Statistics

DAVID FREEDMAN
ROBERT PISANI
ROGER PURVES

UNIVERSITY OF CALIFORNIA · BERKELEY

Statistics

W · W · NORTON & COMPANY · INC · NEW YORK

Cartoons by Dana Fradon

FIRST EDITION

Library of Congress Cataloging in Publication Data
Freedman, David, 1938–
  Statistics.
  Includes index.
  1. Mathematical statistics. I. Pisani, Robert,
joint author. II. Purves, Roger, joint author.
III. Title.
QA276.F683   519.5   77–29265
ISBN 0 393 09076 0

2 3 4 5 6 7 8 9 0

To Jerzy Neyman

Contents

PART V. CHANCE VARIABILITY

PART VI. SAMPLING

PART VIII. TESTS OF SIGNIFICANCE

Preface

TO THE READER

We are going to tell you about some interesting problems which have been studied with the help of statistical methods, and show you how to use some of these methods yourself, explaining why the methods work and what to watch out for when other people use them. Mathematical notation only seems to confuse things for most people, so we are going to do it with words, charts, and tables—and hardly any xs or ys. Even when professional mathematicians read technical books, their eyes often skip over the equations despite their best efforts. What they really need is a sympathetic friend who will explain the ideas and draw the pictures behind the equations. We are trying to be that friend for those who read our book.

WHAT IS STATISTICS?

Statistics is the art of making numerical conjectures about puzzling questions.

- How should experiments be designed to measure the effects of new medical treatments?

- What causes the resemblance between parents and children, and how strong is that force?
- How is the rate of inflation measured? The rate of unemployment? How are they related?
- Why does the casino make a profit at roulette?
- How can the Gallup poll predict election results using a sample of only a few thousand people?

These are difficult issues, and statistical methods help quite a lot in analyzing them. These methods were developed over several hundred years by people who were looking for the answers to hard questions. Some of these people will be introduced later on in the book.

AN OUTLINE

Part I is about designing experiments properly, so that meaningful conclusions can be drawn from the results. It draws examples from medicine, education, and urban policy, and analyzes some studies which are badly designed and misleading. And it explains some of the questions to ask when judging the quality of a design. Most studies produce so many numbers that it is impossible to absorb them all: summaries are needed. Descriptive statistics is the art of describing and summarizing data. This branch of the subject is introduced in Part II. The discussion is continued in Part III, where the focus is on analyzing relationships—for instance, how income depends on education.

Much statistical reasoning depends on the theory of probability, discussed in Part IV. The connection is through chance models, which are developed in Part V and used in Part VII to solve some problems about measurement error and genetics. Parts VI and VIII, statistical inference, explain how to make valid generalizations from samples. Part VI deals with estimation and answers the questions: How does the Gallup poll take a sample to estimate the percentage of Democrats in the population? How far off is the estimate likely to be? Using tests of significance, Part VIII explains how to judge whether a sample confirms or denies a hypothesis about the population from which it is drawn, using tests of significance.

Nowadays, inference is the branch of statistics most interesting to professionals in the field. However, nonstatisticians usually find descriptive statistics the more useful branch—and the one which is easier to understand.

The bare bones of the subject are presented in Chapters 1 to 6, 16 to 21, 23 and 26. After digesting these, the reader can browse anywhere; the next chapters to read might be 8, 10, and 13.

EXERCISES

The numbered sections in each chapter usually end with a set of exercises, the answers being at the end of the book. If you work these exercises as they come along and check the answers against the back of the book, you will get practice in your new skills—and find out the extent to which you have mastered them.

Every chapter except 1 and 7 ends with a set of review exercises; the book does not give answers for these exercises. When working them, you may be tempted to flip backward through the pages until you find the relevant formula. However, reading this book backward will prove very frustrating. The review exercises demand much more than formulas. They call for rough guesses and qualitative judgments, and require a good intuitive understanding of what is going on. The way to develop that understanding is to read the book forward.

Why does this book include so many exercises that cannot be solved by plugging into a formula? The reason is that few real-life statistical problems can be solved that way. Blindly plugging into statistical formulas has caused a lot of confusion. So this book teaches a different approach: thinking.

ACKNOWLEDGMENTS

The writing has been supported by the Ford Foundation (1973–1974) and by the Regents of the University of California (1974–1975). We want to thank Earl Cheit and Sanford Elberg (both of the University of California, Berkeley) for help and encouragement at critical times.

Our typist was Julia Rubalcava. She has been unreasonably patient with us and very deft indeed at converting our scribbles into typescript. We are also in debt to Peggy Darland and Beatrice Rizzolo for imposing administrative order on what would otherwise have been chaos.

Dana Fradon did the cartoons, and Dale Johnson the technical drawings. The computer graphics were handled by David Draper, Roberta Heintz, Richard Lockhart, and Rick Persons.

Very helpful reviews of the manuscript were provided by Frank Anscombe (Yale), Diccon Bancroft (Yale), Leo Breiman (UCLA), Merrill Carlsmith (Stanford), Persi Diaconis (Stanford), David Lane (Minnesota), Richard Light (Harvard), Gerry Mendelsohn (University of California, Berkeley), Tom Rothenberg (University of California, Berkeley), Bruce Rothschild (UCLA), Amos Tversky (Jerusalem), and Geoff Watson (Princeton).

Over the years, people in our courses made hundreds of useful comments on successive drafts of the book. Two of the most perceptive student critics were Carole Rohwer and Elton Sherwin.

Finally, we want to thank our editor, Donald Lamm. Somehow, he turned a permanently evolving manuscript into a book.

PART I

Design of Experiments

1

Controlled Experiments

Always do right. This will gratify some people, and astonish the rest.
—MARK TWAIN (UNITED STATES, 1835–1910)

1. THE SALK VACCINE FIELD TRIAL

A new drug is introduced. How should an experiment be designed to test its effectiveness? The basic method is *comparison*. The drug is given to subjects in a *treatment group*, but other subjects are used as *controls*—they aren't treated. Then the responses of the two groups are compared. Experience shows that subjects should be assigned to treatment or control *at random*, and the experiment should be run *double-blind:* neither the subjects nor the doctors who measure the responses should know who was in the treatment group and who was in the control group.[1] These ideas will be developed in the context of an actual field trial.[2]

The first polio epidemic hit the United States in 1916, and during the next forty years polio claimed many hundreds of thousands of victims, especially children. By the 1950s, several vaccines against this disease had been discovered. The one developed by Jonas Salk seemed the most promising. In laboratory trials, it had proved safe, and had caused the production of antibodies against polio. A large-scale field trial was needed to see whether the vaccine would protect children against polio outside the laboratory. In 1954, the Public Health Service decided to organize this kind of experiment. Nearly two million children were involved, and about half a million were vaccinated. About a million were deliberately left un-vaccinated. And another half a million refused vaccination. The field trial was conducted on children in the most vulnerable age groups—grades 1, 2, and 3. It was carried out in selected school districts throughout the country, those where the risk of polio was believed to be the worst.

This is an example of the method of comparison. The treatment (vaccina-tion) is given only to some of the subjects, who form the treatment group: the others do not get the treatment and are used as controls. The responses of the two groups can then be compared to see if the treatment makes any differ-ence. Here, the treatment group and control group were of different sizes, but that did not matter. The investigators were going to compare the rates at which children got polio in the two groups—cases per hundred thousand. Looking at rates instead of absolute numbers adjusts for the differences in the size of the two groups.

There is a troublesome question of medical ethics here. Shouldn't all the children have been given the protection of the vaccine? One answer is that with new drugs, even after extensive laboratory testing, it is often unclear whether the benefits outweigh the risks. A field trial is needed to find out what the treatment does when it is used in the real world. Now giving the vaccine to a large number of children might seem to provide decisive evidence, even without controls. For instance, if the incidence of polio in 1954 had dropped sharply from 1953, that would seem to be proof of the effectiveness of the Salk vaccine. But it really wouldn't be. Polio is an epidemic disease, whose incidence varies a lot from year to year. In 1952, there were about 60,000 cases; in 1953, there were only half as many. Without controls, low incidence in 1954 could have meant one of two things: either the vaccine worked, or there was no epidemic that year.

The only way to find out whether the vaccine worked was to leave some children unvaccinated. Of course, children could be vaccinated only with their parents' permission. So one possible design was this: the children whose parents consented would form the treatment group and get the vaccine. The other children would form the control group. But it was known that higher-income parents would consent to treatment much more readily than lower-income parents. And this would have created a bias against the vaccine—because children of higher-income parents are more vulnerable to polio than children of lower-income parents. This seems paradoxical at first, but polio is a disease of hygiene. Children who live in less hygienic surroundings tend to contract very mild cases of polio early in childhood, while still protected by antibodies from their mother. After being infected, they generate their own antibodies, which protect them against more severe infection later. Children who live in more hygienic surroundings are less likely to contract these early mild infections, do not develop antibodies, and are less likely to be protected against severe infection later.

The statistical lesson is that to avoid bias, the treatment group and control groups should be as similar as possible—except for the treatment. That makes it possible to conclude that any difference in response between the two groups is due to the treatment, rather than some other factor. If the two groups differ with respect to some factor other than the treatment, the effects of this other factor might be *confounded* (mixed up) with the effects of the treatment. Separating these effects is often difficult or even impossible. Confounded effects are a major source of bias.

Coming back to the Salk vaccine field trial, several designs were proposed. The National Foundation for Infantile Paralysis (NFIP) wanted to vaccinate all grade 2 children whose parents would consent, leaving the children in grades 1 and 3 as controls. And this NFIP design was accepted by many school districts. However, it had two serious flaws. First, polio is a contagious disease, spreading through contact. So the incidence could be much higher in grade 2 than in grades 1 or 3. This would have biased the study against the vaccine. Or the incidence could have been much lower in grade 2, biasing the study in favor of the vaccine. Second, children in the treatment group (where parental consent was needed) were bound to have different family backgrounds from those in the control group (where parental consent was not required).

With the NFIP design, the treatment group would include too many children from the higher-income families, making this group more vulnerable to polio than the control group. Here was a definite bias—against the vaccine.

Many school districts saw these flaws in the NFIP design, and as a result adopted a different design. To make a fair comparison, the control group had to be chosen from the same population as the treatment group: children whose parents consented to vaccination. Otherwise, the effect of family background would have been confounded with the effect of the vaccine. The next question was how to assign the children to treatment and control. It seems that delicate human judgment was called for to make the treatment group and the control group as similar as possible with respect to the relevant variables—like parents' income, or the children's general health, and personality, and social habits. Experience shows, however, that such judgments often result in substantial bias. It is better to use a carefully designed chance procedure. For the Salk trial, the procedure used was equivalent to tossing a coin for each child, who then had a 50% chance to be in treatment and a 50% chance to be in control. Such a procedure is objective and impartial. And the laws of chance guarantee that, with enough subjects, the treatment group and control group will resemble each other very closely with respect to all the important variables—whether or not these have been identified. When an impartial chance procedure is used to assign the subjects to treatment or control, the experiment is said to be *randomized controlled*.

Another basic precaution was the use of a *placebo*. Children in the control group were given an injection of salt dissolved in water, so that during the experiment the subjects did not know whether they were in treatment or in control. This ensured that their response would be to the vaccine, rather than the idea of treatment. It may seem unlikely that subjects could be protected against polio just by the strength of an idea. However, in a number of studies hospital patients suffering from severe postoperative pain have been given a pill which was made of a completely neutral substance—but was described as a pain killer. About one third of the patients experienced prompt relief.[3]

Still another precaution of the same sort: diagnosticians had to decide whether the children contracted polio during the experiment. Many forms of polio are quite difficult to diagnose, and in borderline cases the diagnosis could easily be affected by knowledge of whether the child was vaccinated or not. So the diagnosticians too were not told whether the child was in treatment or in control. This kind of experiment is said to be *double-blind:* the subjects do not know whether they got the treatment, and neither do those who evaluate the response to treatment. This part of the Salk vaccine trial was a double-blind randomized controlled experiment, which is about the best design there is.

How did it turn out? Table 1 shows the rate of polio cases per hundred thousand subjects for the treatment group and the control group in the randomized controlled experiment. The rate is much lower for the treatment group, and this is decisive proof of the effectiveness of the Salk vaccine.

Table 1. The results of the Salk vaccine trial of 1954—size of groups and rate of polio cases per 100,000 in each group. (The numbers are rounded.)

The double-blind randomized controlled experiment			*The NFIP design*		
	Size	*Rate*		*Size*	*Rate*
Treatment	200,000	28	Grade 2 (vaccine)	225,000	25
Control	200,000	71	Grades 1 and 3 (control)	725,000	54
No consent	350,000	46	Grade 2, no consent	125,000	44

Source: Thomas Francis, Jr., *Am. J. of Public Health,* 1955.

The randomized controlled double-blind design reduces bias to a minimum, and that is why it should be used whenever possible. It also has an important technical advantage. To see why, let us play devil's advocate for a moment and assume that the Salk vaccine really had no effect. On this hypothesis, the difference between the polio rates for the treatment and control groups is just due to chance. How likely is that?

With the NFIP design, the results are affected by many factors which (from the point of view of the investigators) are random: which families volunteer, which children are in grade 2, and so on. However, the investigators do not have nearly enough information to estimate the chances of any of these outcomes, so they cannot figure the odds against the difference in polio rates being due just to these accidental factors. With the randomized controlled design, chance enters in a planned and simple way: with the assignment to treatment or control. To spell this out, the "devil's advocate" hypothesis is that the vaccine has no effect. On this hypothesis, some children are fated to contract polio, most are not. Assignment to treatment or control has nothing to do with it—it is preordained. Each child has a 50–50 chance to be assigned to treatment or control—it just depends on the toss of a coin. So each of the children who is fated to get polio has an equal chance to be assigned to treatment or control. On this hypothesis, the number of polio cases in each of the two groups must be about the same; any difference must be due to the chance variability in coin tossing. Now statisticians understand this kind of chance variability very well. They can figure the odds against it making a difference as large as the observed one. This calculation will be presented in Chapter 26, and the odds are astronomical—a billion to one against.

Comparing the results for the two designs, the NFIP design really was somewhat biased against the vaccine. The randomized controlled design shows that the vaccine cuts the polio rate from 71 to 28 per hundred thousand, a 61% reduction; the apparent reduction in the NFIP design, from 54 to 25 per hundred thousand, is only 54%. The impact of this bias may not seem too serious. However, the next two sections will show by example that bad designs really can lead investigators astray.

2. GASTRIC FREEZING

In 1958, Wangensteen introduced a new technique for treating ulcers, called *gastric freezing*.[4] The patient is anesthetized, a balloon is placed in his stomach, and coolant is pumped through the balloon, freezing the stomach for about an hour. This stops the digestive process and the ulcer begins to heal. Wangensteen tried the method on twenty-four patients, and all were cured. For a time his technique became quite popular, as the standard treatment required drastic surgery. However, many doctors were quite suspicious of Wangensteen's results, in part because his experiment included no controls.

In 1963, Ruffin organized a double-blind randomized controlled experiment to evaluate gastric freezing.[5] This experiment was conducted at five different hospitals. There were 82 patients in treatment and 78 in control, assigned at random. Patients in the treatment group were subjected to a true gastric freeze, using Wangensteen's procedure. Patients in the control group were subjected to a sham freeze. The procedure for the sham was the same as for the true freeze, except that a shunt was built into the balloon, returning the coolant before it could freeze the stomach. The follow-up evaluations were conducted over a two-year period by doctors who did not know whether the patients had been given the true freeze or the sham. Ruffin summarized the results as follows.

Most patients were improved (47% in the treatment group and 39% in the control group) or symptom-free (29% in both groups) during the first six weeks. However, as time passed, most patients in both groups relapsed and became clinically worse (. . . at 24 months, 39% in the control group and 45% in the treatment group). At no time was there a significant difference in the two groups at any period of follow-up observation.

The results of this study demonstrate conclusively that the freezing procedure was no better than the sham in the treatment of duodenal ulcer. . . . It is reasonable to assume that the relief of pain and subjective improvement reported by early investigators was probably due to the psychologic effect of the procedure.

The importance of random assignment of patients to treatment and the double-blind method in clinical trials has been emphasized repeatedly, but these features are still too frequently ignored. Only by strict adherence to such principles and resisting the urge to publish until data have been gathered by these rigorous methods will false leads be kept to a minimum and erroneous conclusions avoided.

Wangensteen's experiment was badly designed—no controls. It exaggerated the value of the gastric freeze. Ruffin's experiment was well designed—randomized controlled double-blind. It proved the gastric freeze to be worthless.

3. THE PORTACAVAL SHUNT

In acute cases of cirrhosis of the liver, the patient may start to hemorrhage and bleed to death. One treatment involves surgery to redirect the flow of blood through what is called a *portacaval shunt*. The operation to create the shunt is long and hazardous. Do the benefits of this surgery outweigh the risks? Over fifty studies have been carried out to assess the effect of this surgery.[6] The results are summarized in Table 2 below.

Table 2. A study of studies. The conclusions of 51 studies on the portacaval shunt are related to their designs. The well-designed studies show the surgery to have little or no value. The poorly designed studies exaggerate the value of the surgery.

Design	Degree of enthusiasm		
	Marked	Moderate	None
No controls	24	7	1
Controls, but not randomized	10	3	2
Randomized controlled	0	1	3

Source: Grace, Muench, and Chalmers, *J. Gastroenterology,* 1966.

There were 32 studies without controls: 75% of these studies were markedly enthusiastic about the shunt, concluding that the benefits definitely outweighed the risks. In 15 studies there were controls, but assignment to treatment or control was not randomized. Only 67% were markedly enthusiastic about the shunt. But the 4 studies that were randomized controlled show the surgery to be of little or no value. The badly designed studies exaggerated the value of this risky surgery.

One explanation is that in an experiment without controls, or an experiment where patients are assigned to treatment or control according to clinical judgment, there is a natural tendency to treat only the patients who are in relatively good shape. This biases the study in favor of the treatment. For the experiments summarized in Table 2, in all three types of study about 60% of the patients in the treatment group were still alive three years after the operation. The percentage of controls in the randomized controlled experiments who survived the experiment by three years was also about 60%. But only 45% of the controls in the nonrandomized experiments survived for three years. The most dangerous thing of all was to be a control in a bad experiment. This shows the bias in the nonrandomized experiments.

4. SUMMARY

1. Statisticians often use the *method of comparison*. They want to know the effect of a *treatment*, like the Salk vaccine, on a *response*, like getting polio. To find out, they compare the responses of a *treatment group*, which gets the treatment, with those of a *control group*, which doesn't. Usually, it is hard to judge the effect of a treatment properly without comparing it to something else.

2. If the treatment group is just like the control group, apart from the treatment, then an observed difference in the responses of the two groups is likely to be due to the effect of the treatment. However, if the treatment group is different from the control group with respect to other factors as well, the observed difference may be due in part to these other factors. The effects of these other factors are *confounded* with the effect of the treatment.

3. The best way to try and make sure that the treatment group is like the control group is to assign subjects to treatment or control at random. This is done in *randomized controlled experiments*.

4. Whenever possible, in a well-designed experiment the control group is given a *placebo*, which is neutral but which resembles the treatment. This is to make sure that the response is to the treatment itself rather than to the idea of treatment.

5. A well-designed experiment is run *double-blind* whenever possible. The subjects do not know whether they are in treatment or in control. Neither do those who evaluate the responses. This guards against bias, either in the responses or in the evaluations.

2

Observational Studies

That's not an experiment you have there, that's an experience.
—SIR R. A. FISHER (ENGLAND, 1890–1962)

1. INTRODUCTION

The word *control* is used by statisticians in two different senses:

• A *control* is a subject who did not receive the treatment.
• An experiment is *controlled* when the investigators determine which subjects will be the controls and which will get the treatment—for instance, by tossing a coin.

Statisticians distinguish carefully between controlled experiments and *observational studies*. In an observational study, the investigators do not determine which subjects are to be in control, and which in treatment. For instance, studies of the effects of smoking are necessarily observational—nobody is going to smoke for ten years just to please a statistician. However, the treatment-control terminology is still used. The investigators compare a group of smokers (the treatment group) with a group of nonsmokers (the control group) to determine the effect of smoking—the "treatment." The smokers usually come off badly in this comparison. Heart disease, lung cancer, automobile accidents, and suicide are all more common among smokers than nonsmokers. So there is a strong *association* between smoking and heart disease, lung cancer, and the rest. Does smoking cause cancer? Is smoking the guilty factor? Smoking probably isn't good for you, but the verdict must be: not proved. It may be that some hereditary factor predisposes people both to smoking and to cancer. That would explain the observed association: the effects of the hereditary factor are confounded with the effects of smoking. And then there would be no point to giving up smoking, because that wouldn't change the hereditary factor or reduce the risk of cancer.[1]

This argument is hypothetical. However, many problems can be studied only observationally. And all observational studies have to deal with the problems of confounding. What are the right questions to ask about this? What analytical methods can be used? These questions will be discussed in the context of actual observational studies.

2. THE BALTIMORE HOUSING STUDY

For many years, social scientists have studied the effect of housing on health and social attitudes. This research is related to important policy questions: Should slums be cleared or rehabilitated? Should the state provide low-cost public housing or rent subsidies? What kinds of building codes are needed? In the early 1950s, the School of Hygiene and Public Health at Johns Hopkins University began a study on the effect of public housing.[2]

Three hundred families who lived in the slums were the controls in the Hopkins study; three hundred families living in Lafayette Courts were the treatment group. Lafayette Courts was a public housing project operated by the Baltimore Housing Authority. The treatment was the provision of public housing, which was better and cheaper than slum housing. The six hundred families were followed for three years, from 1955 to 1958. The health and social attitudes of the family members were measured a number of times in questionnaires and interviews; performance of the children in school was determined from their records. The two groups turned out to be quite similar in most respects, with two exceptions. The treatment group was quite a bit happier with its physical environment. And the mortality in the treatment group was much lower. Only two people in the treatment group died during the study, compared with ten controls.

Are the differences between the two groups due to treatment? To make a judgment, it is necessary to find out where the controls came from. When Lafayette Court opened, several thousand families applied for admission. Those families all lived in slums nearby. The Housing Authority chose about eight hundred as tenants. The Hopkins investigators then chose the treatment group from among the successful applicants, and the control group from the unsuccessful ones. This is an observational study. The investigators did not determine which families got the treatment—that was done by the Housing Authority. The treatment group looked like desirable tenants, the control group didn't. It is a fair guess that the decision was based on factors closely related to the ones being studied. So a difference in responses could be the result of the selection process rather than the result of treatment. There is no way to tell, these factors are hopelessly confounded.

There is another major problem in comparing the treatment group and the control group. To begin with, the Hopkins investigators chose four hundred families for the treatment group, and six hundred for the controls, using their judgment to match the groups. But during the three years of the study, they lost one hundred families from the treatment group, and two hundred from the control group. Many of these families just got tired of cooperating with the study. Others left the Baltimore area. Thirty-five treatment families moved out of Lafayette Courts, and were dropped for that reason. About a hundred control families succeeded in escaping from the slums into public housing or decent private housing. That eliminated the difference between them and the treatment families as far as housing was concerned. So the Hopkins investigators dropped them. At the end of the study, the investigators dropped another hundred-odd controls, using their judgment to match what was left of the treatment group and the control group. The effect of this complicated procedure is impossible to determine. But there is a big difference between the controls who escaped from the slums and those who didn't. Dropping those who escaped is bound to bias any comparison between the treatment group and the controls. The design of the Hopkins study makes it impossible to draw any meaningful conclusions about the effect of housing. And no amount of statistical analysis can rescue it.

What should have been done? The investigators should have tried to get the Housing Authority to agree to the following procedure. Select a thousand families who would be suitable as tenants; put their names into a hat and draw five hundred at random; those drawn would be admitted to the housing project and form the treatment group; the others form the control group. If the treatment group and control group had been selected at random in this way, and both groups followed for the period of the study, whether they stayed in their original housing or not, it would have been possible to make some meaningful comparisons at the end of the study.

In an observational study, always find out in detail how the controls were selected. Ask yourself whether the treatment group and control group really were similar, apart from the treatment. If there were other factors whose effects were confounded with the effect of the treatment, find out what adjustments were made for them. Ask yourself whether these adjustments seem reasonable.

3. SEX BIAS IN GRADUATE ADMISSIONS

One source of trouble in observational studies is that the subjects can differ among themselves in crucial ways which aren't noticed at the time. Sometimes, these differences can be found afterward, and adjustments made for them, by comparing smaller and more homogeneous subgroups. This process is called *controlling for* the confounding factor—a third sense of the word *control*. Here is an example.

An observational study on sex bias in admissions to the Graduate Division at the University of California, Berkeley, was carried out in the fall quarter of 1973.[3] There were 8,442 men who applied for admission to graduate school that quarter, and 4,321 women. About 44% of the men and 35% of the women were admitted. (Taking percents adjusts for the difference in numbers of male and female applicants: 44 out of every 100 men were admitted, and 35 out of every 100 women.) Assuming that the men and women were on the whole equally well qualified (and there is no evidence to the contrary), the difference in admission rates looks like a very strong piece of evidence to show that men and women are treated differently in the admissions procedure. Given equally qualified men and women, the university seems to prefer the men by 44 to 35.

" YES, ON THE SURFACE IT WOULD APPEAR TO BE SEX-BIAS BUT LET US ASK THE FOLLOWING QUESTIONS..."

Admissions to graduate work are made separately for each major. By looking at each major separately, it should have been possible to identify the ones which discriminated against the women. At this point, a puzzle appeared. Major by major, there did not seem to be any bias against women. Some majors favor men, but others favored women. On the whole, if there was any bias it ran against the men. What was going on?

Over a hundred majors were involved, and that's too many to look at. However, the six largest majors had over five hundred applicants each, and together accounted for over one third of the total number of applicants to the campus. The pattern for them was typical of the whole campus. Table 1 shows the number of male and female applicants, and the percentages admitted, for each of these majors.

Table 1. Admissions data for the graduate programs in the six largest majors at University of California Berkeley, fall, 1973.

	Men		Women	
Major	Number of applicants	Percent admitted	Number of applicants	Percent admitted
A	825	62	108	82
B	560	63	25	68
C	325	37	593	34
D	417	33	375	35
E	191	28	393	24
F	373	6	341	7

Note: University policy does not allow these majors to be identified by name.
Source: The Graduate Division, University of California, Berkeley.

In each major, the percentage of female applicants who were admitted is roughly equal to the percentage of male applicants. The only exception is major A, which appears to discriminate against men: it admitted 82% of the women, and only 62% of the men. The department that looks most biased against women is E. It admitted 28% of the men and 24% of the women. This difference only amounts to 4%. When all six majors are taken together, they admitted 44% of the male applicants, and only 30% of the females—the difference is 14%.

This is paradoxical indeed, but here is the explanation:

• The first two majors were easy to get into. Over 50% of the men applied to these two.

• The other four majors were much harder to get into. Over 90% of the women applied to these four.

So the men were applying to the easy majors, the women to the harder ones. There was an effect due to the choice of major, confounded with the effect due to sex. When the choice of major is controlled for, as in Table 1, there is little difference in the admissions rates for men or women. In many observational studies, it is possible to control for confounding factors by a similar process: making separate comparisons for homogeneous subgroups.

The main conclusion is that a certain kind of analysis is invalid. When admissions are done by departments, it is wrong to compare campus-wide admissions rates for men and women: a proper comparison has to control for choice of major, since this has a strong confounding effect. Is there a sex bias in graduate admissions at Berkeley? The evidence presented here does not settle the question. It does show that if a bias exists, it is rather subtle. To document it would require proof that the female applicants were on the whole better qualified than the men—or that admissions are deliberately kept low in majors which have high female application rates.

Technical note: Table 1 compares twelve admissions rates and is hard to grasp for that reason. A statistician might summarize Table 1 by computing one overall admissions rate for men and another for women, but adjusting for the sex difference in application rates. The procedure would be to take some kind of average admission rate, separately for the men and women. An ordinary average will not do, because it ignores the differences in size among the departments. Instead, most statisticians would take a *weighted average* of the admission rates, the weights being the total number of applicants (male and female) to each department.

Table 2. Total number of applicants, from Table 1.

Major	Total number of applicants
A	933
B	585
C	918
D	792
E	584
F	714
	4,526

The weighted average admission rate for men is

$$\frac{933 \times 62 + 585 \times 63 + 918 \times 37 + 792 \times 33 + 584 \times 28 + 714 \times 6}{4,526}$$

This works out to 39%. Similarly, the weighted average admission rate for women is

$$\frac{933 \times 82 + 585 \times 68 + 918 \times 34 + 792 \times 35 + 584 \times 24 + 714 \times 7}{4,526}$$

This works out to 43%. In these formulas, the weights are the same for the men and women: they are the totals from Table 2. The admission rates are different for men and women; they are the rates from Table 1. The final comparison: the weighted average admission rate for men is 39%, while the weighted average admission rate for women is 43%.

4. THE INCAP STUDY

What are the effects of malnutrition and what can be done about them? These are critical policy questions in many parts of the world today where growing populations press harder and harder against limited supplies of food. Studies of these questions are now under way. We are going to discuss one in detail. It was conducted by INCAP, a research institute administered by the World Health Organization in Guatemala.[1]

The basic method is comparison. To see the effect of malnutrition, investigators must compare people who eat a proper diet with those who don't. To use statistical methods, the difference must be quantified, expressed in numbers. Many serious effects of malnutrition may be too subtle to quantify. But its impact on physical growth is so obvious that it is easy to measure. For example, the rural children of Guatemala get about two thirds as much to eat as children in the United States. By age 7, Guatemalan children are, on the average, 5 inches shorter and 11 pounds lighter than children from the United States, and the gap widens as the children get older. This growth deficiency is thought to be due mainly to the protein deficiency in diet.

The difference between diets in Guatemala and the United States is staggering, but so are the differences in many other factors which may be relevant to growth: race, prevalence of disease, and standards of hygiene and health care. The effects of these factors are confounded with the effects of diet. How much of the observed differences in growth should be attributed to diet, and how much to these other factors? The effect of race can be ruled out almost completely. Upper-class city children in Guatemala are racially indistinguishable from rural children, but grow according to North American norms. Of course, they live in an environment that is basically similar to that of the United States, and they get about as much to eat as children in the United States.

To see the effects of malnutrition in rural Guatemala, it would be better to compare children who don't get enough to eat there with children who do. But this overlooks a brutal reality: almost all of these children go hungry. INCAP had to take a different approach. In 1969, the investigators set up an experimental program in four Guatemalan villages to measure the effect of supplementing diets on the physical and psychological development of children. These four villages were closely matched on a variety of characteristics.

• *Size*. Each village consisted of about 150 families, with about 600 inhabitants over the age of 7 and 100 under age 7.

• *Culture*. There are two main cultural groups in rural Guatemala: the Indios, who follow some of the old Mayan customs; and the Europeanized Ladinos. The INCAP villages were exclusively Ladino.

• *Farming*. The villagers raise corn and beans and have a small cash crop like tomatoes.

• *Poverty*. The average annual income for a family of five is roughly equivalent to U.S. $200.

• *Illiteracy*. Only about 30% of those over age seven can read.

The villagers live by subsistence agriculture. The staple food is corn tortillas; many of the villagers can afford to add beans to their regular diet, many cannot. There are no paved streets, gutters, sewage systems, water lines, electric wires. Many of the houses are made of branches, mud walls being added during the rainy seasons. Permanent walls are a sign of wealth. So is a surfaced floor, or even a separate kitchen. Social life is as impoverished as the physical surroundings. The villagers have few group ceremonies, no art, and no music. In Guatemalan terms, these villages are not exceptional.

INCAP assigned two of the villages to treatment and two to control:

• The treatment villages get a drink called *atole,* which is rich in calories and protein.

• The control villages get a drink called *fresco,* which is low in calories and contains no proteins.[5] This is the placebo.

Both drinks contain the vitamins and trace elements thought to be missing from the usual diet. The drinks are served at special cafeterias built by INCAP, and the villagers take as much each day as they please.

In many ways, the study is a burden on the villagers. They were asked what they wanted in compensation, and they chose free medical care. So INCAP maintained a dispensary staffed by a nurse in each village. A doctor makes rounds, and villagers who need it are given free hospital care in Guatemala City. In typical Guatemalan villages, medical care is not available.

INCAP set out to run a controlled experiment. The investigators expected all the people in the treatment group to take about the same amount of supplement, and similarly for the controls. However, it turned out differently. There was enormous variation in consumption from villager to villager. And on the average the controls consumed much more by volume than the treatment group, so both groups took roughly the same amount of supplementary calories. In the end, INCAP had an observational study. The analytical procedure was to relate the responses of each child in physical and mental growth to the amount of supplement taken by that child, separately within each village.

There were many confounding factors to take care of in the family backgrounds of children: for example, diet and health conditions in the home, or body size and social status of parents. INCAP made painstaking efforts to measure all these confounding variables and to control for them using methods like those discussed in the previous section.

What did INCAP find? The results[5a] lend support to the theory that the difference in growth between Guatemalan and North American children is largely due to the difference in protein consumption. In the control villages, the supplement provided calories but no protein, and there was no relationship between the growth of children and the amount of supplement they took. In the treatment villages, there was a definite relationship: on the average, the children grew an extra 0.04 inch per pound of supplemental protein taken, after adjusting for all the confounding variables INCAP could find. This rate of 0.04 inches per pound may seem small, but it is enough to account for a large part of the observed five-inch difference between the average heights of Guatemalan and American children at age 7. By that age, the North American child has eaten about 100 pounds of protein more than the Guatemalan. Using INCAP's rate, the American child ought to be taller by

100 pounds × 0.04 inches per pound = 4 inches.

Of course, the proof is not conclusive. The weakness is in the design of the experiment. A key step in the argument was that in the control villages, the supplement did not affect growth. The most likely explanation is that fresco offers no proteins. But it could also be due to some hidden factor which makes the control villages different from the treatment villages. With any observational study, there is always the possibility that the effects are due to a confounded factor. That is why statisticians prefer controlled experiments.

EXERCISE SET A

1. A new teaching method is tried on a large class. At the end of the course, a final is given. The average score is 73 out of 100 and only 3% of the class failed. Does this show the new method was successful?

2. From Table 1 on p. 6, the NFIP controls had a much lower rate of polio than the controls in the randomized experiment. Why?

3. The Salk vaccine field trials were conducted only in certain experimental areas (school districts), selected by the Public Health Service in consultation with local officials.[6] In these areas, there were about 3 million children in grades 1, 2, or 3, and there were about 11 million children in those grades in the United States. In the experimental areas, the incidence of polio was about 25% higher than in the rest of the country. Did the Salk vaccine field trials cause children to get polio instead of preventing it? Explain briefly.

4. The University of Washington initiated an experimental physical-fitness program. To evaluate the effects, all those who registered for the program at the beginning of its first year of operation were tested at registration and retested at the end of the year. Physical fitness improved remarkably. As a result, it was decided to run the program for a second year. It was not considered necessary to test new participants at the beginning of the second year since the results from the previous year were available as benchmarks. At the end of the second year, all the participants were tested. The tests showed a marked deterioration in physical fitness by comparison with the scores at the beginning of the first year.[7] Does this mean that the second year of the program was a failure? If not, what is an alternative explanation?

5. A hypothetical university has two departments, A and B. There are 2,000 male applicants, of whom half apply to each department. There are 1,100 female applicants: 100 apply to department A and 1,000 to department B. Department A admits 60% of the men who apply and 60% of the women. Department B admits 30% of the men who apply and 30% of the women. Since the percentage of men admitted equals the percentage of women admitted for each department, this is so for both together. True or false? Explain. (Hint: For each department, figure out the number of men it admits and the number of women.)

6. In October, 1976, a nationwide vaccination program was started against swine flu. The first shots were given to the group most at risk—the elderly and infirm. During the first week of the program, 24,000 persons aged 65 and over were given shots, and three of these persons died. As a result, eight states suspended the vaccination program.[8] What would a statistician say?

7. Many studies have shown that there is a strong association between delinquency and family size. By comparison with the general population, a high percentage of delinquents come from families with six or more children. It follows mathematically that among children from such large families, a higher percentage will become delinquent than from smaller families.[9] This association remains even when race, religion, and family income are controlled for. This suggests that family size is a contributing factor to delinquency.

   One study found that by comparison with the general population, a high percentage of delinquents are middle children—that is, neither first-born nor last born.[10] This association remained even when race, religion, and family income were controlled for. Being a middle child, therefore, seems to be a contributing factor to delinquency. Or is it? Answer yes or no, and explain.

8. People who are arrested and charged with crimes are held in jail until trial, unless they can post bail or they are *paroled* by the court. (*Parole* means release on one's own recognizance. The person does not post bail, but just promises to appear for trial.) One study showed that in 1960, over 50% of all those charged with a crime in New York were held in jail until trial. And those detained were two or three times as likely to be convicted or imprisoned as those released, even after controlling for family background and type of crime.

(a) Does this show that being detained increased the likelihood of convictions or imprisonment? If not, what is an alternative explanation?

In the same year, the Vera Foundation initiated a randomized controlled experiment on bail procedures.[11] The foundation selected a group of about a thousand people who were detained pending trial. These were mainly indigents who could not raise bail but who were eligible for parole according to the foundation's criteria, mainly for having roots in the community such as a family or a job; persons accused of serious crimes were automatically excluded. The foundation chose five hundred out of these thousand people at random for the treatment group, and the other five hundred formed the control group. For those in the treatment group, the foundation wrote letters to the court, recommending parole and giving some facts on family background and job history. In the treatment group, about 60% were paroled, compared to 15% in the control group.

(b) Does this show that treatment increased the likelihood of parole? Explain.

Of those paroled in the treatment group, only 40% were convicted at their trial. On the other hand, of those detained in the control group, about 80% were convicted.

(c) Is this a fair comparison to measure the effect of the letter on the likelihood of conviction, or should Vera have compared the parolees with the parolees, and those detained with those detained? Explain.

From 1961 to 1964, the Vera Foundation continued the experiment, but without controls. About 2,000 were paroled on Vera recommendations during this period, and only 15 failed to show up for trial, a default rate well under 1%. In fact, the main point of the experiment was to prove that poor people could be released on their own recognizance, without posting bail, and would still show up for trial. City officials were convinced, and the Vera program was institutionalized as a regular part of the New York court system.

(d) Would the default rate be expected to go up, go down, or stay about the same after the experimental phase was over? Explain.

The answers to these exercises are on pp. A-25–A-26.

5. REVIEW EXERCISES

1. A high percentage of the seeds of the Coast redwood tree fail to germinate. A sample of these seeds was subjected to chemical analysis, and found to contain a lot of tannic acid. Does this finding implicate tannic acid as a cause for failure to germinate?

2. From Table 1 on p. 6, those children whose parents refused to participate in the randomized controlled Salk trial got polio at the rate of 46 per 100,000. On the other hand, those children whose parents consented to participation got polio at the slightly higher rate of 49 per 100,000 in the treatment group and control group taken together. Suppose that this field trial were being repeated the following year. On the basis of the figures, some parents refuse to allow their children to participate in the experiment and be exposed to this higher risk of polio. What does statistical theory say?

3. One of the leading causes of death in the United States is coronary artery disease, in which the main arteries to the heart break down. The conventional treatment for this disease involved drugs and special diets to reduce the patient's blood pressure and eliminate fatty deposits in the arteries. Several studies, involving a few hundred patients, reported that only 68% of patients getting this treatment survived for three years or more. In 1972, Dr. Daniel Ullyot and his associates introduced a radical new treatment, in which the diseased arteries were replaced by veins transplanted from the patients' own legs. About one hundred patients had been treated this way, and 98% survived three years or more.[12] A newspaper article described this as "spectacular." Comment briefly.

4. An experiment was carried out to determine the effect of providing free milk to school children in a certain district (Lanarkshire, England).[13] Some children in each school were chosen for the treatment group and got free milk; others were chosen for controls and got no milk. Assignment to treatment or control was done at random, to make the two groups comparable in terms of family background and health. However, it was feared that just by chance, there would still be small differences between the two groups. So the teachers were allowed to use their judgment in switching children between treatment and control, with the object of equalizing the groups. Was it wise to let the teachers use their judgment this way? What kinds of bias might have been introduced?

5. In 1964, the Public Health Service studied the effects of smoking on health, in a sample of 42,000 households.[14] For men and for women in each age group, they found that those who had never smoked were on average somewhat healthier than the current smokers, but the current smokers were on average much healthier than the former smokers.
   (a) Why did they study men and women and the different age groups separately?
   (b) The lesson seems to be that you shouldn't start smoking, but once you've started, don't stop. Comment briefly.

6. Dr. Josephine Lo and her associates treated 31 patients suffering from severe headaches as a result of spinal punctures. Conventional treatment did not help. She found that 30 out of the 31 patients experienced "complete and permanent relief" after one to five acupuncture treatments.[15] Is this good evidence for the effectiveness of acupuncture?

7. (Hypothetical.) A study is carried out to determine the effect of party affiliation on voting behavior in a certain city. The city is divided up into wards. In each ward, the percentage of registered Democrats who vote is higher than the percentage of registered Republicans who vote. Does it follow that for the city as a whole, the percentage of registered Democrats who vote is higher than the percentage of registered Republicans who vote? (Hint: See Exercise 5 on p. 18.)

The following two exercises are designed as warm-ups for the next chapter. Do not use a calculator when working them. Just remember that "percent" really means "per hundred." For example, 41 people out of 398 is just over 10%. The reason: 41 out of 398 is like 40 out of 400, that's 10 out of 100, and that's 10%.

8. Say whether each of the following is closest to 1%, 10%, 25%, 50%.
    (a) 39 out of 398.
    (b) 99 out of 407.
    (c) 57 out of 209.
    (d) 99 out of 197.

9. Among the Statistics 2 students at The University of California, Berkeley, in the fall of 1976, 46 students out of 446 reported family incomes ranging from $5,000 to $10,000 a year.
    (a) About what percentage had family incomes in this range?
    (b) Guess the percentage that had family incomes in the range $5,000 to $6,000 a year.
    (c) Guess the percentage that had family incomes in the range $6,000 to $7,000 a year.
    (d) Guess the percentage that had family incomes in the range $7,000 to $9,000 a year.

6. SUMMARY

1. In an *observational study*, the investigators do not assign the subjects to treatment or control. Typically some of the subjects have the conditions whose effects are being studied; this is the treatment group. The others form the control group. For example, in a study on smoking, the smokers form the treatment group and the nonsmokers are the controls.

2. In an observational study, the effects of treatment are confounded with the effects of the factors that got the subjects into treatment or control in the first place. Observational studies can, therefore, be quite misleading about cause-and-effect relationships.

3. When looking at a study, ask the following questions. Was there any control group at all? If so, how were subjects assigned to treatment or control: through a process under the control of the investigator (a controlled experiment), or through a process outside the control of the investigator (an observational study)? If it was a controlled experiment, was the assignment

made using a chance mechanism (randomized controlled) or did it depend on the judgment of the investigator?

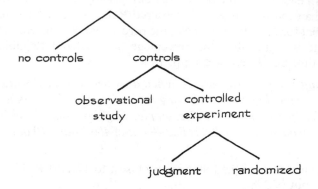

4. With observational studies, and even with nonrandomized controlled experiments, try to find out in detail how the subjects came to be in treatment or in control. How are the treatment groups similar? different? What factors are confounded with treatment—that is, what could explain a difference in response between the two groups, apart from treatment? What adjustments were made to take care of confounding? Were they sensible?

5. In an observational study, a confounding factor can sometimes be *controlled for,* by comparing smaller groups which are relatively homogeneous with respect to the factor. For instance, in a study on admissions, it was advisable to look at applicants to each department separately.

PART II

Descriptive Statistics

3

The Histogram

Grown-ups love figures. When you tell them that you have made a new friend, they never ask you any questions about essential matters. They never say to you, "What does his voice sound like? What games does he love best? Does he collect butterflies?" Instead, they demand: "How old is he? How many brothers has he? How much does he weigh? How much money does his father make?" Only from these figures do they think they have learned anything about him.

—The Little Prince[1]

1. INTRODUCTION

In the United States, how are incomes distributed? How much worse off are minority groups like blacks? Some information is provided by government statistics, obtained from the Current Population Survey. The mechanics of this survey will be discussed in detail in Part VI; each month, the survey provides a representative cross section of American families, about forty thousand of them. In March of each year, these families are asked to report their incomes for the previous year. We are going to look at the results from March, 1974. Of course, these data have to be summarized somehow. Nobody in his right mind wants to look at 40,000 numbers. Statisticians prefer to summarize this kind of data by means of a graph rather than a table. They use a graph called a *histogram*. The histogram for the income data is shown in Figure 1 on the next page.

The object of this section is to explain how to read this kind of graph. First of all, it does not have a vertical scale: unlike most other graphs, a histogram does not need a vertical scale. The next thing to look at is the

Figure 1. A histogram. This graph shows the distribution of families by income in the United States in 1973.

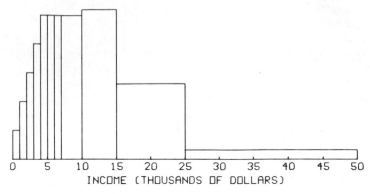

Source: Current Population Survey.[2]

horizontal scale. This shows income in thousands of dollars. The graph consists of a set of blocks. The bottom edge of the first block covers the interval from $0 to $1,000, the bottom edge of the second goes from $1,000 to $2,000, and so on, until the last block, which covers the interval from $25,000 to $50,000. These invervals are called *class intervals*. The graph is drawn so that the area of each block is proportional to the number of families with incomes in the corresponding class interval.

A histogram represents numbers by area, not height.

In this way, histograms are different from bar graphs.

To see how the blocks work, look more closely at Figure 1. About what percentage of the families earned between $10,000 and $15,000? Now the block over this interval amounts to something like one fourth of the total area of all the blocks. So about one fourth, or 25%, of the families had incomes in that range.

Take another example. Were there more families with incomes between $10,000 and $15,000 or with incomes between $15,000 and $25,000? The block over the first interval is taller, but the block over the second interval is wider. The areas of the two blocks are about the same, so the number of families earning $10,000 to $15,000 is about the same as the number earning $15,000 to $25,000.

For a third and final example: Take the percentage of families with incomes under $7,000. Is this closest to 10%, 25%, or 50%? By eye, the area under the histogram between $0 and $7,000 is about a quarter the total area, so the percentage is closest to 25%.

The horizontal axis in Figure 1 stops at $50,000. What about the families earning more than that? The histogram simply ignores them. Of course, in 1973 only 1% of American families had incomes above that level, so not many data are lost.

At this point, a good way to learn more about histograms is to do some exercises. To help you judge the sizes of the blocks in the first exercises, Figure 2 shows the same histogram as Figure 1, but with a vertical scale supplied.

Figure 2. The histogram from Figure 1 with a vertical scale supplied.

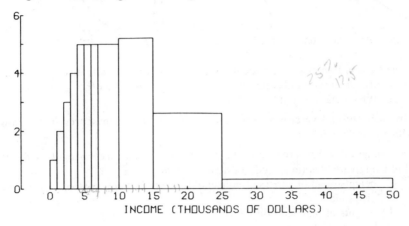

EXERCISE SET A

1. About 1% of the families in Figure 2 had incomes between $0 and $1,000. Estimate the percentage who had incomes
   (a) between $1,000 and $2,000.
   (b) between $2,000 and $3,000.
   (c) between $4,000 and $5,000.
   (d) between $4,000 and $7,000.
   (e) between $7,000 and $10,000.

2. In Figure 2, were there more families earning between $6,000 and $7,000, or between $7,000 and $8,000?

3. In Figure 2, were there more families earning between $10,000 and $11,000, or between $15,000 and $16,000?

4. The histogram below shows the distribution of final scores in a certain class.
   (a) Did anybody score below 20?
   (b) Which block represents the people who scored between 60 and 80?
   Ten percent scored between 20 and 40.
   (c) About what percentage scored between 40 and 60?
   (d) About what percentage scored over 60?

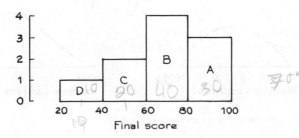

5. Below are three rough sketches of histograms for test scores in different classes. The scores range from 0 to 100; a passing score was 50. For each class, was the percent who passed about 50%, well over 50%, or well under 50%?

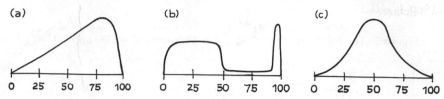

(a)   (b)   (c)

6. One class in Exercise 5 was peculiar, for there were two quite distinct groups of students: one group did rather poorly on the test, and the other group did very well. Which class was it?

7. In class (b) of Exercise 5, were there more people with scores in the range 40–50, or 90–100?

8. Six rough sketches are shown below. Four of them are histograms for the following variables (in a study of a small town):
   (a) heights of all members of households where both parents are less than 24 years old.
   (b) heights of married couples.
   (c) heights of all people.
   (d) heights of all automobiles.
   Match the variables with their histograms. Explain your reasoning.

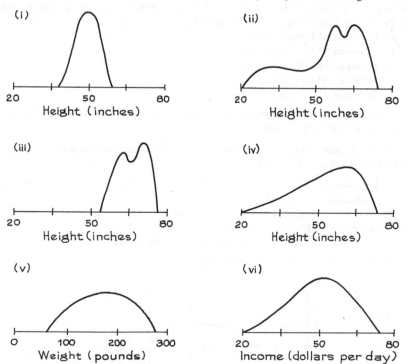

(i)   (ii)

(iii)   (iv)

(v)   (vi)

The answers to these exercises are on p. A-26.

2. DRAWING A HISTOGRAM

The object of this section is to explain how to draw a histogram. The method is not difficult but there are a couple of wrong turns to avoid.

The starting point in drawing a histogram is a distribution table, like Table 1. This table shows the percentage of families with incomes in each class interval. Such a table is obtained by going back to the original data—the income of 40,000 representative American families—and counting. Nowadays this sort of work is done by computer, and in fact Table 1 was drawn up by a computer at the Bureau of the Census. The computer has to be told what to do with families that fall right on the boundary between two class intervals. This is called an *endpoint convention*. The endpoint convention followed in Table 1 is indicated by the caption: the left endpoint is included in the class interval, the right endpoint is excluded. In the first line of the table, for example, $0 is included and $1,000 is excluded: to be counted in this interval a family should earn $0 or more, but less than $1,000. The family which earns $1,000 exactly goes in the next interval.

Table 1. Distribution of families by income in the United States in 1973. Class intervals include the left endpoint, but not the right endpoint.

Income level	Percent
$0–$1,000	1
$1,000–$2,000	2
$2,000–$3,000	3
$3,000–$4,000	4
$4,000–$5,000	5
$5,000–$6,000	5
$6,000–$7,000	5
$7,000–$10,000	15
$10,000–$15,000	26
$15,000–$25,000	26
$25,000–$50,000	8
$50,000 and over	1

Source: Current Population Survey.[2]

The first step in drawing the histogram is to put down the horizontal axis. On their first try, some people get

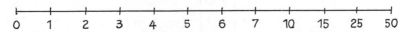

This is a mistake. The interval from $7,000 to $10,000 is three times as long as the interval from $6,000 to $7,000, and the axis must show this. Also, it is a good idea to label the axis. So the horizontal axis should look like:

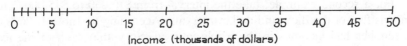

Income (thousands of dollars)

The next step is to draw the blocks. It's tempting to make their heights equal to the percents in the table. Figure 3 shows what happens if you make this mistake. The graph gives much too rosy a picture of the income distribution. For example, Figure 3 says that there are many more families with incomes over $25,000 than there are with incomes under $7,000. The United States is wealthy, but not that wealthy.

Figure 3. Don't plot the percents.

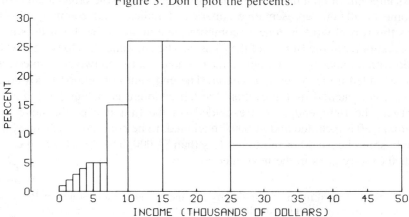

INCOME (THOUSANDS OF DOLLARS)

The source of the trouble is that some class intervals are longer than others. This means that the percents in the distribution table (Table 1) are not on a par with one another. The 8% who earn $25,000 to $50,000, for example, are spread over a much larger range of incomes than the 26% who earn $15,000 to $25,000. Plotting percents directly overlooks this, and makes the blocks over the longer class intervals too big.

There is a simple way to compensate for the different lengths of the class intervals. The idea is to use one-thousand dollar intervals as the common unit. For example, the class interval from $7,000 to $10,000 contains three of these intervals: $7,000 to $8,000, $8,000 to $9,000, and $9,000 to $10,000. From Table 1, 15% of the families had incomes in the entire interval. So within each of the three smaller intervals, there will only be about 5% of the families. This 5, not the 15, is what should be plotted above the interval $7,000 to $10,000.

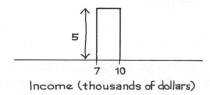

Income (thousands of dollars)

For a second example, take the interval from $10,000 to $15,000. This contains five of the thousand-dollar intervals. According to Table 1, 26% of the families had incomes in the whole interval. So within each of the five smaller intervals there will be about 5.2% of the families: $26 \div 5 = 5.2$. This gives the height of the block over the interval $10,000 to $15,000.

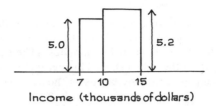

Income (thousands of dollars)

The work is done for two of the lines in Table 1. To complete the histogram, do the same thing for the rest of the class intervals. This is how Figure 2 was plotted.

The histogram represents the distribution as if the percents were spread evenly over each class interval. Often, this is a good first approximation.

> To figure out the height of a block over a class interval, divide the percent by the length of the interval.

The procedure is straightforward, but the units on the vertical scale are a little complicated. For example, to get the height of the block over the interval $7,000 to $10,000, we divide 15% by 3 thousand dollars. So the units for the answer had to be percent divided by thousand dollars—percent per thousand dollars. The unit "percent per thousand dollars" is a bit intimidating at first. But think about the "per" just as you would when reading that there are 50,000 people per square mile in Tokyo: in each square mile of the city, there are about 50,000 people. It is the same with histograms. The height of the block over the interval $7,000 to $10,000 is 5% per thousand dollars: in each thousand dollar interval between $7,000 and $10,000, there are about 5% of the families. Figure 4 shows the complete histogram with these units on the vertical scale.

Figure 4. Distribution of families by income in the United States in 1973.

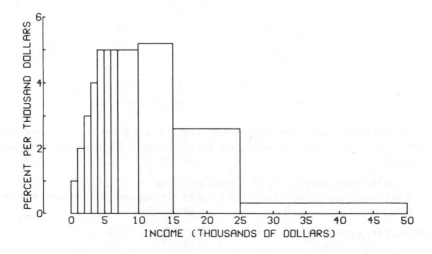

EXERCISE SET B

1. The distribution of family income for the United States in 1973 by race is shown
   below (rounded off). For example, 1% of the white families made less than
   $1,000 a year; so did 3% of the nonwhite families. (The class intervals include
   the left endpoint, but not the right.) Draw a histogram for the white families and
   for the nonwhite families separately but using the same scales.

Income levels	Percent White families	Percent Nonwhite families
$0–$1,000	1	3
$1,000–$2,000	1	6
$2,000–$3,000	3	8
$3,000–$4,000	4	9
$4,000–$5,000	4	8
$5,000–$6,000	4	7
$6,000–$7,000	5	7
$7,000–$10,000	15	17
$10,000–$15,000	26	19
$15,000–$25,000	28	13
$25,000–$50,000	9	3
$50,000 and over	1	negligible

Source: Current Population Survey.[2]

2. The table below gives the distribution of educational level for persons aged 25
   and over in the United States in 1960 and 1970. "Educational level" means
   the number of years of schooling completed. The class intervals include the left
   endpoint, but not the right; for example, from the second line of the table, in
   1970 about 9% of the people had completed 5 to 8 years of schooling, 8 not
   included; in 1960, about 14% of the people were in this category. Draw a
   histogram for the 1970 data. You can interpret "16 or more" as 16 to 17 years of
   schooling; very few people completed more than 17 years of school. Why does
   your histogram have spikes at 8, 12, and 16 years of schooling?

Educational level (years of schooling)	Percent (1970)	Percent (1960)
0–5	5	8
5–8	9	14
8–9	13	18
9–12	17	19
12–13	34	25
13–16	10	9
16 or more	11	8

Source: *Statistical Abstract*, 1976, Table 199.

3. Redraw the histogram for the 1970 data, combining the first two class intervals
   into one (0–8 years, with 14% of the people). Does this change the histogram
   much?

4. Draw the histogram for the 1960 data, and compare it to the 1970 histogram.
   What happened to the educational level of the population between 1960 and
   1970? Did it go up, go down, or stay about the same?

The answers to these exercises are on p. A-26.

3. THE DENSITY SCALE

When reading areas off a histogram, it is convenient to have a vertical scale to refer to. Statisticians use the *density scale*. This scale shows the percentage of cases per unit on the horizontal axis. The income histogram in the previous section was drawn using this density scale (although we did not bother to name it). For that histogram, the vertical scale showed percent of families per $1,000 of income. Figure 5 shows another example of a histogram drawn with a density scale. This is a histogram of educational level for persons aged 25 and over in the United States in 1970. "Educational level" means years of schooling completed.

Figure 5. Distribution of persons aged 25 and over in the United States in 1970 by educational level.

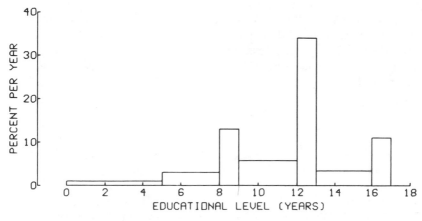

Source: *Statistical Abstract,* 1976, Table 199.

The endpoint convention followed in this histogram is a bit fussy. The block over the interval 8 to 9 years, for example, represents all the people who finished eighth grade, but not ninth grade; people who dropped out part way through ninth grade are included.

The units on the horizontal axis of the histogram are years, so the units on the vertical axis are percent per year. For example, the height of the histogram over the interval 9 to 12 years is nearly 6% per year. In other words, about 6% of the population finished the first year of high school, another 6% finished the second year, and another 6% finished the third year.

In section 1, we described how area in a histogram represents percent: if one block covers a larger area than another, it represents a larger percent of the cases. But what does the height of a block represent? To answer this, look at Figure 5, focusing for a moment on the horizontal axis. Imagine the people lined up on this axis, with each person stationed at his educational level. Some parts of the axis—years—will be more crowded than others. The height of the histogram shows this crowding. The histogram is highest over the interval 12 to 13 years, so the crowding is greatest there. This interval has all the people with high-school degrees and no further education. There are two other peaks, one at 8 to 9 years—finishing elementary school—and another at 16 to 17 years—finishing college. By comparison with the rest of the histogram, the three peaks show how people tend to stop their schooling at one of the three possible graduations rather than dropping out in between.

At first, it may be difficult to keep apart the notion of the crowding in an interval, represented by the height of the block, and the number in an interval, represented by the area of the block. An example will make this clear. Look at the blocks over the intervals 8 to 9 years and 9 to 12 years in Figure 5. The first block is taller, so this interval is more crowded. However, the block over 9 to 12 years has a larger area, so this interval has more people. Of course, there is more room in the second interval—it's three times as long. The two intervals are like Holland and the United States. Holland is more crowded, but the United States has more people.

The density scale shows crowding, but it can also be used to figure per-cents. For an example, take the interval from 13 to 16 years in Figure 5—the people who got through their first year of college but didn't graduate. On the vertical scale, the height of the block over this interval is about 3% per year. In other words, each of the three one-year intervals 13 to 14, 14 to 15, and 15 to 16 holds around 3% of the people. So the whole three-year interval must hold about

$$3 \times 3\% = 9\%$$

of the people. About 9% of the population aged 25 and over in 1970 got through the first year of college, but failed to get a degree.

Example 1. The sketch below shows one block of the family-income histogram for a certain city. About what percent of the families in the city had incomes between $15,000 and $25,000?

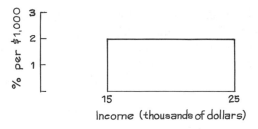

Income (thousands of dollars)

Solution. The height of the block is 2% per thousand dollars: each thousand-dollar interval between $15,000 and $25,000 contains about 2% of the families in the city. There are ten thousand-dollar intervals between $15,000 and $25,000. So the answer is

$$10 \times 2\% = 20\%.$$

A fifth of the families in the city had incomes between $15,000 and $25,000.

With any histogram, percents are represented by areas. But with the density scale, something special happens. Look back at the block sketched in Example 1. This represents the percent of families with incomes between $15,000 and $25,000. The percent was worked out as

10 thousand dollars × 2% per thousand dollars = 20%.

The dollars cancel out. Graphically, we are figuring the area of the block using "base × height" and getting the answer in percent. So when the density scale is used on the vertical axis, the size of the area over an interval is equal to the percentage of cases in that interval. The total area under the histogram represents all the cases, and must be 100%.

> With the density scale on the vertical axis, the area under the histogram over an interval equals the percentage of cases in that interval.[3] The total area is 100%.

Example 2. Someone has sketched a histogram for the weights of some people, using the density scale. What is wrong with it?

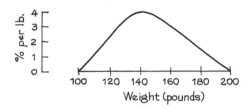

Solution. The area is 200%, and should only be 100%.

(The area can be calculated as follows. The histogram is almost a triangle, whose height is 4% per pound and whose base is 200 − 100 = 100 lbs. The area is

½ × base × height = ½ × 100 × 4% = 200%.)

EXERCISE SET C

1. A histogram of monthly wages for part-time employees is shown below. Nobody earned more than $1,000 a month. The block over the class interval from $200 to $500 is missing. How tall must it be?

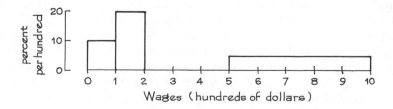

2. Three people plot histograms for the weights of subjects in a study, using the density scale. Only one is right. Which one, and why?

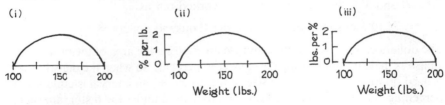

3. In a Public Health Service Study of 1964, a histogram was plotted showing the number of cigarettes per day smoked by each subject (male, current smokers), as shown below.[4] The density is marked in parentheses. The class intervals include the right endpoint, not the left.

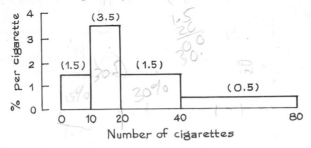

   (a) The percentage who smoked ten cigarettes or less per day is closest to:
       1.5%   15%   30%   50%.
   (b) The percentage who smoked more than a pack a day, but not more than two packs is closest to:
       1.5%   15%   30%   50%.
       (There are twenty cigarettes in a pack.)
   (c) The percent who smoked more than a pack a day is closest to:
       1.5%   15%   30%   50%.
   (d) The percent who smoked more than three packs a day is closest to:
       0.25 of 1%   0.50 of 1%   5%   10%.
   (e) The percent who smoked 15 cigarettes per day is closest to:
       0.35 of 1%   0.5 of 1%   1.5%   3.5%   10%.

The answers to these exercises are on p. A-27.

4. VARIABLES

The Current Population Survey covers many other variables besides income. A *variable* is a characteristic which changes from person to person in a study. Interviewers for the survey use a battery of questions: How old are you? How many people are there in your family? What is your family's total income? Are you married? Do you have a job? The corresponding variables would be: age, family size, family income, marital status, and employment status. Some questions are answered by giving a number: the corresponding variables are said to be *quantitative*. Age, family size, and family income are examples of quantitative variables. Some questions cannot be answered with numbers, and the corresponding variables are *qualitative:* marital status (single, married, divorced, widowed, separated) and employment status are examples.

Quantitative variables may be *discrete* or *continuous*. This is not a hard-and-fast distinction, but it is a useful one.[5] For a discrete variable, the values can only differ by fixed amounts. Family size is discrete. Two families can differ in size by 0 or 1 or 2, and so on. Nothing in between is possible. Age, on the other hand, is a continuous variable. This doesn't refer to the fact that a person is continuously getting older; it just means that the difference in age between two people can be arbitrarily small—a year, a month, a day, an hour, whatever. Finally, the terms *qualitative, quantitative, discrete,* and *continuous* are also used to describe data—qualitative data are data collected on a qualitative variable, and so on.

Section 2 showed how to plot a histogram starting with a distribution table. Often the starting point is the raw data, which usually consists of a list of cases (individuals, families, schools, etc.) and for each case, a record of what the variable was found to be. In order to draw the histogram, a distribution table must be prepared. The first step is to choose the class intervals. With too many or too few, the histogram will not be informative. There is no rule, it is a matter of judgment or trial and error. It is common to start with ten or fifteen class intervals and work from there. In this book, the class intervals will always be given.[6]

When plotting a histogram for a continuous variable, you also have to decide on the endpoint convention; that is, what to do with cases that fall right on the boundary. With discrete variables, there is a convention which gets around this nuisance: the class intervals are centered at the values of the variable. For instance, family size can be 2 or 3 or 4, and so on. (The Census does not recognize one person as a family.) The corresponding class intervals in the distribution table would be

Center	Class interval
2	1.5 to 2.5
3	2.5 to 3.5
4	3.5 to 4.5

and so on. Since a family cannot have 2.5 members, there is no problem about which interval it is to be counted in. Figure 6 shows the histogram for family size.

Figure 6. Histogram showing distribution of families by size. With a discrete variable, the class intervals are centered at the possible values.

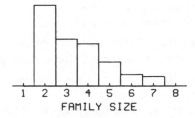

FAMILY SIZE

Source: *Statistical Abstract*, 1976, Table 58.

EXERCISE SET D

1. Classify each of the following variables as qualitative or quantitative: if quantitative, as discrete or continuous.
    (a) occupation   (b) region of residence   (c) weight   (d) height
    (e) number of automobiles owned.

2. In 1960, the Public Health Service interviewed a representative cross section of Americans aged 18 to 79. One of the questions asked of the women was "How many children have you had?" The results are shown below.
    (a) Is this variable discrete or continuous?
    (b) Draw a histogram for these data. You may take "9 or more" as 9—very few women had more than 9 children.

Distribution of women by number of children born.

Number of children	Percent of women
0	22
1	17
2	21
3	16
4	10
5	5
6	3
7	2
8	2
9 or more	3

Source: Microdata tape supplied by the National Center for Health Statistics.

The answers to these exercises are on p. A-27.

5. CONTROLLING FOR A VARIABLE

In the 1960s, many women began using oral contraceptives, "the pill." Since the pill alters the body's hormone balance, it is important to see what the side effects are. Research on this question is carried out by the Contraceptive Drug Study at the Kaiser Clinic in Walnut Creek, California.[7] Over 20,000 women in the Walnut Creek area belong to the Kaiser Foundation Health Plan, paying a monthly insurance fee and getting medical services from Kaiser. One of these services is a routine checkup called the "multiphasic." During the period 1969–1971, about 17,500 women aged 17 to 58 took the multiphasic and became subjects for the Drug Study.

Investigators try to relate the findings of the multiphasic to pill use by the subjects. Following the method of comparison, they look at the multiphasic findings for two different groups of women:

   • "users" who take the pill (the treatment group);
   • "nonusers" who don't take the pill (the control group).

This is an observational study, for it is the women themselves who have decided whether to take the pill or not.

One issue taken up by the Drug Study was the effect of the pill on blood pressure.[8] In comparing the blood pressures of users and nonusers, about 3,500 women are excluded, because they were pregnant, post-partum, or taking hormonal medication other than the pill; these factors affect blood pressure.

It might seem natural to compare the histograms of the blood pressures for the users and nonusers among the remaining women. However, this comparison would be misleading. Blood pressure tends to go up with age, and the nonusers are on the whole older than the users. For example, about 70% of the nonusers were over thirty; only 48% of the users were over thirty. The effect of age is confounded with the effect of the pill, and the two work in opposite directions. To make the full effect of the pill visible, it is necessary to make a separate comparison for each group: this controls for age.[9] The results turn out to be quite similar for each age group, so we will look only at the women aged 25 to 34. Figure 7 shows the histograms for the users and nonusers in this age group.

The two histograms in the top panel of Figure 7 have very similar shapes. However, the user histogram is higher to the right of 120 mm., lower to the left. High blood pressure (above 120 mm.) is more prevalent among users, low blood pressure less prevalent. Now imagine that 5 mm. were added to the blood pressure of each nonuser. That would shift the nonuser histogram 5 mm. to the right, as shown in the bottom panel of Figure 7. The two histograms are now so intertwined that the difference between them could easily be due to chance. As far as the histograms are concerned, it is as if using the pill adds about 5 mm. to the blood pressure of each woman.

Figure 7. The effect of the pill. The top panel shows histograms for the systolic blood pressures of the 1,747 users and the 3,040 nonusers aged 25 to 34 in the Contraceptive Drug Study.[10] The bottom panel shows the histogram for the nonusers shifted to the right by 5 mm.

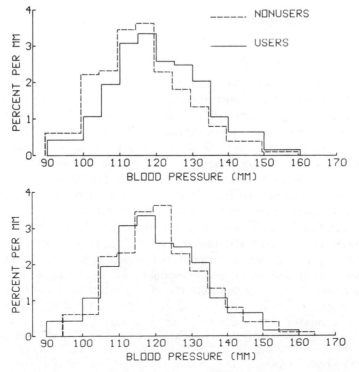

This conclusion must be treated with caution. The results of the Contraceptive Drug Study suggest that if a woman went on the pill, her blood pressure would go up by around 5 mm. But the proof is not complete. It cannot be, because of the design. The Drug Study is an observational study, not a controlled experiment. Part I showed that observational studies can be misleading about cause-and-effect relationships. There could be some factor other than the pill or age, as yet unidentified, which is affecting the blood pressures. For the Drug Study, this is a bit farfetched. The physiological mechanism by which the pill affects blood pressure is well established. The Drug Study data discussed here document the size of the effect.

EXERCISE SET E

1. As a sideline, the Drug Study compared blood pressures for women having different numbers of children. Here are rough sketches of the histograms for women with two or four children. Which group has the higher blood pressures? Does having children cause the blood pressures of the mothers to change? Or could the change be due to some other factor, whose effects are confounded with the effect of the number of children?

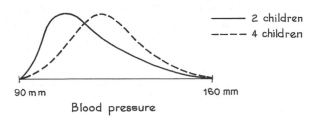

The answer to this exercise is on p. A-27.

6. CROSS-TABULATION

The previous section explained how to control for the effect of age: it was a matter of doing the comparison separately for each age group. The comparison was made graphically, through the histograms in Figure 7. Many investigators prefer to make the comparison in tabular form, using what is called a *cross-tab* (short for *cross-tabulation*). A cross-tab for blood pressure by age and pill use is shown in Table 2. Such tables are a bit imposing, and the eye naturally tends to skip over them until some of the numbers are needed. But all Table 2 amounts to is a distribution table for blood pressures, made separately for the users and nonusers of each age group.

Look at the columns for the age group 17–24. There were 1,024 users and 1,206 nonusers. About 1% of the users had blood pressure below 90 mm.; the corresponding percentage of nonusers was negligible—that is what the dash means. (Each column of percents should total up to 100, but may be off by 1% or so due to rounding off.) To see the effect of the pill on the blood pressures of women aged 17 to 24, it is a matter of looking at the percents in the columns for users and nonusers in the age group 17 to 24. To see the effect of age, look first at the nonusers column in each age group and see how the percents shift toward the high blood pressures as age goes up. Then do the same thing for the users.

Table 2. Distributions of systolic blood pressures, cross-tabulated by age and pill use, for women in the Contraceptive Drug Study, excluding those who were pregnant or taking hormonal medication other than the pill. Class intervals include the left endpoint, but not the right.— means negligible. Percents may not add to 100 due to rounding.

Blood pressure (millimeters)	Age 17–24 Non-users	Users	Age 25–34 Non-users	Users	Age 35–44 Non-users	Users	Age 45–58 Non-users	Users
	(%)	(%)	(%)	(%)	(%)	(%)	(%)	(%)
under 90	—	1	1	—	1	1	1	—
90–95	1	—	1	—	2	1	1	1
95–100	3	1	5	4	5	4	4	2
100–105	10	6	11	5	9	5	6	4
105–110	11	9	11	10	11	7	7	7
110–115	15	12	17	15	15	12	11	10
115–120	20	16	18	17	16	14	12	9
120–125	13	14	11	13	9	11	9	8
125–130	10	14	9	12	10	11	11	11
130–135	8	12	7	10	8	10	10	9
135–140	4	6	4	5	5	7	8	8
140–145	3	4	2	4	4	6	7	9
145–150	2	2	2	2	2	5	7	9
150–155	—	1	1	1	1	3	2	4
155–160	—	—	—	1	1	1	1	3
160 and over	—	—	—	—	1	2	2	5
Total percent	100	98	100	99	100	100	99	99
Total number	1,206	1,024	3,040	1,747	3,494	1,028	2,172	437

EXERCISE SET F

1. Using the cross-tab in Table 2, answer the following:
   (a) What percentage of users aged 17 to 24 have blood pressure of 140 mm. or more?
   (b) What percentage of nonusers aged 17 to 24 have blood pressures of 140 mm. or more?
   (c) What can be concluded from comparing (a) and (b)?

2. Draw histograms for the blood pressures of the users and nonusers aged 17 to 24. What can be concluded?

3. Compare the histograms of blood pressures for nonusers aged 17 to 24 and for nonusers aged 25 to 34. What can be concluded?

The answers to these exercises are on p. A-27.

7. SELECTIVE BREEDING

In 1927, the psychologist Charles Spearman published his book *The Abilities of Man,* presenting a theory of human intelligence. Briefly, Spearman held that scores on tests of different intellectual abilities (like reading comprehension, arithmetic, or spatial perception) varied together in a way which could only be explained as follows: the score on each test is the weighted sum of two independent components, a general intelligence factor which Spearman called "g," and an ability factor specific to each test. This theory attracted a great deal of attention.

Robert Tryon, a professor of psychology at the University of California, Berkeley, decided to check the theory on an animal population, where extraneous variables are much easier to control.[11] Tryon used rats, which are easy to breed in the laboratory. As a test of intelligence, he used maze-running. The rats were put into a maze. As they ran it, they made errors by going into blind alleys. The "intelligence test" consisted of nineteen runs through the maze; the animal's score was the total number of errors it made. So the bright rats are the ones with low scores, the dulls are the ones with high scores. Tryon started out with 142 rats, and the distribution of their "intelligence" scores is sketched in Figure 8.

Figure 8. Tryon's experiment. Distribution of intelligence in original population.

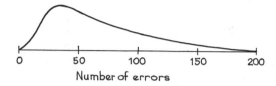

The next step in the experiment was to breed selectively for intelligence. In each generation, the "maze-bright" rats (the ones making only a small number of errors) were bred with each other; similarly, the "maze-dull" animals (with high scores) were bred together. Seven generations later, Tryon had 85 rats in the maze-bright strain, and 68 in the maze-dull strain. And there was a clear separation in their scores. Figure 9 shows the distribution of intelligence for the two groups, and the histograms barely overlap. (In fact, Tryon went on with his selective breeding past the seventh generation, but he was not able to get more of a separation than the one shown in Figure 9.)

Figure 9. Tryon's experiment. After seven generations of selective breeding, there is a clear separation into "maze-bright" and "maze-dull" strains.

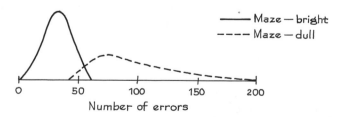

The two strains created by Tryon are still used (for other experiments) by the Berkeley psychology department, and they have been kept separated in breeding. Many hundreds of generations have gone by since the experiment, but rats from the maze-bright population still outperform the maze-dulls—at maze-running.

So Tryon did succeed in selectively breeding for a mental ability. This was strong evidence for the view that some mental abilities are at least in part genetically determined. The experiment tells us something very interesting about mental abilities. But what did it say about Spearman's theory? Tryon found that the maze-bright rats did no better than the maze-dulls on any other test of animal intelligence, such as discriminating between geometric shapes, or between intensities of light. This seems to be strong evidence against Spearman's theory of a general intelligence factor—at least for rats.

On the other hand, Tryon did find intriguing psychological differences between the two rat populations. The "brights" seemed to be unsociable introverts, well adjusted to life in the maze, but neurotic in their relationships with other rats; the "dulls" were quite the opposite.

8. REVIEW EXERCISES

1. The figure below shows a histogram for the heights of a representative sample of men. The shaded area represents the percentage of men whose heights were between __ and __. 66 - 72

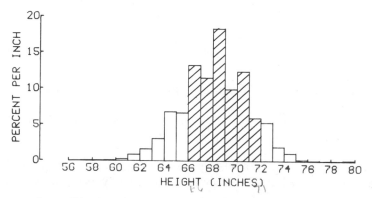

Source: Microdata tape supplied by the National Center for Health Statistics.

2. The distribution by age (for men and women together) for the U.S. population in 1970 is shown below. (The class intervals include the left endpoint, not the right. For instance, from the second line of the table, 18% of the people were aged 5 years or more but had not yet turned 14. The interval for "75 and over" can be ended at 85.) Draw the histogram.

Age	Percent of population	Age	Percent of population
0–5	8	35–45	11
5–14	18	45–55	11
14–18	8	55–65	9
18–21	5	65–75	6
21–25	6	75 and over	4
25–35	12		

Source: *Statistical Abstract,* 1973, Table 34.

Use your histogram to answer the following questions.
  (a) Were there more people aged 5 to 15, or 15 to 25?
  (b) The percentage of people aged 30 and over is closest to: 25%, 50%, or 75%.
  (c) The percentage of 18 year olds in the population is closest to: 2%, 5%, or 8%.

3. Data from the 1970 Census can be used to find the distribution of occupied housing units (this includes apartments) by number of rooms. The Census does this separately for "owner-occupied" and "renter-occupied" units, with the results shown below. Draw a histogram for each of these two distributions. (You may assume that the interval "7 or more" means 7 to 10; very few units have more than 10 rooms.)

Number of rooms in unit	Owner-occupied (percent)	Renter-occupied (percent)
1		3
2	3	7
3		21
4	14	30
5	28	21
6	27	11
7 or more	28	7

Source: *Statistical Abstract*, 1976, Table 1278.

  (a) What is the difference between owner-occupied and renter-occupied units?
  (b) For owner-occupied units, the Census collapsed 1, 2, and 3 rooms into one class interval. Why?

4. The figure below is a histogram showing the distribution of blood pressure for all 14, 148 women in the drug study (see pp. 38–40).

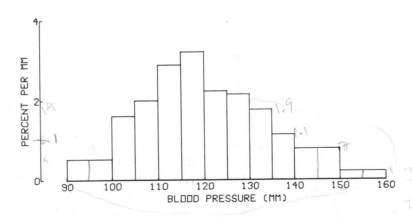

Use the histogram to answer the following questions:
  (a) Is the percentage of women with blood pressure above 130 mm. closest to 25%, 50%, or 75%?
  (b) Which interval has more women: 97–98 mm. or 102–103 mm.?

(c) Where is the millimeter which is most crowded of all?

(d) On the interval 125 mm. to 130 mm., the height of the histogram is about 2.1% per mm. What percentage of the women had blood pressures in this class interval?

(e) Is the percentage of women with blood pressures between 90 mm. and 160 mm. closest to 1%, 50%, or 99%?

(f) Which interval is more crowded: 135–140 mm. or 140–150 mm.?

(g) In which interval are there more women: 135–140 mm. or 140 mm–150 mm.?

5. Someone has sketched one block of a family-income histogram for a wealthy suburb. About what percentage of the families in this suburb had incomes between $40,000 and $50,000 a year?

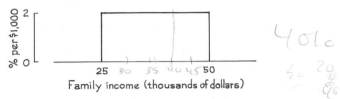

6. Three people have tried to sketch the histogram for blood pressures of the subjects in a certain study, using the density scale. Only one is right. Which one, and why?

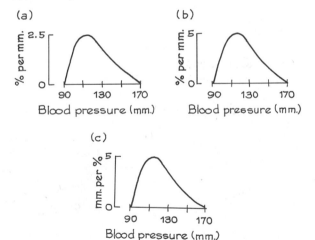

(a)

(b)

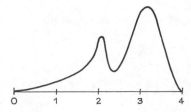

(c)

7. In a survey carried out at the University of California, Berkeley, a sample of students were interviewed and asked what their grade-point average was. A histogram of the results is shown below. How should the spike at 2 be interpreted? (GPA ranges from 0 to 4, and 2 is a bare pass.)

8. (Hypothetical.) In one study, 100 people had their heights measured to the nearest eighth of an inch. A histogram for the results is shown below. Two of the following lists have this histogram. Which ones, and why?
   (i)   25 people, 67 inches tall; 50 people, 68 inches tall; 25 people, 69 inches tall.
   (ii) 10 people, 66¾ inches tall; 15 people, 67¼ inches tall; 50 people, 68 inches tall; 25 people, 69 inches tall.
   (iii) 30 people, 67 inches tall; 40 people, 68 inches tall; 30 people, 69 inches tall.

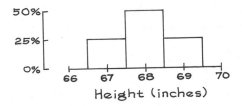

9. The table below shows the distribution of adults by the last digit of their age, as reported in the Census of 1880 and the Census of 1970.[12] You might expect each of the ten possible digits to turn up for 10% of the people, but this is not the case. For example, in 1880, 16.8% of all persons reported an age ending in 0—like 30 or 40 or 50. In 1970, this percentage was only 10.6%.
   +(a) Draw histograms for these two distributions.
    (b) In 1880, there was a strong preference for the digits 0 and 5. How can this be explained?
    (c) In 1970, the preference was much weaker. How can this be explained?
    (d) Are even digits more popular, or odd ones, in 1880? 1970?

Digit	1880	1970
0	16.8	10.6
1	6.7	9.9
2	9.4	10.0
3	8.6	9.6
4	8.8	9.8
5	13.4	10.0
6	9.4	9.9
7	8.5	10.2
8	10.2	10.0
9	8.2	10.1

Source: United States Census.

10. In Chicago, city engineers are hired on the basis of a competitive civil-service examination. In 1966, there were 223 applicants for 15 jobs. The exam was held on March 12, and the test scores are shown below (arranged in increasing order). So is a histogram. (The height of each bar shows the number of people with that score.) On the basis of this evi-

dence, the examiners were charged with rigging the exam.[13] Why?

26,27,27,27,27   29,30,30,30,30   31,31,31,32,32   33,33,33,33,33
34,34,34,35,35   36,36,36,37,37   37,37,37,37,37   39,39,39,39,39
39,39,40,41,42   42,42,42,42,43   43,43,43,43,43   43,43,44,44,44
44,44,44,45,45   45,45,45,45,45,   46,46,46,46,46   46,47,47,47,47
47,47,48,48,48   48,48,48,48,48   49,49,49,49,50   50,51,51,51,51

51,52,52,52,52   52,53,53,53,53   53,54,54,54,54   54,55,55,55,56
56,56,56,56,57   57,57,57,58,58   58,58,58,58,58   58,59,59,59,59
60,60,60,60,60   60,61,61,61,61   61,61,62,62,62   63,63,64,65,66
66,66,67,67,67   67,68,68,69,69   69,69,69,69,69   69,71,71,72,73
74,74,74,75,75   76,76,78,80,80   80,80,81,81,81   82,82,83,83,83

83,84,84,84,84   84,84,84,90,90   90,91,91,91,92   92,92,93,93,93
93,95,95

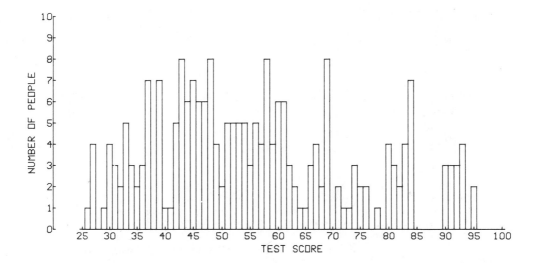

9. SUMMARY

1. A *histogram* represents percents by area. It consists of a set of blocks. The area of each block represents the percentage of cases in the corresponding *class interval*.

2. To plot a histogram using the *density scale,* take the height of each block to be the percentage of cases in the corresponding class interval divided by the length of that interval.

3. When a histogram is drawn with the density scale on the vertical axis, the total area is 100%. The area between two values gives the percentage of cases falling in that interval.

4. A *variable* is a characteristic of the subjects in a study. It can be either *qualitative* or *quantitative*. A quantitative variable can be either *discrete* or *continuous*.

5. A confounding factor is sometimes controlled for by *cross-tabulation*.

4

The Average and the Standard Deviation

It is difficult to understand why statisticians commonly limit their enquiries to Averages, and do not revel in more comprehensive views. Their souls seem as dull to the charm of variety as that of the native of one of our flat English counties, whose retrospect of Switzerland was that, if its mountains could be thrown into its lakes, two nuisances would be got rid of at once.

—SIR FRANCIS GALTON (ENGLAND, 1822–1911)[1]

1. INTRODUCTION

A histogram can be used to summarize large amounts of data. Often, an even more drastic summary is possible, giving just the center of the histogram and measuring the spread around the center. (*Center* and *spread* are ordinary words here, without any special technical meaning.) Figure 1 shows a rough sketch of two histograms with the center and spread indicated. Both histograms have the same center, but the second one is more spread out—there is more area farther away from the center. To make statistical use of these ideas, they have to be given a definite interpretation, and there are several ways to go about this. The center is usually indicated by the *average* or by the *median*.[2] The corresponding ways to measure spread around the center are the *standard deviation* and the *interquartile range*.

Histograms like those in Figure 1 can be summarized quite well by the center and the spread. However, some histograms just shouldn't be sum-

Figure 1. Center and spread. The centers of the two histograms are the same, but the second histogram is more spread out.

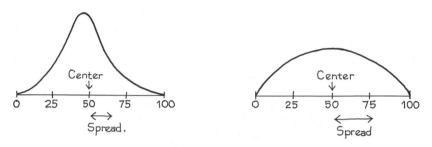

marized this way. For instance, Figure 2 gives the distribution of elevation on the earth's surface. Elevation is shown along the horizontal axis, in miles above (+) or below (−) sea level. The area under the histogram between two elevations gives the percentage of the surface area of the earth which is between these two elevations. There are two definite peaks in this histogram, for the bulk of the surface area of the earth is near one of two elevations: the sea floors, around three miles below sea level; and the continental plains, around sea level. Reporting only the center and spread of this histogram would ignore its most interesting feature: the two peaks.[3]

Figure 2. Distribution of the surface area of the earth by elevation above (+) or below (−) sea level.

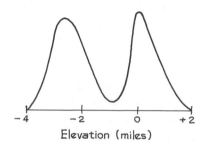

2. THE AVERAGE

The object of this section is to review the idea of the average, in the context of the Health Examination Survey. The difference between *cross-sectional* and *longitudinal* surveys will also be discussed. In 1960–1962, the Public Health Service examined a representative cross section of 6,672 Americans aged 18 to 79 in Cycle I of the Health Examination Survey.[4] The objective was to get base-line data about

- physiological variables like height, weight, and blood pressure;
- psychological variables, like nervousness;
- demographic variables, like age, education, and income;
- prevalence of diseases, especially of the heart.

Some of the results were quite disturbing.

• About 13% of those in the sample suffered from heart disease, according to the diagnostic criteria used by the Public Health Service.

• With another 12%, heart disease was suspected but could not be confirmed.

• About 10% of the men and about 24% of the women reported having had a nervous breakdown or having felt that one was impending.

Subsequent analysis focused on the interrelationships among the variables: what kinds of people are likely to have heart trouble? nervous breakdowns? But the point here is just to take a quick look at the sample, reviewing the idea of the average.

> The average of a list of numbers equals their sum, divided by how many there are.

For instance, the list 9, 1, 2, 2, 0 has 5 entries, the first being 9, and its average is

$$\frac{9 + 1 + 2 + 2 + 0}{5} = \frac{14}{5} = 2.8$$

What did the men and women in the sample look like?

• The average height of the men was 5 feet 8 inches, and their average weight was 168 pounds.

• The average height of the women was 5 feet 3 inches, and their average weight was 142 pounds.

They're pretty chubby!

How are height and weight related to age? Figure 3 shows the average heights and weights separately for the men and women in the seven age groups[5] studied by the Public Health Service; these averages are joined by straight lines in the graph. Figure 3 exemplifies the tremendous power of the average in summarizing data—twenty-eight histograms are compressed into the four curves. But this compression is achieved only by smoothing away all the individual differences. To be more specific, the average height for the 18-to-24-year-old men was 5 feet 9 inches. But about 10% of them were taller than 6 feet; and 10% were shorter than 5 feet 4 inches. These two groups of men have been balanced out in the average and hidden behind the single figure of 5 feet 9 inches.

In Figure 3, the average height of men appears to decrease after age 30, dropping about three inches in fifty years. Similarly for women. Does this show that an average person gets shorter at this rate? Not really, because the Health Examination Survey is *cross-sectional,* not *longitudinal.* In a cross-sectional study, different subjects are compared to one another at one point in time. In a longitudinal study, subjects are followed over time, and compared with themselves at different points in time.

The people aged 25 to 34 in Figure 3 are completely different from

Figure 3. Age-specific average heights and weights for men and women in
the Health Examination Survey. The panel on the left shows height, the
panel on the right shows weight.

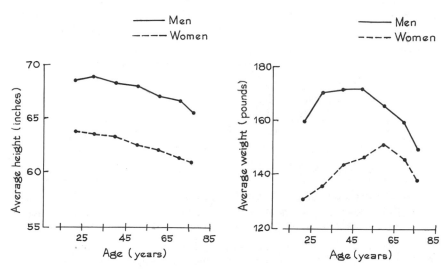

Source: Microdata tape supplied by the National Center for Health Statistics.

those aged 75 to 79. The first group was born around 1930, the second around
1880. How does this affect the picture? There is good evidence to suggest that
Americans have been getting taller. On the average, the mature height of a
person born in 1930 appears to be about an inch more than the mature height
of one born in 1880. This is called the *secular trend* in height. It is confounded
with the effect of aging in Figure 3. We guess that the average person loses
about two inches of height between ages 30 and 80. The other inch is lost due
to the secular trend: the 75 to 79 year olds were born around 50 years before
the 25 to 34 year olds, and are an inch shorter for that reason.

Why do people get shorter as they get older? Here is one possible
explanation. About two inches of height is made up of air spaces between the
bones in the body. The body settles in on itself with age, so these air spaces
get smaller and smaller.

EXERCISE SET A

1. (a) The numbers 3 and 5 are marked by crosses on the horizontal line below.
   Find the average of these two numbers and mark it by an arrow.

   (b) Repeat (a) for the list 3, 5, 5.

(c) Two numbers are shown below by crosses on a horizontal axis. Draw an arrow pointing to their average.

2. A list has ten entries. Each entry is either a 1, or a 2, or a 3. What must the list be if the average is 1? If the average is 3? Can the average be 4?

3. Which of the following two lists has a bigger average? Or are they the same? Try to answer without calculating.
   (i)   10, 7, 8, 3, 5, 9   (ii) 10, 7, 8, 3, 5, 9, 11.

4. (a) For each of the lists below: draw the histogram (the histogram for the first list is shown in Figure 5 on p. 53), find the average, and draw an arrow pointing to the average on the horizontal axis.
      (i) 1, 2, 2, 3   (ii) 2, 3, 3, 4   (iii) 2, 4, 4, 6   (iv) −1, −2, −2, −3.

   (b) Each list is related to list (i). For example, list (ii) can be obtained from list (i) by adding 1 to each entry. How are lists (iii) and (iv) related to list (i)? How do these relationships carry over to the histograms? To the averages?

5. Ten people in a room have an average height of 5 feet 6 inches. An eleventh person, who is 6 feet 5 inches tall, enters the room. Find the average height of all eleven people.

6. Twenty-one people in a room have an average height of 5 feet 6 inches. A twenty-second person, who is 6 feet 5 inches tall, enters the room. Find the average height of all twenty-two people. Compare with Exercise 5.

7. Twenty-one people in a room have an average height of 5 feet 6 inches. A twenty-second person enters the room. How tall would he have to be to raise the average height by 1 inch?

8. An instructor gives a quiz with three questions, each worth 1 point: 40% of the class score 3 points, 30% score 2 points, 20% score 1 point, and 10% score 0:

Score	Percent
3	40
2	30
1	20
0	10

   (a) If there were ten people in the class, what would the average score be?
   (b) If there were twenty people in the class, what would the average score be?
   (c) Suppose you are not told the number of people in the class. Can you still figure out the average score?

9. Average hourly earnings are computed each month by the Bureau of Labor Statistics from payroll data submitted by commercial establishments. The Bureau figures the total wages paid out (to nonsupervisory personnel), and divides by the total hours worked. During recessions, average hourly earnings typically go up. When the recession ends, average hourly earnings often start going down. How can this be?

The answers to these exercises are on pp. A-27–A-28.

3. THE AVERAGE AND THE HISTOGRAM

This section will indicate how to read the average, by eye, from a histogram. It will also show the relationship between the average and the median. To begin with an example, Figure 4 shows a histogram for the weights of the 3,581 women in the Health Examination Survey. The average (142 pounds) is marked with a vertical line. It is natural to guess that 50% of the women were above average in weight, and 50% were below average. However, this guess is somewhat off. In fact, only 43% were above average, and 57% were below. And in other situations, the percentages can be even farther from 50%.

Figure 4. Histogram for the weights of the 3,581 women in the Health Examination Survey. The average is marked by the dashed line. Only 43% of the women are above average in weight.

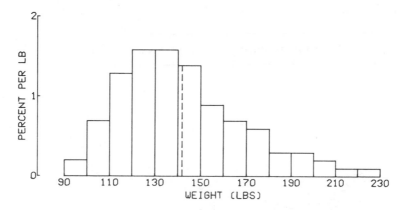

Source: Microdata tape supplied by the National Center for Health Statistics.

How is this possible? To find out, it is easiest to start with some hypothetical data—for instance, the list 1, 2, 2, 3. The histogram for this list is shown in Figure 5. It is *symmetric* about the value 2—that is, imagine drawing a vertical line through the value 2, and folding the histogram in half around that line; the two halves would match up. And, as can be checked by arithmetic, the average equals 2. This is always the case: if the histogram is symmetric around a value, that value equals the average; furthermore, half the area under the histogram lies to the left of that value, and half to the right.

Figure 5. Histogram for the hypothetical data 1, 2, 2, 3. The histogram is symmetric around 2, the average: 50% of the area is to the left of 2, and 50% to the right.

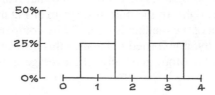

What happens when the value 3 on the list 1, 2, 2, 3 of Figure 5 is increased say to 5 or 7? As shown in Figure 6 the rectangle over that value moves off to the right, destroying the symmetry of the histogram. The average for each histogram is marked with an arrow, and the arrow shifts to the right following the shaded rectangle. To see why the arrow shifts, imagine the histogram cut out of a board, and glued to a stiff, weightless number line. Place it across a taut wire, at the average, as illustrated in the bottom panel of Figure 6. The histogram will balance there. In this balance, areas are weighted by their distance from the average, so that a small area far away from the average can balance a large area close to the average. It is just like a seesaw: the smaller child sits farther away from the center in order to balance a larger child sitting closer to the center. That is why the percentage of cases on each side of the average can differ from 50%. To summarize this part of the discussion, the average locates the center of a histogram in the following precise manner:[6]

A histogram balances when supported at the average.

The *median* of a histogram is the value such that half the area is to the left of it and half to the right. In the third histogram of Figure 6, the median is 2. The area to the right of the median is far away by comparison with the area to the left. Consequently, if you tried to balance this histogram at the median, it would tip to the right. More generally, the average is to the right of the

Figure 6. The average. The top panel shows three histograms for hypothetical data. Averages are marked with an arrow. As the shaded box moves to the right, it pulls the average along with it. The area to the left of the average gets up to 75%. The bottom panel shows the same three histograms cut out of a board and glued to a stiff, weightless number line. They balance when supported at the average.

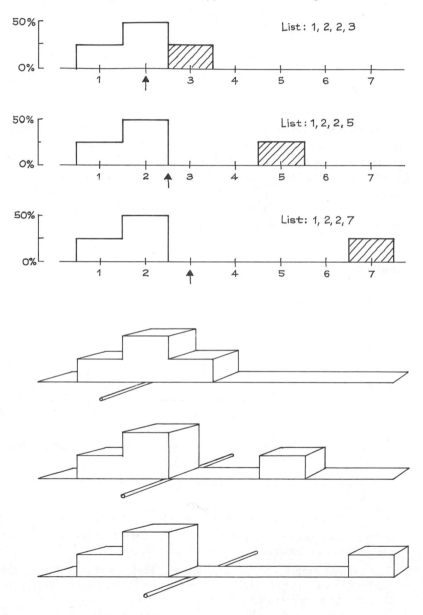

median whenever the histogram has a long right-hand tail, as illustrated in Figure 7. The weight histogram (Figure 4 on p. 53) had a long right-hand tail; that is why the average was bigger than the median.

Figure 7. The tails of a histogram.

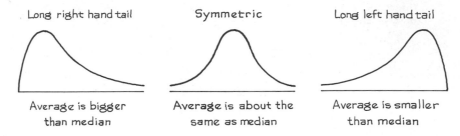

Long right hand tail	Symmetric	Long left hand tail
Average is bigger than median	Average is about the same as median	Average is smaller than median

The median is often used to summarize demographic data. For instance, median family income in the United States in 1973 was about $12,000.[7] The income histogram has a long right-hand tail (Figure 1 on p. 26), so the average income was higher—by about $1,500. Statisticians often use the median rather than the average when dealing with long-tailed distributions, the reason being that in some cases the average might pay too much attention to a small percentage of cases in the extreme tail of the distribution.

EXERCISE SET B

1. Below are rough sketches of histograms for three lists. Guess whether the average of each list is closest to 25, 40, 50, 60, or 75.

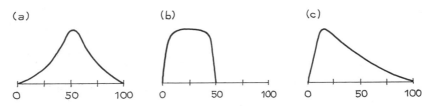

(a)   (b)   (c)

2. For each histogram in Exercise 1, say whether the median is equal to the average, or is to the left of it, or to the right.

3. Look back at the cigarette histogram on p. 36. The median is closest to: 10, 20, 30, 40.

4. For this cigarette histogram, the average is closest to: 15, 20, 25.

5. For registered students at universities in the United States, which is larger: average age or median age?

6. (This exercise prepares for the next section.) For each of the following lists of numbers, say whether the entries are on the whole closest in size to: 1, 5, or 10. No calculations are needed.
    (a) 1.3, 0.9, 1.2, 0.8   (b) 13, 9, 12, 8   (c) 7, 3, 6, 4   (d) 7, −3, −6, 4.

The answers to these exercises are on p. A-29.

Technical note: The median of a list is defined by the property that half or more of the entries are at the median or above, and half or more are at the median or below.

Example. Find the median for each of the following lists:

(a) 1, 5, 7   (b) 1, 2, 5, 7   (c) 1, 2, 2, 7, 8   (d) 8, −3, 5, 0, 1, 4, −1.

Solution. For list (a), the median is 5: two entries out of three are 5 or more, and two are 5 or less. For list (b), any value between 2 and 5 would serve as a median; if pressed, most statisticians would choose 3.5 (which is halfway between 2 and 5) as "the" median. For list (c), the median is 2: four entries out of five are 2 or more, and three are 2 or less. To find the median of list (d), arrange it in increasing order:

$$-3, \quad -1, \quad 0, \quad 1, \quad 4, \quad 5, \quad 8.$$

There are seven entries on this list: four are 1 or more, and four are 1 or less. So 1 is the median.

4. THE ROOT-MEAN-SQUARE

The next main topic in the chapter is the *standard deviation,* which is used to measure the overall size of the deviations from the average. This section presents a mathematical preliminary, the operation of taking the *root-mean-square*. This operation can be applied to any list of numbers, to measure their overall size—neglecting signs. For example, take the list:

$$0, \quad 5, \quad -8, \quad 7, \quad -3.$$

On the whole, how big are these five numbers? Their average is 0.2, but this gives a poor impression of their overall size: it only means that to a large extent, the positives and negatives canceled each other out. The simplest way around this would be to wipe out the signs and then take the average. However, statisticians do something else: they apply the root-mean-square operation to the list. The phrase "root-mean-square" itself says how to do the computation, provided you remember to read it backwards.

Step 1. SQUARE all the entries, getting rid of the signs.
Step 2. Take the MEAN (average) of the squares.
Step 3. Take the square ROOT of the mean.

This can be expressed as an equation, with root-mean-square abbreviated to r.m.s.:

$$\text{r.m.s. size of a list} = \sqrt{\text{average of (entries)}^2}$$

Example. For the list 0, 5, −8, 7, −3 find the average, the average neglecting signs, and the r.m.s. size.

Solution.

$$\text{average} = \frac{0 + 5 - 8 + 7 - 3}{5} = 0.2$$

$$\text{average neglecting signs} = \frac{0 + 5 + 8 + 7 + 3}{5} = 4.6$$

$$\text{r.m.s. size} = \sqrt{\frac{0^2 + 5^2 + (-8)^2 + 7^2 + (-3)^2}{5}}$$

$$= \sqrt{29.4} \approx 5.4$$

The r.m.s. size is a little bigger than the average neglecting signs. It always turns out like that, except in the trivial case when all the entries are of the same size. The root and the square do not cancel each other out, due to the intervening operation of taking the mean.

There doesn't seem to be much to choose between the 5.4 and the 4.6 as a measure of the overall size for the list in the example. And there isn't. Then why do statisticians use the r.m.s. operation? The reason is that it fits in much better with the algebra that they have to do.[8] Whether this explanation is appealing or not, don't worry. Nobody likes the r.m.s. at first, but everyone gets used to it very quickly.

EXERCISE SET C

1. (a) Calculate the average and the r.m.s. size of the numbers on the list
        1, −3, 5, −6, 3.
   (b) Do the same for the list −11, 8, −9, −3, 15.

2. Guess whether the r.m.s. size of each of the following lists of numbers is closest to 1, or 10, or 20. No arithmetic is required.
       (a) 1, 5, −7, 8, −10, 9, −6, 5, 12, −17.
       (b) 22, −18, −33, 7, 31, −12, 1, 24, −6, −16.
       (c) 1, 2, 0, 0, −1, 0, 0, −3, 0, 1.

3. For which of the following three lists is the r.m.s. size the smallest? the largest? No arithmetic is required.
       (i)   −6, −3, −1, 0, 5, 5.
       (ii) −3, −2, 0, 1, 1, 3.
       (iii) −4, −4, 0, 2, 3, 3.

4. (a) Find the r.m.s. size of the numbers on the list 7, 7, 7, 7.
   (b) Repeat, for the list 7, −7, 7, −7.

5. Each of the numbers 103, 96, 101, 104 is almost 100 but is off by a certain amount. Find the r.m.s. size of the amounts off.

6. The list 103, 96, 101, 104 has an average. Find it. Each number in the list is off the average by some amount. Find the r.m.s. size of the amounts off.

7. A computer is programmed to predict test scores, compare them with actual scores, and find the r.m.s. size of the prediction errors. Glancing at the print-

out, you see the r.m.s. size of the prediction errors is 3.6, and the following results for the first ten students:

    predicted score: 90 90 87 80 42 70 67 60 83 94

    actual score: 88 70 81 85 63 77 66 49 71 69

Is anything wrong?

The answers to these exercises are on p. A-29.

5. THE STANDARD DEVIATION

As the quote at the beginning of the chapter suggests, it is often very helpful to think of the way a list of numbers spreads out around its average. This spread is usually measured by a quantity called the *standard deviation, abbreviated to SD.* This quantity may be unfamiliar, but, as the name indicates, it says something about the size of the deviations of the individuals from their average. It is a sort of average deviation. The program is first to see how to interpret the SD in the context of real data, and second to see how to calculate the SD in the context of hypothetical data where the arithmetic is easy to do.

There were 1,747 pill users aged 25 to 34 in the Contraceptive Drug Study (see pp. 38–40). Their average blood pressure was nearly 121 mm., with an SD of about 12.5 mm. These two quantities summarize quite a lot of the information that would be contained in a full list of all 1,747 blood pressures. The fact that the average was 121 mm. tells us that most of the blood pressures were somewhere around 121 mm. But there were deviations from the average. Some of the blood pressures were higher than average, some lower. How big were these deviations? That is where the SD comes in.

> The SD says how far away numbers on a list are from their average. Most entries on the list will be somewhere around one SD away from the average. Very few will be more than two or three SDs away.

The SD of 12.5 mm indicates that many of the women had blood pressures which deviated from the average by 5, 10, 15, or 20 mm.—5 mm. is less than half an SD, and 20 mm. is between one and two SDs. Few of the women had blood pressures deviating from the average by more than 25 mm. (= two SDs). To be more quantitative: 68% of the users aged 25 to 34 had blood pressure deviating from the average by one SD or less, and 95% had blood pressures deviating from the average by two SDs or less. There was only one user in 1,747 whose blood pressure deviated from the average by more than five SDs. The histogram of blood pressures is shown in Figure 8 (next page). The average is marked by a vertical line, and the region within one SD of the average is shaded.

Figure 8. The SD. This histogram is for the blood pressures of the 1,747 users, aged 25 to 34, in the Contraceptive Drug Study. The average of 121 mm. is marked with a dashed vertical line. The region within one SD (= 12.5 mm.) of the average is shaded: 68% of the women differed from the average with respect to blood pressure by less than one SD.

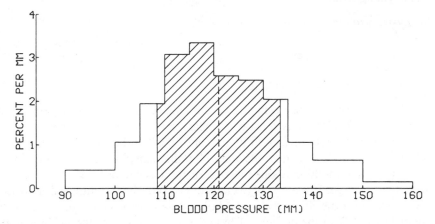

The next example is on the 3,581 women in the Health Examination Survey discussed in section 2 above (pp. 49–51). Their average height was 63 inches, with an SD of 2.5 inches. Figure 9 shows the histogram of heights; the average and the region within one SD of the average are marked as before. With respect to height, 68% of these women differed from the average by less than one SD, and 95% differed from the average by less than two SDs. No woman differed from the average by more than five SDs.

Figure 9. The SD. This histogram is for the heights of the 3,581 women in the Health Examination Survey. The average of 63 inches is marked with a dashed vertical line. The region within one SD (= 2.5 inches) of the average is shaded; 68% of the women differed from the average with respect to height by less than one SD.

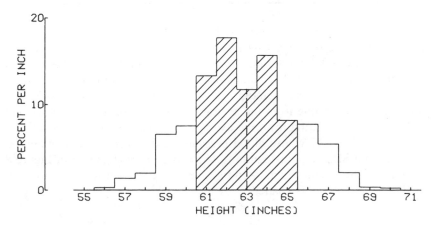

For Figure 8, the basic data is a list of all 1,747 blood pressures. This list has an average of 121 mm. and an SD of 12.5 mm.; 68% of the entries on the list are within one SD of the average, and 95% are within two SDs. For Figure 9, the basic data is a list of all 3,581 heights. This list has an average of 63 inches, and an SD of 2.5 inches. Again, 68% of the entries are within one SD of the average, and 95% are within two SDs.[9] For many sets of data, it comes out the same way.[10]

> Roughly 68% of the entries on a list (two in three) are within one SD of the average, the other 32% are further away. Roughly 95% (19 in 20) are within two SDs of the average, the other 5% are further away. This is so for many lists, but not all.

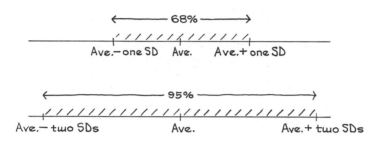

The qualification "roughly" must be kept in mind. For instance, 73% of the women in the Health Examination Survey differed from the average by one SD or less with respect to weight.

Where do the figures of 68% and 95% come from? That will be seen in the next chapter.

About two in three of the Public Health Service women differ from the average by less than one SD, with respect to height.

EXERCISE SET D

1. In another Public Health Service study (Cycle II, on children aged 6 to 11), it was found that for boys aged 11 the average height was about 146 cm., with an SD of about 7 cm. Just roughly, what percentage of boys aged 11 in the study had heights between 139 cm. and 153 cm.? Between 132 and 160 cm.?

2. Each of the following lists has an average of 50. For which one is the spread of the numbers around the average the greatest? the least?
   (i)   0, 20, 40, 50, 60, 80, 100.
   (ii) 0, 48, 49, 50, 51, 52, 100.
   (iii) 0, 1, 2, 50, 98, 99, 100.

3. Each of the following lists has an average of 50. For which one is the spread of the numbers around the average the greatest? the least?
   (i)   47, 49, 50, 51, 53.
   (ii) 46, 48, 50, 52, 54.
   (iii) 46, 49, 50, 51, 54.

4. Each of the following lists has an average of 50. For each one, guess whether the SD is around 1, around 2, or around 10. This does not require any calculations. The SD says how far off the average the entries are, on the whole. So just ask yourself whether the amounts off are on the whole more like 1, or 2, or 10 in size.
   (a) 49, 51, 49, 51, 49, 51, 49, 51, 49, 51.
   (b) 48, 52, 48, 52, 48, 52, 48, 52, 48, 52.
   (c) 48, 51, 49, 52, 47, 52, 46, 51, 53, 51.
   (d) 54, 49, 46, 49, 51, 53, 50, 50, 49, 49.
   (e) 60, 36, 31, 50, 48, 50, 54, 56, 62, 53.

5. The SD of the ages of the people in the Health Examination Survey is closest to:
   5 years   15 years   25 years   10 inches   20 pounds.
   Explain. (This survey was discussed on p. 49 above.)

6. "Educational level" means the number of years of schooling completed. The SD for the educational level of the people in the Health Examination Survey is closest to:
   1 year   4 years   9 years   3 pounds   6 mm.
   Explain.

7. Below are rough sketches of the histograms for three lists. Match the sketch with the description. Some descriptions will be left over. Give your reasoning in each case.
   (i)   ave ≈ 3.5, SD ≈ 1.   (iv) ave ≈ 2.5, SD ≈ 1.
   (ii) ave ≈ 3.5, SD ≈ 0.5.   (v)   ave ≈ 2.5, SD ≈ 0.5.
   (iii) ave ≈ 3.5, SD ≈ 2.   (vi) ave ≈ 4.5, SD ≈ 0.5.

(a)
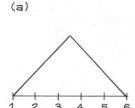

(b)
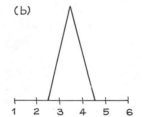

(c)
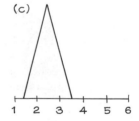

8. Make up a list of ten numbers so that the SD is as large as possible and
   (a) every number is either 1 or 5.
   (b) every number is either 1 or 9.
   (c) every number is either 1 or 5 or 9, and at least two of them are 5.
   This exercise does not require any arithmetic.

9. Repeat Exercise 8, this time making the SD as small as possible.

The answers to these exercises are on pp. A-29–A-30.

6. COMPUTING THE STANDARD DEVIATION

To find the standard deviation of a list, take the entries one at a time. Each deviates from the average by some amount (perhaps 0):

$$\text{deviation from average} = \text{entry} - \text{average}.$$

The SD is just the r.m.s. size of the deviations from the average. As a formula,

$$\boxed{\text{SD} = \sqrt{\text{average of (deviations from average)}^2}}$$

Example. Find the SD of the list 20, 10, 15, 15.

Solution. The first step is to find the average:

$$\text{average} = \frac{20 + 10 + 15 + 15}{4} = 15.$$

The second step is to find the deviations from the average by subtracting the average from each entry. These deviations are

$$5 \quad -5 \quad 0 \quad 0.$$

The last step is to take the r.m.s. size of the deviations:

$$\text{SD} = \sqrt{\frac{5^2 + (-5)^2 + 0^2 + 0^2}{4}}$$

$$= \sqrt{\frac{25 + 25 + 0 + 0}{4}}$$

$$= \sqrt{\frac{50}{4}} = \sqrt{12.5} \approx 3.5$$

This completes the calculation.

Do not confuse the r.m.s. size of a list, which measures the size of the numbers on the list, with the SD, which measures the size of the deviations from the average. The SD is the r.m.s., not of the numbers on the original list, but of their deviations from the average.

The SD comes out in the same units as the numbers which enter the calculation. For example, if the units are inches, as for the heights of the women in the Health Examination Survey, the SD of these heights will be in inches. The intermediate squaring step in the calculation alters the units to inches squared, but the final step of taking the square root returns the units to the original inches.[11]

EXERCISE SET E

1. Guess which of the following two lists has the larger SD. Check your guess by calculating the SD for both lists.
   (i)   9, 9, 10, 10, 10, 12.
   (ii) 7, 8, 10, 11, 11, 13.

2. Someone is telling you how to calculate the SD of the list 1, 2, 3, 4, 5.
   "The average is 3, so the deviations from average are
   $$-2 \quad -1 \quad 0 \quad 1 \quad 2.$$
   Dropping the signs, the average deviation is
   $$\frac{2 + 1 + 0 + 1 + 2}{5} = 1.2$$
   And that's the SD."
   What do you think of this method?

3. Someone is telling you how to calculate the SD of the list 1, 2, 3, 4, 5.
   "The average is 3, so the deviations from average are
   $$-2 \quad -1 \quad 0 \quad 1 \quad 2.$$
   The 0 doesn't count, so the r.m.s. deviation is
   $$\sqrt{\frac{4 + 1 + 1 + 4}{4}} \approx 1.6$$
   And that's the SD."
   What do you think of this method?

4. (a) For each list below, work out the average, all the deviations from average, and the SD.
   (i)   1, 3, 4, 5, 7.
   (ii) 6, 8, 9, 10, 12.
   (b) How is list (ii) related to list (i)? How does this relationship carry over to the average? the deviations from average? the SD?

5. Repeat Exercise 4 for the following two lists:
   (i)   1, 3, 4, 5, 7.
   (ii) 3, 9, 12, 15, 21.

6. Repeat Exercise 4 for the following two lists:
   (i)   5, −4, 3, −1, 7.
   (ii) −5, 4, −3, 1, −7.

7. (a) In 1975, Governor Brown of California proposed that all state employees be given a flat raise of $70 a month. What would this do to the average monthly salary of state employees? to the SD?
   (b) What would a 5% increase in the salaries, across the board, do to the average monthly salary? to the SD?

8. What is the r.m.s. size of the list 17, 17, 17, 17, 17? the SD?

9. For the list 107, 98, 93, 101, 104, which is smaller—the r.m.s. size or the SD? No calculations are necessary.

10. Can the SD ever be negative?

11. For a list of positive numbers, can the SD ever be larger than the average?

12. A list has ten numbers. Each number is either 1 or 2 or 3.
    (a) The average is 2 and the SD is 0. What is the list?
    (b) The SD is 1. What is the list?
    (c) Can the SD be bigger than 1?

13. An instructor gives a quiz with a total of 20 points. The grading policy is such that the scores are necessarily multiples of 5. He gets the following distribution:

Score	Percent
20	40
15	30
10	20
5	10

    (a) Suppose there are 10 people in the class. Can you work out the average and SD for the scores?
    (b) Suppose there are 20 people in the class. Can you work out the average and SD for the scores?
    (c) Suppose you don't know how many people there are in the class. Can you work out the average and SD for the scores?

14. The men in the Health Examination Survey had an average height of 68 inches with an SD of 3 inches. Tomorrow, one of these men will be chosen at random. you have to guess his height. What should you guess? Should you expect to be off by around ½ inch, 1 inch, or 3 inches?

15. As in Exercise 14, but tomorrow a whole series of men will be chosen at random. After each man appears, his actual height will be compared with your guess to see how far off you were. The r.m.s. size of the amounts off should be _____ .

The answers to these exercises are on p. A-30.

Technical note: There is an alternative way to compute the SD, which is more efficient in some cases (for instance, it is often used in computer programs):[12]

$$SD = \sqrt{\text{average of (entries)}^2 - \text{(average of entries)}^2}.$$

7. USING A STATISTICAL CALCULATOR

You have been asked to guess the SD for a few lists and to calculate it for some others in order to get a clear idea of what is involved. But nowadays, an investigator seldom has to do this kind of arithmetic. Large data sets are dealt with by electronic computers, and small ones are easily done on a desk calculator. It is a matter of feeding the list into the machine and then pushing a button.[13]

Most statistical calculators produce not SD, but the slightly larger number SD⁺. (The distinction between SD and SD⁺ will be explained in more detail in Part VII.) To find out what your machine is doing, put in the list −1, 1: if the machine gives you 1, it's working out the SD; if it gives you 1.41 . . . , it's working out the SD⁺. If your machine is giving you the

SD⁺, and you want the SD, you have to multiply by a conversion factor. This depends on the number of entries on the list. With 10 entries, the conversion factor is $\sqrt{9/10}$. With 20 entries, it is $\sqrt{19/20}$. In general,

$$SD = \sqrt{\frac{\text{number of entries} - \text{one}}{\text{number of entries}}} \times SD^{+}.$$

8. REVIEW EXERCISES

1. Find the average and SD of the list 12, 1, 4, 10, 8.

2. Here is a list of numbers:

| 0.7 | 1.6 | 9.8 | 3.2 | 5.4 | 0.8 | 7.7 | 6.3 | 2.2 | 4.1 |
| 8.1 | 6.5 | 3.7 | 0.6 | 6.9 | 9.9 | 8.8 | 3.1 | 5.7 | 9.1 |

(a) Without doing any arithmetic, guess whether the average is closest to 1 or 5 or 10.

(b) Without doing any arithmetic, guess whether the SD is closest to 1 or 3 or 6.

(c) What would happen to the average and SD if 10 were added to each number on the list?

(d) What would happen to the average and SD if a minus sign were put in front of every number on the list?

3. Below are rough sketches of the histograms for three lists.

(a) In scrambled order, the averages are 40, 50, 60. Match the histograms with the averages.

(b) The SD of histogram (iii) is closest to: 5, 15, 50.

(c) The SD for histogram (i) is a lot smaller than that for histogram (iii). True or false?

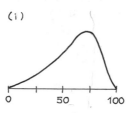 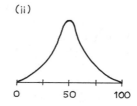 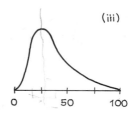

4. In a Public Health Service study on children aged 6 to 11, the average height of the 11-year-old boys was 146 cm.[14] The average height of the 11-year-old children was 147 cm. The average height of the 9-year-old boys was 135 cm. There are 2.54 cm. to the inch.

(a) What was the average height of the 11-year-old boys in inches?

(b) On the average, are boys taller than girls at age 11?

(c) Estimate the average height of the 10-year-old boys.

5. A study on college students found that the men had an average weight of about 66 kg. with an SD of about 9 kg. The women had an average weight of about 55 kg. with an SD of 9 kg.

(a) Find the averages and SDs, in pounds. (1 kg. = 2.2 lbs.)

(b) Just roughly, what percentage of the men weighed between 57 kg. and 75 kg.?

(c) If you took the men and women together, would the SD of their weights be quite a bit less than 9 kg., about 9 kg., or quite a bit bigger than 9 kg.?

6. You are the statistical adviser to a scientist investigating air pollution. He has a computerized data file, showing the amount of carbon monoxide in the air (parts per million) measured every hour for a week at several hundred locations scattered all across California. He is about to compute the average and SD. Is this a meaningful way to summarize the data?

7. A study is made of the age at entrance of college freshmen. The SD turned out to be one of the following: 1 month, 1 year, 5 years. Which was it, and why?

8. Both of the following lists have the same average of 50. Which one has the smaller SD, and why? No computations are necessary.

(i) 50, 40, 60, 30, 70, 25, 75.

(ii) 50, 40, 60, 30, 70, 25, 75, 50, 50, 50.

9. For persons aged 25 and over in the United States in 1975, would the average or the median be higher for income? for years of schooling completed?

10. For the men in the Health Examination Survey, the average income in each age group increased until age 50 or so and then gradually declined. Does this show that for a typical man, his income will increase until age 50 or so and then start decreasing? If not, how else can the HES data be explained?

11. Incoming students at a certain law school have an average LSAT (Law School Aptitude Test) score of 680 with an SD of 25. Tomorrow, one of these students will be picked at random. You have to guess his score now; after he is picked, your guess will be compared with his actual score to see how far off you were. Each point off will cost you a penny. (For example, if you guess 650 and his score is really 750, you will have to pay $1.)

(a) Is your best guess 680, 705, or 730?

(b) About how much can you expect to lose: a nickel, a quarter, or a dollar?

12. As in Exercise 11, but a whole series of students are chosen. The r.m.s. size of your losses should be around _____.

9. SUMMARY

1. A typical list of numbers can be summarized by its *average* and *standard deviation* (SD).

2. Average of a list $= \dfrac{\text{sum of entries}}{\text{number of entries}}$.

ENTERING
HILLSVILLE
FOUNDED   1802
ALTITUDE   620
POPULATION   3,700

TOTAL   6,122

Drawing by Dana Fradon; © 1977 The New Yorker Magazine, Inc.

3. The average locates the center of a histogram in the sense that the histogram balances when supported at the average.

4. Half the area under a histogram lies to the left of the *median*, and half to the right. The median is an alternative way to locate the center of a histogram.

5. The *r.m.s. size* of a list measures how big the entries are, neglecting signs.

6. r.m.s. size of a list = $\sqrt{\text{average of (entries)}^2}$.

7. The SD measures distance from the average. Each number on a list is off the average by some amount. The SD is a sort of average size for these amounts off.

8. The SD is the r.m.s. size of the deviations from the average:

$$SD = \sqrt{\text{average of (deviations from the average)}^2}.$$

9. Roughly 68% of the entries on a list of numbers are within one SD of the average, and about 95% are within two SDs of the average. This is so for many lists, but not all.

5

The Normal Approximation
for Data

1. THE NORMAL CURVE

The normal curve was devised around 1720 by the mathematician Abraham de Moivre in order to solve problems connected with games of chance. (His work will be discussed in Parts IV and V.) Around 1870, the Belgian mathematician Adolph Quetelet had the idea of using this curve as an ideal histogram, to which histograms for data could be compared. The normal

curve has a formidable-looking equation:

$$y = \frac{100\%}{\sqrt{2\pi}} e^{-x^2/2} \qquad \text{where } e = 2.71828 \ldots .$$

This equation involves three of the most famous numbers in the history of mathematics: $\sqrt{2}$, π, and e. This is just to show off a little, but it is easy to work with the normal curve through diagrams and tables, without ever using the equation. A graph of the curve is shown in Figure 1.

Figure 1. The normal curve.

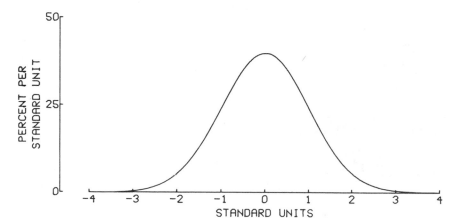

Two features of this graph will be important. First, it is symmetric about 0. That is, the part of the curve to the right of 0 is a mirror image of the part to the left. Second, the total area under the curve equals 100%. The curve is always above the horizontal axis; it appears to stop between 3 and 4, but that's only because it gets so low there. Only about 6/100,000 of the area is outside the interval from −4 to 4.

It will be helpful to find areas under the normal curve between specified values. For instance,

   • the area under the normal curve between −1 and +1 is about 68%;
   • the area under the normal curve between −2 and +2 is about 95%.

Finding these areas is a matter of looking things up in a table or pushing a button on a desk calculator. The method will be explained in section 3.

Most of the histograms for data that have been discussed so far are similar in shape to the normal curve, provided they are drawn to the same scale. Making the horizontal scales match up involves what are called *standard units*.[1] A value is converted to standard units by seeing how many SDs it is above or below the average. Values above the average are given a plus sign; values below the average get a minus sign.

> Standard units say how many SDs above or below the average a value is.

Standard units are discussed in section 2, but here is an example.

Example 1. The average height of the women in the Health Examination Survey was 63 inches, with an SD of 2.5 inches. Convert the following to standard units: 63 inches, 65.5 inches, 58 inches. Find the height which is −1.2 in standard units.

Solution. To answer the first question: 63 inches is the average, so it is 0 SDs away from the average and is 0 in standard units. The second height, 65.5 inches, is 2.5 inches above the average, that is 1 SD above the average. In standard units, 65.5 inches is +1. The third height of 58 inches is 5 inches below the average—that is, 2 SDs below average. In standard units, 58 inches is −2.

To answer the second question: this height is 1.2 SDs below the average, that is, 1.2 × 2.5 inches = 3 inches below the average. So it must equal

$$63 \text{ inches} - 3 \text{ inches} = 60 \text{ inches}.$$

This completes the example.

Figure 2 on p. 72 compares the histogram for the heights of the women in the Health Examination Survey with the normal curve. The horizontal axis for the histogram is drawn so that, when converted to standard units, it matches the horizontal axis for the normal curve (the lower of the two horizontal axes). For example, 63 inches is directly above 0, because 63 inches is 0 in standard units. Similarly, 65.5 inches is directly above +1, and 58 inches is directly above −2.

There are also two vertical axes in Figure 2. The histogram is drawn relative to the inside one, in percent per inch. The normal curve is drawn relative to the outside one, in percent per standard unit. How are the two scales related? For example, take the top values on the scales: 50% per standard unit matches 20% per inch. The reason: there are 2.5 inches to the standard unit, so

$$50\% \text{ per standard unit} = 50\% \text{ per } 2.5 \text{ inches} = 20\% \text{ per inch}.$$

Similarly, 25% per standard unit matches 10% per inch. Any other pair of values can be dealt with in the same way.

Areas under the histogram can be found using either pair of axes: measuring horizontal distances along the inches scale and vertical distance along the percent-per-inch scale, or measuring horizontal distances along the standard-units scale and vertical distances along the percent-per-standard-unit scale. Using either method, the total area under the histogram is 100%, just as the total area under the normal curve is 100%.

The last chapter said that for many lists, roughly 68% of the entries on the list are within one SD of the average. The 68% comes from the normal curve. Take the heights of the women in the Health Examination Survey, for

Figure 2. Comparing a histogram for heights with the normal curve. The horizontal scale for the normal curve matches the standard-units scale for the heights. The histogram follows the normal curve quite well. In particular, the area under the histogram between 60.5 inches and 65.5 inches (the percentage of women within one SD of average with respect to height) is about equal to the area between −1 and +1 under the normal curve—68%.

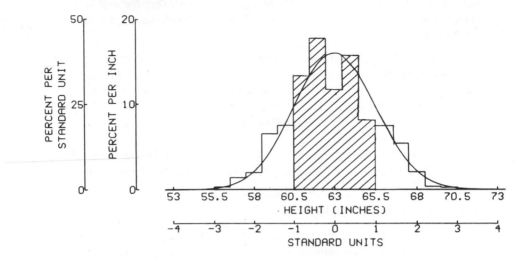

example. The percentage of women whose heights are within one SD of the average is essentially the area under the histogram within one SD of the average, that is, the area under the histogram between −1 and +1 in standard units. This area is shaded in Figure 2. As this figure shows, the histogram follows the normal curve fairly well, although it is quite a bit more jagged-looking. Parts of it are higher than the normal curve, and parts of it are lower. But the highs balance out the lows, and the shaded area under the histogram is just about equal to the area under the normal curve between −1 and +1. This normal area is about 68%. That is where the 68% comes from.

For many lists, roughly 95% of the entries on the list fall within two SDs of the average. Again, the 95% comes from the normal curve. Take the blood pressures of the users aged 25 to 34 in the Drug Study, for example. The percentage of these women whose blood pressures are within two SDs of the average is essentially equal to the area under the histogram between −2 and +2 in standard units. This area is shaded in Figure 3 which also shows a normal curve for comparison. Again, the histogram follows the normal curve fairly well. So the shaded area under the histogram must be about equal to the area under the normal curve between −2 and +2. This normal area is about 95%. And that is where the 95% comes from.

For many lists, then, the percentage of entries falling in an interval can be estimated as follows. First, the interval must be converted to standard units; the method will be reviewed in section 2. Second, the area above this interval under the normal curve must be found. The method will be explained in section 3. Finally, section 4 will show how to put the two steps together. This procedure is called the *normal approximation*. The approximation consists in replacing the original histogram by the normal curve before finding the area.

Figure 3. Comparing a histogram for blood pressure with the normal curve. The horizontal scale for the normal curve matches the standard-units scale for the blood pressures. The histogram follows the normal curve quite well. In particular, the area under the histogram between 96 mm. and 146 mm. (the percentage of women within two SDs of average with respect to blood pressure) is about equal to the area between −2 and +2 under the normal curve—95%.

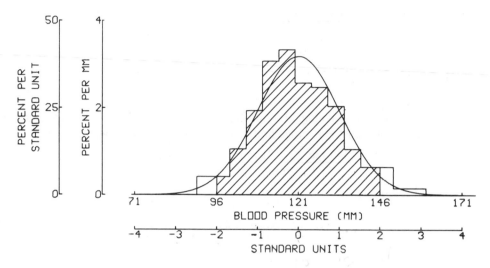

EXERCISE SET A

1. On a certain exam, the scores averaged out to 50, with an SD of 10.
   (a) Convert each of the following scores to standard units: 60, 45, 75.
   (b) Find the scores which in standard units are: 0, +1.2, −2.8.

2. (a) Convert each entry on the following list to standard units (that is, using the average and SD for that list): 13, 9, 11, 7, 10.
    (b) Find the average and SD for the converted list.

The answers to these exercises are on p. A-31.

2. STANDARD UNITS

It will be useful to convert intervals into standard units. The method is just to convert their endpoints, as illustrated by the following examples.

Example 1. A list of blood pressures averages out to 121 mm., with an SD of 12.5 mm. Convert the (open-ended) interval to the left of 146 mm. to standard units.

Solution. Draw a horizontal line, shading the given interval from 146 mm. on indefinitely far to the left.

146 mm

Then mark the average on the line:

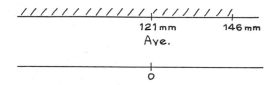

Then draw a second line directly below the first line, and put 0 directly below the average:

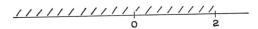

Think of one SD, equaling 12.5 mm., as a yardstick. This yardstick fits into the space between the average and 146 mm. exactly twice, because

$$\frac{146 \text{ mm.} - 121 \text{ mm.}}{12.5 \text{ mm.}} = \frac{25 \text{ mm.}}{12.5 \text{ mm.}} = 2.$$

So put 2 directly below 146 mm.

Finally, shade in the same part of the line as in the first diagram:

This completes the procedure. In standard units, the interval to the left of 146 mm. becomes the interval to the left of 2.

Example 2. A list of heights averages out to 63 inches, with an SD of 2.5 inches. Convert the (open-ended) interval to the right of 60 inches to standard units.

Solution.

Example 3. A list of weights averages out to 160 pounds, with an SD of 30 pounds. Convert the interval from 145 pounds to 205 pounds to standard units.

Solution.

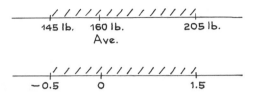

EXERCISE SET B

1. A list of blood pressures averages out to 121 mm., with an SD of 13 mm. Convert each of the following intervals to standard units:
   (a) to the right of 147 mm.   (b) to the left of 95 mm.
   (c) from 95 mm. to 147 mm.

2. A list of heights averages out to 63 inches, with an SD of 2.5 inches. Convert each of the following intervals to standard units:
   (a) to the left of 63 inches.   (b) to the left of 60 inches.
   (c) between 60 inches and 63 inches.

3. A list of weights averages out to 160 pounds, with an SD of 30 pounds. Convert each of the following intervals to standard units:
   (a) from 100 pounds to 220 pounds.   (b) from 145 pounds to 220 pounds.
   (c) from 145 pounds to 175 pounds.

The answers to these exercises are on p. A-31.

3. FINDING AREAS UNDER THE NORMAL CURVE

At the end of the book, there is a table giving areas under the normal curve. For example, to find the area under the normal curve between -1.20 and 1.20 on the horizontal axis, enter the table at 1.20 in the column marked z and read off the entry in the column marked $A(z)$. This entry is about 77%, so the area under the normal curve between -1.20 and 1.20 is about 77%.

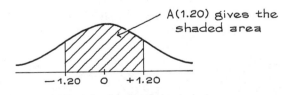

But you are also going to want to find areas of the form

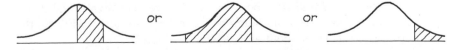

The method for finding these areas is indicated by example.

Example 1. Find the area between 0 and 1 under the normal curve.

Solution. First, make a sketch of the normal curve, and then shade in the area to be found.

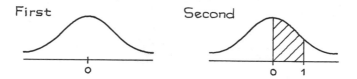

The table will give you the area between −1 and +1. That's about 68%. By symmetry, the area between 0 and 1 is one half the area between −1 and +1, that's $\frac{1}{2} \times 68\% = 34\%$.

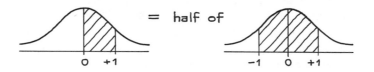

Example 2. Find the area between 0 and 2 under the normal curve.

Solution. It is tempting to think that this must be double the area between 0 and 1. It isn't because the normal curve isn't a rectangle:

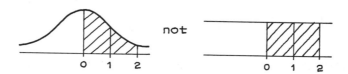

The procedure is the same as in Example 1. The area between −2 and 2 can be found from the table—it's about 95%. The area between 0 and 2 is half that, by symmetry: $\frac{1}{2} \times 95\% \approx 48\%$.

Example 3. Find the area between −2 and +1 under the normal curve.

Solution. The area between −2 and 1 is the sum of the areas between −2 and 0, and between 0 and 1.

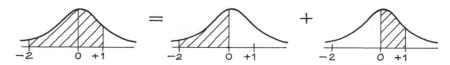

The area between −2 and 0 is the same as the area between 0 and 2, by symmetry, and is about 48%. The area between 0 and 1 is about 34%. So the area between −2 and 0 is about 48% + 34% = 82%.

Example 4. Find the area to the right of 1 under the normal curve.

Solution. The table gives the area between -1 and 1, which is 68%. So the area outside this interval is 32%.

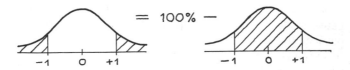

By symmetry, the area to the right of 1 is half this, or 16%.

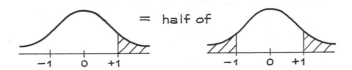

Example 5. Find the area to the left of 2 under the normal curve.

Solution. The area to the left of 2 is the sum of the area to the left of 0, and the area between 0 and 2.

The area to the left of 0 is half the total area, by symmetry; that's $\frac{1}{2} \times 100\% = 50\%$. The area between 0 and 2 is about 48%, the sum is $50\% + 48\% = 98\%$.

Example 6. Find the area between 1 and 2 under the normal curve.

Solution. The diagram below relates this area to other areas which can be read from the table.

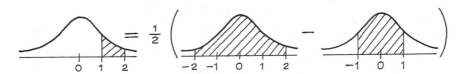

The area between -2 and 2 is about 95%, the area between -1 and 1 is about 68%, half the difference is $\frac{1}{2} \times (95\% - 68\%) = \frac{1}{2} \times 27\% \approx 14\%$.

There is no set procedure to use. It is a matter of drawing pictures which relate the area you want to areas that can be read from the table.

EXERCISE SET C

1. Find the area under the normal curve
   (a) to the right of 1.25.   (b) to the left of −0.40.
   (c) to the left of 0.80.   (d) to the right of −0.85.
   (e) between −1.35 and 1.35.   (f) between 0.40 and 1.30.
   (g) between −0.30 and 0.90.   (h) between −1.10 and −0.35.
   (i) outside −1.5 to 1.5.

2. Fill in the blanks:
   (a) The area between ± _____ under the normal curve equals 68%.
   (b) The area between ± _____ under the normal curve equals 75%.

3. Solve for z:

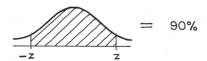

$$= \ 90\%$$

4. A certain curve (not the normal) is sketched below. The total area under it is 100%, and the area between 0 and 1 is 39%.
   (a) If possible, find the area to the right of 1.
   (b) If possible, find the area between 0 and 0.5.

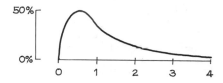

5. A certain curve (not the normal) is sketched below. It is symmetric around 0, and the total area under it is 100%. The area between −1 and 1 is 58%.
   (a) If possible, find the area between 0 and 1.
   (b) If possible, find the area to the right of 1.
   (c) If possible, find the area to the right of 2.

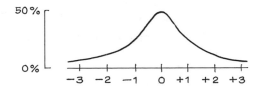

The answers to these exercises are on p. A-32.

4. THE NORMAL APPROXIMATION FOR DATA

The method for the normal approximation will be explained here by example. The diagrams look so simple that you may not think it's worth drawing them. However, it is easy to lose track of the area that is wanted. Please draw the diagrams.

Example 1. The blood pressures of the users aged 25 to 34 in the Drug Study averaged out to 121 mm., with an SD of 12.5 mm. Use the normal curve to estimate the percentage of these women with blood pressures between 96 mm. and 133.5 mm.

Solution. This percentage is given by the area under the blood-pressure histogram, between 96 mm. and 133.5 mm.

Step 1. Draw this interval.

Step 2. Convert the interval to standard units.

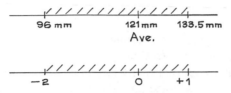

Step 3. Sketch in the normal curve, and find the area above the shaded standard-units interval obtained in step 2. The percentage is approximately equal to the shaded area, which is almost 82%.

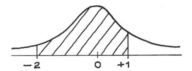

Using the normal curve, it is estimated that about 82% of the blood pressures were between 96 mm. and 133.5 mm. This is only an approximation, but it is pretty good: 83% of the women had blood pressures in that interval. See Figure 4.

Figure 4. The normal approximation consists in replacing the original histogram by the normal curve, before computing areas.

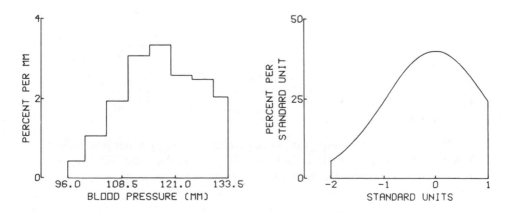

Example 2. The heights of the women in the Health Examination Survey averaged out to 63 inches, with an SD of 2.5 inches. Use the normal curve to estimate the percentage of heights above 60 inches.

Solution.

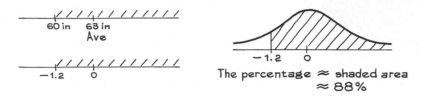

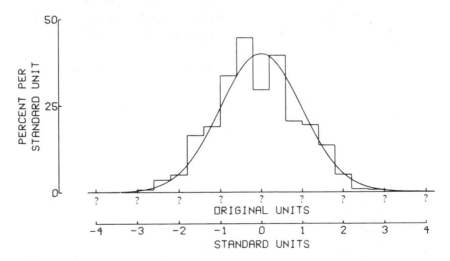

The percentage ≈ shaded area ≈ 88%

Using the normal curve, it is estimated that about 88% of the women were more than 60 inches = 5 feet tall. In fact, 86% were.

It is a remarkable fact that many histograms follow the normal curve. For such histograms, the average and SD are good summary statistics: if a histogram follows the normal curve, it looks something like the sketch in Figure 5. Then, filling in the question marks tells almost all there is to say about the histogram—because its shape is just about like that of the normal curve. The average and SD fill in the question marks.

Figure 5. The average and SD. These statistics summarize histograms which follow the normal curve by locating the center and indicating the spread around the center.

However, it is also important to remember that many other histograms do not follow the normal curve. In such cases, the average and SD are poor summaries. More about this in the next section.

EXERCISE SET D

1. For the men in Cycle I of the Health Examination Survey, the average height was about 68.2 inches, with an SD of about 2.8 inches. Using the normal curve, estimate the percentage of men with heights
    (a) below 71 inches.   (b) between 62.6 inches and 71 inches.
    (c) above 62.6 inches.

2. For these men, the average weight was about 168 pounds, with an SD of about 28 pounds. Using the normal curve, estimate the percentage of men with weights
    (a) below 168 pounds.   (b) between 150 pounds and 180 pounds.
    (c) above 205 pounds.

3. In one law school class, the entering students averaged about 700 on the LSAT, with an SD of about 40. The histogram of LSAT scores followed the normal curve.
    (a) About what percentage of the class scored below 750?
    (b) One student was 0.5 SDs above average on the LSAT. About what percentage of the students scored lower than him?

The answers to these exercises are on p. A-32.

5. PERCENTILES

The average and SD can be used to summarize data following the normal curve. They are less satisfactory for other kinds of data. Take the distribution of white family income in the United States in 1973, shown in Figure 6.

Figure 6. Distribution of white families by income: United States in 1973.

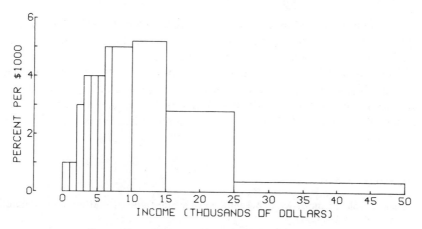

Source: *Current Population Reports,* Series P-60, No. 97.

The average income for the families in Figure 6 was about $14,000 with an SD of about $10,000. So the normal approximation suggests that about 8% of the families had negative income:

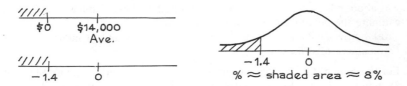

The reason for this blunder is that the histogram in Figure 6 does not follow the normal curve at all well. It is quite unsymmetric, with a long right-hand tail. To summarize such histograms, statisticians use *percentiles* (Table 1).

Table 1. Selected percentiles for income of white families in the United States in 1973.

1	$1,000
10	$4,400
25	$7,800
50	$12,600
75	$20,000
90	$25,000
99	$50,000

Source: *Current Population Reports*, Series P-60, No. 97.

The 1st percentile of the income distribution was $1,000, meaning that about 1% of the white families had incomes of $1,000 or less, and about 99% had incomes above that level. The 10th percentile was $4,400: about 10% of the families had incomes below that level, and 90% were above. The 50th percentile is usually called the median, discussed in Chapter 4. By definition, the *interquartile range* equals

75th percentile − 25th percentile.

This is sometimes used as a measure of spread when the SD would pay too much attention to a small percentage of cases in the tail of a distribution.

For reasons of their own, statisticians call de Moivre's curve "normal." This often makes nonstatisticians think that other curves are abnormal. Not so. Many histograms follow the normal curve very well, and many others— like the income histogram—do not. Later in the book, we will present a mathematical theory which helps to explain when histograms should follow the normal curve.

EXERCISE SET E

1. Using Table 1, about what percentage of white families had incomes between $25,000 and $50,000?

2. Using the data in Exercise 1 on p. 32, is the 25th percentile of the income distribution for nonwhite families closest to $1,000, $4,000, or $10,000?

3. A histogram is sketched below.
    (a) How is it different from the normal curve?
    (b) The interquartile range is closest to: 15, 25, 50.

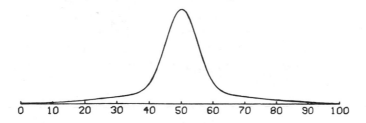

The answers to these exercises are on p. A-32.

6. PERCENTILES FOR THE NORMAL CURVE

When a histogram does follow the normal curve, the normal table can be used to estimate its percentiles. This method is indicated by Examples 1 and 2.

Example 1. Estimate the 95th percentile of the normal curve.

Solution. You are being asked to solve the following equation for z:

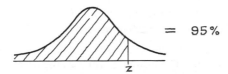

The normal table cannot be used directly, because it does not show areas to the left of z, only areas between $-z$ and z. But the equation is easy to transform. The area to the right of z is 5%, so the area to the left of $-z$ is 5% too, and the area between $-z$ and z must be 90%. The transformed equation is

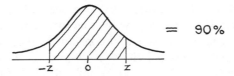

From the table, $z \approx 1.65$. (See Exercise 3 on page 78.)

Example 2. Among all applicants to a certain law school in 1976, the LSAT (Law School Aptitude Test) scores averaged 650, with an SD of 60, and followed the normal curve. Estimate the 95th percentile of the score distribution.

Solution. The problem asks for the score such that: 95% of the students scored lower, and 5% scored higher. The method is to get the score in standard units, and then work backward to original units. In standard units, the score must be about the 95th percentile of the normal curve, which from the previous example is 1.65. So the score is 1.65 SDs above average. That's $1.65 \times 60 \approx 100$ points above average. The 95th percentile of the score distribution is $650 + 100 = 750$.

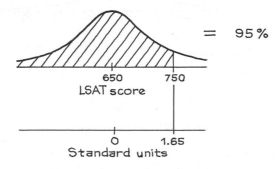

EXERCISE SET F

1. Find the 90th percentile for the normal curve.

2. For the law school in Example 2, estimate the 90th percentile of the scores.

3. For the law school in Example 2, estimate the 20th percentile of the scores.

4. For Berkeley freshmen, the average GPA (grade point average) is around 3.0, with an SD of about 0.5. The histogram follows the normal curve. Estimate the 30th percentile of the GPA distribution.

The answers to these exercises are on p. A-32.

7. REVIEW EXERCISES

1. In the 1960s and 1970s, there has been a slow but steady decline in the Scholastic Aptitude Test scores (SAT). For the verbal test, the average in 1967 was about 465; by 1974, the average was down to about 445. The SD, however, has remained nearly constant at 100, and histograms for the scores follow the normal curve. A 20-point drop may not seem like much, but it has a large effect on the tails of the distribution.

   (a) Estimate the percentage of students scoring over 600 in 1967.

   (b) Estimate the percentage of students scoring over 600 in 1974.

   (It does not seem possible to explain this decline on the basis of changes in the test, or in the population of students taking the test.[2])

2. Men and women have about the same distribution of scores on the verbal SAT; on the mathematical SAT, men have a distinct edge.[3] In 1974, for instance, the men averaged about 500 on the mathematical SAT, while the women averaged about 460. Both histograms follow the normal curve, with an SD of about 100.

    (a) Estimate the percentage of men getting over 600 on this test in 1974.

    (b) Estimate the percentage of women getting over 600 on this test in 1974.

3. One of the men who took the mathematical SAT in 1974 will be picked at random, and you have to guess his test score. You will be given a dollar if you guess it right to within 50 points.

    (a) What should you guess?

    (b) What is your chance of winning?

4. You are looking at a computer printout of a list of 100 numbers which have been converted to standard units. Looking at the first ten entries, you see

    −6.2   3.5   1.2   −0.13   4.2   −5.1   −7.2   −11.3   1.8   6.3

Is anything wrong?

5. Among applicants to one law school in 1976, the average LSAT score was about 600, with an SD of about 100. The LSAT scores followed the normal curve.

    (a) About what percentage of the applicants scored over 700?

    (b) Estimate the 90th percentile for the scores.

6. For men who were current smokers in a Public Health Service study, the average number of cigarettes smoked per day was about 27, with an SD of about 20. The histogram was shown on p. 36 above. If you used the normal approximation to estimate the percentage who smoked 2 cigarettes a day or less, would the estimate be about right, quite a bit too high, or quite a bit too low? Check your guess by working out the normal approximation for the percentage, and then reading it off the histogram.

7. True or false? Explain or give examples.

    (a) The median and the average of any list are always close together.

    (b) Half of a list is always below average.

    (c) With a large, representative sample, the histogram is bound to follow the normal curve quite closely.

    (d) If two lists of numbers have exactly the same average of 50 and the same SD of 10, then the percentage of entries between 40 and 60 must be exactly the same for both lists.

8. An investigator has a computer file showing family incomes for 1,000 subjects in a certain study. These range from $2,800 a year to $78,600 a year. By accident, the highest income in the file gets changed to $786,000.

    (a) Does this affect the average? If so, by how much?

    (b) Does this affect the median? If so, by how much?

9. For about 700 Statistics 2 students at University of California, Berkeley, in the fall of 1975 the average number of college mathematics courses taken (other than Statistics 2) was about 1.1, with an SD of about 1.5. Would a sketch of the histogram for the data look like (i) or (ii) or (iii)?

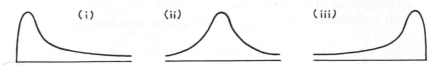

10. The following list of test scores has an average of 50 and an SD of 10:

39, 41, 47, 58, 65, 37, 37, 49, 56, 59, 62, 36, 48,
52, 64, 29, 44, 47, 49, 52, 53, 54, 72, 50, 50.

(a) Use the normal approximation to estimate the number of scores within 1.25 SDs of the average.
(b) How many scores really were within 1.25 SDs of the average?
(c) Compare your answers to (a) and (b).

"LOOK, FRED! THIS SEEMS TO BE THE SAME THING, SUMMARIZED."

8. SUMMARY

1. The normal curve is symmetric about 0, and the total area under it is 100%.

2. *Standard units* say how many SDs above (+) or below (−) the average a value is.

3. Many histograms have roughly the same shape as the normal curve.

4. If a list of numbers follows the normal curve (that is, its histogram has about the same shape as the normal curve) the percentage of entries falling in a given interval can be estimated by converting the interval to standard units and then finding the area under the normal curve above the resulting standard-units interval. This is called the *normal approximation*.

5. In particular, a histogram which follows the normal curve can be reconstructed fairly well from its average and SD; so the average and SD are good summary statistics.

6. All histograms, whether or not they follow the normal curve, can be summarized using *percentiles*.

6

Measurement Error

Jesus: I am come to bear witness unto the truth.
Pilate: What is truth?

1. INTRODUCTION

If the same thing were measured several times, in an ideal world the same result would be obtained each time. In practice, the results are found to differ. Each result is contaminated by *chance error,* and this error changes from measurement to measurement. One of the first scientists to deal with this problem was Tycho Brahé (1546–1601), the Danish astronomer. But it was probably noticed first in the market place, as merchants weighed out spices and measured off lengths of silk. Even today, merchants are troubled by chance error. For instance, live pigs are sold by weight. In some stock-yards, the pig steps on a scale and is weighed three times. The three weights, and their average, are automatically printed out on a paper tape, and the pig is sold by his average weight. Each of the three weights is contaminated by a chance error, some of which cancels out in the average.

There are several questions to ask about chance errors in measurements: Where do they come from? How big are they likely to be? How much is likely to cancel out in the average? The first question is easy to dispose of: in most cases, nobody knows. The second question will be dealt with later in this chapter, and the third will be answered in Part VII.

2. CHANCE ERROR

This section will discuss the idea of chance error in the context of precision weighing done at the National Bureau of Standards in Washington.[1] But first, a brief explanation of why the bureau does this kind of work. When a customer buys something by weight in a store, it is weighed on a scale. To assure honest weight, the scales are checked periodically by county weights-and-measures officials, using county standard weights. In this way, the scale only serves to compare merchandise with these standard weights. The county standards too must be *calibrated* (checked against external standards) periodically. This is done at the state level. And the state standards are calibrated against national standards, by the National Bureau of Standards in Washington. This chain of comparisons ends at the International Prototype Kilogram (for short, The Kilogram). This platinum-iridium weight is held at the International Bureau of Weights and Measures in Paris. By international treaty (The Treaty of the Meter, 1875), "one kilogram" was defined to be the weight of this object, under standard conditions.[2] All other weights are determined relative to The Kilogram. For instance, something weighs a pound if it weighs just a bit less than half as much as The Kilogram. More precisely,

The Pound = 0.4539237 of The Kilogram.

To say that a package of butter weighs a pound is to say that it has been connected by some long and complicated series of comparisons to The Kilogram in Paris, and has been found to weigh 0.4539237 times as much.

When the Treaty of the Meter was concluded, each participating nation got a national prototype kilogram, whose exact weight had been determined as accurately as possible relative to The Kilogram. They were distributed by lot, and the United States got Kilogram #20. The values of all the U.S. national standards are determined relative to it.

Figure 1. The U.S. national prototype kilogram, K_{20}

Source: *National Bureau of Standards Bulletin*, 1905.

In the United States, then, accuracy in weighing at the supermarket depends on the accuracy of the calibration work done at the Bureau. The basic issue is reproducibility: if a measurement is repeated, how much will it change? The Bureau gets at this issue by making repeated determinations of the values of some of their own weights. We will discuss the results for one such weight, called NB 10 because it is owned by the National Bureau and its nominal value is ten grams (the weight of two nickels). This weight is made out of chrome-steel alloy. It was acquired by the Bureau around 1940, and they have weighed it about once a week ever since. We are going to look at one hundred of these weighings, performed by Mr. Almer and Mrs. Jones at the Bureau in 1962–1963. These measurements were made on the same apparatus, in the same room. Every effort was made to follow the same procedure each time. All the factors known to affect the results, like air pressure or temperature, were kept as constant as possible.

The first five weighings gave

9.999591 grams,
9.999600 grams,
9.999594 grams,
9.999601 grams,
9.999598 grams.

At first glance, these numbers all seem to be the same. But look more closely. It is only the first four digits that are solid, at 9.999. The last three digits are shaky, they change from measurement to measurement. This kind of variability is typical of careful measurement work. The last few decimal places are a bit unreliable, they change if the measurement is repeated.[3] This is chance error at work.

Coming back to NB 10, it does seem to weigh just a bit less than ten grams. So instead of writing out the 9.999 each time, the bureau just reports the amount by which NB 10 fell short of ten grams. For the first weighing, this was

$$0.000409 \text{ grams.}$$

All the 0s are distracting, so the Bureau works not in grams but in *micrograms:* a microgram is the millionth part of a gram. In these units, the first five measurements on NB 10 are easier to read; they are

$$409 \quad 400 \quad 406 \quad 399 \quad 402.$$

All one hundred measurements are shown in Table 1. Look down the table. You can see that they run around 400 micrograms, but some are more, some are less. The smallest is number 94 at 375 micrograms; the largest is number 86 at 437 micrograms. And there is a lot of variability in between. To keep things in perspective, one microgram is the weight of a large speck of dust; 400 micrograms is the weight of a grain or two of salt. This really is precision weighing!

Even so, the different measurements can't all be right. The exact amount by which NB 10 falls short of ten grams is very unlikely to equal the first number in the table, or the second, or any of them. In fact, despite the effort of making these hundred measurements, the exact weight of NB 10 remains

Table 1. One hundred measurements on the weight of NB 10, made by Almer and Jones at the National Bureau of Standards. Each measurement was some number of micrograms below ten grams. These numbers are shown in the column headed "Result."

Meas. no.	Result	Meas. no.	Result	Meas. no.	Result	Meas. no.	Result
1	409	26	397	51	404	76	404
2	400	27	407	52	406	77	401
3	406	28	401	53	407	68	404
4	399	29	399	54	405	79	408
5	402	30	401	55	411	80	406
6	406	31	403	56	410	81	408
7	401	32	400	57	410	82	406
8	403	33	410	58	410	83	401
9	401	34	401	59	401	84	412
10	403	35	407	60	402	85	393
11	398	36	423	61	404	86	437
12	403	37	406	62	405	87	418
13	407	38	406	63	392	88	415
14	402	39	402	64	407	89	404
15	401	40	405	65	406	90	401
16	399	41	405	66	404	91	401
17	400	42	409	67	403	92	407
18	401	43	399	68	408	93	412
19	405	44	402	69	404	94	375
20	402	45	407	70	407	95	409
21	408	46	406	71	412	96	406
22	399	47	413	72	406	97	398
23	399	48	409	73	409	98	406
24	402	49	404	74	400	99	403
25	399	50	402	75	408	100	404

unknown, and perhaps unknowable. All we can really say is that it is around 400 micrograms below ten grams—give or take something.

It may still seem quite strange that the Bureau weighs the same weight over and over again. To see how this replication is useful, imagine that a scientific laboratory sends a ten-gram weight off to the Bureau for calibration. Of course, one measurement can't be the last word, for it is contaminated by a chance error. To make intelligent use of the calibration, the lab has to know how big this chance error is likely to be. There is a direct way to find out: sending the same weight back for a second weighing. If the two results only differ by a few micrograms, then the lab can be reasonably sure that the chance error in each one only amounts to a few micrograms. So each measurement is right, give or take a few micrograms. On the other hand, if the two results differ by several hundred micrograms, then the chance error in each one is likely to be that large, and each measurement is likely to be off by several hundred micrograms. The repeated weighings on NB 10 save everyone the bother of sending in their ten-gram weights more than once. There is no need for the Bureau's customers to ask for replicate calibrations, because the Bureau has already done the work on NB 10.

> No matter how carefully it was made, a measurement could have come out a bit differently than it did. If it is repeated, it would come out a bit differently. By how much? Before relying on any measurement, this question must be faced. The best way to face it is by replicating the measurement.

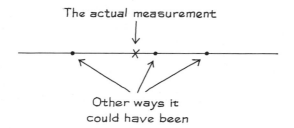

Table 1 shows a hundred replicate measurements on NB 10. Ordinarily, the Bureau would only report the SD of the data, 6 micrograms. What information does this SD convey? It suggests that each measurement on NB 10 was thrown off by a chance error of size around 6 micrograms. Chance errors of size 5 or 10 micrograms were fairly common. Chance errors of size 50 or 100 micrograms must have been extremely rare. The inference is that in calibrating other ten-gram weights by the same process, the chance errors will be similar in size to those for NB 10—that is, about 5 or 10 micrograms.

To see why the SD gives the likely size of the chance error in any single measurement, imagine taking more and more measurements on NB 10. The average of the measurements will tend toward the exact value for the weight. This represents an idealized measurement, made without error.

The individual measurements, the exact value, and chance error are connected by an equation:

> individual measurement = exact value + chance error.

The chance error throws each individual measurement off the exact value by an amount which changes from measurement to measurement. These changes are reflected in the variability of the results. That is why the variability in the measurements is a guide to the likely size of the chance error.

The average of all 100 measurements reported in Table 1 was 405 micrograms below ten grams. This is likely to be very close to the exact weight. The first measurement deviated from the average of the 100 measurements by 4 micrograms. So it must have deviated from the exact weight by nearly 4 micrograms: the chance error in it was nearly 4 micrograms in size. Similarly, the second measurement deviated from the average of the 100 measurements by 5 micrograms, and the chance error in it must have been around 5 micrograms in size. The typical deviation from the average of the 100 measurements was around 6 micrograms in size—the standard deviation. Consequently, the typical chance error must have been around 6 micrograms in size. That is why the likely size of the chance error in a single measurement—its deviation from the exact value—can be estimated by the SD.

> The SD of a series of repeated measurements is an estimate of the likely size of the chance error in a single measurement.

3. OUTLIERS

How well do the measurements reported in Table 1 fit the normal curve? The answer is, not very well at all. Measurement number 36 is three SDs away from the average; number 94 is five SDs away; and number 86 is over five standard deviations away from the average—a minor miracle. Such extreme measurements are called *outliers*. They do not result from blunders: as far as the Bureau could tell, nothing went wrong when these three observations were made. However, these outliers make a very substantial contribution to the SD. Consequently, the percentage of results falling closer to the average than one of these inflated SDs is 86%—quite a bit larger than the 68% predicted by the normal curve. See Figure 2 on the next page.

When these three outliers are discarded, the remaining ninety-seven results average out to 404 micrograms below ten grams, with an SD of only 4 micrograms. So the average did not change much, but the SD dropped by about a third. These ninety-seven measurements follow the normal curve reasonably well, as shown in Figure 2. The measurement process used on

Figure 2. Outliers. The top panel compares the histogram for all 100 measurements on the check-weight NB 10 with the normal curve. (The units are micrograms below ten grams.) The normal curve does not fit at all well. The second panel shows a normal curve fitted to the data with three outliers removed. The fit is good. The data can, therefore, be described as follows. Most of the numbers follow the normal curve, but a small percentage of them are much further away from the average than the normal curve suggests.

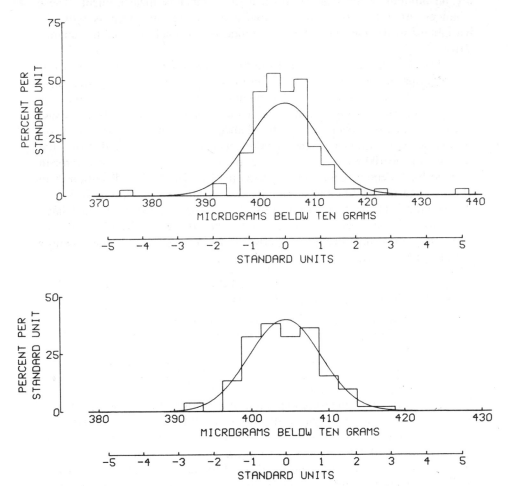

NB 10 can be described this way. Most of the measurements follow the normal curve with an average of about 404 micrograms below ten grams, and an SD of about 4 micrograms. But in a small percentage of cases, the measurements are quite a bit further away from the average than the normal curve suggests. The overall SD of 6 micrograms is a compromise between the SD of the main part of the histogram—4 micrograms—and the three outliers, representing deviations of 18, −30, and 32 micrograms.

In careful measurement work, a small percentage of outliers is expected. The only unusual aspect of the NB 10 data is that the National Bureau of

Standards reported its outliers; many investigators don't. Here is what the Bureau has to say about ignoring outliers; for official prose, the tone is rather tart.[4]

A major difficulty in the application of statistical methods to the analysis of measurement data is that of obtaining suitable collections of data. The problem is more often associated with conscious, or perhaps unconscious, attempts to make a particular process perform as one would like it to perform rather than accepting the actual performance. . . . Rejection of data on the basis of arbitrary performance limits severely distorts the estimate of real process variability. Such procedures defeat the purpose of the . . . program. Realistic performance parameters require the acceptance of all data that cannot be rejected for cause.

Why are outliers usually ignored? If investigators see one, they are faced with a dilemma. Either they have to ignore the evidence of their own eyes, or admit that the measurement process doesn't follow the normal curve. The prestige of the normal curve is so high that the first course is the one most often chosen—a triumph of theory over experience.

4. BIAS

Suppose a butcher always weighs steak with his thumb on the scale. That causes an error in the measurement, but little has been left to chance: this error always makes the steak weigh too much. To take another example, suppose a drapery store measures material with a cloth tape measure which has stretched from 36 inches to 37 inches in length. So every "yard" of cloth they sell to a customer has an extra inch tacked onto it. This isn't a chance error—because it always works in the same direction. The butcher's thumb and the stretched tape are two examples of *bias,* or *systematic error,* in measurement processes.

> Bias affects all measurements the same way, pushing them in the same direction. Chance errors change from measurement to measurement, sometimes up and sometimes down.

If there were no bias in a measurement procedure, the long-run average of a series of repeated measurements would give the exact value of whatever it is that is being measured—all the chance error will have cancelled out. However, when bias is present, the long-run average will itself be either too high or too low. Each measurement is off by the bias, as well as a chance error:

individual measurement = exact value + bias + chance error.

Usually, there is no way to detect bias just by looking at the measurements themselves; instead, they have to be compared to an external standard or to theoretical predictions. In the United States, all weight measurements depend on the connection between K_{20} and The Kilogram. These two weights have been compared a number of times, and it is estimated that K_{20} is a tiny bit lighter than The Kilogram—by 19 parts in a billion. All weight calculations at the Bureau are revised upward by 19 parts in a billion, to compensate. However, this factor is very likely to be just a shade off—after all, it was the result of some measurements. If so, all weights measured in the United States are systematically off, by the same (infinitesimal) percentage. This is another example of bias.

5. REVIEW EXERCISES

These exercises cover all of Parts I and II.

1. An experienced scientist, who is using the best equipment available, only needs to measure things once—provided he didn't make a mistake. After all, if he measured them twice, he'd get the same results both times. Discuss.

2. In the Health Examination Survey, there were 6,672 subjects. The sex of each subject was recorded at two different stages of the survey. In seventeen cases, there was a discrepancy: the subject was recorded as male at one interview, female at the other. How would you account for this?

3. You send a yardstick to a local laboratory for calibration, asking that the procedure be repeated three times. They report the following values:

    $35^{59}/_{64}$ inches $36^{1}/_{64}$ inches $36^{4}/_{64}$ inches.

    If you sent the yardstick back for a fourth calibration, you would expect to get 36 inches give or take:

    $^{4}/_{64}$ inch or so $^{6}/_{64}$ inch or so $^{8}/_{64}$ inch or so.

4. You send a weight to a local laboratory for calibration. They report its value to be 16.007 ounces. If you sent it back for a second calibration, you would expect to get:
    (i)   16.007 ounces exactly.
    (ii) 16.007 ounces, give or take 0.001 ounces or so.
    (iii) 16.007 ounces, give or take 0.01 ounces or so.
    (iv) a complaint.
    (v)   can't say.

5. The bends are caused by rapid changes in air pressure, resulting in the formation of nitrogen bubbles in the blood. The symptoms are: acute pain, and sometimes paralysis leading to death. In World War II, pilots used to get the bends during certain battle maneuvers. It proved feasible to simulate these conditions (but in milder form) in a pressure chamber. As a result, pilot trainees were tested under these conditions once, at the beginning of their training. If they got the bends (only mild cases were induced), they were excluded from the training, on the grounds that they were more likely to get the bends under battle conditions. This procedure was severely criticized by the statistician J. Berkson, and he persuaded the Air Force to replicate the test—that is, repeat it several times for each trainee.
    (a) Why might Berkson have suggested this?
    (b) Give another example where replication is helpful.

6. A carpenter is using a tape measure to get the length of a board.
   (a) What are some possible sources of bias?
   (b) Which is more subject to bias, a steel tape or a cloth tape?
   (c) Would the bias in a cloth tape change over time?

7. True or false? Explain.
   (a) Bias is a kind of chance error.
   (b) Chance error is a kind of bias.
   (c) Measurements are usually contaminated by both bias and chance error.

8. Nineteen students in Statistics 20, at the University of California, Berkeley, in the winter of 1976 were asked to measure the thickness of a table top, using a vernier gauge reading to one-thousandth of an inch. Each person made two measurements, shown below (in inches). For instance, the first person got 1.317 inches and 1.320 inches for the two measurements.
   (a) Did the students work independently of one another?
   (b) Some friends of yours do not believe in chance error. How could you use this data to convince them?

Person	Measurements (inches) 1st	2nd	Person	Measurements (inches) 1st	2nd
1	1.317	1.320	10	13.26	13.25
2	13.26	13.25	11	1.333	1.334
3	1.316	1.335	12	1.315	1.317
4	1.316	1.328	13	1.316	1.318
5	1.318	1.324	14	1.321	1.319
6	1.329	1.326	15	1.337	1.343
7	1.332	1.334	16	1.349	1.336
8	1.342	1.328	17	1.320	1.336
9	1.337	1.342	18	1.342	1.340
			19	1.317	1.318

9. Fill in the blanks, using the options below, and give examples to show that you picked the right answers.
   (a) The SD of a list is 0. This means _____ .
   (b) The r.m.s. size of a list is 0. This means _____ .
   Options:
   (i)   there are no numbers on the list.
   (ii) all the numbers on the list are the same.
   (iii) all the numbers on the list are 0.
   (iv) the average of the list is 0.

10. The scores on the final in one course followed the normal curve, with an average of about 50 and an SD of about 10. About what percentage of the students had scores between 35 and 70?

11. To measure the effect of exercise on the risk of heart disease, investigators decided to compare the incidence of the disease for two large groups of London busmen: drivers and conductors. The conductors got a lot more exercise as they walked around all day collecting fares. The age

distributions for the two groups were very similar, and all the subjects had been on the same job for a period of at least ten years. The incidence of heart disease was substantially lower among the conductors.[5] To what extent does this show that exercise reduces the risk of heart disease?

12. In one course, the final scores averaged 50, with an SD of 20, and followed the normal curve. One student will be chosen at random, and you have to guess his score. You will win $1 if your guess is right to within 10 points. What should you guess, and how likely are you to win?

13. In one course, a histogram for the scores on the final looked like the sketch below. True or false: Because this isn't like the normal curve, there must have been something wrong with the test. Explain.

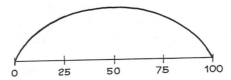

14. In one large class, the male students averaged 5 feet 9 inches in height, with an SD of 3 inches; the female students averaged 5 feet 3 inches in height, with an SD of 3 inches. There are 150 males in the class, and 150 females. Taking the men and the women together:
    (a) the average height would be _____ ;
    (b) would the SD of height be quite a bit more than 3 inches, just about 3 inches, or quite a bit less than 3 inches?
    Explain your reasoning.

15. Repeat Exercise 14, when there are 200 men in the class, and 100 women. (The separate averages and SDs for the men and women stay the same.)

6. SUMMARY

1. No matter how carefully it was made, a measurement could have turned out a bit differently from the way it did. This reflects *chance error*. Before relying on a measurement, the likely size of the error must be estimated. The way to do that is to replicate the measurement.

2. The likely size of the chance error in a single measurement can be estimated by the SD of a sequence of repeated measurements made under the same conditions.

3. *Bias*, or *systematic error*, causes the measurements made by a process to be systematically too high or systematically too low. The equation is:

individual measurement = exact value + bias + chance error.

The chance error changes from measurement to measurement.

4. Even in careful measurement work, a small percentage of *outliers* can be expected.

5. The average and SD can be unduly affected by outliers. Then the histogram will not follow the normal curve at all well.

7

Plotting Points and Lines

Q. What did the dot say to the line?
A. Come to the point.

1. READING POINTS OFF A GRAPH

This chapter reviews some of the ideas about plotting points and lines which will be used in Part III. You can either read this chapter now, or return to it if you run into difficulty in Part III. If you read the chapter now, the first four sections are the most important, the last section is more difficult.

Figure 1 shows a horizontal axis (the x-axis) and a vertical axis (the y-axis). The point shown in the figure has an x-coordinate of 3, because it is in line with 3 on the x-axis. It has a y-coordinate of 2, because it is in line with 2 on the y-axis. This point is written $x = 3, y = 2$. Sometimes, it is abbreviated even more, to (3, 2). The point shown in Figure 2 is $(-2, -1)$: it is directly below -2 on the x-axis, and directly to the left of -1 on the y-axis.

The idea of representing points by pairs of numbers is due to René Descartes (France, 1596–1650). In his honor, the x- and y-coordinates are often called "cartesian coordinates."

Figure 1.   Figure 2.

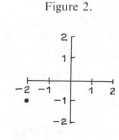

EXERCISE SET A

1. Figure 3 shows five points. Write down the *x*-coordinate and *y*-coordinate for each point.

2. As you move from point A to point B in Figure 3, your *x*-coordinate goes up by _____: your *y*-coordinate goes up by _____.

3. One point in Figure 3 has a *y*-coordinate 1 bigger than the *y*-coordinate of point E. Which point is that?

Figure 3.

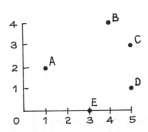

The answers to these exercises are on p. A-33.

2. PLOTTING POINTS

Figure 4 shows a pair of axes. To plot the point (2, 1): find the 2 on the *x*-axis; the point will be directly above this, as in Figure 5; find the 1 on the *y*-axis; the point will be directly to the right of this, as in Figure 6.

Figure 4.   **Figure 5.**   Figure 6.

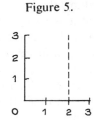   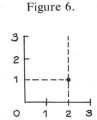

EXERCISE SET B

1. Draw a pair of axes and plot each of the following points:
   (1, 1)   (2, 2)   (3, 3)   (4, 4).
   What can you say about them?

2. Three out of the following four points lie on a line. Which is the maverick? Is it above or below the line?
   (0, 0)   (0.5, 0.5)   (1, 2)   (2.5, 2.5).

3. The table below shows four points. In each case, the y-coordinate is computed from the x-coordinate by the rule $y = 2x + 1$. Fill in the blanks, then plot the four points. What can you say about them?

x	y
1	3
2	5
3	- 7
4	- 9

on a line

4. Figure 7 below shows a shaded region. Which of the following two points is in the region: (1, 2) or (2, 1)?

5. Do the same for Figure 8.

6. Do the same for Figure 9.

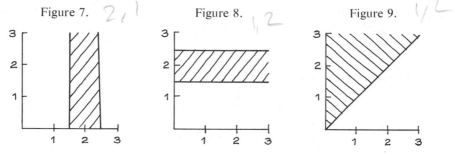

Figure 7.   2, 1   Figure 8.   1, 2   Figure 9.   1, 2

The answers to these exercises are on p. A-33.

René Descartes (France, 1596–1650)

3. SLOPE AND INTERCEPT

Figure 10 shows a line. Take any point on the line—for instance, point A. Now move up the line to any other point—for instance, point B. Your x-coordinate has increased by some amount, called the *run*. In this case, the run was 2. At the same time your *y*-coordinate has increased by some other amount, called the *rise*. In this case, the rise was 1. Notice that in this case, the rise was half the run. Whatever two points you take on this line, the rise will be half the run. The ratio rise/run is called the *slope* of the line:

$$\text{slope} = \text{rise/run}.$$

The slope is the rate at which *y* increases with *x*, along the line. To interpret it another way, imagine the line as a road going up a hill. The slope measures the steepness of the grade. For the line in Figure 10, the grade is 1 in 2—quite steep for a road. In Figure 11, the slope of the line is 0. In Figure 12, the slope is −1. If the slope is positive, the line is going uphill. If the slope is 0, the line is horizontal. If the slope is negative, the line is going downhill.

Figure 10. Slope is 1/2.   Figure 11. Slope is 0. Figure 12. Slope is −1.

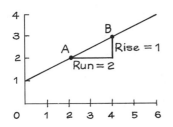 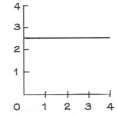 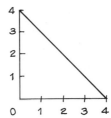

The *intercept* of a line is its height at *x* = 0. Usually, the axes cross at 0. Then, the intercept is where the line crosses the *y*-axis. In Figure 13, the intercept is 2. Sometimes, the axes don't cross at 0, and then you have to be a little bit careful. In Figure 14, the axes cross at (1, 1). The intercept of the line in Figure 14 is 0—that would be its height at *x* = 0.

Often, the axes of a graph show units. For example, in Figure 15 the units for the *x*-axis are inches, the units for the *y*-axis are degrees centigrade. Then the slope and intercept have units too. In Figure 15, the slope of the line is 2.5 degrees per inch; the intercept is −5 degrees.

Figure 13.   Figure 14.   Figure 15.

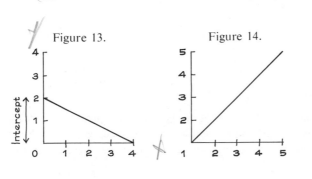 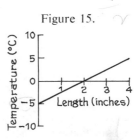

EXERCISE SET C

1. Figures 16 to 18 show lines. For each line, find the slope and intercept. Note: the axes do not cross at 0 in each case.

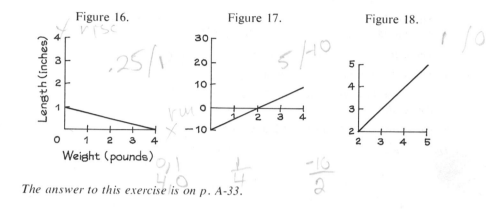

Figure 16.

Figure 17.

Figure 18.

The answer to this exercise is on p. A-33.

4. PLOTTING LINES

Example. Plot the line which passes through the point (2, 1) and has slope ½.

Solution. First draw a pair of axes and plot the given point (2, 1), as in Figure 19. Then move any convenient distance off directly to the right from the given point: Figure 20 shows a run of 3. Make a construction point at this new location. Since the line slopes up, it passes above the construction point. How far? That is, how much will the line rise in a run of 3? The answer is given by the slope. The line is rising at the rate of half a vertical unit per horizontal unit, and in this case there is a run of three horizontal units, so the rise is $3 \times ½ = 1.5$:

$$\text{rise} = \text{run} \times \text{slope}.$$

Make a vertical move of 1.5 from the construction point, and mark a point at this third location, as in Figure 21. This third point is on the line. Put a ruler down and join it to the given point (2, 1).

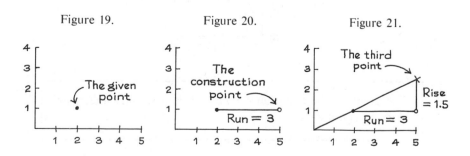

Figure 19.

Figure 20.

Figure 21.

EXERCISE SET D

1. Draw lines through the point (2, 1) with the following slopes:
       (a) +1   (b) −1   (c) 0.

2. Start at the point (2, 1) in Figure 21. If you move over 2 and up 1, will you be on the line, above the line, or below the line?

3. The same, but move over 4 and up 2.

4. The same, but move over 6 and up 5.

5. Draw the line with intercept 2 and slope −1. Hint: This line goes through the point (0, 2).

6. Draw the line with intercept 2 and slope 1.

The answers to these exercises are on p. A-34.

5. THE ALGEBRAIC EQUATION FOR A LINE

Example. Here is a rule for computing the *y*-coordinate of a point from its *x*-coordinate: $y = \frac{1}{2}x + 1$. The table below shows the points with *x*-coordinates of 1, 2, 3, 4. Plot the points. Do they fall on a line? If so, find the slope and intercept of this line.

Solution. The points are plotted in Figure 22. They do fall on a line. Any point whose *y*-coordinate is related to its *x*-coordinate by the same equation $y = \frac{1}{2}x + 1$ will fall on the same line. This line is called the *graph* of the equation. The slope of the line is ½, the coefficient of *x* in the equation. The intercept is 1, the constant term in the equation.

Figure 22.

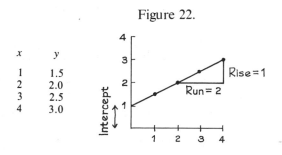

x	y
1	1.5
2	2.0
3	2.5
4	3.0

The graph of the equation $y = mx + b$ is a straight line, with slope *m* and intercept *b*.

Example. Figure 23 shows a line. What is the equation of this line? What is the height of this line at $x = 1$?

Solution. This line has slope -1 and intercept 4, so its equation is $y = -x + 4$. Substituting $x = 1$ gives $y = 3$; so the height of the line is 3 when x is 1.

Figure 23.   Figure 24.   Figure 25.

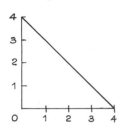   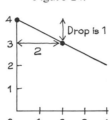   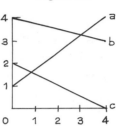

Example. Plot the line whose equation is $y = -\frac{1}{2}x + 4$.

Solution. The intercept of this line is 4; plot the point $(0, 4)$ as in Figure 24. The line must go through this point. Make any convenient horizontal move—say 2. The slope is $-\frac{1}{2}$, so the line must drop 1. Mark the point which is 2 over and 1 down from the first point in Figure 24. Then join these two points by a straight line.

EXERCISE SET E

1. Plot the graphs of the following equations:
   (a) $y = 2x + 1$ (b) $y = \frac{1}{2}x + 2$.
   In each case, say what the slope and intercept are, and give the height of the line at $x = 2$.

2. Figure 25 shows three lines. Match the lines with the equations:
   $$y = \tfrac{3}{4}x + 1 \qquad y = -\tfrac{1}{4}x + 4 \qquad y = -\tfrac{1}{2}x + 2.$$

3. Plot four different points whose y-coordinates are double their x-coordinates. Do these points lie on a line? If so, what is the equation of the line?

4. Plot the points $(1, 1)$, $(2, 2)$, $(3, 3)$, and $(4, 4)$ on the same graph. These points all lie on a line. What is the equation of this line?

5. For each of the following points, say whether it is on the line of Exercise 4, or above, or below:
   (a) $(0, 0)$ (b) $(1.5, 2.5)$ (c) $(2.5, 1.5)$.

6. True or false:
   (a) If y is bigger than x, then the point (x, y) is above the line of Exercise 4.
   (b) If $y = x$, then the point (x, y) is on the line of Exercise 4.
   (c) If y is smaller than x, then the point (x, y) is below the line of Exercise 4.

The answers to these exercises are on p. A-34–A-35.

PART III

Correlation and Regression

8

Correlation

Like father, like son.

1. THE SCATTER DIAGRAM

The methods discussed in Part II are useful for dealing with one variable at a time. Other methods are needed for studying the relationship between two variables.[1] The statistical methods currently used for studying such relationships were invented by Sir Francis Galton (1822–1911), while he was thinking about the following problem: As everyone knows, children resemble their parents. What Galton wanted to know was the strength of this resemblance—how much of a difference the parents made. Statisticians in Victorian England were fascinated by this question, and gathered huge amounts of data in pursuit of the answer. We are going to look at the results of one study carried out by Galton's disciple Karl Pearson (1857–1936) at the turn of the century on the resemblances between family members.[2] One of the things he did was to measure the heights of 1,078 fathers and their sons at maturity, one son per father.

A list of 1,078 pairs of heights would be impossible to grasp. But the relationship between the two variables (father's height and son's height) can

Figure 1. Scatter diagram for the heights of 1,078 fathers and their sons. Families where the height of the son was equal to the height of the father would be plotted along the line $y = x$. But most of the families are scattered around the line, showing how the son's height differed from his father's.

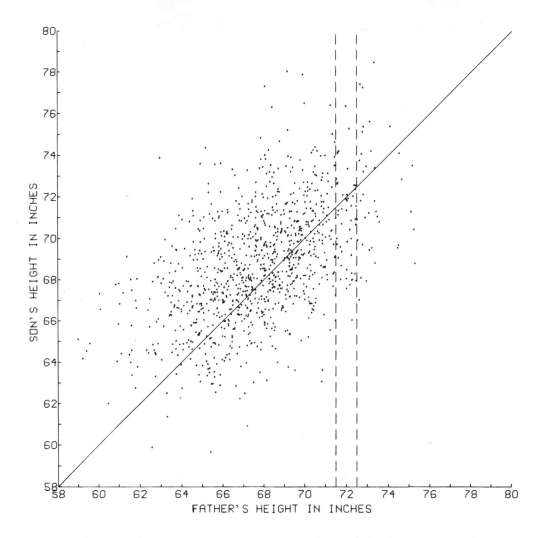

be brought out in a graph called the *scatter diagram* (see Figure 1). Each dot on the diagram represents one father-son pair. The x-coordinate of the dot, measured along the horizontal axis, gives the height of the father; the y-coordinate, measured along the vertical axis, gives the height of the son.

Figure 2a illustrates the mechanics of plotting scatter diagrams, which are discussed in more detail in Chapter 7. The scatter diagram in Figure 1 is a cloud shaped something like a football, with points straggling off the edges.

When making a rough sketch of such a scatter diagram, it is only necessary to show the main oval portion, as in Figure 2b.

Figure 2a. A point on a scatter diagram.   Figure 2b. Rough sketch.

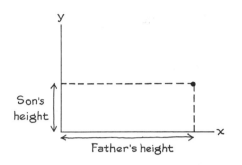
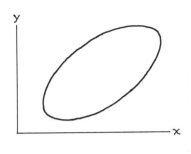

The swarm of points in Figure 1 slopes upward to the right, the y-coordinates of the points tending to increase with their x-coordinates. As a statistician might say, there is a *positive association* between the heights of fathers and sons. In everyday language, as a rule the taller fathers have taller sons. This confirms the obvious. But the scatter diagram says more: it shows how weak the association is. Take a father who is 72 inches tall. It would be natural to guess 72 inches for the height of his son. Such a family is plotted in Figure 3a. Similarly, if a father is 64 inches tall, then 64 inches is the natural guess for the height of his son; if the father is 68 inches tall, 68 inches is the natural guess. These families are all plotted in Figure 3a, and fall on the 45° line. This line expresses the idea that the son's height should be the same as his father's height: it consists of just the points for which the y-coordinate equals the x-coordinate, and has the equation $y = x$. If you think that a son's

Figure 3a. Son's height the same as father's height.

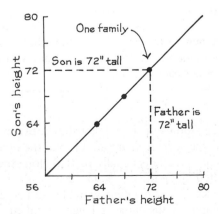

height should be close to his father's height, then you think their point on the scatter diagram should be close to this line, like the points in Figure 3b.

Figure 3b. Son's height close to father's height.

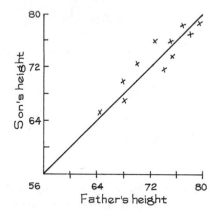

However, in the actual scatter diagram there is a lot more spread around the line. This spread shows the weakness of the relationship between father's height and son's height. Knowing the height of the father helps in guessing the height of the son, but there is still a lot of room for error—reflected by the spread in the ys for any fixed x.

For instance, how closely can you guess the height of a son whose father is known to be 6 feet tall? Look at the dots in the narrow vertical strip over 72 inches in Figure 1. These dots represent all the father-son pairs where the father is 72 inches tall, to the nearest inch. There is still a lot of variability in the heights of these sons, as reflected by the vertical scatter in the strip. In other words, there is still a lot of room for error in trying to guess the height of a son, even if you happen to know his father's height.

> If there is a strong association between two variables, then knowing one helps a lot in predicting the other. But when there is a weak association, information about one variable does not help much in guessing the other.

In social-science studies of the relationship between two variables, it is usual to label one as the *independent* variable and the other as the *dependent* one. The independent variable is usually plotted along the x-axis, the dependent variable is plotted along the y-axis. Ordinarily, the independent variable is thought to influence the dependent variable, rather than the other way· around. Thus, in Figure 1 the father's height is taken as the independent variable and plotted along the x-axis: father's height influences son's height.

Sir Francis Galton (England, 1822–1911)

From Biometrika, 1903.

Before going on, it would be a good idea to work the exercises of this section. They are easy, and they will really help you understand the rest of this chapter. If you have trouble with them, review Chapter 7.

EXERCISE SET A

1. Some hypothetical data are shown below along with the scatter diagram. Fill in the blanks.

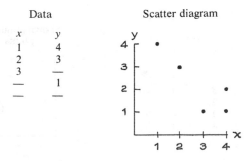

2. Below is a scatter diagram for some hypothetical data.
   (a) The average of the x-values is closest to: 1, 1.5, 3.
   (b) The average of the y-values is closest to: 1, 1.5, 3.
   (c) Which shows more spread, the x-values or the y-values?
   (d) Write down a data table like the one in Exercise 1 for this scatter diagram.
   (e) Check your answers to (a), (b), (c) by calculating the average and SD for the x-values, as well as the average and SD for the y-values.

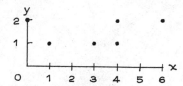

3. In the diagram below—
   (a) The average of the x-values is closest to: 1, 1.5, 2.
   (b) The SD of the x-values is closest to: 0.1, 0.5, 1.
   (c) The average of the y-values is closest to: 1, 1.5, 2.
   (d) The SD of the y-values is closest to: 0.5, 1.5, 3.0.

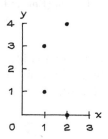

4. Draw the scatter diagram for each of the following hypothetical data sets. The variable labeled "x" should be plotted along the x-axis, the one labeled "y" along the y-axis. Mark each axis fully. In some cases, you will have to plot the same point more than once. By convention, the number of times such a multiple point appears is indicated inside a small circle next to it, as in the diagram below.

(a)		(b)	
x	y	x	y
1	2	3	5
3	1	1	4
2	3	3	1
1	2	2	3
		1	4
		4	1

Showing multiple points

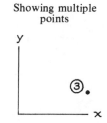

5. Use Figure 1 to answer the following questions:
   (a) How tall was the shortest father? How tall was his son?
   (b) Four fathers are just about tied for tallest. What is their height, and how tall were their sons?
   (c) Take the families where the father was 72 inches tall, to the nearest inch. How tall was the tallest son? the shortest son?
   (d) How many families are there where the sons are just about 76 inches tall? How tall are the fathers?
   (e) The average height of the fathers was closest to:

   64 inches   68 inches   72 inches.

   (f) The SD of the heights of the fathers was closest to:

   3 inches   6 inches   9 inches.

6. Students named A, B, C, D, E, F, G, H, and I took a midterm and a final in a certain course. A scatter diagram for the scores is shown below.
   (a) Which students scored the same on the midterm as on the final?
   (b) Which students scored higher on the final?
   (c) The average score on the final was closest to: 25, 50, 75.
   (d) The SD of the scores on the final was closest to: 10, 25, 50.
   (e) For the students who scored over 50 on the midterm, the average score on the final was closest to: 25, 50, 75.
   (f) True or false: On the whole, students who did well on the midterm also did well on the final.
   (g) True or false: There is a strong positive association between midterm scores and final scores.

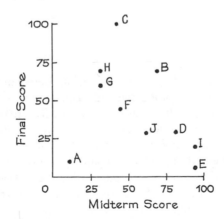

7. The scatter diagram on the next page shows scores on the midterm and final in a certain course.
   (a) The average midterm score was closest to: 25, 50, 75.
   (b) The SD of the midterm scores was closest to: 5, 10, 20.
   (c) The SD of the final scores was closest to: 5, 10, 20.
   (d) Which exam was harder—the midterm or the final?

(e) In which exam was there more spread in the scores?

(f) True or false: There was a strong positive association between midterm scores and final scores.

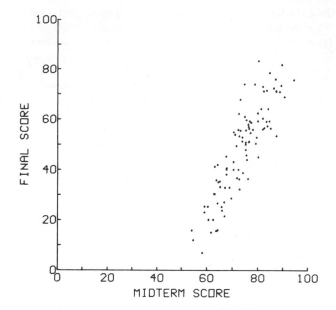

The answers to these exercises are on pp. A-35–A-36.

2. THE CORRELATION COEFFICIENT

Suppose you are looking at the relationship between two variables, and have already plotted the scatter diagram. The graph looks like a cloud of points. How can it be summarized numerically? The first step would be to mark a point showing the average of the x-values and the average of the y-values (Figure 4a). This point is called the *point of averages*.[3] It locates the center of the cloud. The next step would be to measure the spread of the cloud from side to side. This can be done using the SD of the x-values—the horizontal SD. Most of the points will be within two horizontal SDs on either side of the point of averages (Figure 4b). In the same way, the SD of the y-values—the vertical SD—can be used to measure the spread of the cloud from top to bottom. Most of the points will be within two vertical SDs above or below the point of averages (Figure 4c).

So far, the summary statistics are

• average of x-values, SD of x-values;

• average of y-values, SD of y-values.

Figure 4. Summarizing a scatter diagram.

(a) The point of averages   (b) The horizontal SD   (c) The vertical SD

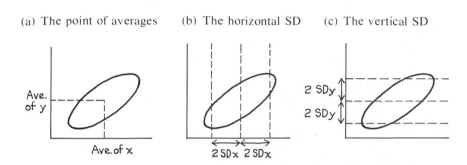

These statistics tell us the center of the cloud, and how spread out it is, both horizontally and vertically. But there is still something missing. The summary says nothing about the association between the two variables.

Look at the scatter diagrams in Figure 5. Both clouds have the same center and show the same spread both horizontally and vertically. However,

Figure 5. Summarizing a scatter diagram. The correlation coefficient measures clustering around a line.

(a) Correlation near one   (b) Correlation near zero
     means tight clustering   means loose clustering

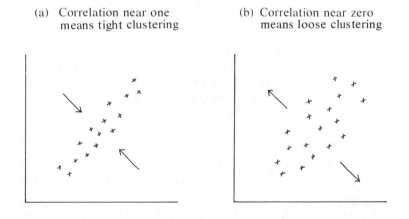

in the first cloud, the points are tightly clustered around a line: there is a strong linear association between the two variables. In the second cloud, the clustering is much looser. The strength of the association is different in the two diagrams. To measure this, one more summary statistic is needed—the correlation coefficient.

> The correlation coefficient is a measure of linear association or clustering around a line. The relationship between two variables can be summarized by giving five statistics:
> - the average of the x-values, the SD of the x-values
> - the average of the y-values, the SD of the y-values
> - the correlation coefficient.

The formula for computing the correlation coefficient will be presented in section 5, but right now we want to focus on the graphical interpretation. Figure 6 shows six scatter diagrams for hypothetical data, each with fifty points. The diagrams were generated by computer. In all six pictures, the average is 3 and the SD is 1 for both variables. The computer has printed the value of the correlation coefficient over each diagram. The one at the top left shows a correlation of 0, and the cloud is completely formless. As x increases, y shows no tendency to increase or decrease, it just straggles around. This diagram shows no association between the variables, no clustering around a line. The next one has a correlation coefficient of 0.4; a linear pattern is just beginning to emerge—there is some weak association between the variables. The next one has a correlation coefficient of 0.6, with a stronger linear pattern, a stronger association between x and y. And so on, through the last one. The closer the correlation coefficient is to 1, the stronger is the linear association between the variables and the more tightly clustered are the points around a line.

A correlation of $+1$, which does not appear in the figure, is often referred to as a *perfect correlation*. It means that all the points lie exactly on a line, so there is a perfect linear relationship between the variables. Correlations are always $+1$ or less.

For example, the correlation between IQs of identical twins raised together is 0.95. Do such twins tend to have nearly the same IQ? If they did, their points on the scatter diagram would be very close to the line $y = x$. The lower right scatter diagram in Figure 6 has a correlation coefficient of 0.95 but still shows spread around the line. A scatter diagram for the twin IQs would look much the same. The twins have similar IQs, but they are far from identical. So a correlation of 0.95 still leaves a fair amount of room for spread around the line. Most social-science correlations are much lower, 0.30 to 0.70 being the usual range. As Figure 6 shows, the corresponding relationships are very rough indeed.

For another example, in the Health Examination Survey, the correlation between income and education was 0.4 for the men aged 25 to 34, rising to 0.6 for the men aged 35 to 44.[4] As the corresponding scatter diagrams in Figure 6 indicate, the relationship between income and education is stronger for the older men. But it is still quite rough.

Be careful. A correlation of 0.80 does not mean that 80% of the points are tightly clustered around a line, nor does it indicate twice as much linearity as a correlation coefficient of 0.40. Right now, there is no direct way to interpret the exact numerical value of the correlation coefficient; this will be done later in Chapters 10 and 11.

Figure 6. The correlation coefficient—six positive values. The diagrams are scaled so that the average equals 3 and the SD equals 1, horizontally and vertically. The clustering around a line is measured by the correlation coefficient.

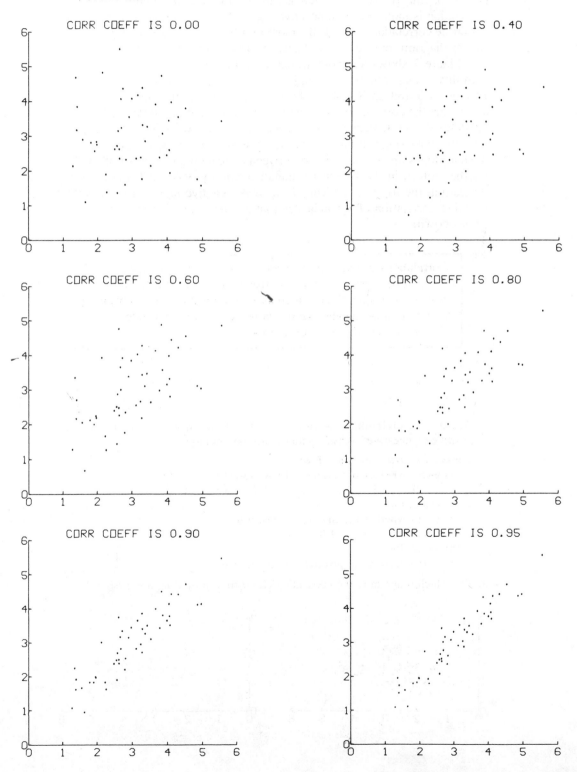

So far, only positive association has been discussed. The Health Examination Survey found that women with more education tended to have fewer children. This is *negative association*. An increase in education is accompanied on the whole by a decrease in the number of children. Negative association is indicated by a negative sign is the correlation coefficient. The closer the correlation is to -1, the more tightly clustered are the points in the scatter diagram, but now they cluster around a line which slopes down.

Figure 7 shows six more scatter diagrams for hypothetical data, each with fifty points. They are scaled just like Figure 6, each variable having an average of 3 and an SD of 1. The only difference is that the clouds slope down, and the correlations are negative. A correlation of -0.90, for instance, indicates the same degree of clustering as one of $+0.90$. With the negative sign, the clustering is around a line which slopes down; with a positive sign, the line slopes up. The correlation between education and number of children for the women in the Health Examination Survey was -0.3, matching the diagram at the top left of Figure 7—a weak negative association. A perfect negative correlation of -1 indicates that all the points lie on a line, which slopes down.

> Correlations are always between -1 and 1, but can take any value in between. A positive correlation means that the cloud slopes up: as one variable increases, so does the other. A negative correlation means that the cloud slopes down: as one variable increases, the other decreases.

EXERCISE SET B

1. Would the correlation between the age of a second-hand car and its price be positive or negative? Why? (Antiques are not included.)

2. For each scatter diagram below:
   (a) guess whether the average of x is closest to
         1.0   1.5   2.0   2.5   3.0   3.5   4.0
   (b) same, for y
   (c) guess whether the SD of x is closest to
         0.25   0.5   1.0   1.5
   (d) same, for y
   (e) is the correlation positive, negative, or 0?

3. For which diagram is the correlation closer to 0, forgetting about signs?

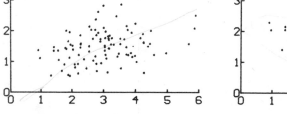
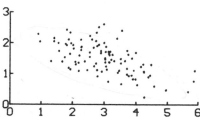

Figure 7. The correlation coefficient—six negative values. The diagrams are scaled so the average equals 3 and the SD equals 1, horizontally and vertically. The clustering around a line is measured by the correlation coefficient.

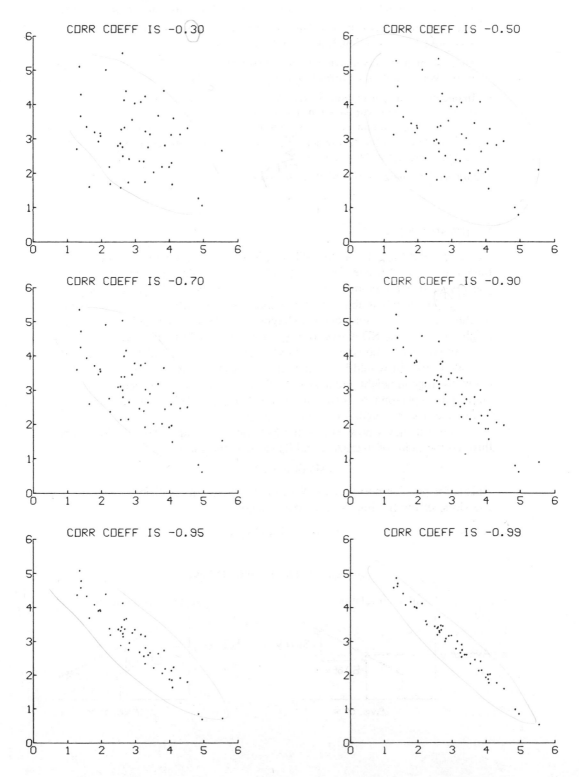

4. The correlation between the heights of the fathers and sons in Figure 1 is closest to: -0.3, 0, 0.5, or 0.8.

5. If you took only the fathers who were taller than 6 feet and their sons in Figure 1, the correlation between the heights would be closest to: -0.3, 0, 0.5, or 0.8.

6. The correlation between the heights of husbands and wives in the USA in 1976 was closest to: -0.9, -0.3, 0.3, or 0.9. Explain.

7. Suppose men always married women who were exactly 8% shorter than they were. What would the correlation between their heights be?

8. In one class, the correlation between scores on the final and on the midterm was 0.60, while the correlation between the scores on the final and homework scores was 0.30. The relationship between the final scores and the midterm scores is twice as linear as the relationship between the final scores and the homework scores. True, false, or meaningless?

The answers to these exercises are on p. A-36.

3. THE SD LINE

The closer the correlation coefficient is to 1, the more tightly clustered the points are around a line. What is this line? It is called the *SD line,* and it goes through all the points which are an equal number of SDs away from the average, for both variables. For example, take a scatter diagram showing heights and weights. Someone who happened to be one SD above average in height and also one SD above average in weight would be plotted on the SD line; but someone who is one SD above average in height and 1.5 SDs above average in weight would be off the line. Similarly, a person who is two SDs below average in height and also two SDs below average in weight would be on the line; a person who is two SDs below average in height and only one SD below average in weight would be off the line.

Figure 8 shows how to plot the SD line on a graph. The SD line goes through the point of averages, and its slope is the ratio

$$(\text{SD of } y)/(\text{SD of } x).$$

This is for positive correlations. When the correlation coefficient is negative, the slope of the SD line is reversed: it becomes[5]

$$-(\text{SD of } y)/(\text{SD of } x).$$

Figure 8. Plotting the SD line.

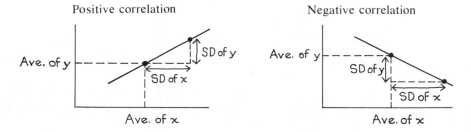

Positive correlation   Negative correlation

EXERCISE SET C

1. One study on male college students found their average height to be 69 inches, with an SD of 3 inches. Their average weight was 140 pounds, with an SD of 20 pounds, and the correlation was 0.60. If one of these people is 72 inches tall, how heavy would he have to be to fall on the SD line?

2. Using the same data as in Exercise 1, say whether each of the following students was on the SD line:
   (a) height 75 inches, weight 180 pounds.
   (b) height 66 inches, weight 130 pounds.
   (c) height 66 inches, weight 120 pounds.

3. For the scatter diagram shown below, say whether it is the solid line or the dashed line which is the SD line.

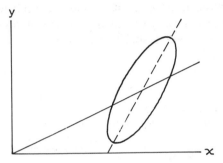

4. One study measured the heights of fathers and sons at different ages. The SD lines plotted below are for:
   (a) fathers at age 30, sons at age 10;
   (b) fathers at age 40, sons at age 20.
   Which one is dashed? solid?

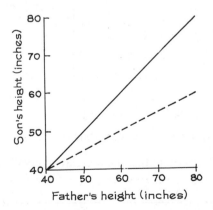

The answers to these exercises are on p. A-36.

4. SOME EXCEPTIONAL CASES

The correlation coefficient does not always give a true indication of the clustering. There are two main exceptional cases: outliers and nonlinearity. In

Figure 9a the dots show a perfect correlation of +1. The outlier, marked by a cross, brings the correlation down almost to 0. As a result of this anomaly, some people get carried away in pursuit of outliers. It is important to remember that in any scatter diagram, there will be some points more or less detached from the main part of the cloud. These points should be rejected only for good reason.

In Figure 9b the correlation coefficient is very close to 0, even though the points show strong association. The reason is that the graph does not look at all like a straight line. An example would be the association between weight and age for adult men (Figure 3 on p. 51). The correlation is very close to 0, but there is a definite pattern: weight rises with age, then falls.

> Whenever possible, look at the scatter diagram to check for outliers and nonlinear association. The correlation coefficient only measures linear association, rather than association in general.

Figure 9. Nonlinearity. The correlation coefficient can be misleading in the presence of outliers or of nonlinear association.

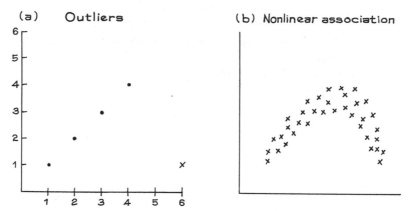

5. COMPUTING THE CORRELATION COEFFICIENT

The procedure for computing the correlation coefficient can be stated as follows:

> Convert each variable to standard units. The average of the products gives the correlation coefficient.

This can be put in a formula, abbreviating x for the first variable, y for the second variable, and r for the correlation coefficient:

$$r = \text{average of } (x \text{ in standard units}) \times (y \text{ in standard units}).$$

Example. Compute r for the hypothetical data in Table 1 below.

Table 1. Data.   Table 2. Computing r.

x	y		x in standard units	y in standard units	Products
1	5		-1.5	-0.5	0.75
3	9		-0.5	0.5	-0.25
4	7		0.0	0.0	0.00
5	1		0.5	-1.5	-0.75
7	13		1.5	1.5	2.25

Note: The first row of Table 1 represents two measurements on one subject in the study; the two numbers are the x- and y-coordinates of the corresponding point on the scatter diagram. Similarly for the other rows.

Solution. The work can be laid out as in Table 2.

Step 1. Convert the x-values to standard units, as in Chapter 5. This is quite a lot of work. First, you have to find the average and SD of the x-values:

$$\text{average of } x\text{-values} = 4, \quad \text{SD} = 2.$$

Then, you have to subtract the average from each x-value, and divide by the SD:

$$\frac{1-4}{2} = -1.5 \quad \frac{3-4}{2} = -0.5 \quad \frac{4-4}{2} = 0 \quad \frac{5-4}{2} = 0.5 \quad \frac{7-4}{2} = 1.5$$

Put the results in the first column of the table (see Table 2). These numbers tell you how far above or below average the x-values are, in terms of their SD. For instance, the value 1 is 1.5 SDs below average.

Step 2. Convert the y-values to standard units, and put the results in the second column of the table.

Step 3. For each row of the table, work out the product

$$(x \text{ in standard units}) \times (y \text{ in standard units}).$$

Put the products in the third column of the table.

Step 4. Take the average of the products:

$$r = \text{average of } (x \text{ in standard units}) \times (y \text{ in standard units})$$

$$= \frac{0.75 - 0.25 + 0.00 - 0.75 + 2.25}{5} = 0.40$$

If you plotted the scatter diagram for this data (Figure 10, p. 126), the points would slope up, but would only show loose clustering around a line.

One feature of the correlation coefficient can be seen from the procedure given in the example.

> The correlation coefficient is a pure number, without units. It does not depend on the units chosen for either variable.

The basic reason is that the original units cancel out when converting to standard units. In a similar way, adding the same number to all the values of one variable, or multiplying all the values of one variable by the same positive number, will not change r.

The scatter diagram for Table 1 is shown in Figure 10a. The products are shown at the corresponding dots. Horizontal and vertical lines have been drawn through the point of averages, dividing the scatter diagram into four quadrants. If a point is in the lower left region, both variables are below average and are negative in standard units; the product of the two negatives is positive. In the upper right quadrant, the product of two positives is positive. In the remaining two quadrants, the product of a positive value and a negative value (in standard units) is negative. The average of these products, positive and negative together, gives the correlation coefficient. If r is positive, then points in the two positive quadrants will predominate, as in Figure 10b. If r is negative, points in the two negative quadrants will predominate, as in Figure 10c. This is how r works as a measure of association.

Figure 10. How the correlation coefficient works.

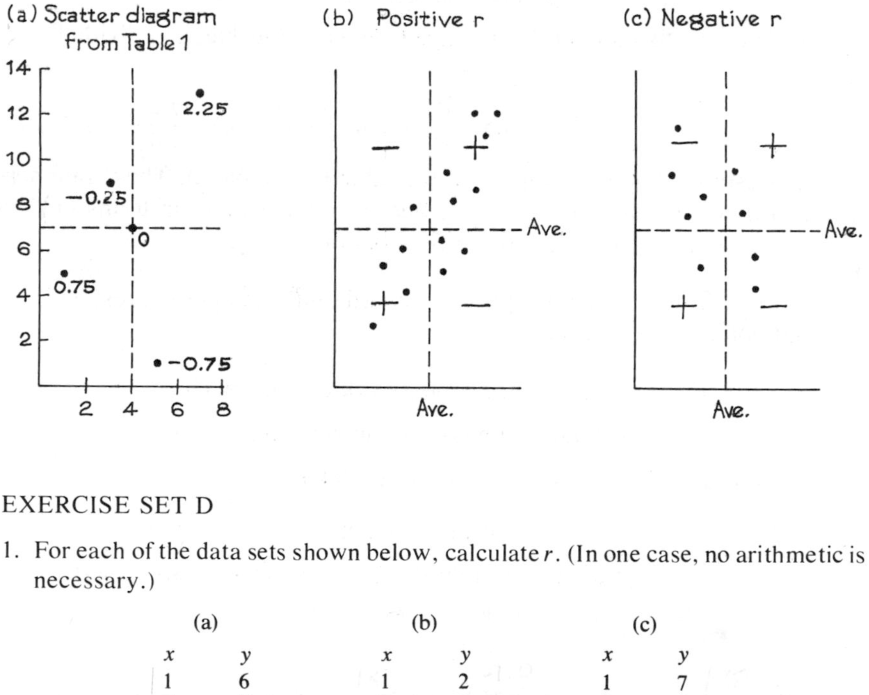

EXERCISE SET D

1. For each of the data sets shown below, calculate r. (In one case, no arithmetic is necessary.)

(a)		(b)		(c)	
x	y	x	y	x	y
1	6	1	2	1	7
2	7	2	1	2	6
3	5	3	4	3	5
4	4	4	3	4	4
5	3	5	7	5	3
6	1	6	5	6	2
7	2	7	6	7	1

2. Modify the data in Table 1 on p. 125 by switching the two columns. Does this change r? Explain or calculate.

3. Modify the data in Table 1 by adding 3 to each number in the first column. Does this change *r*? Explain or calculate.

4. Modify the data in Table 1 by doubling each number in the first column. Does this change *r*? Explain or calculate.

5. Interchange the last two values (1 and 13) for the second variable in Table 1. Does this change *r*? Explain or calculate.

6. Below are three scatter diagrams. Do they have the same correlation? Try to answer without calculating.

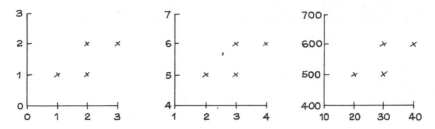

7. Someone hands you the scatter diagram shown below. He has forgotten to label the axes. Can you still calculate *r*? If so, what is it?

8. Find the scatter diagram in Figure 6 on p. 119 with a correlation of 0.95. In this diagram, the percentage of points where both variables are simultaneously above average is closest to: 5%, 25%, 50%, 75%, 95%.

9. Repeat Exercise 8, for a correlation of 0.00.

10. Using Figure 7 on p. 121, repeat Exercise 8 for a correlation of −0.95.

The answers to these exercises are on pp. A-36–A-37.

Technical note: There is another way to compute *r*, which is more efficient in some cases (for instance, it is often used in computer programs):[6]

$$r = \frac{\text{Cov}(x, y)}{(\text{SD of } x) \times (\text{SD of } y)}$$

where

Cov(*x*, *y*) = (average of products *xy*) − (ave of *x*) × (ave of *y*).

6. CHANGING SDs

So far, we have looked at scatter diagrams where the SDs were fixed, and the correlation coefficient varied. When the SDs change, it is more difficult. The appearance of a scatter diagram depends on the SDs. For instance, look at Figure 11. In both diagrams, *r* is 0.70. However, the first one appears to be much more tightly clustered around the SD line. That is because its SDs are much smaller. The point is that *r* does not measure clustering in absolute terms, but only relative to the SDs. You can see this from the formula for *r*. This involved converting the variables to standard units—that is, measuring deviations from average relative to the SDs.

Figure 11. The effect of changing SDs. The two scatter diagrams have the same correlation coefficient of 0.70. The first one is much more tightly clustered around the SD line than the second, but only because its SDs are smaller.

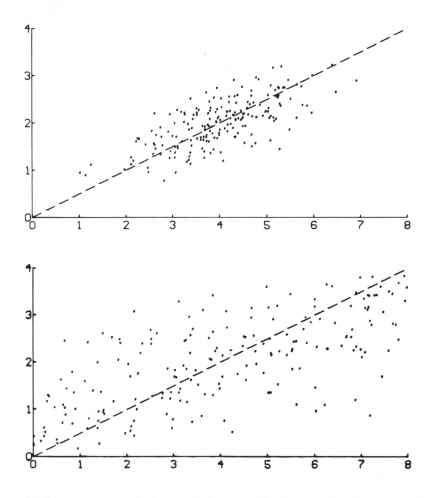

To interpret a correlation coefficient graphically, imagine the scatter diagram drawn so the vertical and horizontal SDs cover the same distance on the page as those in Figure 6. If *r* for your scatter diagram is 0.40, it will probably show about the same amount of clustering around the diagonal as the one with an *r* of 0.40 in Figure 6 at the top right. If *r* is 0.90, it will look like the corresponding one in Figure 6, at the bottom left. In general, your scatter diagram will match the ones in Figure 6 or Figure 7 that have similar values for *r*.

EXERCISE SET E

1. For each of the following diagrams, say whether *r* is closest to 0.3 or 0.7. In (c), take the dots and crosses together.

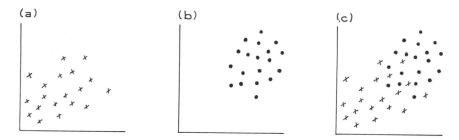

(a)   (b)   (c)

2. The correlation coefficient between height and weight for male college students in one study was about 0.60. For female college students, it was about the same. If you took both the men and the women together, would the correlation between height and weight be just about 0.60, quite a bit lower, or quite a bit higher? Explain.

The answers to these exercises are on p. A-37.

Technical note: If *r* is close to 1, then the points are tightly clustered around the SD line: a typical point is only a small fraction of a vertical SD above or below the line. If *r* is close to 0, then the points are widely scattered around the SD line: a typical point is above or below the line by an amount roughly comparable in size to the vertical SD.

Figure 12. The correlation coefficient: *r* measures the vertical distance of a typical point above or below the SD line, relative to the vertical SD. When *r* is near 1, this ratio is near 0. When *r* gets smaller, this ratio gets bigger.

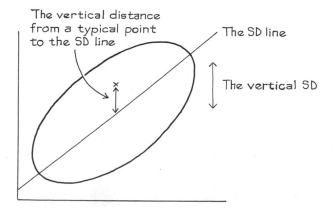

The connection between the correlation coefficient and the typical distance above or below the SD line can be expressed mathematically as follows: the r.m.s. vertical distance to the SD line equals

$$\sqrt{2(1 - r)} \times \text{ the SD in the vertical direction.}$$

There are similar formulas for the horizontal direction.

Take, for example, a correlation of 0.95. Then

$$\sqrt{2(1 - r)} = \sqrt{0.1} \approx 0.3$$

So the spread around the SD line is about 30% of an SD. That is why the lower right scatter diagram in Figure 6 shows a fair amount of spread around the SD line, even though a correlation of 0.95 looks so close to 1.

7. REVIEW EXERCISES

1. A study of the IQs of husbands and wives obtained the following results:

for husbands, average IQ ≈ 100,   SD ≈ 15
for wives,   average IQ ≈ 100,   SD ≈ 15
correlation ≈ 0.6

One of the following is a scatter diagram for the data. Which one? Say briefly why you reject the other three.

(a)

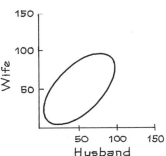

(b)

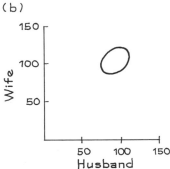

(c)

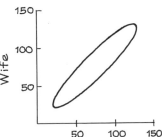

(d)

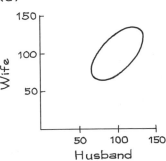

2. For a representative sample of cars, would the correlation between weight and gasoline economy (miles per gallon) be positive or negative? Explain.

3. If the correlation coefficient is +0.90, then 90% of the points are highly correlated. True, false, or meaningless?

4. Figure 13 shows six scatter diagrams for hypothetical data. The correlation coefficients, in scrambled order, are:

     −0.85   −0.38   −1.00   +0.06   +0.62   0.97

Match the scatter diagrams with the correlation coefficients.

Figure 13. Guess r for each diagram.

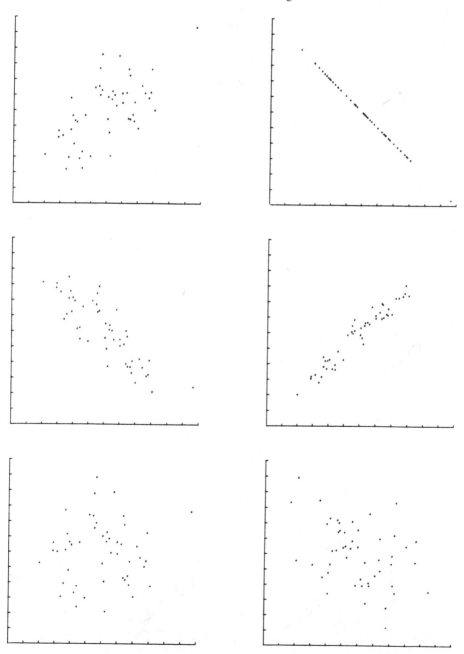

5. Three data sets are collected, and the correlation coefficient is computed in each case. The variables are
   (a) grade point average in freshman year and in sophomore year
   (b) grade point average in freshman year and in senior year
   (c) length and weight of two-by-four boards.
   In scrambled order, the correlations are: 0.30, 0.60, and 0.95. Match the correlations with the data sets. Explain your reasoning.

6. The scatter diagram for test scores on two IQ tests (Stanford-Binet L and M) is sketched below. You are trying to predict the score on form M from the score on form L. Which of the following is true and why?
   (i)   It is easier to do when the score on form L is 75.
   (ii) It is easier to do when the score on form L is 125.
   (iii) It is about the same for both.

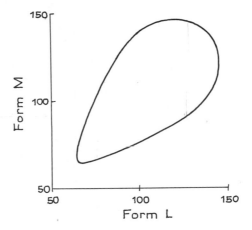

7. The correlation between the ages of husbands and wives in the United States in 1976 was
   exactly −1   close to −1   close to 0   close to +1   exactly +1.
   Why?

8. If women always married men who were five years older, what would the correlation between their ages be? Explain.

9. Find the correlation coefficient for each of the three data sets shown below. (In one case, no arithmetic is necessary.)

(a)		(b)		(c)	
x	y	x	y	x	y
1	5	1	1	1	2
1	3	1	2	1	2
1	5	1	1	1	2
1	7	1	3	1	2
2	3	2	1	2	4
2	3	2	4	2	4
2	1	2	1	2	4
3	1	3	2	3	6
3	1	3	2	3	6
4	1	4	3	4	8

10. In Exercise 9(b), would *r* change if you
    (a) interchanged the two columns?
    (b) switched the 2 and the 3 at the bottom of the right-hand column of the table?
    (c) simultaneously doubled all the entries in the first column, and added 17 to all the entries in the second column?
    Explain or calculate.

11. A teaching assistant gives a quiz to his section. There are ten questions on the quiz and no part credit is given. After grading the papers the TA writes down for each student the number of questions the student got right and the number wrong. The average number of correct answers is 6.4 with an SD of 2.0; the average number of wrong answers is 3.6 with the same SD of 2.0. The correlation coefficient is:·
     0   −0.50   +0.50   −1   +1   can't tell without the data.
    Explain.

12. The subjects in Cycle II of the Health Examination Survey were children aged 6 to 11. At each age, the correlation between height and weight was about 0.50. For all the children together, would the correlation between height and weight be around 0.50, quite a bit more than 0.50, or quite a bit less than 0.50? Explain.

13. A longitudinal study of human growth has been under way since 1929, at the Berkeley Institute of Human Development.[7] The scatter diagram below shows the heights of 64 boys, measured at ages 4 and 18.
    (a) The average height at age 4 was closest to:
          38 inches   42 inches   44 inches.
    (b) The SD of height at age 18 was closest to:
          0.5 inches   1.0 inches   2.5 inches.
    (c) The correlation coefficient is closest to:
          0.50   0.80   0.95.
    (d) Which is the SD line—solid or dashed?

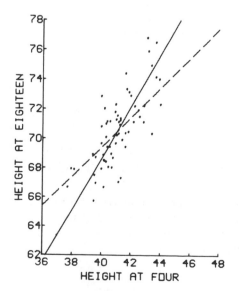

14. Fifteen students in an elementary statistics course (taught at the University of California, Berkeley, in fall, 1976) were asked to count the dots on a diagram like the one shown below. (In fact, there were 85 dots on the diagram.) Each student counted the dots twice, and the results are shown below. Make a scatter diagram for the counts; each student is represented by one point on the diagram, showing his first count and his second count. Mark both axes fully. Choose the scale so you can see the pattern in the points. Use your scatter diagram to answer the following questions:

   (a) Did all the students work independently?
   (b) True or false: those students who counted high the first time also tended to be high the second time.
   (c) True or false: there is almost no relationship between the two variables.

1st count	2nd count
91	85
81	83
86	85
83	84
85	85
85	84
85	89
84	83
91	82
91	82
91	82
85	85
85	85
87	85
90	85

8. SUMMARY

1. Data concerning the relationship between two variables can be represented visually by a *scatter diagram*.

2. When the scatter diagram is tightly clustered around a line, there is a strong relationship between the variables: one is easy to predict from the other.

3. A scatter diagram can be summarized by means of five statistics:
   • the average of the x-values, the SD of the x-values,
   • the average of the y-values, the SD of the y-values,
   • the *correlation coefficient r*.

4. Positive association is indicated by a plus-sign and negative association by a minus-sign in the correlation coefficient.

5. The correlation coefficient measures clustering around a line, relative to the SDs.

6. The correlation coefficient ranges from -1, when all the points lie on a line which slopes down, to $+1$, when all the points lie on a line which slopes up.

7. In a series of scatter diagrams with the same SDs, the closer r is to ± 1 the tighter the clustering around a line. But the actual numerical value of r is harder to interpret. For instance, an r of 0.60 does not indicate twice as much clustering as an r of 0.30.

8. The points cluster around a line called the *SD line*. This line goes through the *point of averages*. When r is positive, the slope of this line is

$$(SD \text{ of } y)/(SD \text{ of } x).$$

When r is negative, the slope is

$$-(SD \text{ of } y)/(SD \text{ of } x).$$

9. In the presence of outliers, or of nonlinear association, the correlation coefficient can give a misleading impression about the appearance of the scatter diagram.

10. To calculate the correlation coefficient, convert each variable to standard units, and then take the average product.

11. The correlation coefficient is a pure number, without units. It is unaffected by:

- interchanging the two variables
- adding the same number to all the values of one variable
- multiplying all the values of one variable by the same positive number.

9

More about Correlation

"Very true," said the Duchess: "flamingoes and mustard both bite. And the moral of that is—'Birds of a feather flock together.'"
"Only mustard isn't a bird," Alice remarked.
"Right, as usual," said the Duchess: "what a clear way you have of putting things!"
—Alice in Wonderland

1. ASSOCIATION IS NOT CAUSATION

The Salk vaccine against polio was discussed in Part I. Before the introduction of this vaccine, investigators looked at the relationship between the incidence of polio and the number of soft drinks sold. For each week of the year, they tabulated the number of soft drinks sold that week, and the number of new cases of polio reported. These data points showed strong positive correlation. During weeks when more soft drinks were sold, there were more new cases of polio; when fewer soft drinks were sold, there were fewer such cases.

Do soft drinks cause polio? If so, prohibiting their sale would have reduced the incidence of the disease. Clearly, the answer was no, and nobody was fooled by the correlation. Polio epidemics were most severe in the summer just when soft-drink sales are at their highest. So there was a third factor driving both variables—season. The correlation coefficient was just picking up this third factor (see Figure 1). Controlling for this third factor, by looking at the seasons separately, would make the correlation go away. Nothing in the formula warns you about such third factors.

Figure 1. A misleading correlation. Soft-drink sales are correlated with the incidence of polio.

Correlation measures association. But association is not the same as causation.

Part I explained the difference between observational studies and controlled experiments. The same kind of distinction is useful here. In a laboratory experiment, the investigator usually varies the independent variable on his own initiative, and watches the effect on the dependent variable. For example, Robert Hooke (England, 1653–1703) was able to determine the relationship between the length of a spring and the load placed on it. He just hung weights of different sizes on the end of a spring, and watched what happened. When the load was increased, the spring got longer. When the load was reduced, the spring got shorter. In this experiment, weight was the independent variable; Hooke could vary that at will. Length was the dependent variable. Hooke did not choose its value, but watched how it responded to weight. Since the weight was under the direct control of the experimenter, there is no question here about what was causing what. The weight caused the spring to get longer.

Social science studies of the relationship between two variables are usually observational. For example, if an investigator wants to know how weight depends on height among children of a certain age, weight would be the dependent variable and height the independent variable. Height is usually thought to influence weight, rather than the other way around. However, such a study is bound to be observational. Nobody but Procrustes would entertain the notion of varying a person's height to observe the effect on weight.[1]

If the investigator can control the independent variable, setting it at whatever values he pleases, then association does argue strongly for causation. If the investigator just observes the values of the independent and dependent variables for some set of subjects, association does not establish causation. The polio study in Figure 1 is an example. With observational studies, caution is always in order.

EXERCISE SET A

1. Many studies have found an association between cigarette smoking and heart disease. Recently, the Framingham study found an association between coffee drinking and heart disease. Should you conclude that coffee drinking causes heart disease? Or can you explain the association between coffee drinking and heart disease in some other way?

2. (a) Should the correlation between blood pressure and age be positive or negative? What about the correlation between income and age? The subjects are a representative sample of American men aged 18 to 54.

   (b) A positive correlation is observed between blood pressure and income. Does this point to a causal connection, or can it be explained another way?

3. For the men aged 25 to 34 in the Health Examination Survey of 1960, the relationship between educational level (years of schooling completed) and height can be summarized as follows:[3]

   average educational level ≈ 11 years,   SD ≈ 3 years
   average height ≈ 69 inches, SD ≈ 3 inches, $r ≈ 0.2$

   Does going to school longer make a man taller? If not, how can you explain the correlation of 0.2?

4. Many economists believe that there is a trade-off between unemployment and inflation: low rates of unemployment will cause high rates of inflation, while higher rates of unemployment will reduce the rate of inflation. The relationship between the two variables is shown below for the United States in the decade 1960–1969. There is one point for each year, with the rate of unemployment that year shown on the x-axis, and the rate of inflation shown on the y-axis. The points fall very close to a smooth curve known as the *Phillips Curve*. Is this an observational study or a controlled experiment? If you plotted the points for the 1970s or the 1950s, would you expect them to fall along the curve?

The Phillips Curve for the 1960s

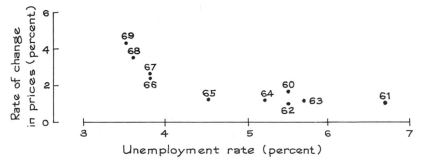

Source: *Economic Report of the President*, 1975.

5. Two different investigators are working on a growth study. The first measures the heights of 100 children, in inches. The second prefers the metric system, so he converts the results to centimeters (multiplying by the conversion factor 2.54 centimeters per inch).

(a) If no mistakes are made in the conversion, what is the correlation between the two sets of measurements?
(b) What happens to r if mistakes are made in the arithmetic?
(c) What happens to r if he goes out and measures the same children again, using metric equipment?

6. A sociologist wants to describe the relationship between religion of family head and family size. Is it proper to use the correlation coefficient?

The answers to these exercises are on p. A-37–A-38.

2. THE SKEELS-SKODAK STUDY

Is intelligence determined by heredity or by environment? This nature-nurture controversy has been raging for at least three centuries, since the English philosopher John Locke (1632–1704) published his *tabula rasa* theory, according to which the mind is a blank tablet at birth, to be written on by circumstance. In this century, attempts have been made to investigate this problem in a quantitative way, and some insight is provided by studies of adopted children. One of the best-known of these studies was conducted by Howard Skeels and Marie Skodak, psychologists at an Iowa adoption agency. In the 1930s, they undertook a longitudinal study of children placed by this agency, from adoption (at an age of around six months) until adolescence.[4] In 100 cases, they were able to test the child's intelligence four times (roughly at the ages of 2, 4, 7, and 14). In 63 of these cases, they also managed to test the intelligence of the child's natural mother at the time of placement. These data could be reported as in Table 1.

Table 1. The Skeels-Skodak data.

Case	Natural mother's IQ	Foster mother's educational level	Child's IQ at test			
			I	II	III	IV
1	unknown	12	121	114	115	105
2	100	10	120	115	109	106
3	71	10	131	109	113	95
.	.	.	.	.	.	.
.	.	.	.	.	.	.
.	.	.	.	.	.	.
100	76	12	122	107	128	101

For each of the four tests, then, it is possible to correlate the IQs of the children with the IQs of their natural mothers, focusing on the 63 cases where the information is available. This correlation is 0 at test I (when the children were about two years old), rising to about 0.4 at test IV, when the children were around fourteen. These correlations are plotted in Figure 2. There was absolutely no contact between the natural mothers and the children after adoption, so the increasing correlation cannot be explained as the result of nurture.

How about the correlation between the IQs of these children and the IQs of their foster mothers? Unfortunately, the foster mothers weren't tested. However, their educational level (years of school completed) was known (see

Table 1), and this is a good proxy for IQ. At each test, the IQs of the children can be correlated with the educational level of their foster mothers. The results are plotted in Figure 2: the correlation is extremely low.

Figure 2. The Skeels-Skodak study.

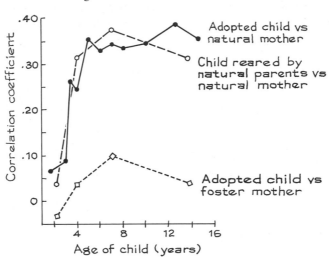

Figure 2 also shows some results from another study (conducted by Dr. Marjorie Honzik at the Institute of Human Development, University of California, Berkeley), the correlations between the IQs of children and the educational level of their natural mothers, for a sample of children who were raised at home by their natural parents. This curve is remarkably similar to the Skeels-Skodak curve for the children who were separated from their natural mothers. This similarity suggests that the contact between the mother and child has little to do with increasing correlation. It is a strong piece of evidence for the hereditary theory of intelligence.

Skeels and Skodak themselves were very cautious about interpreting their results. One question they asked was this. Did the environment make some difference which the correlation coefficient has ignored? This may be expected, for as they pointed out,

On the whole the foster families were above the average of their communities in economic security and educational and cultural status. They were highly regarded by the town's business, professional, and religious leaders, and had usually demonstrated a long-time interest in children through church or community activities.

On the other hand, the natural parents came "from relatively inferior socio-economic backgrounds." And as it turned out, the average IQ of the natural mothers was 86, while the average IQ of their children was 106! This 20-point improvement can be attributed in large measure to the superior environment provided by the foster parents.[5]

Why doesn't the correlation coefficient pick up this change? Because it happened across the board. It is as if the change in environment added about

20 points to all the IQs for the children in Table 1. Adding 20 points to all the numbers in a column is a change of scale, which does not affect the correlation—so *r* won't tell you that it happened.

> The correlation coefficient by itself can be a misleading summary of data concerning a relationship. It is usually advisable to report the averages, SDs, and the number of data points, in addition to *r*.

Here is another question raised by Skeels and Skodak, about the correlation between the IQs of the children and their natural mothers: Are these pure measures of the effect of heredity? They seem to be, but in fact they aren't. They also reflect any correlation between the environment (provided by the foster parents) and heredity (the natural parents). This correlation must depend on the way the foster parents picked the children for adoption. As Skeels and Skodak explain,

The primary factors in matching were the stipulations of the foster parents regarding religion, sex and hair color in that order.

That would seem to preclude any connection between environment and heredity. However, the correlation between the foster mother's educational level and the natural mother's IQ turned out to be around 0.25. Despite the fact that the children never saw their natural mothers after the age of 6 months, the effects of environment are confounded with the effects of heredity in the correlations between the IQs of the children and of their natural mothers.

3. ECOLOGICAL CORRELATIONS

In 1955, Doll published a landmark article on the relationship between cigarette smoking and lung cancer.[6] A key piece of evidence was a scatter diagram showing the relationship between the rate of cigarette smoking (per capita) and the rate of deaths from lung cancer in eleven countries. The correlation between these eleven pairs of rates was 0.7, and this was taken as showing the strength of the relationship between smoking and cancer. However, it is not countries which smoke and get cancer, but people. To measure the strength of the relationship for people, it is necessary to have data relating smoking and cancer for individual people. In 1955, this wasn't available. That is why Doll used rates for countries—it was all he could get.

However, correlation coefficients based on rates or averages are often misleading. Here is an example.[7] From 1970 Census data, it is possible to compute the correlation between income and education, for men aged 35 to 54 in the United States. This correlation is about 0.4. The Census Bureau divides the United States up into nine geographical regions. For each region, it is possible to compute the average income and average education for the men living in that region. Then, it is possible to compute the correlation coefficient between these nine pairs of averages, and it works out to 0.7. If you used the correlation for the regions to estimate the correlation for the men, you would be way off. The reason is that within each region, there is a lot of spread around the averages. Replacing each region by the average eliminates the spread, and gives a misleading impression of tight clustering. Figure 3 presents this graphically, for three regions.

Correlations based on rates or averages are called *ecological* correlations.[8] They are often used in political science and sociology. So watch out.

Figure 3. Correlations based on rates or averages are often too big. The panel on the left represents income and education for individuals in three geographic regions, labeled A, B, C. Each individual is marked by the letter showing his region of residence. The correlation is moderate. The panel on the right shows the averages for each region: the correlation between the averages is almost 1.

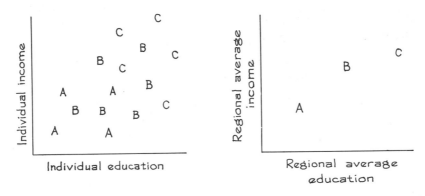

EXERCISE SET B

1. The following table is adapted from Doll and shows per-capita consumption of cigarettes in various countries in 1930, and the death rates from lung cancer for men in 1950. (In 1930, hardly any women smoked: and it seems fair to allow some long period of time for the effects of smoking to manifest themselves.)

Country	Cigarette consumption	Deaths per million
Australia	480	180
Canada	500	150
Denmark	380	170
Finland	1,100	350
Great Britain	1,100	460
Holland	490	240
Iceland	230	60
Norway	250	90
Sweden	300	110
Switzerland	510	250
USA	1,300	200

(a) Plot a scatter diagram for this data.
(b) Does the scatter diagram show that the higher cigarette consumption was in 1930 in one of these countries, on the whole the higher the death rate from lung cancer in 1950?

(c) Does it show that death rates from lung cancer tend to be higher among those persons who smoke more?

2. A sociologist is studying the relationship between suicide and literacy in nineteenth-century Italy.[9] He has data for each province, showing the percentage of literates, and the suicide rate, in that province. The correlation is 0.6. Does this give a fair estimate of the strength of the association between literacy and suicide?

The answers to these exercises are on p. A-38.

4. REVIEW EXERCISES

These exercises cover the previous chapter as well.

1. For persons aged 25 and over in Cycle I of the Health Examination Survey, the relationship between age and educational level (years of schooling completed) can be summarized as follows:

    average age $\approx$ 46 years, SD $\approx$ 14 years
    average ed. level $\approx$ 10 years, SD $\approx$ 4 years, $r \approx -0.3$

    How can you explain the negative correlation? (This study was discussed on pp. 49–51.)

2. If the correlation coefficient is -0.80, below-average values of the dependent variable tend to be associated with below-average values of the independent variable. True or false? Explain.

3. If the dependent variable is usually less than the independent variable, the correlation coefficient will be negative. True or false? Explain.

4. In 1910, Hiram Johnson entered the California gubernatorial primaries. For each county, data are available to show the percentage of native-born Americans in that county, as well as the percentage of the vote for Johnson. A political scientist calculated the correlation between these percentages.[10] It is 0.5. Is this a fair measure of the extent to which "Johnson received native, as opposed to immigrant, support?" Why?

5. When studying one variable, you can use a graph called a _____. When studying the relationship between two variables, you can use a graph called a _____.

6. In each case, guess which correlation is higher, and say why.
    (a) height at age 4 and height at age 18, height at age 16 and height at age 18.
    (b) height at age 4 and height at age 18, weight at age 4 and weight at age 18.
    (c) height and weight at age 4, height and weight at age 18.

7. At the University of California, Berkeley, Statistics 2 is a large lecture course with small discussion sections led by teaching assistants. In the fall of 1975, two of the lecturers taught the course jointly, and gave a common

final. They had eleven assistants. At the second-to-last lecture, the students filled out anonymous questionnaires, rating the teaching effectiveness of the instructors, their teaching assistant (by name), and the course, on the scale

1	2	3	4	5
poor	fair	good	very good	excellent

For each assistant, it was then possible to compute
- his average rating, by the students in his section;
- the average rating of the course, by students in his section;
- the average score on the final, for the students in his section.

The results are shown below; the assistants are identified only by a letter. Draw a scatter diagram for each pair of variables (there are three pairs), and if you have a statistical calculator find the correlations. (The correlations are based on averages. However, since the questionnaires were anonymous, it was not possible to link up student ratings with scores on an individual basis. Student ability may be a confounding factor; however, controlling for pretest results turns out to make no difference in the analysis.[11])

True or false? Discuss.
(a) On the average, those sections that liked their TA more did better on the final.
(b) There was almost no relationship between the section's average rating of the assistant and the section's average rating of the course.
(c) There was almost no relationship between the section's average rating of the course and the section's average score on the final.

Assistant	Ave. rating of assistant	Ave. rating of course in that section	Ave. score on final in that section
A	3.3	3.5	70
B	2.9	3.2	64
C	4.1	3.1	47
D	3.3	3.3	63
E	2.7	2.8	69
F	3.4	3.5	69
G	2.8	3.6	69
H	2.1	2.8	63
I	3.7	2.8	53
J	3.2	3.3	65
K	2.4	3.3	64

8. Exercise 8 on p. 97 reported an experiment in which 19 students each measured the thickness of a table top twice. Two students made gross errors; plot the scatter diagram for the remaining 17 students, and if you have access to a statistical calculator, find the correlation coefficient for these 17 pairs of numbers. True or false?
   (a) There is almost no relationship between the first and second measurements.
   (b) Those students who measured high the first time also tended to measure high the second time.

5. SUMMARY

1. Correlation measures association. But association does not necessarily show causation. It may only show that both variables are simultaneously influenced by some third variable.

2. If the independent variable is controlled by the experimenter, then an association between the two variables is good evidence for a cause-and-effect connection. Otherwise, be careful.

3. The correlation coefficient by itself can be a misleading summary of data concerning a relationship. It is usually advisable to report the averages, SDs, and the number of data points in addition to r.

4. Correlations based on rates or averages can be quite misleading.

10

Regression

You've got to draw the line somewhere.

1. INTRODUCTION

How is weight related to height? For example, there were 411 men aged 18 to 24 in Cycle I of the Health Examination Survey. Their average height was 5 feet 8 inches = 68 inches, with an overall average weight of 158 pounds. But those men who were one inch above average in height had a somewhat higher average weight. Those men who were two inches above average in height had a still higher average weight. And so on. On the average, how much of an increase in weight is associated with each unit increase in height? The best way to get started is by looking at the scatter diagram for these heights and weights (Figure 1). The object is to see how weight depends on height, so height is taken as the independent variable and plotted horizontally. The diagram shows vertical stripes because heights were measured to the nearest 0.1 inch. So, for instance, there are people of height 68.0 inches or 68.1 inches in the diagram—but nothing in between.

The summary statistics are[1]

average height $\approx$ 68 inches, SD $\approx$ 2.5 inches
average weight $\approx$ 158 pounds, SD $\approx$ 25 pounds, $r \approx 0.36$

The scales on the vertical and horizontal axes have been chosen so that one SD of height and one SD of weight cover the same distance on the page. This makes the SD line (dashed) rise at 45° across the page, as shown in the figure. There is a fair amount of scatter around the line, as indicated by the low correlation.

Figure 1. Scatter diagram for heights and weights. Each point represents the height and weight of one of the 411 men aged 18 to 24 in Cycle 1 of the Health Examination Survey. The points in the vertical strip represent all the men who are one SD above average in height, to the nearest inch. Those who are also one SD above average in weight would be plotted along the dashed SD line, and there is one such man. Most of these men are below the SD line, so they are only part of an SD above average in weight. The solid *regression* line estimates the average weight at each height.

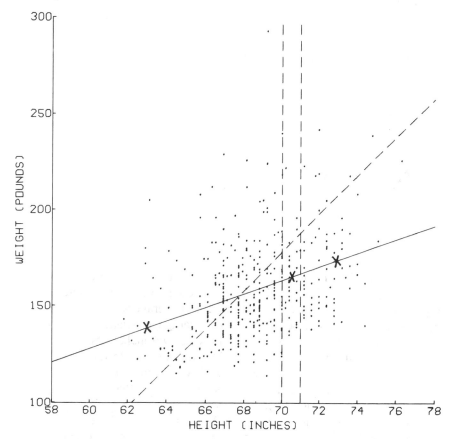

The vertical strip in Figure 1 shows the men who were one SD above average in height (to the nearest inch). Those men who were also one SD above the average in weight would be plotted along the SD line, and the diagram shows one such man. Most of the points in the strip are well below the SD line; only a few are above it. In other words, most of the men who were one SD above average in height were quite a bit less than one SD above average in weight. So the average weight of these men is only part of an SD above the overall average weight. What part? That is where the correlation of 0.36 comes in. Associated with an increase of one SD in height there is an increase of only 0.36 SDs in weight, on the average.

To be more specific, take the men who were one SD above average in height. Their height was

overall average height + SD of height = 68″ + 2.5″ = 70.5″.

Their average weight is estimated as being above the overall average weight by r × SD of weight. In pounds, that's 0.36 × 25 lb. = 9 lb. So their average weight is estimated as 158 lbs. + 9 lbs. = 167 lbs. The point

70.5 inches, 167 pounds

is marked by a cross in Figure 1.

What about the men who were two SDs above average in height? Their height equals

overall average height + 2 × SD of height = 68″ + 2 × 2.5″ = 73″.

Their average weight is estimated as being above the overall average weight by r × 2 SD of weight. In pounds, that's 0.36 × 2 × 25 lb. = 18 lb. So their average weight is estimated as 158 lb. + 18 lb. = 176 lb. The point

73 inches, 176 pounds

is also marked by a cross in Figure 1.

What about the men who were two SDs below average in height? Their height equals

average height − 2 SD of height = 68″ − 2 × 2.5″ = 63″.

The regression estimate for their average weight is below the overall average weight by r × 2 SD of weight. In pounds, that's 0.36 × 2 × 25 lb. = 18 lb. So their average weight is estimated as 158 lb. − 18 lb. = 140 lb. This point too (63 inches, 140 pounds) is marked by a cross in Figure 1.

All the points (height, estimate for weight) fall on the solid line shown in Figure 1. This line is called the *regression line*. It goes through the point of averages: the estimate is that men of average height will also be of average weight.

> The regression line is to a scatter diagram as the average is to a list. The regression line estimates the average value for the dependent variable corresponding to each value of the independent variable.

Along the regression line, associated with each increase of one SD in height there is an increase of only 0.36 SDs in weight. To interpret this, imagine grouping the men by height. There is a group which is one SD below average in height, another group which is average in height, still another group which is one SD above average in height, and so on. Moving from group to group, the average weight will also go up, but only by around 0.36 SDs.

This way of using the correlation coefficient to estimate the average value of the dependent variable for each value of the independent variable is called the *regression method*. It can be stated as follows:

> Associated with each increase of one SD in the independent variable there is an increase of only r SDs in the dependent variable, on the average.

Two different SDs are involved here: the SD of x, to gauge changes in x; and the SD of y to gauge changes in y. It is easy to get carried away by the rhythm: if x gives up by one SD, so does y. But it doesn't. On the average, y only goes up by r SDs.

Figure 2. Regression estimate. When the independent variable goes up by an SD, the average value of the dependent variable only goes up by r SDs.

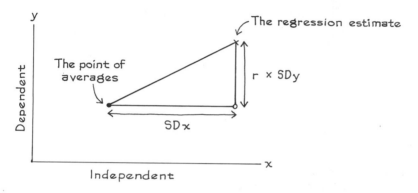

Why is r the right factor? To prove this, a complicated mathematical argument is needed. But two cases are easy to see directly. First, suppose r is 0. Then there is no association between x and y: an increase in x doesn't change y. Therefore, a one-SD increase in x is accompanied by a zero-SD increase in y, on the average. Second, suppose r is 1. Then all the points lie on the SD line. So a one-SD increase in x is accompanied by a one-SD increase in y.

With in-between values of r, the best thing to do is to look at a picture. Figure 3 (next page) shows a scatter diagram with a correlation coefficient of 0.50. The x-values and y-values average out to 4, with an SD of 1. The 45° line in the figure is the SD line. Along this line, a one-SD increase in x is matched by a one-SD increase in y. Now look at the points in the vertical strip over 5, where the x-values are one SD above average. The average of the y-values in this strip is marked by a cross. This is halfway between the horizontal line through the average of y and the sloping SD line. In other words, a one-SD increase in x is accompanied by a half-SD increase in y, on the average. Again, r is the right factor.

Figure 3. A scatter diagram with an *r* of 0.50. The average of the *x*-values and *y*-values is 4, with an SD of 1. The points in the vertical strip over 5 have *x*-values which are one SD above average. A cross marks the average of their *y*-values. This cross is halfway between the horizontal line through the average of the *y*-values and the sloping SD line. So a one-SD increase in *x* is accompanied by a half-SD increase in *y*, on the average.

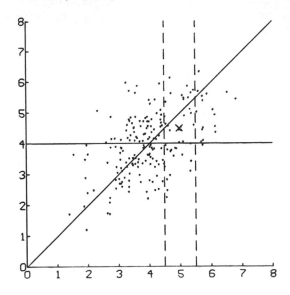

EXERCISE SET A

1. For men aged 35 to 54 in the United States in 1970, the relationship between education (years of schooling completed) and personal income can be summarized as follows:[2]

   average education ≈ 11 yrs., SD ≈ 4 yrs.
   average income ≈ $10,000, SD ≈ $8,000, *r* ≈ 0.4

   Estimate the average income of those men who have finished elementary school but have not gone to high school (so they have 8 years of education).

2. For the men aged 18 to 24 in the Health Examination Survey, the relationship between height and systolic blood pressure can be summarized as follows:

   average height ≈ 68 inches, SD ≈ 2.5 inches
   average blood pressure ≈ 120 mm., SD ≈ 15 mm., *r* ≈ −0.2

   Estimate the average blood pressure of the men who were 6 feet tall.

3. In a certain class, midterm scores average out to 60 with an SD of 15, as do scores on the final. The correlation between midterm scores and final scores is about 0.50. Estimate the average final scores for the students whose midterm score was

   (a) 60   (b) 75   (c) 30.

4. A longitudinal study obtained the following results for boys:

   average height at age 4 ≈ 41 inches, SD ≈ 1.5 inches
   average height at age 18 ≈ 70 inches, SD ≈ 2.5 inches, *r* ≈ 0.80

(The scatter diagram is on p. 133). Estimate the average height at age 18 of the boys whose height at age 4 was

    (a) 41 inches   (b) 44 inches   (c) 40 inches.

Plot your regression estimates as in Figure 1. Do they lie on a line?

5. For all the men in the Health Examination Survey (ages 18 to 79),

        average height $\approx$ 68 inches, SD $\approx$ 2.8 inches
        average weight $\approx$ 168 pounds, SD $\approx$ 28 pounds, $r \approx 0.40$

Estimate the average weight of the men whose height was

    (a) 68 inches   (b) 66 inches   (c) 24 inches   (d) 0 inches.

Comment on your answers to (c) and (d).

6. Can you use the information in Exercise 5 to estimate the average weight of the men who were 70 inches tall and 30 years old?

7. The men aged 45 to 54 in the Health Examination Survey (Exercise 5) had an average height of 68 inches, equal to the overall average height. True or false: Their average weight should be around 168 pounds. Explain.

8. For the men aged 18 to 24 in the Health Examination Survey (p. 146), the ones who were 70.5 inches tall averaged about 167 pounds in weight. True or false: The ones who weighed about 167 pounds must have averaged about 70.5 inches in height.

9. Suppose r is -1. Can you explain why a one-SD increase in x is matched by one-SD decrease in y?

The answers to these exercises are on pp. A-38–A-40.

2. THE GRAPH OF AVERAGES

How good is the regression method? Figure 4 (p. 152) is the *graph of averages* for the heights and weights of the men aged 18 to 24 in the Health Examination Survey.[3] The men who were for instance 63 inches tall (to the nearest inch) had an average weight of 155 pounds. This is represented by the point (63 inches, 155 pounds) in the figure. There were 8 such men, as indicated by the number next to the point. For each specific height, the graph of averages shows the average weight among the men of that height in the sample. This graph is very close to a straight line in the middle, where most of the people are. But at the ends, it is quite bumpy. For example, the men who were 64 inches tall averaged only 138 pounds in weight, 17 pounds less than the men who were 63 inches tall. In this instance, a one-inch increase in height was accompanied by a seventeen-pound decrease in weight. There is no real reason for this, it is the result of chance variation. The men were chosen for the sample at random. By the luck of the draw, there were a few too many heavy men of height 63 inches, and not enough of height 64 inches. The regression line smooths away this kind of chance variation.

> The regression line is a smoothed version of the graph of averages. If the graph of averages happens to be a straight line, the regression line and the graph of averages must be one and the same.

Figure 4. The graph of averages. This graph shows the average weight at each height for the 411 men aged 18 to 24 in the Health Examination Survey. The regression line smooths this graph.

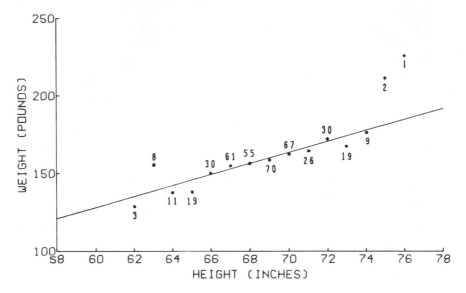

In some situations, the regression line smooths away too much. If there is a nonlinear association between the two variables, as in Figure 5, the regression line will pass it by. Then, it is better to use the graph of averages. An example is the relationship between age and weight (Figure 3, p. 51).

Figure 5. Nonlinear association. Regression lines should not be used when there is a nonlinear association between the variables.

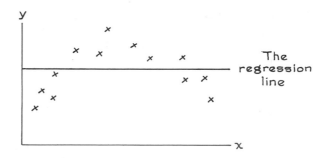

EXERCISE SET B

1. Figure 6 shows six scatter diagrams, each with a solid line and a dashed line. For each diagram, say which is the SD line and which is the regression line.

Figure 6.   Which is the SD line? the regression line?

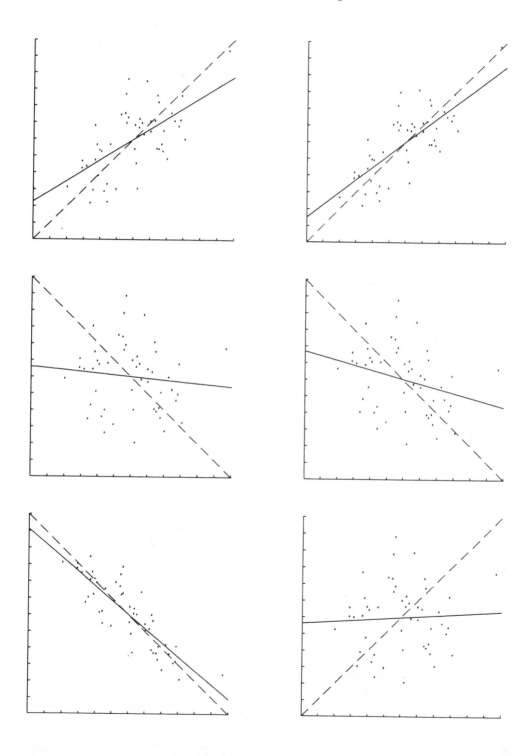

2. Trace the diagram below on a piece of paper, and mark a cross at the average for each of the vertical strips—one of them has been done for you. Then draw in the regression line; the dashed line is the SD line.

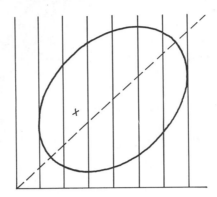

3. Below are some hypothetical data sets. For each one, draw the scatter diagram, plot the point of averages and the graph of averages. Draw the regression line for estimating y from x. Please do not do any calculations: make the best guess you can.

(a)		(b)		(c)		(d)		(e)		(f)	
x	y	x	y	x	y	x	y	x	y	x	y
1	3	1	0	0	0	0	0	0	0	0	2
2	5	1	6	0	2	1	1	0	1	1	3
3	7	2	5	1	2	2	4	1	0	2	0
		3	6					1	1	3	1
		3	8							4	2

4. An economist wishes to study the relationship between income (as the dependent variable) and certain independent variables. Should he use a regression line
   (a) when the independent variable is age, for men aged 18 to 34?
   (b) when the independent variable is age, for men aged 18 to 79?
   (c) when the independent variable is marital status?

The answers to these exercises are on pp. A-40–A-41.

Technical note: In general, the regression line fitted to the graph of averages, with each point weighted according to the number of cases it represents, coincides with the regression line fitted to the original scatter diagram. (This is exact when points with different x-coordinates are kept separate in the graph of averages; otherwise, it is a good approximation.)

3. ESTIMATES FOR INDIVIDUALS

For the men aged 18 to 24 in the Health Examination Survey, the relationship between height and weight was summarized as follows:

average height ≈ 68 inches,   SD ≈ 2.5 inches
average weight ≈ 158 pounds, SD ≈ 25 pounds, r ≈ 0.36

Now suppose you had to guess the weight of one of these men without being told anything else about him. The best guess is the overall average weight of 158 pounds. Next, suppose you are told the man's height: 73 inches, for example. This should help you guess more accurately; this man is tall, and likely to be heavier than average. At this point, you should want to know the average weight for all the 73-inch men. This new average can be estimated by the regression method, as 176 pounds. And this new average is your best guess for the man's weight, knowing his height. The pattern is a general one. If you have to predict the dependent variable from the independent one, the thing to figure out is the new average. In many cases, the regression method gives a sensible way of estimating the new average. Of course, if there is a nonlinear association between the variables, the regression method would not apply.

Usually, people work out regression estimates from a study, and then extrapolate—they use the estimates on subjects they've never seen. In many cases, this makes sense, provided the subjects in the survey are representative of the people about whom the inferences are going to be made. But you have to think about this point each time: the mathematics of the regression method does not protect you here at all.

For example, if back in 1962 someone had to guess weight from height for a randomly selected man aged 18 to 24, not a subject in the survey, it would have been very reasonable to use the regression estimates based on the survey data, because the men in the survey were representative of all the men in that age bracket in the population.

Suppose you had to guess weight from height for men aged 18 to 24 today. Would it be reasonable to use estimates based on the 1962 survey data? There is nothing in the statistics to guarantee this. But it does seem plausible, because the distribution of heights and weights changes only very slowly over time.

Example 1. A law school has made a statistical analysis of the relationship between LSAT scores (ranging from 0 to 800) and first-year scores (ranging from 0 to 100), for students who complete the first year. The results:

$$\text{average LSAT score} = 650, \text{SD} = 80$$
$$\text{average first-year score} = 65, \quad \text{SD} = 8, r = 0.4$$

Predict the first-year score of a student whose LSAT score is 750.

Solution. This student is $100/80 = 1.25$ SDs above average on the LSAT. The regression estimate for his first-year score is: above average by $0.4 \times 1.25 = 0.5$ SDs. That's $0.5 \times 8 = 4$ points. Thus, the estimate is $65 + 4 = 69$.

Some caution is in order here. The law school only has experience with those students it admits. It would be unsafe to use this regression procedure on students who are quite different from the ones usually admitted—for instance, those from a foreign country. (Law schools typically do make regression estimates for students in a category very different from those who complete the first year—those who are denied admission.)

Regression techniques can also be used to make estimates for *percentile ranks*. If your percentile rank on a test is 90%, you did very well: only 10% of the class scored higher, the other 90% scored lower. A percentile rank of 25% is not very good: 75% of the class scored higher, the other 25% scored lower.

Example 2. The law school in Example 1 finds that LSAT scores and first-year scores follow the normal curve. Suppose the percentile rank of one student on the LSAT is 90%, among the first-year students. Predict his percentile rank on the first-year scores.

Solution.

Step 1. Use the normal table to find the number z such that the area to the left of z under the normal curve is the given percentile rank (p. 83).

In standard units, this student scored 1.3 on the LSAT, that is, 1.3 SDs above average.

Step 2. Use the regression method to estimate the dependent variable in standard units. In standard units, his estimated score on the first-year tests is $0.4 \times 1.3 \approx 0.5$. That is, the regression method estimates his score on the first-year tests as 0.5 SDs above average.

Step 3. Find the area under the normal curve to the left of this standard-units estimate.

Area $\approx$ 69%

That is the answer. His percentile rank on the first-year test is predicted as 69%.

In solving this problem, the averages and SDs of the two variables were never used. All that mattered was r. Basically, this is because the whole problem was worked in standard units.

The student in Example 2 was compared with his class in two different competitions, the LSAT and the first-year tests. He did very well on the LSAT—in the 90th percentile. But the regression estimate only puts him in the 69th percentile on the first year tests—still above average, but not as much. On the other hand, for a poor student—say in the 10th percentile of the LSAT—the regression method predicts an improvement. It will put him in the 31st percentile on the first-year tests. This is still below average, but closer.

To go at this more carefully, take all the people in the 90th percentile on the LSAT—good students. Some of them will move up on the first-year tests, some will move down. On the average, however, this group moves down. Now let us take all the people in the 10th percentile of the LSAT—poor students. Again, some will do better on the first-year tests, others worse. On the average, however, this group moves up. This is what the regression method is telling us.

Initially, many people would predict a first-year rank equal to the LSAT rank. This is definitely wrong. To see why, imagine that you had to predict a student's rank in a mathematics class. In the absence of other information, the safest guess is to put him at the median. Now, if you were told that this student was very good in physics, you would probably put him well above the median in mathematics. After all, there is a very strong correlation between physics and mathematics. On the other hand, if all you knew was the student's rank in a pottery class, that wouldn't help you at all in guessing his rank in mathematics. You would be reduced to putting him at the median. There is no correlation between pottery and mathematics.

Now let's return to the problem of predicting first-year rank from LSAT rank. If the two sets of scores were perfectly correlated, then it makes sense to predict that first-year rank will be equal to LSAT rank. At the other extreme, if the correlation were 0, then the LSAT rank wouldn't help at all in predicting first-year rank: back to the median. In fact, the correlation is somewhere between the two extremes, so you have to predict a rank on the first-year tests somewhere between the LSAT rank and the median. The regression method tells you exactly where.

EXERCISE SET C

1. In a certain class, midterm scores average out to 60, with an SD of 15, as do scores on the final. The correlation between midterm scores and final scores is about 0.50. Predict the final score for a student whose midterm score is
   (a) unknown   (b) 60   (c) 75   (d) 30.

2. For the first-year students at a certain law school, the correlation between LSAT scores and first-year scores was 0.60. Estimate the percentile rank on the first-year score for a student whose percentile rank on the LSAT (among his classmates) was
   (a) unknown   (b) 50%   (c) 90%   (d) 30%.
   Compare your answer to (c) with Example 2 on page 156.

3. In the Health Examination Survey, the men aged 75 to 79 average about 3 inches shorter than the men aged 25 to 34. This is one SD of height. The correlation between height and weight was 0.4. True or false: the men aged 75 to 79 must have averaged about 0.4 SDs of weight lighter than the men aged 25 to 34.

4. In Example 2, the regression method predicted that a student in the 90th percentile of the LSAT would only be in the 69th percentile of the first-year tests. True or false: a student in the 69th percentile of the first-year tests should be in the 90th percentile in the LSAT.

5. The scatter diagram below shows the scores on the midterm and final in a certain course. Three lines are drawn across the diagram. People who have the same percentile rank on both tests are plotted along one of these lines. Which one, and why?

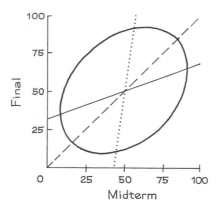

The answers to these exercises are on pp. A-41–A-42.

Technical note: The method discussed in this example estimates median ranks. Assuming normality and $r = 0.4$, the estimate in Example 2 is that of students in the 90th percentile on the LSAT (relative to their classmates), about half will rank above the 69th percentile on first-year scores, and half will rank below. The procedure for estimating average ranks is harder.

4. THE REGRESSION FALLACY

A preschool program attempts to boost children's IQs. The children are tested when they enter the program (the pre-test), and again when they leave (the post-test). On both occasions, the scores average out to nearly 100, with an SD of about 15. The program seems to have no effect. A closer look at the data, however, seems to show something very surprising. The children who were below average on the pre-test show an average gain of about 5 IQ points at the post-test. Conversely, those children who were above average on the pre-test show an average loss of about 5 IQ points at the post-test. What does this prove? Does the program operate to equalize intelligence? Perhaps when the brighter children play with the duller ones, the difference between the two groups tends to be diminished. Is this good or bad?

These speculations may be very interesting, but the sad fact is nothing interesting is going on, good or bad. Here is why. The children cannot be expected to score exactly the same on the two tests. There will be differences between the two scores. Nobody would think these differences mattered, or needed any explanation. But these differences make the scatter diagram for the test scores spread out around the SD line into that familiar football-shaped cloud. And it is just this spread around the line which makes the bottom group come up and the top group come down. There is nothing else to it.

> In virtually all test-retest situations, the bottom group on the first test will on average show some improvement on the second test—and the top group will on average fall back. This is called the *regression effect*.

The *regression fallacy* consists in thinking that the regression effect must be due to something important, not just the spread around the line.

We are now going to see why the regression effect appears whenever there is spread around the SD line. This effect was first noticed by Galton in his study of family resemblances, so we are going to discuss it in that context; the reasoning is perfectly general. Figure 7 (p. 160) shows a scatter diagram for the heights of 1,078 pairs of fathers and sons, as discussed in Chapter 8. The height of the father is taken as the independent variable, the height of the son is the dependent variable. The summary statistics are:[4]

average height of fathers $\approx$ 68 inches, SD $\approx$ 2.7 inches
average height of sons $\approx$ 69 inches, SD $\approx$ 2.7 inches, $r \approx 0.5$

The summary shows that the sons averaged one inch taller than the fathers. On this basis, it is natural to guess that a 72-inch father should have a 73-inch son; similarly, a 64-inch father should have a 65-inch son, and so on. Such fathers and sons are plotted along the dashed line in Figure 7. Of course, not many families are going to be right on the line. In fact, there is a lot of spread around the line. Some of the sons are a lot taller than their fathers. Others are a lot shorter.

For example, take the fathers who are 72 inches tall (to the nearest inch). The corresponding families are plotted in the vertical strip over 72 inches in Figure 7, and there is quite a range in the sons' heights. Some of the points are above the dashed line: the son is taller than 73 inches. But most of the points are below the dashed line: the son is shorter than 73 inches. All in all, the sons of the 72-inch fathers only averaged 71 inches in height. With tall fathers (high score on first test), on the average the sons are shorter (score on second test drops).

Now look at the points in the vertical strip over 64 inches, representing the families where the father was 64 inches tall (to the nearest inch). The height of the dashed line there is 65 inches, representing a son who is 1 inch taller than his 64-inch father. Some of the points fall below the dashed line, but most are above, and the sons of the 64-inch fathers averaged 67 inches in height. With short fathers (low score on first test), on the average the sons are taller (score on second test goes up). The aristocratic Galton termed this "regression to mediocrity."

In fact, the dashed line in Figure 7 is the SD line. It goes through the point corresponding to an average father of height 68 inches, and his average son of height 69 inches. Along the dashed line, each one-SD increase in father's height is matched by a one-SD increase in sons's height. These two facts make it the SD line.

In Figure 7 the cloud is symmetric around the SD line. However, the

Figure 7. The regression effect. The scatter diagram shows the heights of 1,078 fathers and sons. The sons averaged 1 inch taller than the fathers. If the son was 1 inch taller than his father, this family is plotted along the dashed line. The points in the strip over 72 inches correspond to the families where the father was 72 inches tall, to the nearest inch. Most of these points are below the dashed line, and the average height of these sons is only 71 inches. The points in the strip over 64 inches correspond to families where the father was 64 inches tall, to the nearest inch. Most of these points are below the dashed line, and the average height of these sons is 67 inches. The solid regression line picks off the centers of all the vertical strips, and is much flatter than the dashed line.

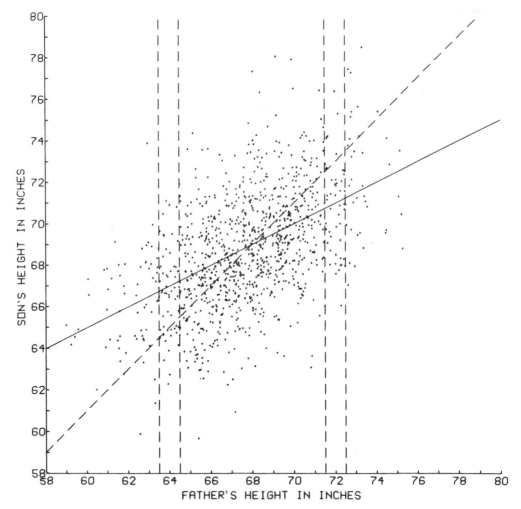

strip at 72 inches isn't. This strip only contains points with unusually big *x*-coordinates. And most of the points in this strip fall below the SD line. Conversely, the strip at 64 inches only contains points with unusually small *x*-coordinates. Most of the points in this strip fall above the SD line. This hidden imbalance is always there in football-shaped clouds, and it is the graphical explanation for the regression effect. This explanation may not seem very romantic. But then, statistics isn't known as a romantic subject.

Figure 7 also shows the regression line for estimating the son's height from his father's height. This solid line rises much less steeply than the dashed SD line, and it picks off the center of each vertical strip of dots—the average y-value in the strip. For instance, take the fathers who are 72 inches tall. They are 4 inches above average in height, which is 4 inches/2.7 inches $\approx$ 1.5 SDs. The regression line estimates their sons should average

$$r \times 1.5 \text{ SDs} = 0.75 \text{ SDs} = 2 \text{ inches}$$

above the overall average height for sons. This overall average is 69 inches, so the regression estimate for the average height of these sons is 71 inches—dead on.

Figure 8 shows the regression effect at its starkest, without the cloud. The dashed SD line rises at a 45° angle. The dots show the average height of the sons corresponding to each value of father's height—the graph of averages. These dots are the centers of the vertical strips in Figure 7. The dots rise much less steeply than the SD line. This is the regression effect. On the

Figure 8. The regression effect. The SD line is dashed, the regression line is solid. The dots show the average height of the sons, for each value of father's height. They rise much less steeply than the SD line. This is the regression effect. The regression line follows the dots very well indeed.

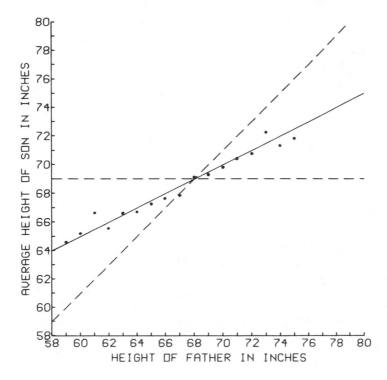

whole, the dots are halfway between the SD line and the horizontal line through the point of averages. This is because the correlation coefficient is one half. So each one-SD increase in father's height is accompanied by a half-SD increase in son's height, not a one-SD increase as the dashed line would suggest. The solid regression line shows the half-SD increase in son's height associated with each one-SD increase in father's height, and it tracks the graph of averages very well indeed.

At first glance, the scatter diagram in Figure 7 is quite chaotic. It was a stroke of genius on Galton's part to see a straight line in this chaos. Since then, many other investigators have found that the graphs of averages in their scatter diagrams were close to straight lines too. That is why the regression line is so useful.

Modern genetics has explained why Galton's graph of averages followed a straight line: more about this in Chapter 25.

EXERCISE SET D

1. As part of their training, air force pilots make two practice landings with instructors, and are rated on their performance. The instructors discuss the ratings with the pilots after each landing. After a statistical analysis, the air force found that pilots who made poor landings the first time tended to do better the second time. Conversely, pilots who made good landings the first time tended to do worse the second time. The conclusion was that criticism tended to make the pilots do better, while praise tended to make them do worse. As a result of this study, the instructors were ordered to criticize all landings, good or bad. Is this warranted by the facts?

2. An instructor standardizes his midterm and final so the class average is 50 with an SD of 10 on both tests. The correlation between the tests is always around 0.50. On one occasion, he took all the students who scored below 30 at the midterm, and gave them special tutoring. They all scored above 50 on the final. Can this be explained by the regression effect?

3. IQ scores are scaled to have an average of about 100, and an SD of about 15, both for men and for women. The correlation between the IQs of husbands and wives is about 0.50. In a large study of families, it was found that for the men whose IQ was 140 the average IQ of their wives was 120. Looking at the wives of IQ 120 in the study, is it fair to expect that the average IQ of their husbands will be greater than 120? Explain your answer briefly.

4. Of the people who were in the 90th percentile of the education distribution in the United States in 1976, what percentage is above the 90th percentile of the income distribution? (The scatter diagram looks like Figure 9 on p. 179.)
   about 50%   well over 50%   well under 50%   can't tell.

5. In Figure 8, are the sons of the 61-inch fathers taller on the average than the sons of the 62-inch fathers, or shorter? What is the explanation?

6. In Pearson's study, the sons of the 72-inch fathers only averaged 71 inches in height. True or false: if we take the 71-inch sons, their fathers will average about 72 inches tall. Explain.

7. A student writes, "The regression effect is that things get closer to average." In Figure 7, the overall average height for the fathers is 68 inches, for the sons it is 69 inches. Take the families where the father is 68 inches to the nearest inch—a vertical strip over 68 inches. Fill in the blanks below, using:
   1/4   1/2   1   2   4   68   69   70   71.
   (a) In this strip, the fathers are just about _____ inches tall.
   (b) In this strip, the sons are _____ inches tall, give or take _____ inches or so.
   (c) Now comment briefly on the quote. Which "things" get closer to the average, and which "things" don't.

The answers to these exercises are on pp. A-42–A-43.

Technical note: In some cases, for instance in the context of a repeated IQ test, the explanation for the regression effect can be expanded a bit. The basic fact is that the two scores are apt to be different. This can be explained in terms of chance variability. Each person may be lucky or unlucky on the first test. But if the score on the first test is very high, that suggests the person was lucky on that occasion, implying that the score on the second test will probably be lower—you wouldn't say, "He scored 140, must have had bad luck that day." The implication is that the score on the second test will be lower. On the other hand, if the score on the first test was very low, the person was probably unlucky to some extent on that occasion, and will do better next time.

We will now present a crude model for this test-retest situation, which brings the regression effect into sharper focus. The basic equation in the model is

observed test score = true score + chance error.

Assume that the distribution of true scores in the population follows the normal curve. Suppose too that the chance error is as likely to be positive as negative, and tends to be about five points in size. So, if someone has a true score of 135 he is just as likely to score 130 or 140 on the test. Someone who has a true score of 145 is just as likely to score 140 as 150.

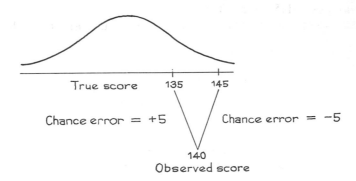

Take the people who scored 140 on the first test. There are two alternative explanations for this observed score:

- true score below 140, coupled with a positive chance error;
- true score above 140, coupled with a negative chance error.

Many more people have true scores of 135 than 145, as the figure shows. Any other pair of values equally spaced around the observed score can be dealt with in a similar way. So the first explanation is more likely.

The model has explained the regression effect. If someone scores above average on the first test, we are forced to estimate that his true score is a bit lower than the observed score. If he takes the test again, we have to predict that his second score will be a bit lower than his first score. On the other hand, if he scores below average on the first test, we estimate that his true score is a bit higher than the observed score; and our prediction for the second score is a bit higher than the first score. It is not the true score which changes from test to test, but our estimate of it.

5. REVIEW EXERCISES

1. Pearson and Lee obtained the following results in a study of about 1,000 families:

   average height of husband ≈ 68 inches, SD ≈ 2.7 inches
   average height of wife ≈ 63 inches, SD ≈ 2.5 inches, r ≈ 0.25

   Estimate the height of a wife when the height of her husband is
   (a) unknown   (b) 68 inches   (c) 72 inches   (d) 64 inches.

2. In a study of the stability of IQ scores, a large group of individuals is tested once at age 18 and again at age 35. The following results are obtained.

age 18: average score ≈ 100, SD ≈ 15
age 35: average score ≈ 100, SD ≈ 15, $r \approx 0.80$

(a) Estimate the average score at age 35 for all the individuals who scored 115 at age 18.

(b) Predict the score at age 35 for an individual who scored 115 at age 18.

3. For each of the scatter diagrams below, say which is the SD line and which is the regression line. Explain briefly.

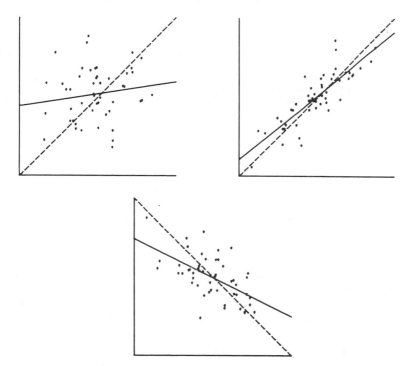

4. True or false: The regression line never rises more steeply than the SD line. Explain briefly.

5. An investigator measuring various characteristics of a large group of athletes found that the correlation between the weight of an athlete and the amount of weight that athlete could lift was 0.60. True or false? Explain briefly.

(a) The average athlete can lift 60% of his body weight.

(b) If an athlete gains 10 pounds, he can expect to be able to lift an additional 6 pounds.

(c) The more an athlete weighs, on the average the more he can lift.

(d) The more an athlete can lift, on the average the more he weighs.

(e) 60% of an athlete's lifting ability can be attributed to his weight alone.

6. A doctor is in the habit of measuring blood pressures twice. He notices that patients who are unusually high on the first reading tend to have somewhat lower second readings. He concludes that patients are more relaxed on the second reading. A colleague disagrees, pointing out that the patients who are unusually low on the first reading tend to have somewhat higher second readings, suggesting they are more nervous. Which doctor is right? Or perhaps both are wrong? Explain.

7. A large study is made on the blood-pressure problem discussed in the previous exercise. It is found that first readings average 130 mm., and second readings average 120 mm. Does this support either doctor's argument? Or is it the regression effect? Explain.

8. In one study, the correlation between the educational level of husbands and wives in a certain town was about 0.50; both averaged 12 years of schooling completed, with an SD of 3 years.[5]
   (a) Estimate the educational level of a woman whose husband has completed 18 years of schooling.
   (b) Estimate the educational level of a man whose wife has completed 15 years of schooling.
   (c) Apparently, well-educated men marry women who are less well educated than themselves. At the same time, such women marry men with even less education. How can this be explained?

9. In a large statistics class, the correlation between midterm scores and final scores is found to be nearly 0.50, every quarter. Estimate the percentile rank on the final for a student whose percentile rank on the midterm is
   (a) unknown   (b) 50%   (c) 5%   (d) 80%.

10. True or false: A person who is in the 40th percentile of the height distribution of people in the United States in 1976 is also likely to be in the 40th percentile of the weight distribution. Explain.

6. SUMMARY

1. The *regression method* estimates that associated with an increase of one SD in the independent variable, there is an increase of only r SDs in the dependent variable, on the average.

2. Plotting the regression estimate for the average of the dependent variable against the value of the independent variable gives a straight line, called the *regression line*. It is a summary of one aspect of the scatter diagram, just as the average is a summary of one aspect of a list. The regression line estimates the average value of the dependent variable for each value of the independent variable.

3. The *graph of averages* is often close to a straight line, but may be a little bumpy. The regression line smooths it out. If the graph of averages is a straight line, then it coincides with the regression line.

4. For any individual in a study, the regression estimate for the average value of the dependent variable from the independent variable can be used as a guess for the actual value of the dependent variable.

5. This procedure is often used to make inferences about individuals who were not subjects in the study. This is legitimate when the people you are generalizing about are like the people in the study. Nothing in the mathematics guarantees this. So watch out.

6. In a typical test-retest situation, the subjects get different scores on the two tests. Take the bottom group on the first test. Some will improve on the second tests, others will do worse; but on the average, the bottom group will show an improvement. Take the top group on the first test. Some do even better the second time, others fall back; but on the average, the top group does worse the second time. This is called the *regression effect,* and it happens whenever the scatter diagram spreads out around the SD line into a football-shaped cloud of points. The same effect can therefore be seen when comparing percentile ranks on almost any pair of variables.

7. The *regression fallacy* consists in thinking that the regression effect must be due to something other than spread around the SD line.

11

The R.M.S. Error for Regression

Such are the formal mathematical consequences of normal correlation. Much bio-metric material certainly shows a general agreement with the features to be expected on this assumption: although I am not aware that the question has been subjected to any sufficiently critical inquiry. Approximate agreement is perhaps all that is needed to justify the use of the correlation as a quantity descriptive of the population; its efficacy in this respect is undoubted, and it is not improbable that in some cases it affords, in conjunction with the means and variances, a complete description of the simultaneous variation of the variates.

—SIR R. A. FISHER (ENGLAND, 1890–1962)[1]

1. INTRODUCTION

The regression method estimates the average value of the dependent variable for each value of the independent variable. However, individuals differ from regression estimates. By how much? The object of this section is to measure the overall size of these differences using the r.m.s. error. For example, take the heights and weights of the 411 men aged 18 to 24 in the Health Examination Survey study discussed on p. 49. The scatter diagram is reproduced in Figure 1. The summary statistics are:

$$\text{average height} \approx 68 \text{ inches}, \quad \text{SD} \approx 2.5 \text{ inches}$$
$$\text{average weight} \approx 158 \text{ pounds}, \text{SD} \approx 25 \text{ pounds}, r \approx 0.36$$

Person A on the diagram is 63 inches tall and weighs 205 pounds. This height of 63 inches is 2 SDs below average. So the regression estimate for his weight is: below average by $2 \times 0.36 = 0.72$ SDs. In pounds, that's

0.72 × 25 = 18 lbs. So the regression estimate for A's weight is

158 lbs. − 18 lbs. = 140 lbs.

However, his actual weight is 205 pounds. The regression estimate is off, by an error or *residual* of 65 pounds:

residual = actual weight − estimated weight
= 205 lbs. − 140 lbs. = 65 lbs.

In the diagram, this residual is the vertical distance of A above the regression line. It is marked by a brace.

Person B on the diagram is 74 inches tall and weighs 140 pounds. The regression estimate for his weight works out to 180 pounds. The residual is 140 lbs. − 180 lbs. = −40 lbs. In the diagram, this residual is the vertical distance of B below the regression line. This residual too is indicated by a brace.

Each point on the scatter diagram has a residual, representing the error made by the regression method for that man. The overall size of these errors

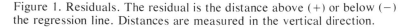

Figure 1. Residuals. The residual is the distance above (+) or below (−) the regression line. Distances are measured in the vertical direction.

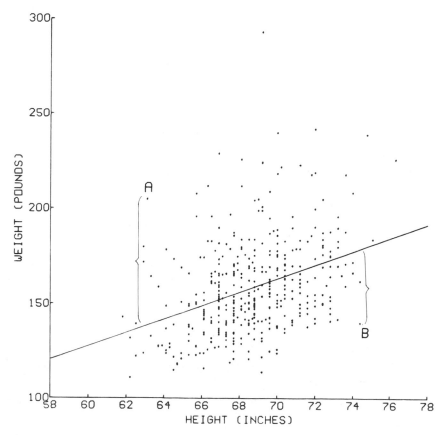

is measured by taking their root-mean-square (p. 57). The result is called the
r.m.s. error of the regression line. In this case, it works out to 23 pounds. The
interpretation: for typical men in the study, the regression estimate of weight
is off by around 23 pounds or so. So A and B are definitely atypical: A is very
short and fat, B is tall and skinny.

In general, the r.m.s. error of the regression line tells how far off the
regression estimates are, on the average, as in Figure 2.

Figure 2. The r.m.s. error. This says how far above or below the regression line the points are, on the average.

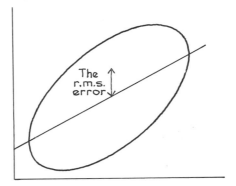

The points on a scatter diagram deviate from the regression line
(up or down) by residuals which are similar in size to the r.m.s.
error. The r.m.s. error is to the regression line as the SD is to
the average.

The analogy between the r.m.s. error and the SD goes even further: For
roughly 68% of the subjects in a study, the regression estimate will be off by
one r.m.s. error or less (Figure 3). For roughly 95% of the subjects, the

Figure 3. The r.m.s. error. About 68% of the points on a scatter diagram
fall inside the strip whose edges are parallel to the regression line, and one
r.m.s. error away (up or down). About 95% of the points are in the wider
strip whose edges are parallel to the regression line, and twice the r.m.s.
error away.

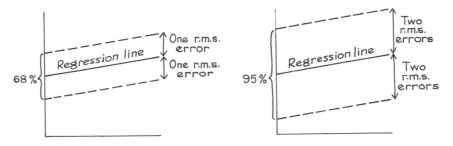

regression estimate will be off by two r.m.s. errors or less. This rule of thumb holds for many data sets, but not all.

For the HES study, the r.m.s. error was 23 pounds. The computer found that the regression estimates were right to within 2 r.m.s. errors, or 46 pounds, for 392 out of 411 men, or 95.4% of them. Here, the rule of thumb works quite well. On the other hand, the regression estimates were right to within one r.m.s. error, or 23 pounds, for 352 out of the 411 men—or 85.6%. Here, the 68% is quite rough indeed.

EXERCISE SET A

1. Below are three scatter diagrams. The regression line has been drawn across each one, by eye. In each case, guess whether the r.m.s. error is closest to 0.2 or 1 or 5.

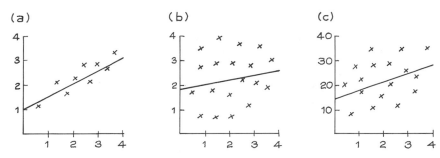

2. A regression line for estimating income has an r.m.s. error of $2,000. It estimates someone's income as $20,000. This estimate is likely to be right give or take: a few hundred dollars, a few thousand dollars, ten or twenty thousand dollars.

3. An admissions officer is trying to choose between two methods of predicting first-year scores. One method has an r.m.s. error of 12. The other has an r.m.s. error of 7. Other things being equal, which should he choose? Why?

4. A regression line for estimating test scores has an r.m.s. error of 8 points.
   (a) About 68% of the time, the regression estimates will be right to within _____ points.
   (b) About 95% of the time, the regression estimates will be right to within _____ points.

The answers to these exercises are on p. A-43.

2. COMPUTING THE R.M.S. ERROR

It would take a long time to compute the r.m.s. error directly. But there is a formula which does it indirectly:

> The r.m.s. error for the regression line estimating y from x can be figured as
> $$\sqrt{1 - r^2} \times \text{the SD of } y.$$

This formula relates two things: the r.m.s. error for the regression line, and the SD of y. These two quantities describe different aspects of a scatter diagram. To begin with the SD of y, Figure 4 shows a horizontal line through the average of y. The SD of y measures the average distance of the points above or below this horizontal line. By contrast, the r.m.s. error measures distances above or below the regression line. The r.m.s. error will be smaller than the SD, because the regression line can get closer to the points than a horizontal line—the regression line is allowed to slope up or down, following the trend. The formula gives the exact factor by which the r.m.s. error will be smaller: $\sqrt{1 - r^2}$.

Figure 4. The SD of the dependent variable.

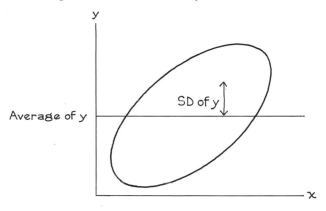

Which SD goes into the formula? The SD of the variable being estimated. If you are estimating weight from height, it is the SD of weight. The r.m.s. error has to come out in pounds, not inches. If you are estimating income from education, it is the SD of income. The r.m.s. error has to come out in dollars, not years.

> The units for the r.m.s. error are the same as the units for the dependent variable, that is, the variable being estimated.

For the height-weight example of section 1: the regression estimates for weight from height have an r.m.s. error of

$$\sqrt{1 - r^2} \times \text{SD of weight} = \sqrt{1 - (0.36)^2} \times 25 \text{ lbs.} \approx 23 \text{ lbs.}$$

The r.m.s. error isn't much smaller than the SD of weight, because weight isn't that well correlated with height.

The formula is hard to prove without algebra. But three special cases are easy to see. First, suppose $r = +1$. Then all the points lie on a straight line (which slopes up). The regression line goes through all the points on the scatter diagram, and all the residuals are 0. So the r.m.s. error should be 0.

And that is what the formula says: the factor works out to

$$\sqrt{1 - r^2} = \sqrt{1 - 1^2} = \sqrt{1 - 1} = \sqrt{0} = 0.$$

The case $r = -1$ is the same, except that the line slopes down. The r.m.s. error should still be 0, and that is what the formula says: the factor works out to

$$\sqrt{1 - r^2} = \sqrt{1 - (-1)^2} = \sqrt{1 - 1} = \sqrt{0} = 0.$$

The third case is $r = 0$. Then there is no linear relationship between the variables. So the regression line does not help in estimating the dependent variable, and its r.m.s. error should equal the SD. That is what the formula says: the factor works out to

$$\sqrt{1 - r^2} = \sqrt{1 - 0^2} = \sqrt{1} = 1.$$

The r.m.s. error measures spread around the regression line in absolute terms: pounds, dollars, and so on. The correlation coefficient, on the other hand, measures in relative terms—relative to the SD—and has no units. The r.m.s. error is related to the SD through the correlation coefficient. This is the third time that r comes into the story:

• r describes the clustering of the scatter diagram around the SD line, relative to the SDs;

• r says how the average value of y depends on x—associated with each one-SD increase in x there is an increase of only r SDs in y, on the average;

• r determines the accuracy of the regression estimates, through the formula for r.m.s. error.

EXERCISE SET B

1. A law school finds the following relationship between LSAT scores and first-year scores (for admitted students):

average LSAT score $\approx$ 650, SD $\approx$ 80
average first-year score $\approx$ 65,   SD $\approx$ 10, $r \approx$ 0.60

(a) Suppose you had to guess the first-year score for one of these students, without any other information. What would you guess?
(b) What would the r.m.s. error be, for the method in (a)?
(c) Suppose you knew the student's LSAT score. What method would you use?
(d) Find the r.m.s. error for the method in (c).
(e) Which is smaller, the answer to (b) or (d)?
(f) For about 32% of the students, your guess in (c) will be off by more than:
    10   80   8   64.

2. In one study of identical male twins, the average height was found to be about 68 inches, with an SD of about 3 inches. The correlation between heights of twins is about 0.95.
(a) You have to guess the height of one of these twins, without any further information. What would you guess?
(b) What would the r.m.s. error be for the method in (a)?

   (c) One twin of the pair is standing in front of you. You have to guess the
       height of the other twin. What would you do? (For instance, suppose the
       twin you see is 6 feet 6 inches tall.)
   (d) Find the r.m.s. error for the method in (c).
   (e) Which is smaller, the answer to (b) or (d)?
   (f) For about 5% of the twins, your guess in (c) will be off by more than:
             3 inches   1.8 inches   0.9 inches.

The answers to these exercises are on p. A-43.

3. PLOTTING THE RESIDUALS

When doing a regression, it is usually advisable to plot the residuals on a
graph. The method is indicated by Figure 5. Each point on the scatter diagram
is transferred to a second diagram, called the *residual plot,* in the following
way. The *x*-coordinate is left alone. But the *y*-coordinate is changed. It is
replaced by the residual at the point—the distance above (+) or below (−) the
regression line.

Next, Figure 6 shows the residual plot for the height-weight scatter diagram of Figure 1. Figures 5 and 6 suggest that the positive residuals balance
out the negative ones, so the average is 0. This can be shown mathematically: The residuals from the regression line average out to 0. Looking
across the residual plot, there is no systematic tendency for the points to drift
up (or down).

> The regression line for the residual plot coincides with the
> *x*-axis.

Basically, the reason is that all the trend up or down has been taken out of the
residuals, and is in the regression line.

Figure 5. Plotting the residuals.

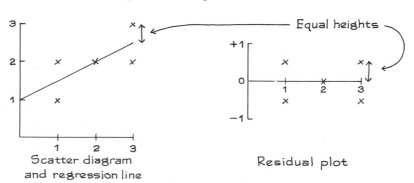

Scatter diagram
and regression line

Residual plot

Figure 6. A residual plot. The scatter diagram at the left shows the heights and weights of the 411 men aged 18 to 24 in the Health Examination Survey, with the regression line. The residual plot is shown at the right. There is no trend or pattern left in the residuals.

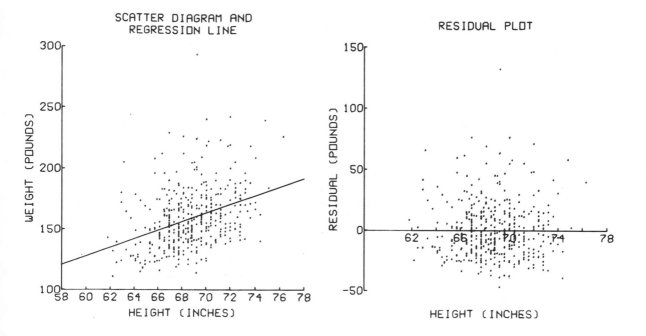

The residual plot in Figure 6 shows no pattern. By comparison, Figure 7 shows a residual plot (for hypothetical data) with a strong pattern. When the residual plot shows this kind of pattern, it was probably a mistake to use a regression line. Sometimes, you can spot this kind of nonlinearity by looking directly at the original scatter diagram. However, the residual plot gives a more sensitive test—because you can make the scale on the vertical axis of the residual plot big enough to examine things carefully.[2] (For an example, work Exercise 2 on p. 197 below.)

Figure 7. A residual plot with a strong pattern. It was a mistake to fit the regression line.

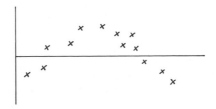

EXERCISE SET C

1. Several different regression lines are used to estimate the price of a stock (from different independent variables). Histograms for the residuals from each line are sketched below. Match the description with the histogram:
   (a) r.m.s. error = $5   (b) r.m.s. error = $15   (c) something's wrong.

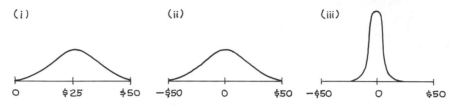

2. Look at the figure below.
   (a) In the scatter diagram, the averages of x and of y are closest to: 3, 4, 5.
   (b) The SDs of x and y are closest to: 0.6, 1.0, 2.0.
   (c) The correlation coefficient is closest to: 0.4, 0.8, 0.95.
   (d) The average of the residuals is closest to: 0, 2, 4.
   (e) The SD of the residuals is closest to: 0.6, 1.0, 2.0.
   (f) Take the points in the scatter diagram whose x-coordinates are between 4.5 and 5.5. Guess the average of their y-coordinates.
   (g) The SD of the y-coordinates of the points in (f) is closest to: 0.6, 1.0, 2.0.

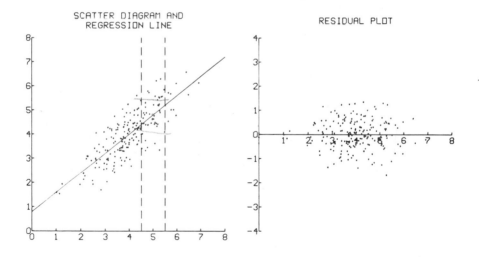

The answers to these exercises are on p. A-43.

4. LOOKING AT VERTICAL STRIPS

Figure 8 repeats the scatter diagram for the heights of the 1,078 fathers and sons in Pearson's study (p. 109). The families where the father is 64 inches

Figure 8. Scatter diagrams for the heights of 1,078 fathers and their sons. Families with 64-inch fathers are plotted in the solid vertical strip: the solid histogram below is for the heights of these sons. Families with 72-inch fathers are plotted in the dashed vertical strip; the dashed histogram below is for the heights of these sons. The dashed histogram is to the right of the solid one—the taller fathers have taller sons, on the whole. But the two histograms have similar shapes, and their SDs are nearly the same.

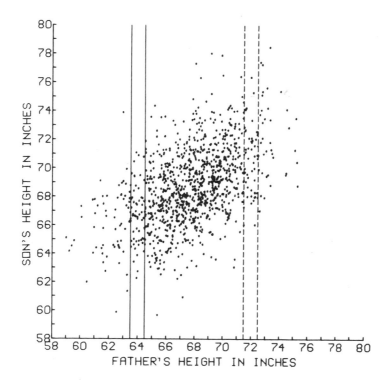

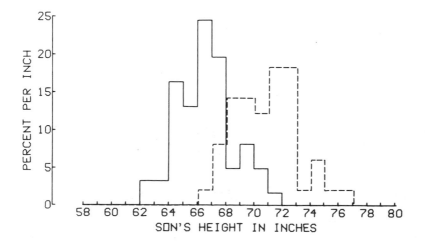

tall, to the nearest inch, are plotted in the solid vertical strip. A histogram for the heights of the sons in this strip is shown at the bottom of the figure: it is the solid one. The families with 72-inch fathers are plotted in the dashed vertical strip. And a histogram for the heights of the sons in this strip is shown too—the dashed one. The dashed histogram is farther to the right than the solid one: on the average, the taller fathers do have taller sons. However, both histograms have similar shapes, and just about the same amount of spread.[3]

When all the vertical strips in a scatter diagram show similar amounts of spread, the diagram is said to be *homoscedastic*. (*Homo* means "same," *scedastic* means "scatter.") The height-weight scatter diagram in Figure 6 is homoscedastic too. The range of weights for given height is greater in the middle of the picture, but that is only because there are more people there. The SD of weight for given height is pretty much the same from one end of the picture to the other. In general, homoscedastic scatter diagrams have oval-shaped residual plots, and the best way to check for homoscedasticity is to look at the residual plot. (*Homoscedasticity* is a terrible word, but statisticians use it.)

When the scatter diagram is homoscedastic, the regression estimates are off by similar amounts all along the regression line. In Figure 8, the regression line for estimating a son's height from his father's height had an r.m.s. error of 2.3 inches. If the father is 64 inches tall, the regression estimate for the son's height is 67 inches, and this estimate is likely to be off by 2.3 inches or so. If the father is 72 inches tall, the regression estimate for the son's height is 71 inches, and this is likely to be off by the same amount, 2.3 inches or so.

By comparison, Figure 9 shows a *heteroscedastic* scatter diagram (*hetero* means "different") of income against education. As education goes up, average income goes up, and so does the spread in income.

When the scatter diagram is heteroscedastic, the regression estimates are off by different amounts in different parts of the scatter diagram. For example, in Figure 9 the r.m.s. error of the regression line is $4,100. However, it is quite a bit harder to predict the incomes of the highly educated men than the incomes of the poorly educated ones. For instance, for men with 8 years of schooling the regression estimates of income are only off by around $3,100 or so; for men with 12 years of schooling, the amounts off go up to $3,500 or so; with 16 years, the errors run around $4,900 or so.

In this case, the r.m.s. error of the regression line only gives you the average size of the error—averaging over all the different *x*-values.

Suppose that a scatter diagram is homoscedastic, and there is no pattern in the residuals. Take the points in a narrow vertical strip. Their *y*-values will be off the regression estimate by amounts similar in size to the r.m.s. error of the regression line.

Figure 9. A heteroscedastic scatter diagram. This scatter diagram plots income against education, for a representative sample of 713 men aged 25 to 29 in 1970 holding full-time jobs.[1]

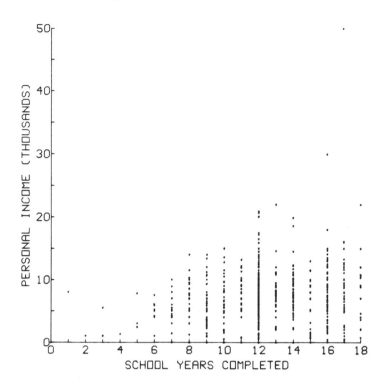

EXERCISE SET D

1. In Figure 8, the r.m.s. error of the regression line was 2.3 inches. If a father is 72 inches tall, the regression estimate for his son's height is 71 inches.
    (a) This estimate is likely to be right give or take _____ inches or so.
    (b) For about _____% of the families where the father was 72 inches tall, the height of the son is in the range

    71 inches − 2.3 inches to 71 inches + 2.3 inches.

    (c) For about 95% of the families where the father was 72 inches tall, the height of the son is in the range 71 inches − _____ to 71 inches + _____ .

2. With the height-weight example of Figure 1: If a man is 73 inches tall, the regression estimate for his weight is 176 pounds, with an r.m.s. error of 23 pounds. Of the men who are about 73 inches tall, approximately what percentage weigh between 176 pounds − 46 pounds and 176 pounds + 46 pounds?

3. In 1937, the Stanford-Binet IQ test was restandardized with two forms (L and M). A large number of subjects took both tests, and the results can be summarized as follows:

Form L average $\approx$ 100, SD $\approx$ 15
Form M average $\approx$ 100, SD $\approx$ 15, $r \approx 0.80$

The r.m.s. error of the regression line is

$$\sqrt{1 - (0.8)^2} \times 15 = 9.$$

   (a) Suppose the scatter diagram looks like (i) below. If someone scores 130 on form L, the regression estimate for the score on form M is 124. True or false: This estimate is likely to be off by 9 or so.
   (b) Repeat, if the scatter diagram looks like (ii).

(i)

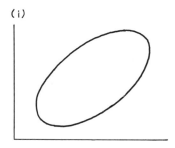

(ii)

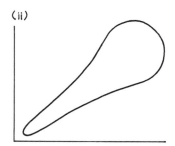

The answers to these exercises are on p. A-43.

5. USING THE NORMAL CURVE

Often, it is possible to use the normal approximation when working inside a vertical strip. For this to be legitimate, the scatter diagram has to be football-shaped, with the dots thickly scattered in the center of the picture and fading off toward the edges. Figure 8 is a perfect example. If the scatter diagram is heteroscedastic, or shows a nonlinear pattern, do not use the method of this section. An example of this kind is Figure 9. With the height-weight data in Figure 1, the normal curve would not work especially well either: the cloud isn't football-shaped, it is stretched out on top and squeezed in at the bottom.

Example. A law school finds the following relationship between LSAT scores and first-year scores (for admitted students):

average LSAT score $\approx$ 650, SD $\approx$ 80
average first-year score $\approx$ 65, SD $\approx$ 10, $r \approx 0.60$

The scatter diagram is football-shaped.
   (a) About what percentage of the students had first-year scores over 75?
   (b) Of the students who scored about 730 on the LSAT, about what percentage had first-year scores over 75?

Solution. Part (a). This is a straightforward normal approximation problem. The LSAT results and *r* have nothing to do with it.

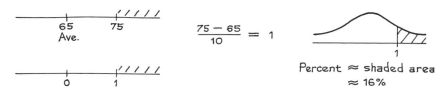

$$\frac{75 - 65}{10} = 1$$

Percent ≈ shaded area
≈ 16%

Part (b). This kind of problem is different from the ones solved in Chapter 10. The new element is spread around the regression line, representing the differences between individuals and averages. The solution may seem a bit intricate at first, so it is presented in the form of an imaginary dialog between a law professor and a statistician.

Law. Which variable is the dependent one?
Stat. First-year scores, because that's what is being estimated. We're looking at the students who scored about 730 on the LSAT. Are they above average, or below?
Law. Above, by 80 points.
Stat. How many SDs is that?
Law. That's one SD.
Stat. Are they going to be above or below average on the first-year tests?
Law. Above, by an SD.
Stat. Not so fast, Remember the regression effect.
Law. Oh. Then they'll be above by *r* SDs. That's

$$0.60 \times 10 \text{ points} = 6 \text{ points.}$$

Stat. So what do you estimate for their average?
Law. The overall average was 65, so the regression estimate is 65 + 6 = 71 points. I see. Now to get the percent I can convert to standard units:

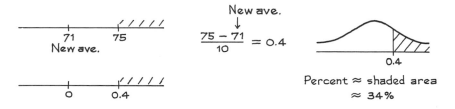

$$\frac{75 - 71}{10} = 0.4$$

Percent ≈ shaded area
≈ 34%

Stat. Hold it. Why did you use 10 for the SD?
Law. That's given. It's the SD for the first-year scores.
Stat. Yes, but that measures the variability in the first-year scores for the whole class. We're just looking at the students who scored 730 on the LSAT.
Law. So?
Stat. So we have a smaller and more homogeneous group. There is going to be less variability in their first-year scores. Then the SD has to be smaller.

Law. OK. How do we get this new SD?

Stat. Let me draw a picture [Figure 10]. The new average is the regression estimate. We're trying to see by how much these individual students differ from the regression estimate for their average. What measures these differences?

Figure 10. The new SD is the r.m.s. error.

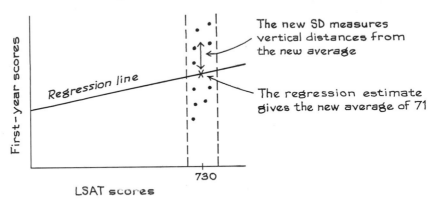

Law. The r.m.s. error.

Stat. Right. The formula is

$$\sqrt{1 - r^2} \times SD = \sqrt{1 - (0.60)^2} \times 10 = \sqrt{0.64} \times 10 = 8.$$

Law. Why did you use the SD for the first-year scores?

Stat. Because we're estimating first-year scores. The error has to come out in first-year points, not LSAT points.

Law. Now what?

Stat. Now convert to standard units:

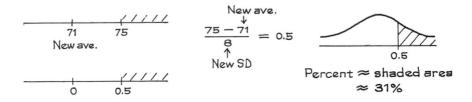

Law. Could you summarize the method for me?

Stat. Sure. There are three steps.

Step 1. Find the new average—that's the regression estimate.
Step 2. Find the new SD—that's the r.m.s. error for the regression line.
Step 3. Convert to standard units, and use the normal curve.

Law. I see. Can I always use this reasoning?

Stat. Whenever the scatter diagram has the football shape. If it shows a nonlinear pattern, or it's heteroscedastic, this method doesn't apply.

Law. Where does heteroscedastic come in?

Stat. The r.m.s. error is sort of an average error. If the scatter diagram is homoscedastic, the residuals tend to be similar in size to the r.m.s. error all along the regression line—whether the LSAT score is 730 or 600 or whatever. If the diagram is heteroscedastic, the residuals at 730 could be much larger, those at 600 could be much smaller. Or the opposite. They just balance out at the r.m.s. error.

Law. What do we do then?

Stat. Then we transform the variables, but I don't want to go into that now.

EXERCISE SET E

1. Pearson and Lee obtained the following results for about a thousand families:

   average height of husband $\approx$ 68 inches, SD $\approx$ 2.7 inches
   average height of wife $\approx$ 63 inches, SD $\approx$ 2.5 inches, $r \approx$ 0.25

   (a) What percentage of the women were over 5 feet 8 inches?
   (b) Of the women who were married to men of height 6 feet, what percentage were over 5 feet 8 inches?

2. From the same study:

   average height of fathers $\approx$ 68 inches, SD $\approx$ 2.7 inches
   average height of sons $\approx$ 69 inches, SD $\approx$ 2.7 inches, $r \approx$ 0.50

   (a) What percentage of the sons were over 6 feet tall?
   (b) What percentage of the 6-foot fathers had sons over 6 feet tall?

3. From the same study:

   average height of men $\approx$ 68 inches, SD $\approx$ 2.7 inches
   average forearm length $\approx$ 18 inches, SD $\approx$ 1 inch, $r \approx$ 0.80

   (a) What percentage of men have forearms which are 18 inches long, to the nearest inch?
   (b) Of the men who are 68 inches tall, what percentage have forearms which are 18 inches long, to the nearest inch?

The answers to these exercises are on pp. A-43–A-44.

6. REVIEW EXERCISES

1. Several regression lines are used to estimate the annual salaries in a certain company, from different independent variables. Residual plots from each regression are shown below. Match the plot with the description. Explain.
   (a) r.m.s. error = $1,000   (b) heteroscedastic   (c) something's wrong.

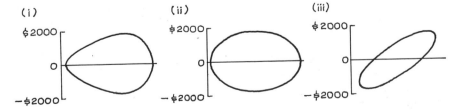

2. The r.m.s. error of the regression line for estimating y from x is:

   (i) SD of y (ii) SD of x (iii) $r \times$ SD of y (iv) $r \times$ SD of x

   (v) $\sqrt{1 - r^2} \times$ SD of y (vi) $\sqrt{1 - r^2} \times$ SD of x.

3. Tuddenham and Snyder obtained the following results for 66 California boys at ages 6 and 18 (see p. 133 for the scatter diagram):[5]

   average height at 18 $\approx$ 5 feet 10 inches, SD $\approx$ 2.5 inches

   average height at 6 $\approx$ 3 feet 10 inches, SD $\approx$ 1.7 inches, $r \approx 0.80$

   (a) Find the r.m.s. error for the regression estimates of heights at 18 from height at 6.

   (b) Find the r.m.s. error for the regression estimates of height at 6 from height at 18.

4. In a certain class, the midterm scores had an average of 50, with an SD of 25. The final scores averaged out to 60, with an SD of 15. The correlation between the midterm and final scores was 0.60. For each student, the final score was estimated from the midterm score using the regression line.

   (a) For about 1/3 of the students, the regression estimate for the final score was off by more than _____ points.

   (b) Estimate the final score for a student whose midterm score was 70.

   (c) This estimate is likely to be off by _____ points or so.

Fill in the blanks in (a) and (c) using the options

          25   15   12   9   6.

You may assume the scatter diagram was football-shaped.

5. A statistical analysis is made of the midterm and final scores in a large course, with the following results:

      average midterm score $\approx$ 65, SD $\approx$ 15

      average final score $\approx$ 60, SD $\approx$ 20, $r \approx 0.50$

The scatter diagram is football-shaped.

   (a) About what percentage of students scored over 80 on the final?

   (b) Of the students who scored 80 on the midterm, about what percentage scored over 80 on the final?

6. A computer program is developed to predict the GPA of college freshmen from their high school GPAs. This program is tried out on a class whose college GPAs are known. The r.m.s. error is 2.0. Is anything wrong? Explain. (GPAs range from 0.0 to 4.0.)

7. Every year, baseball's major leagues honor their outstanding first-year players with the title "Rookie of the Year." From 1949 to 1975, the overall batting average for the Rookies of the Year was about .300, far above the major league batting average of around .250. However, Rookies of the Year don't do so well in their second year: their overall second-season batting average was only around .275. Baseball writers call this "sophomore slump," their explanation being that these players get distracted by outside activities like product endorsements and television appearances. Do you agree?

8. The freshmen at a large university are required to take a battery of aptitude tests. Students who score high on the mathematics test also tend to score high on the physics test. On both tests, the average score is 60, with an SD of 20. Of the students who scored about 80 on the mathematics test:

   (i)   just about half scored over 80 on the physics test.

  (ii)   more than half scored over 80 on the physics test.

 (iii) less than half scored over 80 on the physics test.

Give your reason.

7. SUMMARY

1. The regression line can be used to estimate the dependent variable from the independent one; the difference between the actual value and the estimated value is a *residual*.

2. The *r.m.s. error* of the regression line is the root-mean-square of the residuals. This measures the accuracy of the regression estimates. The estimates are off by amounts similar in size to the r.m.s. error. Often, about 68% of the estimates will be right to within one r.m.s. error; about 95% will be right to within two r.m.s. errors.

3. The r.m.s. error of the regression line can be found by the formula

$$\sqrt{1 - r^2} \times \text{SD of dependent variable.}$$

4. After carrying out a regression, it is advisable to plot the residuals. If the residuals show a pattern, the regression may not have been appropriate.

5. Suppose that a scatter diagram is *homoscedastic*, and there is no pattern in the residuals. Take the points in a narrow vertical strip. Their y-values will be off the regression estimate by amounts similar in size to the r.m.s. error of the regression line. With *heteroscedastic* diagrams, the size of the error depends on the strip.

12

The Regression Line

The estimation of a magnitude using an observation subject to a larger or smaller error can be compared not inappropriately to a game of chance in which one can only lose and never win and in which each possible error corresponds to a loss. . . . However, what specific loss we should ascribe to any specific error is by no means clear of itself. In fact, the determination of this loss depends at least in part on our judgment. . . . Among the infinite variety of possible functions the one that is the simplest seems to have the advantage and this is unquestionably the square. . . . Laplace treated the problem in a similar fashion, but he chose the size of the error as the measure of loss. However, unless we are mistaken this choice is surely not less arbitrary than ours.

—C. F. GAUSS (GERMANY, 1777–1855)[1]

1. SLOPE AND INTERCEPT

Does education pay? Figure 1 shows the relationship between income and education in 1970, for a representative sample of 481 men aged 35 to 54, who were at work full time with professional or managerial jobs. The summary statistics:[2]

average education $\approx$ 14 years, SD $\approx$ 3 years
average income $\approx$ \$16,000, SD $\approx$ \$9,000, $r \approx 0.30$

The regression estimates for average income at each educational level fall along the regression line, shown in the figure. The line slopes up, showing that on the average, income does go up with education.

Figure 1. The regression line. The scatter diagram shows income and education in 1970, for a representative sample of 481 men aged 35 to 54, at work full time, with professional or managerial jobs.

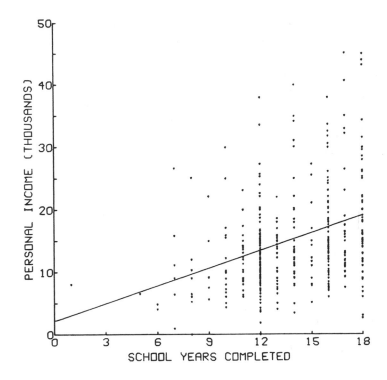

Any line can be described in terms of its slope and intercept (Chapter 7). The *y*-intercept is the height of the line when *x* is 0. And the slope is the rate at which *y* increases, per unit increase in *x*. Slope and intercept are illustrated in Figure 2.

Figure 2. Slope and intercept.

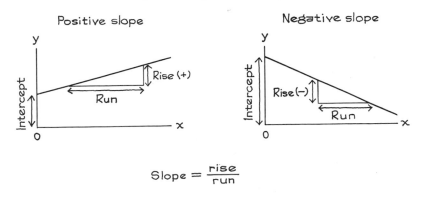

What do the slope and intercept mean for the regression line? To go on with the income-education example: associated with an increase of one SD in education, there is an increase of r SDs in income. That is, three extra years of education are worth an extra $0.30 \times \$9,000 = \$2,700$ of income, on the average. So each extra year is worth $900, on the average. The slope of the regression line is $900 per year. So far, it looks like education does pay off, at the rate of $900 a year.

The intercept of the regression line is its height when $x = 0$—namely, the regression estimate for the average income of the men with 0 years of education. Such men are 14 years below average in education, costing them (Figure 3)

$$14 \text{ years} \times \$900 \text{ per year} = \$12,600.$$

Their average income is estimated at $\$16,000 - \$12,600 = \$3,400$. This is the intercept.

Zero years of education may sound extreme, but there was a man in the study who only finished first grade. And his income was quite respectable, too. His point is in the lower left corner of Figure 1.

Figure 3. Finding the slope and intercept of the regression line.

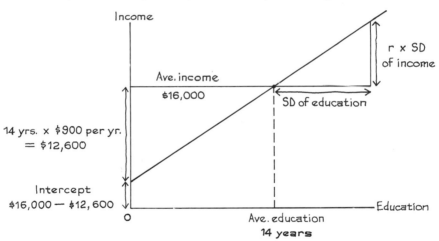

$$\text{Slope} = \frac{r \times \text{SD of income}}{\text{SD of education}}$$

Associated with each unit increase in x there is some average change in y. The slope of the regression line says how much this change is. The formula for the slope is

$$\frac{r \times \text{SD of } y}{\text{SD of } x}.$$

The intercept of the regression line is just the regression estimate for y, when x is 0.

Any line has an equation, in terms of its slope and intercept:

$$y = \text{slope} \times x + \text{intercept}.$$

For the regression line, this equation is called the *regression equation*.

There is nothing new here: the regression equation is just an alternative way of making the regression estimates. We are presenting the regression equation because it is often used in technical papers in the social sciences. The reason is that if someone has to make many regression estimates, it is usually easier to compute the slope and intercept once and for all, and then make the estimates by substituting into the equation.

Example 1. For the men in Figure 1, the regression equation for estimating income from education is

estimated income = ($900 per year) × education + $3,400.

Estimate the income for one of these men who has
  (a) 8 years of schooling—elementary education
  (b) 12 years of schooling—a high-school diploma
  (c) 16 years of schooling—a college degree.

Solution. It is just a matter of substituting into the equation. With 8 years of schooling, the estimated income is

$$\$900 \times 8 + \$3,400 = \$10,600.$$

With 12 years, it's

$$\$900 \times 12 + \$3,400 = \$14,200.$$

With 16 years, it's

$$\$900 \times 16 + \$3,400 = \$17,800.$$

This completes the solution.

This example brings out an interesting point. The correlation between income and education is only 0.30—rather low. However, with enough of a difference in educational level, the difference in incomes becomes quite large: the people with college degrees are doing a lot better than those who only had an elementary education. Even with a low correlation, big changes in x may be accompanied by big changes in y. The slope says the extent to which this happens.

Example 2. For women aged 35–54, working full time, in professional or managerial jobs, the relationship between income and education in 1970 can be summarized as follows:[3]

average education ≈ 14 years, SD ≈ 3 years
average income ≈ $8,000,   SD ≈ $3,000, r ≈ 0.4

  (a) Find the regression equation for estimating income from education.
  (b) Find the regression estimate for the income of a woman whose educational level is: 12 years, 13 years.

Solution. The first step is to find the slope (Figure 3). In a run of one SD, the regression line rises r SDs. So

$$\text{slope} = \frac{0.4 \times \$3,000}{3 \text{ years}} = \$400 \text{ per year.}$$

The interpretation: on the average, each extra year of schooling is worth an extra $400 of income; each year less of schooling costs $400 of income.

The next step is to find the intercept. That is the height of the regression line at $x = 0$. In other words, it is the regression estimate for the income of a woman with no education. Such a woman is 14 years below average; using the slope, she is estimated to be below average in income by

$$14 \text{ years} \times \$400 \text{ per year} = \$5,600.$$

So her income is estimated as

$$\$8,000 - \$5,600 = \$2,400.$$

That is the intercept: the regression estimate for y when $x = 0$.

The regression equation is

estimated income = ($400 per year) × (education) + $2,400.

Substituting 12 years for education gives

($400 per year) × (12 years) + $2,400 = $7,200.

That is the regression estimate for the income of a woman with 12 years of education.

Substituting 13 years for education gives

($400 per year) × (13 years) + $2,400 = $7,600.

That is the regression estimate for the income of a woman with 13 years of education. The extra year of education (from 12 to 13) is worth $400, on the average. In this example, the slope is $400 per year. Associated with each extra year of education, there is an increase of $400 in income, on the average.

The phrase "associated with" sounds like it is talking around some point. And it is. In most regression problems, there is a serious issue. Is a change in the dependent variable caused by a change in the independent variable, or do both changes reflect the common influence of other, more basic, factors? The phrase "associated with" was invented to let statisticians talk about regressions without having to commit themselves on this point.

Often, the slope is used to estimate how much the dependent variable would change, if the independent variable were changed. This inference is valid in a controlled experiment. (There are some examples in Exercise 1 below, and in the next section.) With observational studies, the inference is often shaky—because of confounding. Take Example 2. On the average, the women who finished one year of college (13 years) earned $400 more than women who just finished high school (12 years). This described the actual situation in 1970. But suppose the government sent a representative group of women with high-school degrees on to college for a year. The slope suggests that their income would go up, by an average of $400. But this is too hasty.

Example 2 is based on an observational study, the Census. Those people who went on to college were different from those who just finished high school, with respect to many factors besides education. The effects of intelligence, ambition, and family background are confounded with the effect of education, and inflate the slope. Sending people on to college for a year probably will make their incomes go up, but not by the full $400. To find out by how much, it would be necessary to run a controlled experiment—and that's almost impossible.

> With an observational study, the slope and intercept of the regression line are only descriptive statistics. They say how the average value of the dependent variable is related to given values of the independent variable, in the population being observed. They cannot be relied on to predict how the dependent variable would respond if the investigator changed the values of the independent variable.

Returning for a moment to Examples 1 and 2, each extra year of education was worth much less to the women than to the men. Does this show sex bias? Perhaps, but other explanations are possible. For instance, in the period 1960–1970, women did not stay in the labor force as consistently as men. So they built up less experience, and their job skills were more likely to get out of date.[4]

EXERCISE SET A

1. The International Rice Research Institute in the Philippines developed the hybrid rice IR 8, setting off "the green revolution" in tropical agriculture. Among other things, they made a thorough study of the effects of fertilizer on rice yields. These experiments involve a number of experimental plots (of about 20 square yards in size). Each plot is planted with IR 8, and fertilized with some amount of nitrogen fixed in advance by the investigators. (The amounts range from 0 to about a pound.) When the rice is harvested, the yield can be determined and related to the amount of nitrogen applied. In one such experiment, the correlation between yield and nitrogen was 0.95, and the regression equation was[5]

   estimated rice yield = (20 oz. rice per oz. nitrogen) × (nitrogen) + 240 oz.

   (a) An unfertilized plot can be expected to produce around _____ of rice.
   (b) Each extra ounce of nitrogen fertilizer can be expected to increase the rice yield by _____.
   (c) Estimate the amount of rice yielded when the amount of fertilizer is: 3 ounces of nitrogen, 4 ounces of nitrogen.
   (d) Was this an observational study or a controlled experiment?
   (e) In fact, fertilizer was applied only in the following amounts: 0 ounces, 4 ounces, 8 ounces, 12 ounces, 16 ounces. Would you trust the regression estimate for 3 ounces of nitrogen, even though this particular amount was never applied?
   (f) Would you trust the regression estimate for 100 ounces of nitrogen?

2. In 1970, the relationship between income and education for men aged 35 to 54, with full time jobs, can be summarized as follows—controlling for race:[6]

white $\left\{\begin{array}{l}\text{average education} \approx 11 \text{ years, SD} \approx 3 \text{ years} \\ \quad\text{average income} \approx \$11{,}000, \text{SD} \approx \$7{,}200, \; r \approx 0.40\end{array}\right.$

black $\left\{\begin{array}{l}\text{average education} \approx 9 \text{ years, SD} \approx 4 \text{ years} \\ \quad\text{average income} \approx \$6{,}000, \text{ᐟSD} \approx \$4{,}000, \; r \approx 0.50\end{array}\right.$

Find the regression estimate for the average income of white men with high-school degrees (12 years of education). Repeat for black men.

3. For the men aged 35 to 44 in the Health Examination Survey, the regression equation for estimating height from education is[7]

estimated height = (0.1 inches per year) × (education) + 67 inches.

Estimate the height of a man with: 12 years of education; 16 years of education. Does going to college increase a man's height? Explain.

4. For each of Exercises 2, 3, 5 on pp. 150–151, write down the equation of the regression line. Use this equation to find the regression estimates.

5. Write down the regression equation for estimating the height of a son from the height of his father. (The summary statistics are on p. 159.)

6. Write down the regression equation for estimating the height of a father from the height of his son. (Same data as in Exercise 5.)

The answers to these exercises are on p. A-44.

2. THE METHOD OF LEAST SQUARES

Sometimes the points on a scatter diagram seem to be following a line. The problem is to find the line which best fits the points. Usually, this involves a compromise: moving a line closer to some points will increase its distance from others. To resolve the conflict, two steps are necessary. First, get an average distance from the line to all the points. Second, move the line around until this average distance is as small as possible.

To be more specific, suppose the line will be used to predict y from x. Then, the error made at each point is the vertical distance from the point to the line. The usual way to handle the first step, defining the average distance to the line, is to take the root-mean-square of the errors (p. 57). This measure of average distance is called the *r.m.s. error* of the line. It was first proposed by Gauss, and he was careful to point out that it is only a convention.

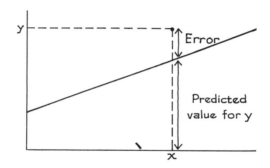

The second problem, how to move the line around to minimize the r.m.s. error, was also solved by Gauss:

> Among all lines, the regression line makes the smallest r.m.s. error in predicting y from x.

For this reason, the regression line is often called the *least-squares line:* the errors were combined by squaring, and the regression line makes the combined error as small as possible.

Now for an example. Hooke's law was discussed in Chapter 9. To review briefly, a spring has length b when unloaded. A weight of x kilograms is tied to the end of the spring. As illustrated in Figure 4, the spring stretches to a new length. According to Hooke's law, the amount of stretch is proportional to the weight x. So the new length of the spring is

$$y = mx + b.$$

In this equation, m and b are constants which depend on the spring. They have to be found using experimental data. The statistical procedure for doing this will now be explained.

Figure 4. Hooke's law: the stretch is proportional to the load.

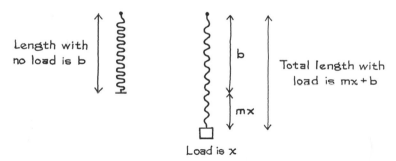

Table 1 shows the results of an experiment on Hooke's law, in which weights of various sizes were loaded on the end of a length of piano wire.[8] The first column shows the load (the independent variable). The second column shows the measured length (the dependent variable). With twenty pounds of load, this spring only "stretched" about one fifth of an inch (10 kg. ≈ 22 lbs., 0.5 cm. ≈ 0.2 in.). Piano wire is not very stretchy.

Table 1. Data on Hooke's law.

Weight (kg.)	Length (cm.)
0	439.00
2	439.12
4	439.21
6	439.31
8	439.40
10	439.50

The correlation coefficient for the data in Table 1 is 0.999, very close to 1 indeed. So the points almost form a straight line (Figure 5), just as Hooke's law predicts. The minor deviations from linearity are probably due to measurement error—neither the weights nor the lengths have been measured with perfect accuracy.

Figure 5. Scatter diagram for Table 1.

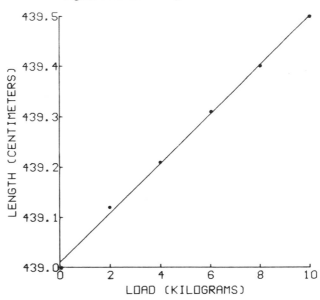

Our goal is to estimate the parameters m and b in the equation of Hooke's law for the piano wire:

$$y = mx + b.$$

The graph of this equation is an ideal straight line, approximated by the scatter diagram in Figure 5. If the points in Figure 5 happened to fall exactly on some straight line, we would take that line as an approximation to the ideal line. Its slope would be an estimate for m, its intercept an estimate for b.

The trouble is that the points do not form a perfect straight line. Many different straight lines could be drawn across the scatter diagram, each having a slightly different slope and intercept. Which line should be used? Hooke's equation predicts length from weight. As discussed above, it is natural to choose m and b so as to minimize the r.m.s. error. That line is the regression line.[9] In other words, m should be estimated as the slope of the regression line, and b as its intercept. These are called the *least squares estimates,* because they minimize root-mean-square error. Doing the arithmetic,

$$m \approx 0.05 \text{ cm. per kg. and } b \approx 439.01 \text{ cm.}$$

The method of least squares estimates the length of the spring under no load to be 439.01 cm. And each kilogram of load causes the spring to stretch by an amount estimated as 0.05 cm. There is no need to hedge this statement, because it is based on a controlled experiment. The investigator puts the weights on, and the wire stretches. He takes the weights off, and the wire comes back to its original length. This process can be repeated as often as is desired. There is no question here about what is causing what.

The least squares estimate for the length of the spring under no load was 439.01 cm. This is a tiny bit longer than the measured length at no load (439.00 cm). A statistician would trust the least squares estimate over the measurement. Why? Because the least squares estimate takes advantage of all six measurements, not just one. Some of the measurement error is likely to cancel out. Of course, the six measurements are tied together by a good theory—Hooke's law. Without the theory, the least squares estimate wouldn't be worth much.

3. DOES THE REGRESSION MAKE SENSE?

A regression line can be put down on any scatter diagram. However, there are two questions to ask about the result. First, was there a nonlinear association between the variables? If so, the regression line may be quite misleading. (This is discussed on pp. 124 and 152.) Even if the association looks linear, there is still a second question to ask. Did the regression make sense? The second question is much harder. To answer it requires some understanding of the mechanism which produced the data. If this mechanism is not understood, fitting a line can be intellectually disastrous. To make up an example, suppose an investigator does not know the formula for the area of a rectangle. Taking an empirical approach, he draws twenty typical rectangles, as shown in Figure 6. He thinks the area ought to depend on the perimeter, so

Figure 6. Twenty typical rectangles.

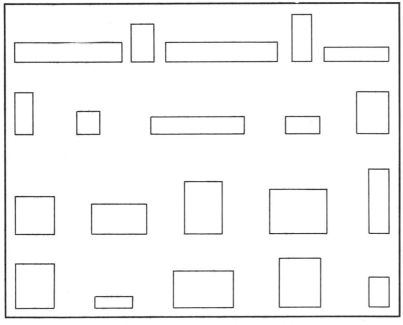

Scale: ⊢————┼————┼————┤ inches
       0   1   2   3

for each rectangle he measures the perimeter—the independent variable—and the area—the dependent variable. A scatter diagram for his results is shown in Figure 7. He works out the correlation coefficient and gets 0.87. This is a good deal higher than many correlations reported in social-science journals, and the investigator concludes that he really understands rectangles. His regression equation is

estimated area = (0.29 inches) × (perimeter) − 0.27 square inches.

(Area is measured in square inches and perimeter in inches.) The regression line is shown in Figure 7.

The arithmetic is all in order. But this investigator went at the problem so crudely that his regression was ridiculous. He should have looked at two other variables, length and width. These two variables determine both area and perimeter:

area = length × width,   perimeter = 2 (length + width).

But our straw-man investigator would never find the relationships just by doing regressions.

Of course, this example is only hypothetical. But many investigators do fit lines to scatter diagrams when they don't really know what's going on. This can make a lot of trouble. When thinking about a regression, ask whether it is more like Hooke's law, or more like area and perimeter.

Figure 7. Scatter diagram of area against perimeter, for the twenty rectangles in Figure 6, with the regression line.

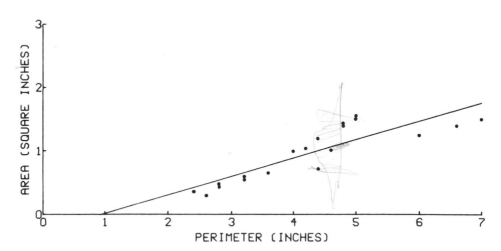

EXERCISE SET B

1. For the piano wire discussed in section 2 (Table 1 and Figure 5), predict the length under the following loads, if possible: 3 kg., 7 kg., 50 kg.

2. The table below shows per-capita disposable income and personal consumption expenditures, in 1958 dollars, yearly from 1950 to 1969.[10] If you have access to a statistical calculator:
   (a) find r and the regression equation for estimating consumption from income
   (b) plot the residuals.

| | Per capita in 1958 dollars | |
Year	Income	Consumption
1950	1,646	1,520
1951	1,657	1,509
1952	1,678	1,525
1953	1,726	1,572
1954	1,714	1,575
1955	1,795	1,659
1956	1,839	1,673
1957	1,844	1,683
1958	1,831	1,666
1959	1,881	1,735
1960	1,883	1,749
1961	1,909	1,756
1962	1,969	1,814
1963	2,015	1,867
1964	2,126	1,948
1965	2,239	2,047
1966	2,335	2,127
1967	2,403	2,164
1968	2,486	2,256
1969	2,534	2,315

The answers to these exercises are on pp. A-44–A-45.

Technical note: Examples 1 and 2 in section 1 presented regression equations for estimating income from education. A regression equation is a good way to describe the relationship between income and education. But it is not legitimate to interpret the equation in a cause-and-effect way: if you send people on to school for another x years, they will earn another y dollars on average. The reason was pointed out before: the effects of many other variables, in personality and family background, are confounded with education.

Many investigators would tackle the problem of controlling for confounded variables by using what is called *multiple regression*. For instance, they might develop some measure for the socioeconomic status of an individual's parents, and fit a multiple regression equation of the form

$$y = a + b \times E + c \times S,$$
where

$$y = \text{estimated income}, \quad E = \text{educational level},$$
$$S = \text{measure of socioeconomic status}.$$

The coefficient b would be interpreted as showing the effect of education, free of the effect of family background.

This often gives sensible results. But it can equally well produce nonsense. Take the hypothetical investigator who was working on the area of rectangles. He could decide to control for the shape of the rectangle by multiple regression, using the length of the diagonal to measure the shape of the rectangle. Of course, this isn't a very good measure of shape, but then nobody knows how to measure socioeconomic status very well either. The investigator would then fit a multiple regression equation of the form

$$\text{estimated area} = a + b \times \text{perimeter} + c \times \text{diagonal}.$$

He might tell himself that b measures the effect of perimeter free of the effect of shape. As a result, he could be even more confused than before. The perimeter and diagonal do determine the area, but only by a complicated nonlinear formula. Multiple regression is a powerful technique, but it is no substitute for understanding.

4. REVIEW EXERCISES

1. (a) The r.m.s. error of the line below is closest to:
          0.1, 0.3, or 1.
   (b) Is it the regression line?

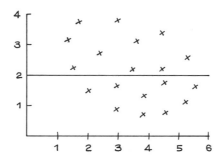

2. An investigator develops a regression equation for estimating the weight of a car (in pounds) from its length (inches). The slope is closest to

   3 pounds per inch   30 pounds per inch   300 pounds per inch
   3 inches per pound   30 pounds per sq. in.   300 cm. per kg.

   Explain.

3. Find the regression equation for estimating final scores from midterm scores, based on the following information:

   average midterm score = 70, SD = 10
   average final score = 55, SD = 20, $r = 0.60$

4. For the piano wire discussed in section 2 (Table 1 and Figure 5), predict the length under the following loads, if possible:
   (a) 1 kilogram   (b) 5 kilograms   (c) 9 kilograms
   (d) 100 kilograms

5. During the first two weeks of March, 1975, thirty-one pieces of real estate were sold in Glendale, California. The selling price was related to the assessed value, with the following results:[11]

> average assessed value ≈ $46,700, SD ≈ $37,100
> average selling price ≈ $50,950, SD ≈ $39,900, $r = 0.998$

(a) The next three pieces of real estate to sell in that city had assessed values of: $25,000, $50,000, $75,000. Estimate the selling prices.
(b) How good do you think these estimates are?
(c) Could you use these data to estimate selling price from assessed values for properties in New York?

Explain your answers to (b) and (c).

6. For men aged 18 to 24 in the Health Examination Survey, the regression equation for estimating height from weight is

> estimated height = (0.04 inches per pound) × (weight) + 62 inches,

where height is measured in inches and weight in pounds. If someone puts on 20 pounds, will they get taller by

> 20 pounds × 0.04 inches per pound = 0.8 inches?

If not, what does the slope mean?

7. For men aged 25 to 34 in the Health Examination Survey of 1960, the relationship between height and income can be summarized as follows:[12]

> average income ≈ $6,000,   SD ≈ $3,000
> average height ≈ 69 inches, SD ≈ 3 inches, $r = 0.2$

What is the regression equation for estimating income from height? Interpret it.

8. A regression equation for estimating weight from height is

> estimated weight = (4 lbs. per in.) × (height) − 130 lbs.

Here, weight is in pounds and height in inches. What would the equation be in the metric system? (1 in. ≈ 2.5 cm., 1 lb. ≈ 0.45 kg.).

9. Working from the same data as in Exercise 8, another investigator wants to get the regression equation for estimating height from weight. Solving, he gets

> estimated height = (1/4 in. per lb.) × (weight + 130 lbs.).

Is this legitimate? If you want to compute anything,

> average height ≈ 68 inches,   SD of height ≈ 2.5 inches
> average weight ≈ 142 pounds, SD of weight ≈ 25 pounds, $r = 0.40$.

10. In a large study (hypothetical) of the relationship between parental income and the IQ of their children, the following results were obtained:

> average income ≈ $12,000, SD ≈ $4,500
> average IQ ≈ 100,   SD ≈ 15, $r ≈ 0.50$.

In analyzing the data, a graph was plotted as follows. For each income group (0–$1,000, $1,000–$2,000, $2,000–$3,000, etc.) the average IQ of children with parental income in that group was calculated and then plotted above the midpoint of the group ($500, $1,500, $2,500, etc.). It was found that the points on this graph followed a straight line very closely. The slope of this line (in IQ points per dollar) would be about:

120   600   1/150   1/300   1/600

can't say from this information.

Explain.

11. One child in the study referred to in Exercise 10 had an IQ of 120, but the information about his parents' income was lost. At $24,000 the height of the line plotted in Exercise 10 corresponds to an IQ of 120. Is $24,000 a good estimate for the parents' income? Or is it likely to be too high? Or too low? Explain.

5. SUMMARY

1. The regression line can be specified by two descriptive statistics: the *slope* and the *intercept*.

2. The slope of the regression line is the average rate at which y changes with x. It is the average change in y, per unit change in x.

3. The formula for the slope of the regression line is: $r \times$ SD of y/SD of x.

4. The intercept of the regression line is just the regression estimate for y, when x is 0.

5. The regression equation is: $y = $ slope $\times x + $ intercept.

6. The equation can be used to make all the regression estimates, by substituting given values of the independent variable for x.

7. Among all lines, the regression line makes the smallest r.m.s. error in estimating the dependent variable from the independent one.

8. Sometimes, two quantities are thought to be connected by a linear relationship (like Hooke's law, for length and weight. The slope and intercept of the line are parameters to be estimated. The *least squares estimates* are the slope and intercept of the regression line, minimizing the root-mean-square error.

9. When the regression line is fitted to data from an observational study, the slope cannot be interpreted as the average change in the dependent variable that would be caused by a unit change in the independent variable. Association is not the same as causation.

10. If the mechanism which produced the data is poorly understood, regression equations can be quite misleading.

PART IV

Probability

13

What Are the Chances?

In the long run, we are all dead.

—JOHN MAYNARD KEYNES (ENGLAND, 1883–1946)

1. INTRODUCTION

People talk loosely about chance all the time, without doing any harm. What are the chances of getting a job? of meeting someone? of rain tomorrow? But for scientific purposes, it is necessary to give the word *chance* a definite, clear interpretation. This turns out to be hard, and mathematicians have struggled with the job for centuries. They have developed several careful and rigorous theories of chance, but it must be admitted that these theories only cover a small range of the cases in which people ordinarily speak of chance.[1] This book will present only one of these theories—the *frequency theory*. This theory works best for processes which can be repeated over and over again, independently and under the same conditions. Games of chance fall into this category, and in fact much of theory was developed to solve gambling problems. One of its great masters was Abraham de Moivre, a French Protestant who fled to England to avoid religious persecution. Part of the dedication to his book, *The Doctrine of Chances,* is reproduced in Figure 1 on the next page.[2]

One simple game of chance involves betting on the toss of a coin. The process of tossing the coin can be repeated over and over again, independently and under the same conditions. The chance of getting heads is 50%: that is to say, in the long run heads will turn up about 50% of the time.

Abraham de Moivre (England, 1667–1754).

Etching by Faber. Reproduced with the permission of the Trustees of the British Museum.

Figure 1. De Moivre's dedication to *The Doctrine of Chances*.

To the Right Honorable the
Lord CARPENTER.
My Lord,
There are many people in the World who are prepossessed with an Opinion, that the Doctrine of Chances has a Tendency to promote Play; but they soon will be undeceived, if they think fit to look into the general Design of this Book; in the mean time it will not be improper to inform them, that your Lordship is pleased to espouse the Patronage of this second Edition: which your strict Probity, and the distinguished Character you bear in the World, would not have permitted, were not their Apprehensions altogether groundless.

Your Lordship does easily perceive, that this Doctrine is so far from encouraging Play, that it is rather a Guard against it, by setting in a clear light, the Advantages and Disadvantages of those Games wherein Chance is concerned. . . .

Another use to be made of this Doctrine of Chances is that it may serve in conjunction with the other parts of the Mathematicks, as a fit Introduction to the Art of Reasoning: it being known by experience that nothing can contribute more to the attaining of that Art, than the consideration of a long Train of Consequences, rightly reduced from undoubted Principles, of which this Book affords many Examples.

Take another example. A die is rolled. The chance of getting an ace— $\boxed{\cdot}$ —is 1 in 6, or $16\frac{2}{3}\%$. The interpretation is similar. If the die is rolled over and over again, repeating the basic chance process independently and under the same conditions, in the long run an ace will show about $16\frac{2}{3}\%$ of the time.

> The chance of something says about what percentage of the time it is expected to happen, when the basic process is repeated over and over again, independently and under the same conditions.

So the theory of chance does not say very much about any particular toss of the coin or roll of the die. But it does make very definite predictions about the long-run behavior of coins and dice, traffic flows and nuclear reactors, sample surveys and measurement procedures.

Because a chance represents a percentage of time:

> Chances are between 0% and 100%.

If something is impossible, it happens 0% of the time. At the other extreme, if something is sure to happen, then it happens 100% of the time. All chances are between these two extremes.

Another consequence:

> The chance of something equals 100% minus the chance of the opposite thing.

For example, if you have a 45% chance to win a game, then you expect to win about 45% of the time. So you must expect to lose the other 55% of the time.

One way to determine chances is empirically. For example, if three dice are thrown, what is the chance that the total number of spots is 9? The empirical way to find out is to throw three dice over and over again, and figure the percentage of the time that a total of 9 spots showed. To a first approximation, this gives the chance. Of course, throwing the dice involves a lot of dull work. And in some cases, the results are so different from what common sense suggests that you might want to write the whole thing off as experimental error—even though it's perfectly right. When working problems about chance, common sense can sometimes lead you far astray. For these reasons, a mathematical theory for calculating chances is needed. This part of the book makes a start on presenting the theory. The dice problem, for instance, will be solved on p. 222.

EXERCISE SET A

1. A computer is programmed to compute various chances. Match the computer's numerical answers with the verbal descriptions (which may be used more than once).

    Numerical answer   *Verbal description*

    (a) −50%   (i)   This is as likely to happen as not.
    (b)   0%   (ii) This is very likely to happen, but it's not certain.
    (c)   10%   (iii) This won't happen.
    (d)   50%   (iv) This may happen, but it's not likely.
    (e)   90%   (v)   This will happen, for sure.
    (f) 100%   (vi) There's a bug in the program.
    (g) 200%

2. A coin will be tossed 1,000 times. About how many heads do you expect to see?

3. A die will be rolled 6,000 times. About how many aces do you expect to see?

4. In poker (five card draw), you have about a .0014 chance to be dealt a full house (three of a kind and a pair). If you play poker 10,000 times, about how many times do you expect to be dealt a full house?

The answers to these exercises are on p. A-45.

2. USING THE LONG-RUN ARGUMENT

A box contains 10 tickets, numbered from 1 through 10:

| 1 | 2 | 3 | 4 | 5 | 6 | 7 | 8 | 9 | 10 |

They are all of the same size, shape, and texture. The box is shaken well to mix them up, and one is drawn out *at random*—so that each ticket has the same chance. The chance of drawing 7, for instance, is 1 in 10 or 10%. Why? Imagine drawing a ticket over and over again at random from this box, replacing it after each draw, so as to repeat the process independently and under the same conditions. In the long run, each ticket should appear about as often as any other—about 1 time in 10.

Example. A box contains red marbles and blue marbles. One marble is drawn at random from the box (each marble has an equal chance to be drawn). If it is red, you win $1. If it is blue, you win nothing. You can choose between two boxes:
   • box A contains 2 blue marbles and 3 red ones
   • box B contains 20 blue marbles and 30 red ones.
Which box offers a better chance of winning, or are they the same?

Solution. Some people prefer box A, because it has fewer blue marbles. Others prefer B, because it has more red marbles. Both views are wrong. The two boxes offer the same chance of winning—3 in 5. To see why, imagine drawing many times at random from box A (replacing the marble after each

Figure 2. De Moivre's solution.

The Probability of an Event is greater or less, according to the number of Chances by which it may happen, compared with the whole number of Chances by which it may either happen or fail.

Wherefore, if we constitute a Fraction whereof the Numerator be the number of Chances whereby an Event may happen, and the Denominator the number of all the Chances whereby it may either happen or fail, that Fraction will be a proper designation of the Probability of it happening. Thus if an Event has 3 Chances to happen, and 2 to fail, the Fraction 3/5 will fitly represent the Probability of its happening, and may be taken as the measure of it.

The same things may be said of the Probability of failing, which will likewise be measured by a Fraction, whose Numerator is the number of Chances whereby it may fail, and the Denominator the whole number of Chances, both for its happening and failing: thus the Probability of the failing of that Event which has 2 Chances to fail and 3 to happen will be measured by the Fraction 2/5.

The Fractions which represent the Probabilities of happening and failing, being added together, their Sum will always be equal to Unity, since the Sum of their Numerators will be equal to their common Denominator: now it being a certainty that an Event will either happen or fail, it follows that Certainty, which may be conceived under the notion of an infinitely great degree of Probability, is fitly represented by Unity. [By ''Unity,'' de Moivre means the number 1.]

These things will easily be apprehended, if it be considered that the word Probability includes a double Idea: first, of the number of Chances whereby an Event may happen: secondly, of the number of Chances whereby it may either happen or fail.

draw, so as not to change the conditions of the experiment). Each of the 5 marbles will appear about the same number of times in the long run—that is, about 1 time in 5. In particular, the red marbles will turn up about 3/5 of the time. So with box A, your chance of drawing a red marble, and winning, is 60%.

Now imagine drawing many times at random with replacement from box B. Each of the 50 marbles will only turn up about 1 time in 50. But now there are 30 red marbles, so you can expect to draw one and win about 30 times in 50. With box B, your chance of winning is 30/50 = 3/5 or 60%. What counts is the ratio

$$\frac{\text{number of red marbles}}{\text{total number of marbles}} .$$

This ratio is the same in both boxes. De Moivre's solution for this example is given in Figure 2.

Many problems take the form of drawing at random from a box. A typical instruction is

"Draw two tickets at random WITH replacement from the box

$$| \boxed{1}\ \boxed{2}\ \boxed{3} |$$ "

This asks you to imagine the following process: shake the box, draw out one ticket at random (equal chance for all three tickets), make a note of the number on it, put it back in the box, shake the box again, draw a second ticket at random (equal chance for all three tickets), make a note of the number on it, and put the second ticket back in the box.

This should be contrasted with the instruction

"Draw two tickets at random WITHOUT replacement from the box

| 1 | 2 | 3 | "

The second instruction asks you to imagine the following process: shake the box, draw out one ticket at random (equal chance for all three tickets), set it aside, draw out a second ticket at random (equal chance for the two tickets left in the box).

Figure 3. The difference between drawing with or without replacement.

Two draws are made at random from the box | 1 | 2 | 3 |. Suppose the

first draw is | 3 |.

WITH replacement . . . the second draw is from

| 1 | 2 | 3 |

WITHOUT replacement . . . the second draw is from

| 1 | 2 |

EXERCISE SET B

1. A box contains 3 red marbles and 2 blue ones; 500 draws are made at random with replacement from this box. About how many red marbles do you expect to see? About how many blues?

2. The situation is the same as in Exercise 1, and you win $1 on a red marble, nothing on a blue.
   (a) How much do you win if you draw 308 red marbles?
   (b) How much do you win if you draw 273 red marbles?
   (c) About how much do you expect to win?

3. As in Exercises 1 and 2, but now you win $1 on red, and lose $1 on blue. About how much money do you expect to make?

4. As in Exercises 1, 2, and 3, but now you win $8 on red, lose $10 on blue. About how much money do you expect to make?

5. A hundred tickets will be drawn at random with replacement from one of the two boxes shown below. On each draw, you will be paid the amount shown on the ticket, in dollars. Which box is better, and why?

(i) | 1 | 2 | (ii) | 1 | 3 |

6. (a) Suppose that a dart player is as likely to hit the board at any point as at any other point. What is the chance of hitting the shaded region A on the dart board below?

   1%   10%   25%   50%   75%   90%   99%.

(b) As in (a), for shaded region B.

The answers to these exercises are on p. A-45.

3. ADDING CHANCES

Suppose that in one class, 10% of the students are freshmen and 20% are sophomores. Then 10% + 20% = 30% of the students are in lower division—freshmen or sophomores. The key point is that nobody can be both a freshman and a sophomore; that is why the percentages can be added. To take an example in the other direction, suppose that in one class, 25% of the students are seniors, and 50% are women. What percentage of the students are seniors or women? From the information given, this question cannot be answered. The percentage could be as small as 50%, or as large as 75%. Adding the percentages isn't legitimate when the two groups overlap: it double-counts the people who happen to be in both groups.

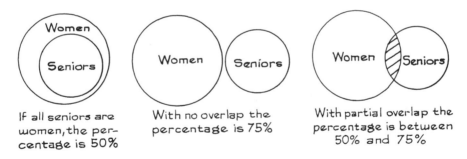

If all seniors are women, the percentage is 50%

With no overlap the percentage is 75%

With partial overlap the percentage is between 50% and 75%

A similar principle applies when figuring chances, but first a definition:

• Two things are said to be *incompatible* when the occurrence of one prevents the occurrence of the other.

• Two things are *compatible* when one does not prevent the other from happening.

The next two examples will illustrate these definitions.

Example 1. A card is dealt off the top of a well-shuffled deck. One outcome is that it is a heart; another, that it is a spade. Are these two outcomes compatible, or incompatible?

Solution. If the card is a heart, it can't be a spade. These two outcomes are incompatible.

Example 2. Someone is about to throw a pair of dice. One die is white with black spots; the other is black with white spots.

Are the following two outcomes compatible, or incompatible?

• The white die lands ace

• The black die lands ace [black die face showing one]

Solution. A white ace does not prevent a black ace. So these two outcomes are compatible.

[white die face] [black die face]

We can now state a general principle for figuring chances; it is called the *addition rule*.

> If two things are incompatible, the chance that at least one of the two will happen is found by adding the two chances. However, it is not legitimate to add the chances of compatible outcomes.

Example 3. A card is dealt off the top of a well-shuffled deck. There is 1 chance in 4 for it to be a heart. There is 1 chance in 4 for it to be a spade. What is the chance for it to be in a major suit (hearts or spades)?

Solution. As in Example 1, these two outcomes—hearts or spades—are incompatible: each one prevents the other. So it is legitimate to add the chances. The chance of getting a card in a major suit is $1/4 + 1/4 = 1/2$.

Example 4. Someone throws a pair of dice. True or false: The chance of getting at least one ace is $1/6 + 1/6 = 1/3$.

Solution. This is false. As in Example 2, imagine one of the dice to be white, the other black. A white ace is compatible with a black ace—one doesn't prevent the other. So it isn't legitimate to add the chances.

To see why adding the chances isn't legitimate here, imagine throwing the pair of dice 600 times. On about 100 throws, the white die will land ace. On about 100 throws, the black die will land ace. But the number of throws with at least one ace will be smaller than 200—because there will be some throws where both dice land aces. These throws would be double-counted in the sum $100 + 100 = 200$. The chance of getting at least one ace looks hard to figure right now, but it is definitely smaller than 1/3. The reason is overlap.

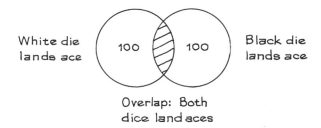

Overlap: Both
dice land aces

Note on terminology: In more technical language, outcomes are called *events,* and incompatible outcomes are said to be *mutually exclusive.*

EXERCISE SET C

1. In poker (five card draw) there is about a 2.1% chance of being dealt three of a kind, like

There is about a 4.6% chance of being dealt two pair, like

What is the chance of getting either three of a kind or two pair?

2. A number is drawn at random from a box. There is a 20% chance for it to be 10 or less. There is a 10% chance for it to be 50 or more. True or false: the chance for it to be between 10 and 50 (exclusive) is 70%.

3. A die will be rolled three times. True or false: the chance of getting at least one ace is $1/6 + 1/6 + 1/6 = 1/2$.

4. A box contains 10 tickets numbered 1 through 10. Five draws will be made at random with replacement from this box. True or false: there are 5 chances in 10 of getting ⑦ at least once.

5. Two cards will be dealt off the top of a well-shuffled deck. The chance that the first one will be a heart is 1/4. The chance that the second will be a spade is 1/4. True or false: the chance that the first one will be a heart or the second one will be a spade equals $1/4 + 1/4$.

The answers to these exercises are on p. A-45.

4. INDEPENDENCE

To begin with an example from demography, in 1970 just over half the population of the United States was female—51%. Also, 42% of the population was aged 35 and over.[3] To find the percentage of the population consisting of women aged 35 and over, it is natural to take

$$51\% \text{ of } 42\% = 0.51 \times 42\% \approx 21\%.$$

But this is a little bit off: the correct answer is 22%. The calculation assumed that the percentage of women was the same across all age groups, and it isn't. Women live longer than men, so the percentage of women is higher in the older age groups: of the people aged 35 and over, 53% are women. (Among those below 35 years of age, less than half are women.) The right calculation is

$$53\% \text{ of } 42\% = 0.53 \times 42\% \approx 22\%.$$

The question of multiplication also comes up when calculating chances. In probability theory, outcomes are said to be *independent* if the chance that one happens remains the same, regardless of how the others turn out. If the chances change, the outcomes are said to be *dependent*.

Example 1. Someone is going to toss a coin twice. If the coin lands heads on the second toss, you win a dollar. Suppose he has just done the first toss.
   (a) If this first toss is heads, what is your chance of winning the dollar?
   (b) If this first toss is tails, what is your chance of winning the dollar?
   (c) Are the tosses independent?

Solution. If the first toss was heads, you have a 50–50 chance to get heads the second time. The chance is the same if the first toss was tails. The chances for the second toss stay the same, however the first toss turns out. This is independence.

Example 2. A box contains three black tickets numbered 1, 2, 3, and three white tickets numbered 1, 2, 3. One ticket will be drawn at random; you have to guess the number on the ticket.

   (a) You catch a glimpse of the ticket as it is drawn out of the box. You cannot make out the number, but you do see that the ticket is black. What is the chance that the number on it will be 2?
   (b) Same, but you see that the ticket is white.
   (c) Are color and number independent?

Solution. If the ticket is black, it has one chance in three to show the number 2. Similarly if it is white. Knowing the color does not change the chances for the number. So color and number are independent.

To see whether two outcomes are dependent or independent, pretend you know the first one, and then calculate the chance of the second one.

If the chance for the second one depends on the first one, the outcomes are dependent.

Example 3. One draw will be made at random from the box

$$\boxed{1}\,\boxed{2}\,\boxed{2}\,\boxed{\blacksquare}\,\boxed{\blacksquare}\,\boxed{\blacksquare}$$

Are color and number independent?

Solution. Pretend you catch a glimpse of the ticket as it is drawn out of the box—enough to catch the color, not enough to make out the number. If you see the ticket is black, there are two chances in three for the number to be 1. If the ticket is white, there is only one chance in three for the number to be 1. The chances for the number change if you know the color. So number and color are dependent.

Example 4. Two draws will be made at random with replacement from the box

$$\boxed{1}\,\boxed{2}\,\boxed{3}\,\boxed{4}\,\boxed{5}$$

(a) Suppose the first draw is $\boxed{1}$. What is the chance of getting a $\boxed{4}$ on the second draw?

(b) Suppose the first draw is $\boxed{2}$. What is the chance of getting $\boxed{4}$ on the second draw?

(c) Are the draws independent?

Solution. Whether the first draw is $\boxed{1}$ or $\boxed{2}$ or anything else, the chance of getting $\boxed{4}$ on the second draw stays the same—one in five. The reason is that the first ticket is replaced, so the second draw is always made from the same box:

$$\boxed{1}\,\boxed{2}\,\boxed{3}\,\boxed{4}\,\boxed{5}$$

The draws are independent.

Example 5. Two draws will be made at random without replacement from the box

$$\boxed{1}\,\boxed{2}\,\boxed{3}\,\boxed{4}\,\boxed{5}$$

Are the draws dependent or independent?

Solution. You have to reason this out the same way. Pretend you know how the first draw turned out. Do the chances for the second draw depend on the first draw? The answer is, yes. If the first draw turns out to be $\boxed{1}$, for example, then the second draw is from the box

$$\boxed{2}\,\boxed{3}\,\boxed{4}\,\boxed{5}$$

There is no chance for the second draw to be $\boxed{1}$. On the other hand, if the first draw turns out to be $\boxed{3}$, then the second draw is from the box

$$\boxed{1}\,\boxed{2}\quad\boxed{4}\,\boxed{5}$$

Now there is one chance in four for the second to be $\boxed{1}$. The draws are dependent.

When drawing at random with replacement, the draws are independent. Without replacement, the draws are dependent.

What does independence of the draws mean? To answer this question, focus on any bet which can be settled on one draw—for instance, that the draw will be 3 or more. Then the chance of winning this bet stays the same from draw to draw—and it doesn't matter how the earlier draws turn out. In this sense, with an honest roulette wheel the plays are independent. By contrast, if the house starts using a magnet when things go badly, the plays would be dependent.

This section discussed independent outcomes; the previous section was about compatible outcomes. How are these two ideas related?

- Independent outcomes are always compatible with each other.
- Dependent outcomes may be compatible, or incompatible; there is no general rule.

We will now introduce a second rule for calculating chances, called the *product rule*.

When two things are independent, the chance that both will happen can be found by multiplying the chances. However, it is not legitimate to multiply the chances of dependent outcomes.

Example 6. A coin is tossed twice. What is the chance of getting heads on the first toss and heads on the second?

Solution. The outcomes are independent, so the chance of getting heads on the first toss and on the second toss can be figured as

$$\text{chance of heads on first toss} \times \text{chance of heads on second toss}$$
$$= 1/2 \times 1/2 = 1/4.$$

What is the logic behind the multiplication? Imagine a long series of weekends spent repeating the basic chance process—tossing a coin twice. On Saturday, you make the first toss, and on Sunday, the second. About 50% of the weekends get off to a good start, with a head on Saturday. Of these promising weekends, only 50% end well, with a head on Sunday. So, the percentage of weekends where you get a head on Saturday and a head on Sunday is 50% of 50% = 25%.

Example 7. A coin will be tossed three times. What is the chance of getting three heads?

Solution. There are three outcomes to think about: getting a head on the first toss, getting a head on the second, getting a head on the third. These are independent, so the chance that they all happen is $1/2 \times 1/2 \times 1/2$, or 1 in 8.

The reasoning is like Example 6—but you need three-day weekends.

Example 8. A box contains one red marble, one green, and one blue. Someone reaches in with his left hand, and draws one marble at random; then, he reaches in with his right hand, and draws a second marble; the draws are made without replacement. True or false:

    (a) With probability 1/3, he gets the red marble in his left hand.

    (b) With probability 1/3, he gets the green marble in his right hand.

    (c) With probability $1/3 \times 1/3 = 1/9$ he gets the red marble in his left hand, and the green marble in his right hand.

Solution. Statements (a) and (b) are true, but (c) is false. The draws are dependent, because they are made without replacement; so the multiplication was not legitimate.

(Here is a correct method for finding the chance. Imagine the person repeating the experiment many times, drawing twice at random without replacement each time. One time in three, he will draw the red marble first. Now there are two marbles left in the box, so on half of these times, the next draw will be green. The chance is 1/2 of $1/3 = 1/6$, not 1/9.)

People often ask, "When do I add the chances, and when do I multiply?" There is no easy answer to this question. It all depends on the chance you are trying to find. Sometimes, you can work out this chance just by looking at it. Other times, it is harder. Then there are two steps to make before doing any arithmetic at all:

    • Try to visualize the chance process that the problem is about— throwing dice, dealing cards, or whatever it is.

    • Identify the outcome whose chance is asked for.

Then, you have to relate this outcome to simpler things whose chances you know. You may want to compute the chance that at least one of these simpler things will happen. Or, you may want to compute the chance that all of these simpler things will happen. If you are dealing with one of these two cases, addition or multiplication may be legitimate.

    • Add the chances when you want to find the probability that at least one of the things will happen—provided no two are compatible.

    • Multiply the chances when you want to find the probability that all of them will happen—provided they are independent.

Sometimes, you may not be able either to add or to multiply. Then, more thinking is needed. (Other strategies for finding chances are dealt with in Chapter 14.)

EXERCISE SET D

1. For each of the following boxes, say whether color and number are dependent or independent.

(a)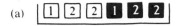

(b)

(c) ![1 2 3 1 2 2]

2. A die is rolled twice. What is the chance of getting two aces?

3. A deck is shuffled and two cards are dealt. There is 1 chance in 4 for the first card to be a club. There is 1 chance in 4 for the second card to be a diamond. True or false: the chance that the first card will be a club and the second card will be a diamond is 1 in 16.

4. A die will be rolled six times. You have the choice
   (i) to win $1 if at least one ace shows:
   (ii) to win $1 if an ace shows on all the rolls.
   Which option offers the better chance of winning? Or are they equivalent?

5. Suppose that in a certain class, there are
   - 80% men and 20% women:
   - 15% freshmen and 85% sophomores.
   (a) How small can the percentage be of sophomore women in the class?
   (b) How large can this percentage be?
   (c) What is the percentage of sophomore women if sex and year are independent?

6. Two draws are made at random without replacement from the box

![1 2 3 4] ;

the first ticket is lost, and nobody knows what was written on it. True or false: In this case the two draws would be independent. Explain.

The answers to these exercises are on pp. A-45–A-46.

5. REVIEW EXERCISES

1. Classify each of the following statements as true, false, or meaningless.
   (a) If something has probability −100%, it can't happen. *meaningless*
   (b) If something has probability 80%, it can be expected to happen about four times as often as its opposite.
   (c) If a coin is tossed twice, the chance of getting exactly one head is 50%. *true*
   (d) If a die is rolled four times, the chance of getting at least one ace is 1/6 + 1/6 + 1/6 + 1/6 = 2/3.

2. Two cards will be dealt off the top of a well-shuffled deck. You have a choice:
   (i) To win $1 if the first is a club.
   (ii) To win $1 if the first is a club, or the second is a diamond (or both).
   Which option is better? Or are they equivalent? Explain.

3. Two cards will be dealt from the top of a well-shuffled deck. You have a choice:
   (i) To win $1 if at least one card is a spade.
   (ii) To win $1 if the first is a spade, or the second is a spade (or both).
   Which option is better? Or are they equivalent? Explain.

4. One hundred draws will be made at random from one of the two boxes shown below. On each draw, you will be paid the number on the ticket, in dollars. Which box do you prefer? About how much do you expect to make, using that box?

   (i)   (ii)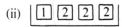

5. In poker (five card draw), there is a 49.8% chance of being dealt one pair or better. A hand which does not even have a pair is called a *bust*. What is the chance of getting a bust?

6. A box has tickets with whole numbers on them. One ticket is drawn at random. The chance it is between 1 and 10 (inclusive) equals 30%; the chance it is between 11 and 20 (inclusive) equals 40%. Find the chance it is between 1 and 20 (inclusive).

7. One ticket will be drawn at random from the box below. Are color and number independent?

8. A coin is tossed six times. Two possible sequences of results are
   (i) H T T H T H   (ii) H H H H H H   H = heads   T = tails.
   (The coin must land H or T in the order given.) Which of the following is correct? Explain.[4]
   (a) Sequence (i) is more likely.
   (b) Sequence (ii) is more likely.
   (c) Both sequences are equally likely.

9. A die is rolled three times. What is the chance that all three rolls show three or more spots?

10. A die is rolled three times. Find the chance of getting three sixes. Find the chance of not getting three sixes.

6. SUMMARY

1. The *frequency theory* of chance applies most directly to chance processes which can be repeated over and over again, independently and under the same conditions—like games of chance.

2. The chance of something says about what percentage of the time it is expected to happen, when the basic process is repeated over and over again.

3. Chances are between 0% and 100%. Impossibility is represented by 0%, certainty by 100%.

4. The chance of something equals 100% minus the chance of its opposite.

5. If two things are incompatible, the chance that at least one will happen equals the sum of the individual chances.

6. If two things are independent, the chance that both happen equals the product of the individual chances.

14

More about Chance

Some of the Problems about Chance having a great appearance of Simplicity, the Mind is easily drawn into a belief, that their Solution may be attained by the meer Strength of natural good Sense; which generally proving otherwise and the Mistakes occasioned thereby being not unfrequent, 'tis presumed that a Book of this Kind, which teaches to distinguish Truth from what seems so nearly to resemble it, will be looked upon as a help to good Reasoning.

—ABRAHAM DE MOIVRE (ENGLAND, 1667–1754)[1]

1. LISTING THE WAYS

This chapter and the next cover special topics in probability, which will not be used later in the book.

When trying to figure chances, it is sometimes very helpful to list all the possible ways that a chance process can turn out. If this is too hard, writing down a few typical ones is a good start. Figuring out their total number is usually possible—and often almost as helpful as listing them. The idea will be illustrated with some examples.

Example 1. A die is thrown. What is the chance of getting an even number of spots (2, 4, or 6)?

Solution. The chance process here consists of throwing the die, and there are six possible ways for it to fall:

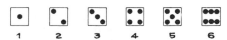

Each of these six ways has the same chance—1 in 6. There are 3 ways to get an even number, so the chance is 3 in 6.

Example 2. A pair of dice are thrown. What is the chance of getting a total of 4 spots?

Solution. The chance process here consists of throwing the two dice. What matters is the number of spots shown by each die. To keep them separate, imagine that one die is white and the other black. One way for the dice to fall is

This means the white die showed two spots, and the black die showed three.

How many ways are there for the two dice to fall? To begin with, the white die can fall in any one of six ways:

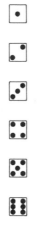

When the white die shows ⬚, say, there are still six possible ways for the black die to fall:

We now have six of the possible ways that the two dice can fall. These ways are shown in the first row of Figure 1. Similarly, the second row shows another six ways for the dice to fall—with the white die showing ⬚. And so on. The table shows there are $6 \times 6 = 36$ possible ways for the dice to fall. They are all equally likely, so each has 1 chance in 36. Counting up, there are 3 ways to get a total of four spots:

So the chance is 3 in 36.

Many people find this argument disturbing, because they feel there should be two outcomes with the white die and the black die showing aces:

But we are not counting position, just the number of spots on each die. People who want to count position are free to do so—provided they are consistent. For instance, they will also need two outcomes where the white die shows an ace and the black die a deuce. In other words, their version of Figure 1 needs 72 outcomes, not 36. And on this basis, the chance of throwing a total of four spots is 6 in 72—which is the same as 3 in 36. This complication is unnecessary.

Figure 1. Throwing a pair of dice. There are 36 ways for the dice to fall, shown in the body of the diagram; all are equally likely.

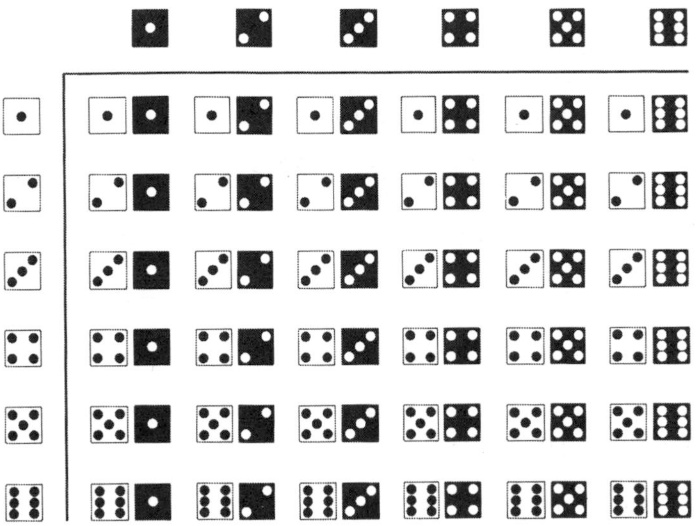

What about three dice? A three-dimensional picture like Figure 1 would be a bit much to absorb. However, the same kind of reasoning can be used, as will be shown by a historical example. In the seventeenth century, Italian gamblers used to bet on the total number of spots rolled with three dice. They believed that the chance of rolling a total of 9 ought to equal the chance of rolling a total of 10. For instance, they said, one combination with a total of 9 spots is

1 spot on one die, 2 spots on another die, 6 spots on the third die.

This can be abbreviated as "1 2 6." There are altogether six combinations for 9:

126   135   144   234   225   333.

Similarly, they found six combinations for 10:

145   136   226   235   244   334.

Thus, argued the gamblers, 9 and 10 should by rights have the same chance. However, experience showed that 10 came up a bit more often than 9.

They asked Galileo for help with this contradiction, and he reasoned as follows. Color one of the dice white, another one grey, and another one black, so they can be kept apart. This won't affect the chances. Now, how many ways can the three dice fall? The white die can land in 6 ways. Corresponding to each of them, the grey die can land in 6 ways, making 6 × 6 possibilities for these two dice. And corresponding to each of these possibilities, there are still 6 for the black die. Altogether, there are $6 \times 6 \times 6 = 6^3$ ways for three dice to land. (With four dice, there would be 6^4; with five dice, 6^5; and so on.)

Now $6^3 = 216$ is a lot of ways for three dice to fall. But Galileo sat down and listed them. Then he went through his list and counted the ones with a total of 9 spots; he found 25. And he found 27 ways to get a total of 10 spots. He concluded that the chance of rolling 9 is $25/216 \approx 11.6\%$, while the chance of rolling 10 is $27/216 \approx 12.5\%$.

What was the trouble with the gamblers' argument? Just that they didn't get down to the different ways for the dice to land. For instance, the triplet 3 3 3 for 9 corresponds to only one way for the dice to land:

But the triplet 3 3 4 for 10 corresponds to three ways for the dice to land:

The gamblers' argument can be corrected as in Table 1.

Table 1. The chance of getting 9 or 10 spots with three dice, by the gamblers' argument with Galileo's correction.

Triplets for 9	Number of ways to roll each triplet	Triplets for 10	Number of ways to roll each triplet
1 2 6	6	1 4 5	6
1 3 5	6	1 3 6	6
1 4 4	3	2 2 6	3
2 3 4	6	2 3 5	6
2 2 5	3	2 4 4	3
3 3 3	1	3 3 4	3
Total	25	Total	27

EXERCISE SET A

1. Looking at Figure 1, make a list of the ways to roll a total of five spots. What is the chance of throwing a total of five spots with two dice?

2. A pair of dice is thrown 1,000 times. What total should appear most often? What totals should appear least often?

3. Two draws will be made at random with replacement from the box

$$\boxed{1}\ \boxed{2}\ \boxed{3}\ \boxed{4}\ \boxed{5}$$

Draw a picture like Figure 1 to represent all possible results. How many are there? What is the chance that the sum of the two draws turns out to equal 6?

The answers to these exercises are on p. A-46.

Galileo (Italy, 1564–1642).

From the Wolff-Leavenworth collection,
George Arents Research Library, Syracuse University.

2. ARE REAL DICE FAIR?

The previous chapter said that when a die is rolled, it is equally likely to
show any of its six faces. This assumes an ideal die which is perfectly fair or
symmetric. It is like ignoring friction in the study of the laws of motion—the
results are a good first approximation. In this light, what does Galileo's
calculation say about real dice?

- For real dice, the 216 possible ways the three dice can land are very
  close to being equally likely.
- If these ways were equally likely, the chance of rolling a total of 9 spots
  would be exactly 25 in 216.
- So for real dice, the chance of rolling a total of 9 spots is just about
  25 in 216.

For loaded dice, the calculations would be badly off. But ordinary dice, coins,
roulette wheels, and the like are very close to fair—in the sense that all the
outcomes are equally likely. (The games of chance based on these fair mechanisms may be quite unfair, as will be seen later.) In a similar way, when you
are asked to draw a ticket at random from a box of tickets, assume each ticket
is equally likely to be drawn. If the tickets are close to the same size, shape,
and texture, and the box is well shaken, this is quite a reasonable approximation to an actual drawing.

3. THE PARADOX OF THE CHEVALIER DE MÉRÉ

In the seventeenth century, French gamblers used to bet on the event
that in four rolls of a die, at least one ace would turn up; an ace is ⚀. In
another game, they bet on the event that in twenty-four rolls of a pair of dice,
at least one double-ace would turn up: a double-ace is a pair of dice which

shows $\boxed{\bullet}\boxed{\bullet}$. The Chevalier de Méré, a French nobleman of the period, thought these two events were equally likely. He reasoned this way about the first game:

- In one roll of a die, I have 1/6 of a chance to get an ace.
- So in four rolls, I have $4 \times 1/6 = 2/3$ of a chance to get at least one ace.

His reasoning for the second game was similar:

- In one roll of a pair of dice, I have 1/36 of a chance to get a double-ace.
- So in twenty-four rolls, I must have $24 \times 1/36 = 2/3$ of a chance to get at least one double-ace.

By this argument, both chances were the same, namely 2 in 3. But experience showed the first event to be a bit more likely than the second. This contradiction became known as the *Paradox of the Chevalier de Méré*.

Blaise Pascal (France 1623–1662).

From the Wolff-Leavenworth Collection, George Arents Research Library, Syracuse University.

Pierre de Fermat (France, 1601–1665).

From the *Oeuvres complètes*, the Library, University of California, Berkeley.

De Méré asked the philosopher Blaise Pascal about the problem, and Pascal solved it with the help of his friend, Pierre de Fermat. Fermat was a judge and a member of parliament, who is remembered today for what he did after hours—mathematical research. Fermat and Pascal saw that de Méré's reasoning was false—he had used the addition rule for compatible outcomes. After all, it is possible to get an ace on both the first and the second roll of a die. In fact, pushing de Méré's argument a little further, it shows the chance of getting an ace in six rolls of a die to be 6/6; but the chance of getting an ace in six rolls certainly isn't 100%.

Now the question is how to calculate the chances correctly. Pascal and Fermat solved this problem, with a typically indirect piece of mathematical reasoning—the kind that always leaves nonmathematicians feeling a bit cheated. Of course, in this problem a direct attack could easily bog down: with four rolls of one die, there are $6^4 = 1{,}296$ outcomes to worry about; with twenty-four rolls of a pair of dice, there are $36^{24} \approx 2.2 \times 10^{37}$ outcomes.

Let's look at the first game first, Pascal might have said to Fermat.[2]

Fermat. OK. The chance of winning is hard to compute, so let's work out the chance of the opposite event—losing. Then

$$\text{chance of winning} = 100\% - \text{chance of losing.}$$

[Ed. note. See p. 205 above.]

Pascal. Fine. The gambler loses when none of the four rolls shows an ace. But how do you work out the chances?

Fermat. It does look complicated. Let's start with one roll. What's the chance that the first roll doesn't show an ace?

Pascal. It has to show something from 2 through 6, so the chance is 5/6.

Fermat. Right. Now, what's the chance that the first two rolls don't show aces? Remember, they're independent.

Pascal. Then we can use the product rule: the chance that the first roll doesn't give an ace and the second doesn't give an ace equals $(5/6) \times (5/6) = (5/6)^2$.

Fermat. What about three rolls?

Pascal. It looks like $(5/6) \times (5/6) \times (5/6) = (5/6)^3$.

Fermat. Right. Now what about four rolls?

Pascal. Must be $(5/6)^4$.

Fermat. Yes, and that's about 0.482, or 48.2%.

Pascal. So there is a 48.2% chance of losing. Now

$$\begin{aligned} \text{chance of winning} &= 100\% - \text{chance of losing} \\ &= 100\% - 48.2\% = 51.8\%. \end{aligned}$$

Fermat. That settles the first game. The chance of winning is a little over 50%. Now what about the second?

Pascal. Well, in one roll of a pair of dice, there is 1 chance in 36 of getting a double-ace, and 35 chances in 36 of not getting a double-ace. By the product rule, in 24 rolls of a pair of dice, the chance of getting no double-aces must be

$$(35/36)^{24}.$$

Fermat. That's about 50.9%. So we have the chance of losing. Now

$$\begin{aligned} \text{chance of winning} &= 100\% - \text{chance of losing} \\ &= 100\% - 50.9\% = 49.1\%. \end{aligned}$$

Pascal. Yes, and that's a bit less than 50%. I guess that's why you win the second game a bit less frequently than the first, just like de Méré said. He must have seen a lot of dice rolled to pick up such a small difference.

EXERCISE SET B

1. A box contains four tickets, one marked with a star, and the other three blank:

   Two draws are made at random with replacement from this box.
   (a) What is the chance of getting a blank card on the first draw?
   (b) What is the chance of getting a blank card on the second draw?
   (c) What is the chance of getting a blank card on the first draw and a blank card on the second draw?
   (d) What is the chance of not getting the star in the two draws?
   (e) What is the chance of getting the star at least once in the two draws?

2. A die is rolled six times. What is the chance of getting at least one ace?

3. A pair of dice is rolled thirty-six times. What is the chance of getting at least one double-ace?

4. In eighteenth-century England, de Moivre reports that people played a game similar to modern roulette. It was called "Royal Oak." There were 32 "points" or numbered pockets on a table. A ball was thrown in such a way that it landed in each pocket with an equal chance, 1 in 32. If you bet £1 on a point and it came up, you got your stake of £1 back, together with winnings of £27. If your point didn't come up, you lost your stake. The players (or "Adventurers," as de Moivre called them) complained that the game was unfair, and they should have won £31 if their point came up. They were right, as will be explained later.[3]

   De Moivre continues:

   "The Master of the Ball maintained they had no reason to complain; since he would undertake that any particular point of the Ball should come up in Two-and-Twenty Throws; of this he would offer to lay a Wager, and actually laid it when required. The seeming contradiction between the Odds of One-and-Thirty to One, and Twenty-two Throws for any [point] to come up, so perplexed the Adventurers, that they begun to think the Advantage was on their side; for which reason they played on and continued to lose."

   What is the chance that the point 17, say, will come up in Two-and-Twenty Throws? (The Master of the Ball laid this wager at even money, so if the chance is over 50%, he shows a profit here too.)

5. In his novel *Bomber*, Len Deighton argues that a World War II pilot had a 2% chance of being shot down on each mission. So in 50 missions he is "mathematically certain" to be shot down: $50 \times 2\% = 100\%$. Is this a good argument? (Hint: To make chance calculations, it is necessary to see how the situation being considered resembles a game of chance. The analogy here is to draw 50 times at random with replacement from the box

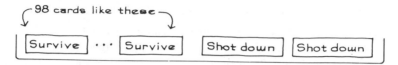

Surviving 50 missions is like drawing the card "survive" 50 times. What is the chance?)

The answers to these exercises are on p. A-46.

4. THE COLLINS CASE

People v. *Collins* is an actual law case in which there was a major statistical issue. The facts were described by the court as follows.[4]

On June 18, 1964, about 11:30 A.M. Mrs. Juanita Brooks, who had been shopping, was walking home along an alley in the San Pedro area of the City of Los Angeles. She was pulling behind her a wicker basket carryall containing groceries and had her purse on top of the packages. She was using a cane. As she stooped down to pick up an empty carton, she was suddenly pushed to the ground by a person whom she neither saw nor heard approach. She was stunned by the fall and felt some pain. She managed to look up and saw a young woman running from the scene. According to Mrs. Brooks the latter appeared to weigh about 145 pounds, was wearing "something dark", and had hair "between a dark blond and a light blond", but lighter than the color of defendant Janet Collins' hair as it appeared at trial. Immediately after the incident, Mrs. Brooks discovered that her purse, containing betweeen $35 and $40, was missing.

About the same time as the robbery, John Bass, who lived on the street at the end of the alley, was in front of his house watering his lawn. His attention was attracted by "a lot of crying and screaming" coming from the alley. As he looked in that direction, he saw a woman run out of the alley and enter a yellow automobile parked across the street from him. He was unable to give the make of the car. The car started off immediately and pulled wide around another parked vehicle so that in the narrow street it passed within six feet of Bass. The latter then saw that it was being driven by a male Negro, wearing a mustache and beard. At the trial Bass identified defendant as the driver of the yellow automobile. However, an attempt was made to impeach his identification by his admission that at the preliminary hearing he testified to an uncertain identification at the police lineup shortly after the attack on Mrs. Brooks, when defendant was beardless.

In his testimony Bass described the woman who ran from the alley as a Caucasian, slightly over five feet tall, of ordinary build, with her hair in a dark blond ponytail, and wearing dark clothing. He further testified that her ponytail was "just like" one which Janet had in a police photograph taken on June 22, 1964.

The prosecutor then had a mathematics instructor at a local state college testify as to the product rule, apparently without paying much attention to the condition of independence. After this testimony, the prosecution assumed the following chances:

Yellow automobile	1/10	Girl with blond hair	1/3
Man with mustache	1/4	Negro man with beard	1/10
Girl with ponytail	1/10	Interracial couple in car	1/1,000

When multiplied together, these chances come to about 1 in 12,000,000. According to the prosecution, this procedure gave the chance "that any [other] couple possessed the distinctive characteristics of the defendants." If no other couple possessed these characteristics, the defendants were guilty.

The jury convicted. On appeal, the Supreme Court of California reversed the verdict, for two reasons. First, it found no evidence to support the assumed values for any of the six chances. Second, it found no evidence to support the assumed independence of the six factors. (Some of them, like "Negro man with beard" and "interracial couple in car" are clearly dependent.) In other words, the basic conditions for using the product rule were not satisfied. There is a third and even more fundamental objection. Probability calculations like the product rule were developed for dealing with games of chance. They have been tested on sequences of similar events. In fact, according to the frequency theory, the chance of an event is the percentage of time it occurs in the long run. The prosecutor was trying to apply this theory to a unique event: something that either happened—or didn't happen—on June 18, 1964, at 11:30 A.M. What does chance mean, in this new context? Was the prosecutor thinking of a series of muggings? or possible variations on Los Angeles? or a sequence of a hypothetical universes? Just what was he really talking about? It was up to the prosecutor to answer these questions, and to show that the theory applied to his situation. He didn't even bother trying. The prosecutor's reasoning is an example of what not to do with statistics.[5]

5. REVIEW EXERCISES

Review exercises may cover material from previous chapters.

1. Three cards will be dealt from the top of a well-shuffled deck. You have a choice:
   (i) To win $1 if all three cards are aces.
   (ii) To win $1 if the first card is an ace, and the second card is an ace, and the third card is an ace.
   Which option is better? Or are they the same?

2. Three cards will be dealt off the top of a well-shuffled deck. You have two options:
   (i) To win $1 if the first card is an ace.
   (ii) To win $1 if the first card is an ace, and the second card is an ace, and the third card is an ace.
   Which option is better? Or are they the same?

3. A hundred draws will be made at random with replacement from one of the two boxes shown below. On each draw, you will be paid the amount shown on the ticket, in dollars. Which box is better? If you use that box, about how much can you expect to win?

(i) $\boxed{\boxed{1}\ \boxed{2}}$ (ii) $\boxed{\boxed{1}\ \boxed{1}\ \boxed{1}\ \boxed{2}}$

4. You have two options:
   (i) A die will be rolled 100 times. Each time it shows an ace, you win $1; on the other rolls, you win nothing.
   (ii) One hundred draws will be made at random with replacement from the box $\boxed{\boxed{1}\ \boxed{0}\ \boxed{0}\ \boxed{0}\ \boxed{0}\ \boxed{0}}$. On each draw, you will be paid the amount shown on the ticket, in dollars.
   Which option is better? Or are they the same?

5. You have two options:
   (i) You toss a coin 100 times; on each toss, if it lands heads you win $1, if it lands tails you lose $1.
   (ii) You draw 100 times at random with replacement from the box $\boxed{\boxed{-1}\ \boxed{1}\ \boxed{1}}$; on each draw, you are paid (in dollars) the number on the ticket, so $\boxed{-1}$ costs you a dollar.
   Which option is better? Or are they the same?

6. Fill in the blanks, using one word from each pair below, to make up two true sentences. Write out both sentences.
   "If two things are ___(i)___, and you want to find the chance that ___(ii)___ will happen, you can ___(iii)___ the chances."
   (i) incompatible, independent.
   (ii) both, at least one.
   (iii) add, multiply.

7. Three draws will be made at random with replacement from the box $\boxed{\boxed{1}\ \boxed{3}\ \boxed{5}\ \boxed{7}}$. How many possible ways are there for the three draws to turn out?

8. One ticket will be drawn at random from each of the two boxes shown below:

   (A) $\boxed{\boxed{1}\ \boxed{2}\ \boxed{3}}$ (B) $\boxed{\boxed{1}\ \boxed{2}\ \boxed{3}\ \boxed{4}}$

   Find the chance that
   (a) the number drawn from A equals the one from B.
   (b) the number drawn from A is larger than the one from B.
   (c) the number drawn from A is smaller than the one from B.
   (Hint: Draw a picture of the outcomes, like Figure 1.)

9. Two draws are made at random with replacement from the box $\boxed{\boxed{1}\ \boxed{2}\ \boxed{3}\ \boxed{4}\ \boxed{5}}$. What is the chance that the sum of the draws will equal 4? (Hint: Draw a picture of the outcomes, like Figure 1.)

10. A die will be rolled three times. Find the chance that the total number of spots will be 3. That the total will be 4. (Hint: Write down some outcomes, like Galileo did.)

11. Two draws are made at random with replacement from the box ⎢ 1 2 3 ⎢ . What is the chance of getting 3 at least once?

12. A die is rolled twelve times. What is the chance of getting at least one ace?

6. SUMMARY

1. When figuring out chances, one helpful strategy is to write down a complete list of all the possible ways that the chance process can turn out. If this is too hard, at least write down a few typical ones, and figure out how many there are.

2. If you are having trouble working out the chance of an outcome, try to figure out the chance of its opposite, and then subtract from 100%.

3. The addition rule should be used only for outcomes which are incompatible. The product rule should be used only for independent outcomes.

4. The mathematical theory of chance only applies in certain cases. Using it where it doesn't apply can lead to ridiculous results.

15

The Binomial Coefficients

Man is a reed, but a reed that thinks.

—BLAISE PASCAL (FRANCE, 1623–1662)

1. INTRODUCTION

This chapter covers a special topic in probability, which will not be used later in the book.

• A coin is tossed four times. What is the chance of getting exactly one head?

• A die is rolled ten times. What is the chance of getting exactly three aces?

• A box contains one red marble and nine green ones. Five draws are made at random with replacement. What is the chance that exactly two draws will be red?

These questions are all very similar, and can be solved by the same technique, using what are called *binomial coefficients,* discovered by Pascal and Newton. The method will be illustrated on the marbles. The problem is to find the chance of getting two reds (no more and no less) in five draws from the box—so the other three draws must be green. One way this can happen is that the first two draws are red and the final three are green. Taking R for red and G for green, this possibility may be expressed by

R R G G G

Isaac Newton (England, 1642–1727).

From the Wolff-Leavenworth Collection, George Arents Research Library,
Syracuse University

Of course, there are many other ways it can happen that two will be red. For example, the second and the fifth draws might be red, while all the rest are green:

G R G G R

To find the chance, we must find all the possible ways, calculate the chance of each, and then use the addition rule to add up all these chances. The first task seems formidable, so postpone it for a moment and turn to the second one. The chance of the pattern R R G G G is

$$1/10 \times 1/10 \times 9/10 \times 9/10 \times 9/10 = (1/10)^2 \, (9/10)^3$$

This follows from the product rule: the draws are independent. On each draw, the chance of red is 1/10, the chance of green is 9/10.

Similarly, the chance of the pattern G R G G R equals

$$9/10 \times 1/10 \times 9/10 \times 9/10 \times 1/10 = (1/10)^2 \, (9/10)^3$$

The pattern R R G G G has the same chance as the pattern G R G G R. In fact, each pattern with two reds and three greens has the same chance, since the two R's will contribute $(1/10)^2$ to the product, and the three G's will contribute $(9/10)^3$ to the product. It follows that the sum of the chances of all the patterns will be just the number of patterns times the common chance $(1/10)^2(9/10)^3$. So, all that is left to finish the calculation is to find out how many patterns there are. This will take a little effort.

Each pattern is specified by writing down in a row two R's and three G's in some order. All that is necessary is to find the number of different ways of doing this. There is a simple formula which gives the number of different ways of arranging two R's and three G's in a row. It is

$$\frac{5 \times 4 \times 3 \times 2 \times 1}{(2 \times 1) \times (3 \times 2 \times 1)}$$

Doing all the cancellations, this is equal to 10. There are 10 different patterns with two R's and three G's. Thus, the chance of drawing exactly two reds is

$$10(1/10)^2(9/10)^3 \approx 7\%$$

The formula gives a *binomial coefficient*. But it is rather messy. Mathematicians get around this by introducing convenient notation. In this case, they put the ordinary exclamation mark (!) to the right of a number to indicate the result of multiplying together the number and all the positive numbers which precede it. For example,

$$1! = 1$$
$$2! = 2 \times 1 = 2$$
$$3! = 3 \times 2 \times 1 = 6$$
$$4! = 4 \times 3 \times 2 \times 1 = 24$$

and so on. The exclamation mark is read "factorial," so that $4! = 24$ reads "four-factorial equals twenty-four."

Using this notation, the binomial coefficient becomes more readable:

$$\frac{5!}{2!3!}$$

Remember what this represents: the number of different ways of arranging two R's and three G's in a row. The number 5 in the numerator is the sum of 2 and 3 in the denominator. Binomial coefficients always take this form. For example, the number of ways to arrange four R's and one G in a row is the binomial coefficient

$$\frac{5!}{4!1!} = 5$$

The patterns are

RRRRG   RRRGR   RRGRR   RGRRR   GRRRR

How many ways are there to arrange five R's and zero G's in a row? There is only one way: R R R R R. Applying the formula mechanically gives

$$\frac{5!}{5!0!}$$

If the formula is to work, the binomial coefficient must equal 1. But we have not yet said what 0! means. It is a convention of mathematics that $0! = 1$. With this convention, the binomial coefficient does equal 1.

Binomial coefficients and factorials get very large very quickly. For instance, the number of ways to arrange ten R's and ten G's in a row is given by the binomial coefficient

$$\frac{20!}{10!10!} = 184,756$$

However, there was a lot of cancellation going on: $10! = 3,628,800$; and $20! \approx 2 \times 10^{18}$, or 2 followed by 18 zeros. (A trillion is 1 followed by 12 zeros.)

EXERCISE SET A

1. Find the number of different ways of arranging one R and three G's in a row. Write out all the patterns.

2. Find the number of different ways of arranging two R's and two G's in a row. Write out all the patterns.

3. A box contains one red ball and five green ones. Four draws are made at random with replacement from the box. Find the chance that:
   (a) a red ball is never drawn.
   (b) a red ball appears exactly once.
   (c) a red ball appears exactly twice.
   (d) a red ball appears exactly three times.
   (e) a red ball appears on all the draws.
   (f) a red ball appears at least twice.

4. A die is rolled four times. Find the chance that:
   (a) an ace (one dot) never appears.
   (b) an ace appears exactly once.
   (c) an ace appears exactly twice.

5. A coin is tossed ten times. Find the chance of obtaining exactly five heads. Find the chance of obtaining between four and six heads inclusive.

6. A coin is tossed four times. What is the chance of getting exactly one head?

7. A die is rolled ten times. What is the chance of getting exactly three aces?

The answers to these exercises are on pp. A-46–A-47.

2. THE BINOMIAL FORMULA

The reasoning of section 1 can be summarized by what is called the *binomial formula*. Suppose a chance process is carried out in stages, as a sequence of trials. (An example would be rolling a die ten times; each roll counts as a trial.) There is an event of interest which may or may not occur at each trial. (The die may or may not land ace.) The problem is to calculate the chance that the event will occur some given number of times. (A die is rolled ten times; find the chance of getting two aces.)

> The chance that an event will occur exactly k times out of n is given by the binomial formula:
>
> $$\frac{n!}{k!(n-k)!}\, p^k(1-p)^{n-k}.$$
>
> When using this formula:
> - n is the number of trials;
> - the exact value of n must be fixed in advance;
> - p is the probability that the event will occur on any particular trial;
> - p must be the same from trial to trial;
> - the trials must be independent.

Example 1. A die is rolled ten times. What is the chance of getting exactly two aces?

Solution. The number of trials is fixed in advance. It is 10. So $n = 10$. The event of interest is rolling an ace. The probability of rolling an ace is the same from trial to trial. It is 1/6. So $p = 1/6$. The trials are independent. So the binomial formula can be used. The answer is

$$\frac{10!}{2!\,8!}\,(1/6)^2\,(5/6)^8 \approx 29\%.$$

Example 2. A die is rolled until it first lands six. If this can be done using the binomial formula, find the chance of getting two aces. If not, why not?

Solution. The number of trials is not fixed in advance. It could be 1, if the die lands six right away. Or it could be 2, if the die lands five then six. Or it could be 3. Or anything. So the binomial formula does not apply.

Example 3. Four draws are made at random without replacement from the box

$$\boxed{\;\boxed{1}\;\boxed{1}\;\boxed{3}\;\boxed{3}\;\boxed{5}\;\boxed{5}\;}$$

If this can be done using the binomial formula, find the chance of drawing exactly two $\boxed{1}$'s. If not, why not?

Solution. The trials are dependent, so the binomial formula does not apply.

Example 4. Ten draws are made at random with replacement from the box in Example 3. However, just before the last draw is made, whatever has gone on, the ticket $\boxed{5}$ is removed from the box. If this can be done using the binomial formula, find the chance of drawing exactly two $\boxed{1}$'s. If not, why not?

Solution. In this example, n is fixed in advance and the trials are independent. However, p changes at the last trial from 2/6 to 2/5. So the binomial formula does not apply.

3. REVIEW EXERCISES

1. A die will be rolled four times. What is the chance of obtaining exactly two sixes?

2. A box contains four red marbles and two green ones. Three draws are made without replacement. Find the chance of getting the two green marbles, if this can be done using the binomial formula. If not, why not?

3. There are ten people in a club.[1] One person makes up a list of all the possible committees with two members. Another person makes up a list of all the possible committees with five members. Which list is longer?

4. A coin is tossed eight times. True or false: The chance it will land heads exactly four times equals 50%.

5. A die will be rolled six times. The chance it lands neither ace nor deuce can be found by one of the following calculations. Which one, and why?

$$(i) \ (2/6)^2 \ (4/6)^4 \qquad (ii) \ \frac{6!}{4!2!} \ (2/6)^4 \ (4/6)^2 \qquad (iii) \ (4/6)^6$$

6. It is claimed that a vitamin supplement helps kangaroos to learn to run a special maze with high walls. To test whether this is true, sixteen kangaroos are divided up into eight pairs. In each pair, one kangaroo is selected at random to receive the vitamin supplement; the other is fed a normal diet. The kangaroos are then timed as they learn to run the maze. In six of the eight pairs, the treated kangaroo learns to run the maze more quickly than its untreated partner. If in fact the vitamin supplement has no effect, so that each animal of the pair is equally likely to be the quicker, what is the probability that six or more of the treated animals would learn the maze more quickly than their untreated partners, just by chance?

7. Of families with six children, what proportion have three boys and three girls? You may assume that the sex of a child is determined as if by drawing at random with replacement from[2]

$$\boxed{\boxed{M} \ \boxed{F}} \quad M = male, \quad F = female.$$

8. A quiz consists of twenty true-false questions. Find the chance that someone who knows the correct answer to ten of the questions, but answers the remaining ones by tossing a coin, will obtain a score of at least 85% on the quiz.

4. SUMMARY

1. The *binomial coefficient* $\frac{n!}{k!(n-k)!}$ is the number of ways to arrange n objects in a row, when k are alike of one kind and $n-k$ are alike of another kind (for instance, red and blue marbles).

2. The chance that an event will occur exactly k times out of n is given by the *binomial formula*

$$\frac{n!}{k!(n-k)!} \ p^k(1-p)^{n-k}.$$

In using this formula,
   • n is the number of trials;
   • the exact value of n must be fixed in advance;
   • p is the chance that the event will occur on any particular trial;
   • p must be the same from trial to trial;
   • the trials must be independent.

PART V

Chance
Variability

16

The Law of Averages

The roulette wheel has neither conscience nor memory.
—JOSEPH BERTRAND (FRENCH MATHEMATICIAN, 1822–1900)

1. WHAT DOES THE LAW OF AVERAGES SAY?

A coin lands heads with chance 50%. After many tosses, the number of heads should equal the number of tails, shouldn't it? Isn't that what the law of averages says? John Kerrich, an English mathematician, found out the hard way. He was teaching at the University of Copenhagen when World War II broke out, and two days before he was scheduled to fly home, the Germans invaded Denmark. Kerrich spent the rest of the war interned at a camp in Jutland, and to pass the time he carried out a series of experiments in probability theory.[1] In one, he tossed a coin ten thousand times. With his permission, some of the results are summarized in Table 1 and Figure 1. What do Kerrich's results say about the law of averages? To find out, let's pretend that at the end of World War II, Kerrich was invited to demonstrate the law of averages to the King of Denmark. He is discussing the invitation with his assistant.[2]

Assistant. So you're going to tell the King about the law of averages.
Kerrich. Right.
Assistant. But what is there to tell? I mean, everyone knows about the law of averages, don't they?

Kerrich.   OK. Tell me what the law of averages says.

Assistant.   Well, suppose you're tossing a coin. If you get a lot of heads, then tails start coming up. Or if you get too many heads, the chance for tails goes up. In the long run, the number of heads and the number of tails even out.

Kerrich.   It's not true.

Assistant.   What do you mean, it's not true?

Kerrich.   I mean, what you said is all wrong. First of all, with a fair coin the chance for heads stays at 50%, no matter what happens. Whether there are two heads in a row or twenty, the chance of getting a head next time is still 50%.

Assistant.   I don't believe it.

Kerrich.   All right. Take a run of four heads, for example. I went through the record of my first 2,000 tosses. In 130 cases, the coin landed heads four times in a row; 69 of these runs were followed by a head, and only 61 by a tail. A run of heads just doesn't make tails more likely next time.

Assistant.   You're always telling me these things I don't believe. What are you going to tell the King?

Kerrich.   Well, I tossed the coin ten thousand times, and I got about five thousand heads. The exact number was 5,067, the difference of 67 is less than 1% of the number of tosses. I have the record here in Table 1.

Assistant.   Yes, but 67 heads is a lot of heads. The King won't be impressed, if that's the best the law of averages can do.

Kerrich.   So what do you suggest?

Assistant.   Toss the coin another ten thousand times. With twenty thousand tosses, the number of heads should be quite a bit closer to the

Table 1. John Kerrich's coin-tossing experiment. The first column shows the number of tosses. The second shows the number of heads. The third shows the difference

number of heads − half the number of tosses.

Number of tosses	Number of heads	Difference	Number of tosses	Number of heads	Difference
10	4	−1	600	312	12
20	10	0	700	368	18
30	17	2	800	413	13
40	21	1	900	458	8
50	25	0	1,000	502	2
60	29	−1	2,000	1,013	13
70	32	−3	3,000	1,510	10
80	35	−5	4,000	2,029	29
90	40	−5	5,000	2,533	33
100	44	−6	6,000	3,009	9
200	98	−2	7,000	3,516	16
300	146	−4	8,000	4,034	34
400	199	−1	9,000	4,538	38
500	255	5	10,000	5,067	67

expected number. After all, eventually the number of heads and the number of tails have to even out, right?

Kerrich.   You said that before, and it's all wrong. Look at Table 1. In a thousand tosses, the difference between the number of heads and the expected number was 2. With two thousand tosses, the difference went up to 13.

Assistant.   That was just a fluke. By toss 3,000, the difference was only 10.

Kerrich.   That's just another fluke. At toss 4,000, the difference was 29. At 5,000, it was 33. Sure, it dropped back to 9 at toss 6,000, but look at Figure 1: the chance error is climbing pretty steadily from 1,000 to 10,000 tosses, and it's going straight up at the end.

Assistant.   So where's the law of averages?

Kerrich.   With a large number of tosses, the size of the difference between the number of heads and the expected number is likely to be quite large in absolute terms.[3] But in percentage terms—relative to the number of tosses—the difference is likely to be quite small. That's the law of averages. Just like I said, 67 is only a small fraction of 10,000.

Assistant.   I don't understand.

Kerrich.   Look. In 10,000 tosses you expect to get 5,000 heads, right?

Assistant.   Right.

Kerrich.   But not exactly. You only expect to get around 5,000 heads. I mean, you could just as well get 5,001 or 4,998 or 5,007. The amount off 5,000 is what we call "chance error."

Figure 1. Kerrich's coin-tossing experiment. A plot of the "chance error"

number of heads − half the number of tosses

against the number of tosses. As the number of tosses goes up, the size of the chance error tends to go up. The horizontal axis is not drawn to scale.

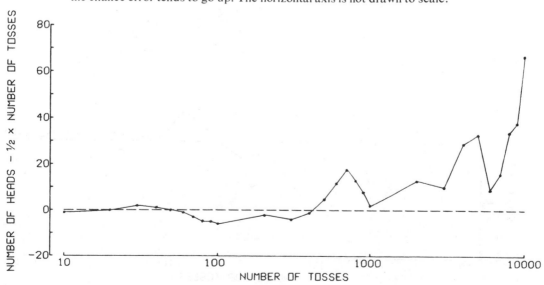

Assistant. Could you be more specific?

Kerrich.   Let me write an equation:

number of heads = half the number of tosses + chance error.

This error is likely to be large in absolute terms, but small in percentage terms—relative to the number of tosses. Look at Figure 2. That's the law of averages, right there.

Assistant. Hmmmm. But what would happen if you tossed the coin another 10,000 times. Then you'd have 20,000 tosses to work with.

Kerrich.   The chance error would be likely to be even bigger with 20,000 tosses, but unlikely to be double what it was with 10,000. In absolute terms, the chance error would be bigger. But in percentage terms, it would be smaller.

Assistant. All right. Tell me again what the law of averages says.

Kerrich.   It says that the number of heads will be around half the number of tosses, give or take a chance error. As the number of tosses goes up, the chance error gets bigger. But relative to the number of tosses, it gets smaller and smaller.

Assistant. Can you give me some idea of how big the chance error is likely to be?

Kerrich.   Well, with 100 tosses, the chance error is likely to be around 5 in size. With 10,000 tosses, the chance error is only likely to be

Figure 2. The chance error expressed as a percentage of the number of tosses. As the number of tosses goes up, this percentage goes down. In other words, the chance error gets smaller relative to the number of tosses. The horizontal axis is not drawn to scale.

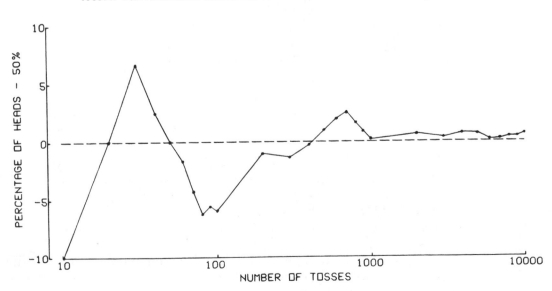

around 50 in size. Multiplying the number of tosses by 100 only multiplies the likely size of the chance error by $\sqrt{100} = 10$.

Assistant. So what you're saying is that as the number of tosses goes up, the difference between the number of heads and half the number of tosses gets bigger; but the difference between the percentage of heads and 50% gets smaller.

Kerrich. That's it, exactly.

EXERCISE SET A

1. A machine has been designed to toss a coin automatically, and keep track of the number of heads. After 1,000 tosses, it has 550 heads. Express the chance error both in absolute terms and as a percent of the number of tosses.

2. After 1,000,000 tosses, the machine in Exercise 1 has 501,000 heads. Express the chance error in the same two ways.

3. A coin is tossed 100 times, landing heads 53 times. However, the last seven tosses are all heads. True or false:The chance that the next toss will be heads is somewhat less than 50%. Explain.

4. A coin will be tossed, and you win a dollar if the number of heads is exactly equal to the number of tails. Which is better for you: ten tosses or a hundred tosses? Or are they the same? Explain.

5. A coin will be tossed, and you win a dollar if the percentage of heads is between 40% and 60%. Which is better for you: ten tosses or one hundred tosses? Explain.

6. A coin is tossed, and you win a dollar if there are more than 60% heads. Which is better for you: ten tosses, or a hundred? Explain.

The next three exercises involve drawing at random from a box. This was described on pp. 207–208 and is reviewed on p. 245.

7. A box contains 20% red marbles and 80% blue marbles. One thousand marbles are drawn at random with replacement. Which of the following is true, and why?
   (i) Exactly 200 marbles are going to be red.
   (ii) About 200 marbles are going to be red, give or take a dozen or so.

8. Repeat Exercise 7, if you draw at random without replacement, when the box contains fifty thousand marbles.

9. A hundred tickets will be drawn at random with replacement from one of the two boxes shown below. On each draw, you will be paid the amount shown on the tickets, in dollars. (If a negative number is drawn, that amount will be taken away from you.) Which box is better? Or are they the same?

   (i) $\boxed{-1}\ \boxed{1}$ (ii) $\boxed{-1}\ \boxed{-1}\ \boxed{1}\ \boxed{1}$

10. (Hard.) Look at Figure 1. If Kerrich kept on tossing, would the graph ever get negative?

The answers to the exercises are on p. A-47.

2. CHANCE PROCESSES

Kerrich's assistant was struggling with the problem of chance variability. Eventually, he came to see that when a coin is tossed a large number of times, the actual number of heads is likely to differ from the expected number. But then, he didn't know how big a difference to anticipate. A method for working out the likely size of the difference will be presented in the next chapter. As it turns out, the mathematics works in many different situations. For example, it can be used to see how much the house should expect to win at roulette (p. 262), or how accurate a sample survey is likely to be (Part VI), or how much averaging reduces measurement error (Part VII).

What is the common element here? The answer is that all these problems are about numbers determined by chance processes.[4] Take the number of heads in Kerrich's experiment. Chance comes in with each toss of the coin. If you repeat the experiment, the tosses turn out differently, and so does the number of heads. Next take the amount of money won or lost at roulette. Spinning the wheel is a chance process, and the amount of money won or lost is fixed by the outcome. Spin the wheel again, and winners become losers. Another example is the percentage of Democrats in a random sample of voters. A chance process is used to draw the sample. So the number of Democrats in the sample is determined by the luck of the draw. Draw another sample, and you would have a slightly different percentage of Democrats. The last example is replicating measurements and finding their average. Chance comes in with measurement error. The average of the series depends on how each measurement turns out—and chance throws the measurements off a bit. If you repeat the series, the average changes a little.

To what extent are such numbers thrown off by chance? For example, take the Gallup poll estimate for the percentage of Democratic votes in the next election. This estimate is based on a random sample of voters, and is therefore subject to chance variability: if the sample were done again, the estimate would change. The question is, by how much? This sort of issue must be faced over and over again in statistics. The general strategy for dealing with it will be presented in this part of the book. There are two main elements to this strategy:

• finding an analogy between the chance process being studied (in the Gallup poll example, this is sampling voters) and drawing numbers at random from a box;

• connecting the chance variability you want to know about (in the estimate for the Democratic vote) with the chance variability in the sum of the numbers drawn from the box.

The analogy between a chance process and drawing from a box is called a *box model*. The point of it is that the chance variability in the sum of the numbers drawn from the box will be easy to analyze mathematically. More complicated processes can then be dealt with through the analogy.

Drawing numbers at random from a box and then adding them up is a simple idea, but not a familiar one. So the next section will go through two examples in detail. The section after will do some examples of box models. In the next chapter, the chance variability in the sum of draws from a box will be analyzed. At that point, there will be some payoff from the theory.

3. THE SUM OF THE DRAWS

The object of this section is to present two examples illustrating the following process. There is a box of tickets, and each ticket has a number written on it. Then, some tickets are drawn at random from the box, and the numbers on these tickets are added up. For the first example, take the box

$$\boxed{1}\ \boxed{2}\ \boxed{3}\ \boxed{4}\ \boxed{5}\ \boxed{6}$$

Imagine drawing twice at random with replacement from this box: you shake the box to mix up the tickets, pick one ticket at random, make a note of the number on it, put it back in the box; then you shake the box again, make a second drawing at random, and note the number before putting the ticket back in the box. (The phrase "with replacement" reminds you to put the ticket back in the box before drawing again. Putting the tickets back enables you to draw over and over again, under the same conditions. Drawing with and without replacement was discussed on pp. 207–208 above.)

Having drawn twice at random with replacement, imagine adding up the two numbers you get. For example, the first draw might be $\boxed{3}$ and the second $\boxed{5}$: then the sum of the draws is 8. Or the first draw might be $\boxed{3}$ and the second $\boxed{3}$ too: then the sum of the draws is 6. There are many other possibilities. So the sum is subject to chance variability: if the draws turn out one way, the sum is one thing; if they turn out differently, the sum is different too.

At first, this example may seem artificial. But it is just like a turn at

Monopoly: you roll a pair of dice, add up the two numbers, and move that many squares. Rolling a die is just like picking a number from the box.

Next, imagine taking a large number of draws—say twenty-five—from the same box:

$$\boxed{\; \boxed{1}\ \boxed{2}\ \boxed{3}\ \boxed{4}\ \boxed{5}\ \boxed{6}\;}$$

Of course, the draws must be made with replacement. About how big is their sum going to be? The most direct way to find out is to make up the box and do the draws. However, we programmed the computer to simulate this experiment.[5] It got 3 on the first draw, 2 on the second, 4 on the third. Here they all are:

3 2 4 6 2 3 5 4 4 2 3 6 4 1 2 4 1 5 5 6 2 2 2 5 5.

The sum of these twenty-five draws is 88.

Of course, if the draws had been different, their sum would have been different. So we had the computer repeat the whole process ten times. Each time, it made twenty-five draws at random with replacement from the box, and took their sum. The results:

88   84   80   90   83   78   95   94   80   89.

Chance variability is easy to see. The first sum is 88, the second drops to 84, the third drops even more to 80. The values range from a low of 78 to a high of 95.

In principle, the sum could have been as small as $25 \times 1 = 25$, or as large as $25 \times 6 = 150$. But in fact, the ten observed values are all between 75 and 100. Would this keep up with more repetitions? Just what is the chance that the sum turns out to be between 75 and 100? This kind of problem will be solved in the next two chapters.

The *sum of the draws* from a box is shorthand for the process discussed in this section:

  • draw tickets at random from a box;
  • add up the numbers on the tickets.[6]

EXERCISE SET B

1. One hundred draws are made from the box $\boxed{\; \boxed{1}\ \boxed{2}\;}$. Forty-seven of the draws turn out to be $\boxed{1}$, and the remaining fifty-three are $\boxed{2}$. How much is the sum?

2. One hundred draws are made at random with replacement from the box [1] [2] .

    (a) How small can the sum be? How large?

    (b) About how many times do you expect the ticket [1] to turn up? The ticket [2] ?

    (c) About how much do you expect the sum to be?

3. One hundred draws are made at random with replacement from the box [1] [2] [9] .

    (a) How small can the sum be? How large?

    (b) About how much do you expect the sum to be?

4. A hundred draws will be made at random with replacement from one of the following boxes. Your job is to guess what the sum will be, and you win $1 if you are right to within 10. In each case, what would you guess? Which box is best for you? Worst?

    (i) [1] [9]   (ii) [4] [6]   (iii) [5] [5]

5. One hundred draws will be made at random from a box which contains two hundred tickets marked "1" and two hundred tickets marked "0." Your job is to guess the sum, and you win $1 if your guess is right to within 5. You have two options:

    (i) the draws are made with replacement.

    (ii) the draws are made without replacement.

In each case, say what you would guess. Which option is better?

6. Fifty draws will be made at random with replacement from one of the two boxes shown below. On each draw, you will be paid in dollars the amount shown on the ticket: if a negative number is drawn, that amount will be taken away from you. Which box is better? Or are they the same? Explain.

    (i) [-1] [2]   (ii) [-1] [-1] [2]

7. You gamble four times at a casino. You win $4 on the first play, lose $2 on the second, win $5 on the third, lose $3 on the fourth. Which of the following calculations tells you how much you come out ahead (more than one may be correct):

    (i) $4 + $5 − $2 − $3.

    (ii) $4 + (−$2) + $5 + (−$3).

    (iii) $4 + $2 + $5 − $3.

    (iv) −$4 + $2 + $5 + $3.

The answers to these exercises are on pp. A-47–A-48.

4. MAKING A BOX MODEL

    The object of this section is to make some box models, as practice for later. The sum of the draws from the box turns out to be the key ingredient for many statistical procedures, so keep your eye on the sum. It will turn out that there are three questions to answer when making a box model:

- what numbers go into the box?
- how many of each kind?
- how many draws?

The purpose of a box model is to analyze chance variability. And chance variability can be seen, in its starkest form, at any gambling casino. So this section will focus on box models for a game of chance—roulette. A Nevada roulette wheel has 38 pockets: one is numbered 0, another is numbered 00, and the others are numbered from 1 through 36. There is a ball inside the wheel. When the wheel is spun, the ball is equally likely to land in any one of the 38 pockets. Before it lands, bets can be placed on the table, as shown in Figure 3.

One bet is *red-and-black*. Except for 0 and 00, which are colored green, the numbers on the roulette wheel alternate red and black. If you bet a dollar on red, say, and a red number comes up, you get the dollar back, together with another dollar in winnings. If a black or green number comes up, the croupier smiles and rakes in your dollar.

Suppose you are at Harrah's Club in Reno, Nevada. You have just put a dollar on red, and the croupier spins the wheel. What is likely to happen? At first, this question may seem hard, because the game is complicated. But let's get down to basics. You will either win a dollar or lose a dollar:

$$+\$1 \qquad -\$1.$$

You win if one of the 18 red numbers comes up, and lose if one of the 18 black numbers comes up. But you also lose if 0 or 00 come up. And that is where the house gets its edge. Your chance of winning is only 18 to 38, and the chance of losing is 20 in 38.

As far as the chances are concerned, betting a dollar on red is just like playing a much simpler game: draw a ticket at random from the box

18 tickets $\boxed{+\$1}$ 20 tickets $\boxed{-\$1}$

The amount shown on the ticket is what you win (when it's positive) or lose (when it's negative). The great advantage of the box game is that all the irrelevant details—the wheel, the table, and the croupier's smile—have been stripped away. And you can see the cruel reality: you have 18 tickets, they have 20. *oh really how*

That clears up one play. But suppose you play roulette ten times, betting a dollar on red each time. What is likely to happen then? You will end up ahead or behind by some amount. This amount is called your *net gain*. The net gain is positive if you come out ahead, negative if you come out behind. The object is to say something about the net gain, and the first thing to do is to connect it to the box. Now, on each of the plays, you win or lose some amount. These ten win-lose numbers are like ten draws from the box, made at random with replacement. (Replacing the tickets keeps the chances the same as for the wheel.) The net gain—the total amount won or lost—is just the sum of these ten win-lose numbers. So your net gain in ten plays is like the sum

Figure 3. A Nevada roulette table.

"Roulette is a pleasant, relaxed, and highly comfortable way to lose your money."
—JIMMY THE GREEK (*San Francisco Chronicle*, JULY 25, 1975)

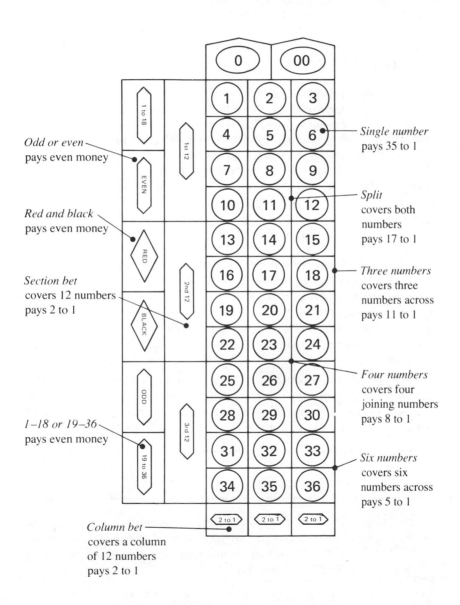

Odd or even
pays even money

Red and black
pays even money

Section bet
covers 12 numbers
pays 2 to 1

1–18 or 19–36
pays even money

Column bet
covers a column
of 12 numbers
pays 2 to 1

Single number
pays 35 to 1

Split
covers both
numbers
pays 17 to 1

Three numbers
covers three
numbers across
pays 11 to 1

Four numbers
covers four
joining numbers
pays 8 to 1

Six numbers
covers six
numbers across
pays 5 to 1

of ten draws made at random with replacement from the box

| 18 tickets +$1   20 tickets −$1 |

This is our first model, so it is a good idea to look at it more closely. Suppose, for instance, that the ten plays came out this way:

R   R   R   B   G   R   R   B   B   R

(R means red, B means black, and G means green—the house numbers, 0 and 00). Table 2 below shows the ten corresponding win-lose numbers, and the net gain.

Table 2. The net gain. This is the cumulative sum of the win-lose numbers.

Win-lose numbers	1	1	1	−1	−1	1	1	−1	−1	1
Net gain	1	2	3	2	1	2	3	2	1	2

Follow the net gain along. When you get a red, the win-lose number is 1, and the net gain goes up by 1. When you get a black or a green, the win-lose number is −1, and the net gain goes down by 1. The net gain is just the cumulative sum of the win-lose numbers—and these are like the draws from the box.

This game had a happy ending: you came out ahead $2. To see what would happen if you kept on playing, read the next chapter.

Example 1. If you bet a dollar on a single number at Nevada roulette, and that number comes up, you get the $1 back, together with winnings of $35. If any other number comes up, you lose the dollar. Gamblers say that a single number *pays 35 to 1*. Suppose you play roulette a hundred times, betting a dollar on the number 17 each time. Your net gain is like the sum of _____ draws made at random with replacement from the box _____. Fill in the blanks.

Solution. What numbers go into the box? To answer this question, think about one play of the game. You put a dollar chip on 17. If the ball drops into the pocket 17, you'll be up $35. If it drops into any other pocket, you'll be down $1. So the box has to contain the tickets $35 and −$1 .

> The tickets in the box show the various amounts that can be won or lost on a single play.

How many tickets of each kind? Keep thinking about one play. You have only 1 chance in 38 of winning; so the chance of drawing +$35 has to be 1 in 38. You have 37 chances in 38 of losing; so the chance of drawing −$1 has to be 37 in 38. The box is

| 1 ticket +$35   37 tickets −$1 |

The chance of drawing any particular value from the box must equal the chance of winning that amount on a single play. The box must be set up to make this so. ("Winning" a negative amount is the mathematical equivalent of what most people call losing.)

How many draws? You are playing a hundred times, so the number of draws has to be one hundred. The tickets must be replaced after each draw, so as not to change the odds.

The number of draws equals the number of plays.

The amounts won or lost on each play are like draws made at random with replacement from the box. The overall net gain is the sum of these individual wins and losses. So, the net gain in 100 plays is like the sum of 100 draws made at random with replacement from the box

| 1 ticket $\boxed{\$35}$ 37 tickets $\boxed{-\$1}$ |

This completes the solution. (The expected net gain will be worked out in Exercise 3 on p. 257 below.)

This section focused on roulette, but the same reasoning can be used in many different situations—as will be seen in the rest of the book.

EXERCISE SET C

1. Consider the following three situations.
    (i)   A box contains one ticket marked "0" and nine marked "1." One ticket is drawn at random. If it shows "1" you win a panda bear.
    (ii) A box contains ten tickets marked "0" and ninety marked "1." One ticket is drawn at random. If it shows "1" you win the panda.
    (iii) A box contains one ticket marked "0" and nine marked "1." Ten draws are made at random with replacement. If the sum of the draws equals 10, you win the panda.
   Assume you want the panda. Which is better—(i) or (ii)? Or are they the same? What about (i) or (iii)?

2. One ticket will be drawn at random from the box

| $\boxed{0}$ $\boxed{1}$ $\boxed{2}$ $\boxed{3}$ $\boxed{4}$ $\boxed{5}$ $\boxed{6}$ $\boxed{7}$ $\boxed{8}$ $\boxed{9}$ |

   What is the chance that it will be 0? That it will be 3 or less? That it will be 4 or more?

3. You are going to play roulette 25 times, putting a dollar on a *split* each time. A split is two adjacent numbers, like 11 and 12 in Figure 3 on p. 249. If either number comes up, you get the dollar back, together with winnings of $17. If neither number comes up, you lose the dollar. So a split pays 17 to 1, and you

have 2 chances in 38 to win. Your net gain in the 25 plays is like the sum of 25 draws made from one of the following boxes. Which one, and why?

(i) | 0 | 00 | 36 tickets numbered | 1 | through | 36 |

(ii) | $17 | $17 | 34 tickets | −$1 |

(iii) | $17 | $17 | 36 tickets | −$1 |

4. In one version of chuck-a-luck, three dice are rolled out of a cage. You can bet that all three show six, and if you are right the house pays 36 to 1. You have 1 chance in 216 to win this bet. Suppose you make this bet ten times, staking $1 each time. Your net gain is like the sum of _____ numbers drawn from the box _____. Fill in the blanks.

5. A quiz has ten true-false questions: a correct answer is worth one point, an incorrect answer is worth nothing. Suppose you guess the answers by tossing a coin. The number of points you get is like the sum of _____ draws made at random with replacement from the box _____. Fill in the blanks.

The answers to these exercises are on p. A-48.

5. REVIEW EXERCISES

1. A coin will be tossed some number of times, and you win $1 if the number of heads is exactly equal to half the number of tosses. Which is better for you: 100 tosses, or 1,000 tosses? Explain.

2. As genetic theory shows, there is very close to an even chance that both children in a two-child family will be of the same sex. Here are two possibilities.
    (i) Fifteen couples have two children each. In ten or more of these families, it will turn out that both children have the same sex.
    (ii) Thirty couples have two children each. In twenty or more of these families, it will turn out that both children are of the same sex.
    Which possibility is more likely, and why?

3. A die will be rolled 6,000 times. Which of the following best describes the situation?
    (i)   The number of 1's will be 1,000 exactly.
    (ii) The number of 1's is very likely to be 1,000 exactly, but there is also some small chance that it will not be equal to 1,000.
    (iii) The number of 1's is likely to be different from 1,000, but the difference is likely to be small by comparison with 6,000.

4. Repeat Exercise 3 for 6,000 draws made at random without replacement from a box of 6,000 tickets, containing equal numbers of digits from 1 through 6.

5. A gambler loses five times running at roulette. He decides to continue playing; he feels that by the law of averages, he is due for a win. A

bystander advises him to quit, on the grounds that his luck is cold. What does statistics say?

6. You are going to play roulette 25 times, betting a dollar on four adjoining numbers each time (such as 23, 24, 26, and 27 in Figure 3 on p. 249.) If any one of the four numbers comes up, you get the dollar back, together with winnings of $8. If none of them comes up, you lose the dollar. So this bet pays 8 to 1, and you have 4 chances in 38 of winning. Your net gain in 25 plays is like the sum of _____ draws from the box _____. Fill in the blanks.

7. A quiz has 16 true-false questions. A correct answer is worth two points, but a point is taken off for each incorrect answer. Suppose you guess the answers by tossing a coin. Your score will be like the sum of _____ draws from the box _____. Fill in the blanks.

8. One hundred draws will be made at random with replacement from the box

$$\boxed{-3} \boxed{-1} \boxed{0} \boxed{1} \boxed{3}$$

   (a) If the sum of the 100 numbers drawn is 30, what is their average?
   (b) If the sum of the 100 numbers drawn is −20, what is their average?
   (c) In general, how can you figure the average of the 100 draws, if you are told their sum?
   (d) You are given a choice between two alternatives:
       (i) You win $1 if the sum of the 100 numbers drawn is between −5 and +5.
       (ii) You win $1 if the average of the 100 numbers drawn is between −0.05 and +0.05.
       Which is better, or are they the same?

6. SUMMARY

1. When a coin is tossed a large number of times, the number of heads is likely to be quite different from half the number of tosses, in absolute terms; but quite close, in relative terms. This is the *law of averages*.

2. The *chance error* in the number of heads is defined by the equation

number of heads = half the number of tosses + chance error.

The error is likely to be large in absolute terms, but only a small percentage of the number of tosses.

3. The law of averages can be restated in percentage terms. With a large number of tosses, the percentage of heads is very likely to be close to 50%, although it is not at all likely to be exactly equal to 50%.

4. The law of averages does not work by compensation. When tossing a coin, a run of heads is just as likely to be followed by a head as by a tail. Similarly, a run of tails is just as likely to be followed by a tail as by a head.

Drawing by Dana Fradon; © 1976 The New Yorker Magazine, Inc.

5. A complicated chance process for generating a number can often be modeled by drawing from a box. The *sum of the draws* is a key ingredient in many statistical procedures.

6. The three basic questions to ask when making a *box model* are:
   • Which numbers go into the box?
   • How many of each kind?
   • How many draws?

7. For gambling problems in which the same bet is made several times, a box model can be set up as follows:
   • The tickets in the box show the amounts that can be won (+) or lost (−) on each play.
   • The number of times each value appears in the box is adjusted to make the chance of drawing it equal the chance of winning it.
   • The number of draws equals the number of plays.
Then, the *net gain* is like the sum of the draws from the box.

17

The Expected Value and Standard Error

If you believe in miracles, head for the Keno lounge.

—JIMMY THE GREEK

1. THE EXPECTED VALUE

A chance process is running. It delivers a number. Then another, And another. You are about to drown in random output. But mathematicians have found a little order in this chaos. The numbers delivered by the process vary around the *expected value,* the amounts off being similar in size to the *standard error*. If the process only runs once, the number it generates should be somewhere around the expected value, give or take a standard error or so.

To be more specific, imagine generating a number through the following chance process: count the number of heads in a hundred tosses of a coin. The expected value is fifty. In one hundred tosses of a coin, you expect to get around fifty heads, give or take a chance error. If you actually did this experiment, you might get the number 57: the chance error would be 7. If you did it again, you would get a different number—perhaps 46: the chance error would be −4. A third repetition would generate still a third number—say 47: the chance error would be −3. Your numbers would vary around fifty, being off by chance amounts similar in size to the standard error, which is 5 (p. 267 below).

The formulas for the expected value and standard error depend on how the chance process generates the number. This chapter presents the formulas that apply to the sum of draws from a box. The formula for the expected value will be introduced with an example: the sum of one hundred draws made at random with replacement from the box

$$\boxed{\;\boxed{1}\;\boxed{1}\;\boxed{1}\;\boxed{5}\;}\;.$$

About how large should this sum be? To answer this question, think how the draws should turn out. The ticket $\boxed{5}$ should come up on around one fourth of the draws, and $\boxed{1}$ on three fourths. With a hundred draws, you can expect to get around twenty-five $\boxed{5}$'s and seventy-five $\boxed{1}$'s to add up. So the sum should be around

$$25 \times 5 + 75 \times 1 = 200$$

That is the expected value.

The formula for the expected value gives a short-cut around this reasoning. It has two ingredients: the number of draws, and the average of the numbers in the box—abbreviated to "average of box."

The expected value for the sum of draws made at random with replacement from a box equals

(number of draws) × (average of box).

To see the logic behind the formula, go back to the example. The average of the box is

$$\frac{1 + 1 + 1 + 5}{4} = 2$$

On the average, each draw adds around 2 to the sum. So, with 100 draws, the sum must be around $100 \times 2 = 200$.

Example 1. Suppose you are going to Las Vegas to play Keno. Your favorite bet is a dollar on a single number. When you win, they give you the dollar back, and two dollars more. When you lose, they keep the dollar. There is one chance in four to win.[1] About how much should you expect to win (or lose) in a hundred plays, if you make this bet on each play?

Solution. The first step is to write down a box model. On each play, your net gain either goes up by $2 (with 1 chance in 4) or goes down by $1 (with 3 chances in 4). So your net gain after a hundred plays is like the sum of a hundred draws (at random with replacement) from the box

$$\boxed{\;\boxed{-\$1}\;\boxed{-\$1}\;\boxed{-\$1}\;\boxed{+\$2}\;}$$

The average of this box is

$$\frac{-\$1 - \$1 - \$1 + \$2}{4} = -\$0.25$$

On the average, each play costs you a quarter. In a hundred plays, you can expect to lose around $25. This is the answer. If you continued on, in a thousand plays you should expect to lose around $250. The more you play, the more you lose. Perhaps you should look for another game.

EXERCISE SET A

1. Find the expected value for the sum of a hundred draws at random with replacement from the box

   (a) $\boxed{0}\ \boxed{1}\ \boxed{1}\ \boxed{6}$ (b) $\boxed{-2}\ \boxed{-1}\ \boxed{0}\ \boxed{2}$ (c) $\boxed{-2}\ \boxed{-1}\ \boxed{3}$

   (d) $\boxed{0}\ \boxed{1}\ \boxed{1}$

2. Find the expected number of squares you move in your first play at Monopoly. (See p. 246.)

3. You are going to play roulette 100 times, betting a dollar on the number 17 each time. Find the expected value for your net gain. (See Example 1 on p. 250.)

4. You are going to play roulette a hundred times, staking $1 on red-and-black each time. Find the expected value for your net gain. (This bet pays even money, and you have 18 chances in 38 of winning: see p. 248.)

5. Repeat Exercise 4 for a thousand plays.

6. A game is *fair* if the expected value for the net gain equals 0, so on the average you neither win nor lose. A generous casino would offer you a bit more than $1 in winnings if you staked $1 on red-and-black in roulette and won. How much should they pay you, to make it a fair game? (Hint: Let x stand for what they should pay you. The box has 18 tickets marked "x" and 20 tickets "−$1." Write down the formula for the expected value in terms of x and set it equal to 0.)

7. If an Adventurer at the Game of the Royal Oak staked £1 on a point and won, how much should the Master of the Ball have paid him, for the Game to be fair? (The rules are explained on p. 226 above.)

The answers to these exercises are on p. A-48

2. THE STANDARD ERROR

Suppose you draw twenty-five times at random with replacement from the box

$$\boxed{0}\ \boxed{2}\ \boxed{3}\ \boxed{4}\ \boxed{6}$$

(There is nothing special about the numbers in the box, they were chosen to make later calculations come out evenly.) Each of the five tickets should appear on about one fifth of the draws, that is, 5 times. So the sum should be around

$$5 \times 0 + 5 \times 2 + 5 \times 3 + 5 \times 4 + 5 \times 6 = 75.$$

This is the expected value for the sum. Of course, each ticket won't appear on exactly one fifth of the draws, just as Kerrich didn't get heads on exactly half the tosses. The sum will be off the expected value by a chance error:

sum = expected value + chance error.

The chance error is the amount above (+) or below (−) the expected value. For example, if the sum came out as 70, the chance error would be −5.

The likely size of the chance error is given by the standard error (usually abbreviated to SE):

> A sum is likely to be around its expected value, but to be off by a chance error similar in size to the standard error.

There is a formula to use in computing the SE for a sum of draws made at random with replacement from a box. It is called a *square root law,* because it involves the square root of the number of draws. All the statistical procedures presented in the rest of the book depend on this formula.

> When drawing at random with replacement from a box of numbered tickets, the standard error for the sum of the draws is
>
> $$\sqrt{\text{number of draws}} \times (\text{SD of the box}).$$

The formula has two ingredients: the square root of the number of draws, and the SD of the numbers in the box—abbreviated to "SD of the box." The SD measures the spread among the numbers in the box. If there is a lot of spread in the box, the SD is big, and it is hard to predict how the draws will turn out. So the standard error must be big too. Now for the number of draws. The sum of two draws is more variable than a single draw; the sum of a hundred draws, still more variable. Each draw adds some extra variability to the sum, because you don't know how it is going to turn out. As the number of draws goes up, the sum gets harder to predict, the chance errors get bigger, and so does the standard error. However, the standard error goes up slowly, by a factor equal to the square root of the number of draws. For instance, the sum of 100 draws is only $\sqrt{100} = 10$ times as variable as a single draw.

Do not confuse the SD and the SE.[2] The SD applies to spread in lists of numbers—for example, the numbers in the box. It is worked out using the method explained on p. 63. The SE applies to chance variability—for instance, in the sum of the draws. The SE for the sum of the draws is computed from the SD of the box, through the square root law.

The SD is for a list   The SE is for a chance process
1 2 3 4 5 6

At the beginning of the section, we looked at the sum of twenty-five draws drawn at random with replacement from the box

$$\boxed{0}\ \boxed{2}\ \boxed{3}\ \boxed{4}\ \boxed{6}$$

The expected value for this sum is 75: it is likely to be around 75, give or take a chance error. How big is the chance error likely to be? To find out, calculate the standard error from the square root law. The average of the numbers in the box is 3, the deviations from average are

$$-3 \quad -1 \quad 0 \quad 1 \quad 3$$

so the SD of the box is

$$\sqrt{\frac{9 + 1 + 0 + 1 + 9}{5}} = \sqrt{\frac{20}{5}} = 2$$

This measures the variability in the box. According to the square root law, the sum of 25 draws is more variable, by the factor $\sqrt{25} = 5$. The SE for the sum of 25 draws is $5 \times 2 = 10$. The likely size of the chance error is 10, and the sum of the draws should be around 75, give or take 10 or so.

To show what this means empirically, we had the computer programmed to do these sums. It drew 25 times at random with replacement from the box $|\boxed{0}\,\boxed{2}\,\boxed{3}\,\boxed{4}\,\boxed{6}|$, and got

$$0\ 0\ 4\ 4\ 0\ 4\ 3\ 2\ 6\ 2\ 2\ 0\ 2\ 6\ 2\ 6\ 4\ 2\ 6\ 3\ 0\ 3\ 6\ 4\ 0$$

The sum of these 25 draws is 71. This is 4 below the expected value, so the chance error is -4. Then it drew another 25 times and took the sum, getting 76. The chance error was $+1$. The third sum was 86, with a chance error of 11. In fact, we had the computer generate a hundred sums, shown in Table 1. These numbers are all around 75, the expected value. They are off by chance errors similar in size to 10, the standard error.

In other words, the sum is likely to be around its expected value, give or take a standard error or so.

Table 1. Computer simulation of one hundred observed values of the sum of twenty-five draws made at random with replacement from the box

$$|\boxed{0}\,\boxed{2}\,\boxed{3}\,\boxed{4}\,\boxed{6}|$$

Repetition	Sum	Repetition	Sum	Repetition	Sum	Repetition	Sum	Repetition	Sum
1	71	21	80	41	64	61	64	81	60
2	76	22	77	42	65	62	70	82	67
3	86	23	70	43	88	63	65	83	82
4	78	24	71	44	77	64	78	84	85
5	88	25	79	45	82	65	64	85	77
6	67	26	56	46	73	66	77	86	79
7	76	27	79	47	92	67	81	87	82
8	59	28	65	48	75	68	72	88	88
9	59	29	72	49	57	69	66	89	76
10	75	30	73	50	68	70	74	90	75
11	76	31	78	51	80	71	70	91	77
12	66	32	75	52	70	72	76	92	66
13	76	33	89	53	90	73	80	93	69
14	84	34	77	54	76	74	70	94	86
15	58	35	81	55	77	75	56	95	81
16	60	36	68	56	65	76	49	96	90
17	79	37	70	57	67	77	60	97	74
18	78	38	86	58	60	78	98	98	72
19	66	39	70	59	74	79	81	99	57
20	71	40	71	60	83	80	72	100	62

The sums in Table 1 show remarkably little spread around the expected value. In principle, they could be as small as 0, or as large as $25 \times 6 = 150$. However, all but one of them are between 50 and 100, that is, within 2.5 SEs of the expected value. Observed values are rarely more than two or three SEs away from the expected value. And the SE for a sum increases very slowly with the number of draws, due to the square root in the formula. The basic reason for this is cancellation. In the example, the average of the box is 3, so the values 0 and 2 are below average, 4 and 6 above average. Below-average values occur about as often in twenty-five draws as above-average ones, and they tend to cancel each other out. So the sum of draws stays relatively close to its expected value. The square root law tells you exactly how much cancellation to anticipate—because it tells you exactly how far from the expected value a sum is likely to get.

EXERCISE SET B

1. One hundred draws are made at random with replacement from the box

    $$\boxed{1}\ \boxed{2}\ \boxed{3}\ \boxed{4}\ \boxed{5}\ \boxed{6}\ \boxed{7}$$

    (a) Find the expected value and standard error for the sum.
    (b) Suppose you had to guess what the sum was going to be. What would you guess? Would you expect to be off by around 2, 4, or 20?

2. You gamble a hundred times on the toss of a coin. If it lands heads, you win $1. If it lands tails, you lose $1. Your net gain will be around _____, give or take _____ or so. Fill in the blanks, using the options

    $$-\$10 \quad -\$5 \quad \$0 \quad +\$5 \quad +\$10.$$

3. The expected value for a sum is 50, with an SE of 5. The chance process generating the sum is repeated ten times. Which of the following is the sequence of observed values, and why?
    (i)   51, 57, 48, 52, 57, 61, 58, 41, 53, 48.
    (ii) 51, 49, 50, 52, 48, 47, 53, 50, 49, 47.
    (iii) 45, 50, 55, 45, 50, 55, 45, 50, 55, 45.

4. Example 1 on p. 256 concerned a hundred plays at Keno, the bet being $1 on a single number at each play. Find the standard error for the player's net gain on these plays.

5. If the computer kept on running, do you think it would eventually generate a sum more than three SEs away from the expected value? Explain.

The answers to these exercises are on p. A-49.

3. USING THE NORMAL CURVE

Mathematicians discovered the normal curve while trying to solve problems of the following general type:

• A large number of draws will be made at random with replacement from a box. What is the chance that the sum of the draws will be in a given interval?

   The logic behind the curve will be discussed in the next chapter; the object of this section is only to sketch the method, which applies whenever the number of draws is reasonably large. Basically, it is a matter of converting to standard units (using the expected value and standard error) and then working out areas under the curve, just as in Chapter 5.

   Now for an example. Suppose the computer is programmed to take the sum of twenty-five draws made at random with replacement from the magic box $\boxed{0}\ \boxed{2}\ \boxed{3}\ \boxed{4}\ \boxed{6}$, and print out the result—repeating this process over and over again. About what percentage of the observed values should be between 50 and 100?

   Each sum will be a number on the horizontal axis somewhere between 0 and $25 \times 6 = 150$:

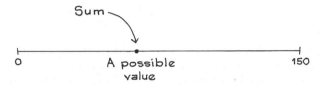

   The problem is really asking for the chance that the sum of 25 draws turns out to be between 50 and 100. This can be represented in a diagram:

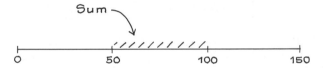

   Now convert to standard units and use the normal curve, as in Chapter 5. In this context, standard units say how many SEs a number is away from the expected value. In the example, 100 becomes 2.5 in standard units: the expected value for the sum is 75 and the SE is 10 (p. 259), so 100 is 2.5 SEs above the expected value. Similarly, 50 becomes -2.5:

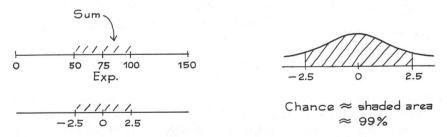

Chance ≈ shaded area ≈ 99%

   The interval from 50 to 100 is the interval within 2.5 SEs of the expected value, so the sum should be there about 99% of the time.

   That finishes the calculation. Now for some data. Table 1 on p. 259 reported one hundred observed values for the sum: according to the calculation 99 of them should be in the interval from 50 to 100, and in fact 99 of them are. To take some less extreme ranges, about 68% of these observed values should be in the interval from $75 - 10$ to $75 + 10$. In fact, 68 are (exclusive of endpoints). Finally, about 95% of the observed values in Table 1 should be in the range 75 ± 20, and 98% of them are. So the theory looks pretty good.

Example 1. In a month, there are 10,000 independent plays on a roulette wheel in a certain casino. To keep things simple, suppose the gamblers only stake $1 on red at each play. Estimate the chance that the house will win more than $250 from these plays. (Red-and-black pays even money, and the house has 20 chances in 38 to win.)

Solution. The problem asks for the chance that the net gain of the house will be more than $250.

Here, the box model isn't specified by the problem. The first thing to do is to write it down. What numbers go into the box? On each play, the house either wins $1 or loses $1, so the tickets are $+\$1$ and $-\$1$. How many of each kind? On each play, the house has 20 chances in 38 to win, and 18 in 38 to lose. So the box is

$$\boxed{18 \text{ tickets } \boxed{-\$1} \quad 20 \text{ tickets } \boxed{\$1}}$$

How many draws? There are 10,000 plays, so there must be 10,000 draws. The net gain for the house is like the sum of 10,000 draws made at random with replacement from the box. This completes the model, and puts us in a position to use the square root law. (The gambler and the house are on opposite sides of this box: 18 tickets are good for the gambler, and 20 are good for the house. This example will show how the house edge mounts up. See p. 248 for a discussion of the model from the gamblers' point of view.)

The next step is to find the expected value for the net gain, so the average of the numbers in the box is needed. This equals their total, divided by 38. The positive numbers contribute $20 to the total, and the negative numbers take $18 away, so the average of the box is

$$\frac{\$20 - \$18}{38} = \frac{\$2}{38} \approx \$0.05$$

On the average, each draw adds around $0.05 to the sum. So the sum of 10,000 draws has an expected value of 10,000 × $0.05 = $500. To translate, the house averages about a nickel on each play, so in 10,000 plays it can expect to win around $500.

The next step is to find the SE for the net gain. This requires the SD of the numbers in the box. Their deviations from average are all just about $1, because the average is so close to 0; so the SD of the box is about $1. (An exact calculation shows the SD to be $0.9986 . . . .) This $1 measures the variability in the box. According to the square root law, the sum of 10,000 draws is more variable, by the factor $\sqrt{10,000} = 100$. The SE for the sum of

10,000 draws is $100 \times \$1 = \$100$. The house can expect to win around $500, give or take $100 or so.

Now the normal curve can be used.

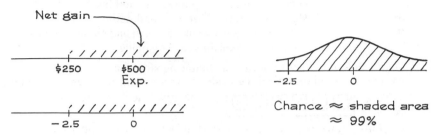

This completes the solution. The key step was showing that the net gain was like the sum of the draws from the box, for that made it possible to use the square root law.

The house has about a 99% chance to win more than $250. This may not seem like much, but you have to remember that the house owns many wheels, there often is a crowd of gamblers playing on each spin of each wheel, and a lot of bets are over a dollar. The house can expect to win about 5% of the money which crosses the table—and the square root law virtually eliminates the risk. For instance, suppose the house runs twenty-five wheels. To be very conservative, suppose each wheel operates under the conditions of Example 1. With these assumptions, the casino's expected winnings go up by a full factor of 25, to $25 \times \$500 = \$12,500$. But their standard error only goes up by the factor $\sqrt{25} = 5$, to $500. Now the casino can be virtually certain—99%—of winning at least $11,000. For the casino, roulette is a volume business, just like groceries are for Safeway.

The square root law is a bit complicated. As a result, it is a common mistake to compute the SE for the sum as "number of draws × $\sqrt{\text{SD of box}}$." This blunder makes the units for the SE come out as the square root of dollars. Naturally, the units for the SE have to be the same as the units for the net gain—$$.

EXERCISE SET C

1. One hundred draws will be made at random with replacement from the box $\boxed{\boxed{1}\ \boxed{1}\ \boxed{2}\ \boxed{2}\ \boxed{2}\ \boxed{4}}$.

   (a) The sum of the draws will be around _____ , give or take _____ or so.
   (b) The smallest the sum can be is _____, the largest is _____.
   (c) The chance that the sum will be bigger than 250 is almost _____%.

2. One hundred draws will be made at random with replacement from the box $\boxed{\boxed{1}\ \boxed{3}\ \boxed{3}\ \boxed{9}}$.

   (a) How large can the sum be? How small?
   (b) How likely is the sum to be in the range from 370 to 430?

3. You can draw either 10 times or 100 times at random with replacement from the box $\boxed{\boxed{-1}\ \boxed{1}}$. How many times should you draw—
   (a) if you win \$1 when the sum is 5 or more, and nothing otherwise?
   (b) if you win \$1 when the sum is −5 or less, and nothing otherwise?
   (c) if you win \$1 when the sum is between −5 and 5, and nothing otherwise?
   No calculations are needed, but explain your reasoning.

4. You are allowed to choose one of the following situations.
   (i) One hundred draws will be made at random with replacement from the box $\boxed{\boxed{1}\ \boxed{1}\ \boxed{5}\ \boxed{7}\ \boxed{8}\ \boxed{8}}$.

   (ii) Fifty draws will be made at random with replacement from the box $\boxed{\boxed{4}\ \boxed{7}\ \boxed{11}\ \boxed{13}\ \boxed{15}}$.

   Which should you choose—
   (a) if you win \$1 when the sum is 550 or more, and nothing otherwise?
   (b) if you win \$1 when the sum is 450 or less, and nothing otherwise?
   (c) if you win \$1 when the sum is between 450 and 550, and nothing otherwise?

5. Suppose that in one week at a certain casino, there are 25,000 independent plays at roulette. On each play, the gamblers stake \$1 on red. The chance that the casino will win more than \$1,000 from these 25,000 plays is closest to: 5%, 50%, 95%.

6. Suppose that one person stakes \$25,000 on one play at red-and-black in roulette. The chance that the casino will win more than \$1,000 from this play is closest to: 5%, 50%, 95%. (There is no formula to use.)

7. Suppose you play once at roulette, staking \$1 on each number (including 0 and 00). So you have staked \$38 in all. What will happen? (There is no formula to use.)

8. A box contains six tickets, numbered 1 through 6; the average of this box is 3.5, with an SD of 1.7. The tickets numbered 1, 2, and 3 are white, but the ones numbered 4, 5, and 6 are black.

$$\boxed{\boxed{1}\ \boxed{2}\ \boxed{3}\ \blacksquare\blacksquare\blacksquare}$$

One person chooses a hundred numbers from this box by the following procedure. He sorts out the white tickets and the black tickets into separate boxes. Then, he takes fifty white tickets at random with replacement. Finally, he takes fifty black tickets at random with replacement. True or false: the standard error for the sum of the 100 draws is $\sqrt{100} \times 1.7$.

The answers to these exercises are on pp. A-49– A-50.

4. CLASSIFYING AND COUNTING

In this chapter, the focus has been on the chance variability in the sum of the draws from a box. The sum is likely to be around its expected value, give or take an SE or so. How does this apply to the coin-tossing game of the previous chapter? For instance, in 10,000 tosses of a coin, around 5,000 heads may be expected—give or take something. But in order to calculate the give-or-take number from the square root law, we have to show that counting the number of heads is like adding numbers drawn from a box. The square root law only applies to the sum of draws from a box. The next two examples will establish the connection.

Example 1. A die is rolled sixty times. Fill in the blanks.
(a) The total number of spots thrown should be around _____, give or take _____ or so.
(b) The number of 6's thrown should be around _____, give or take _____ or so.

By way of illustration, Table 2 shows the results of throwing a die sixty times: the first throw was a 4, the second was a 5, and so on.

Table 2. Sixty throws of a die.

```
4 5 5 2 4    5 3 2 6 3    5 4 6 2 6    4 4 2 5 6
1 5 3 1 2    2 1 2 5 3    3 6 6 1 1    5 1 6 1 2
4 4 2 1 4    4 5 2 6 3    2 4 6 1 6    4 6 1 5 2
```

Solution. Part (a) is familiar. It involves adding. Each throw contributes some number of spots, and then we add these numbers up. So the total number of spots in 60 throws of the die is like the sum of 60 draws from the box

$$\boxed{1}\ \boxed{2}\ \boxed{3}\ \boxed{4}\ \boxed{5}\ \boxed{6}$$

The average of this box is 3.50 with an SD of 1.71. The expected value for the sum is $60 \times 3.50 = 210$, with an SE of $\sqrt{60} \times 1.71 \approx 13$, by the square root law. The total number of spots will be around 210, give or take 13 or so. In fact, the sum of the numbers in Table 2 is 212. The sum was off its expected value by around one sixth of an SE.

Part (b) is new. Instead of adding the 60 numbers in Table 2, we first classify each one: is it a 6, or not? (There are only two classes here: 6's on one hand, everything else on the other.) Then we count the number of 6's. The point to notice is that on each throw, the number of 6's either goes up by 1, or stays the same.

Table 3. Counting the 6's.

Throw	4 5 5 2 4 5 3 2 6 3 5 4 6 2 6 4 4 2 5 6
Running count of 6's	0 0 0 0 0 0 0 0 1 1 1 1 2 2 3 3 3 3 3 4

This can be seen in Table 3, which shows the first twenty throws from Table 2 as well as a running count of the number of 6's. You should follow the count along. It usually stays the same, but every so often it goes up by 1—whenever a 6 is thrown.

To put this more mathematically, at each throw
• either 1 is added to the count (if the throw was 6)
• or 0 is added to the count (if the throw was not a 6).

So the number of 6's in sixty throws of the die is really like the sum of sixty draws from a new box—

$$\boxed{\;\boxed{\cancel{1}\;0}\;\boxed{\cancel{2}\;0}\;\boxed{\cancel{3}\;0}\;\boxed{\cancel{4}\;0}\;\boxed{\cancel{5}\;0}\;\boxed{\cancel{6}\;1}\;}$$

To spell the analogy out: on each throw, the count has 1 chance in 6 to go up by one, and 5 chances in 6 to stay the same. Likewise, on each draw, the sum has 1 chance in 6 to go up by one, and 5 chances in 6 to stay the same. So, counting the number of 6's really is like drawing and adding—from the new box. This puts us in a position to use the square root law.

The new box has five tickets marked "0," and a "1." The average of these six numbers is 1/6, so the sum of 60 draws from the new box has an expected value of $60 \times 1/6 = 10$. To get the SE for the sum, the SD of the new box is needed. Finding the SD usually involves quite a bit of work, but when there are only 0's and 1's in the box, there is a shortcut:

When the box only contains 0's and 1's, the SD of the box equals
$$\sqrt{(\text{fraction of 1's}) \times (\text{fraction of 0's})}.$$

With the new box in the example, one ticket out of six was marked "1," so the fraction of 1's is 1/6. Five tickets out of six were marked "0," so the fraction of 0's was 5/6. The SD of the new box is $\sqrt{(1/6) \times (5/6)} \approx 0.37$. The sum of 60 draws from the new box has an expected value of 10, with an SE of $\sqrt{60} \times 0.37 \approx 3$—by the square root law. In other words, in sixty throws of a die, the number of 6's will be around 10, give or take 3 or so. In fact, in Table 2 there were eleven 6's: the number of 6's was off its expected value by a third of an SE. This completes the example.

Many of the problems in this book can be put in the following form. There is a box of tickets. Some are drawn at random. An operation is performed on these draws to determine a number. The problem asks for the chance that this number will be in a given interval. For instance:

• Sixty draws are made at random with replacement from the box $\boxed{\boxed{1}\;\boxed{2}\;\boxed{3}\;\boxed{4}\;\boxed{5}\;\boxed{6}}$. Estimate the chance that the sum of the draws will be between 200 and 225.

Here, the given interval runs from 200 to 225. More important, the operation performed on the draws in this problem is addition. The draws are treated as quantitative data—numbers—which can be added up.

Now there is another possibility, for instance:

• Sixty draws are made at random with replacement from the box $\boxed{\boxed{1}\;\boxed{2}\;\boxed{3}\;\boxed{4}\;\boxed{5}\;\boxed{6}}$. Estimate the chance of getting between 10 and 20 tickets marked "6."

Here, the given interval runs from 10 to 20. And the operation performed on the draws is completely different—they are classified and counted. For each draw, we answer the question, "Is it a 6 or not?" Then we count the 6's. In other words, the draws are treated as qualitative data. (The distinction between quantitative and qualitative data was discussed on p. 36.)

At this point, there are two possible operations to perform on the draws:
• adding;
• classifying and counting.
But the message of this section is that both kinds of problems can be treated the same way—provided you change the box.

> If you have to classify and count the draws, put 0's and 1's on the tickets. Mark "1" on the tickets that count for you, "0" on the others.

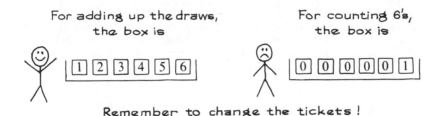

For adding up the draws,
the box is

$\boxed{1}\ \boxed{2}\ \boxed{3}\ \boxed{4}\ \boxed{5}\ \boxed{6}$

For counting 6's,
the box is

$\boxed{0}\ \boxed{0}\ \boxed{0}\ \boxed{0}\ \boxed{0}\ \boxed{1}$

Remember to change the tickets!

Example 2. A coin will be tossed one hundred times. Find the expected value and standard error for the number of heads. Estimate the chance of getting between 40 and 60 heads.

Solution. The first thing is to make a box model. The problem involves classifying the tosses as heads or tails, and then counting the number of heads, so there should be only 0's and 1's in the box. The chances are 50–50 for heads, so the box should be $\boxed{0}\ \boxed{1}$. The number of heads in 100 tosses of a coin is like the sum of 100 draws made at random with replacement from the box $\boxed{0}\ \boxed{1}$. The mathematical reason: each toss of the coin either pushes the number of heads up by one or leaves it alone, with a 50–50 chance; likewise, each draw from the box either pushes the sum up by one or leaves it alone, with the same 50–50 chance. This completes the model.

Since the number of heads is like the sum of the draws, the square root law can be used. The average of the box is 1/2, and the expected number of heads is $100 \times 1/2 = 50$. The SD of the box is 1/2 too, measuring the variability in the box. The sum of 100 draws is more variable, but only by the factor $\sqrt{100} = 10$. So the SE for the sum of 100 draws is $10 \times 1/2 = 5$. The number of heads will be around 50, give or take 5 or so. The range from 40 to 60 heads represents the expected value give or take two SEs. So the chance is around 95%. This completes the solution.

To interpret this 95% probability, imagine counting the number of heads in a hundred tosses of a coin. You might get 44 heads. If you toss again, you might get 54 heads. A third time, the number would change again, perhaps to 48 heads. And so on. In the long run, about 95% of these counts would come out in the range from 40 to 60. Actually, there is no need to imagine this experiment—John Kerrich did it. Table 4 shows the results, with Kerrich's 10,000 tosses broken down into successive groups of one hundred. In fact, 95 out of 100 groups had 40 to 60 heads inclusive, 93 out of 100 had 40 to 60 heads exclusive. The theory looks good.

Table 4. Kerrich's coin tossing experiment, showing the number of heads he got in each successive group of 100 tosses.

Group of 100 tosses	No. of heads	Group of 100 tosses	No. of heads	Group of 100 tosses	No. of heads	Group of 100 tosses	No. of heads
1–100	44	2,501–2,600	44	5,001–5,100	42	7,501–7,600	48
101–200	54	2,601–2,700	34	5,101–5,200	68	7,601–7,700	43
201–300	48	2,701–2,800	59	5,201–5,300	45	7,701–7,800	58
301–400	53	2,801–2,900	50	5,301–5,400	37	7,801–7,900	57
401–500	56	2,901–3,000	51	5,401–5,500	47	7,901–8,000	48
501–600	57	3,001–3,100	51	5,501–5,600	52	8,001–8,100	45
601–700	56	3,101–3,200	48	5,601–5,700	51	8,101–8,200	50
701–800	45	3,201–3,300	56	5,701–5,800	49	8,201–8,300	53
801–900	45	3,301–3,400	57	5,801–5,900	48	8,301–8,400	46
901–1,000	44	3,401–3,500	50	5,901–6,000	37	8,401–8,500	56
1,001–1,100	40	3,501–3,600	54	6,001–6,100	47	8,501–8,600	58
1,101–1,200	54	3,601–3,700	47	6,101–6,200	52	8,601–8,700	54
1,201–1,300	53	3,701–3,800	53	6,201–6,300	45	8,701–8,800	49
1,301–1,400	55	3,801–3,900	50	6,301–6,400	48	8,801–8,900	48
1,401–1,500	52	3,901–4,000	53	6,401–6,500	44	8,901–9,000	45
1,501–1,600	54	4,001–4,100	52	6,501–6,600	51	9,001–9,100	55
1,601–1,700	58	4,101–4,200	54	6,601–6,700	55	9,101–9,200	51
1,701–1,800	50	4,201–4,300	55	6,701–6,800	53	9,201–9,300	48
1,801–1,900	53	4,301–4,400	52	6,801–6,900	52	9,301–9,400	56
1,901–2,000	42	4,401–4,500	51	6,901–7,000	60	9,401–9,500	55
2,001–2,100	56	4,501–4,600	53	7,001–7,100	50	9,501–9,600	55
2,101–2,200	53	4,601–4,700	54	7,101–7,200	57	9,601–9,700	50
2,201–2,300	53	4,701–4,800	47	7,201–7,300	49	9,701–9,800	48
2,301–2,400	45	4,801–4,900	42	7,301–7,400	46	9,801–9,900	59
2,401–2,500	52	4,901–5,000	44	7,401–7,500	62	9,901–10,000	52

It is time to connect the square root law and the law of averages. Suppose a coin is tossed a large number of times. Then heads will come up on about half the tosses:

number of heads = half the number of tosses + chance error.

How big is the chance error likely to be? At first, Kerrich's assistant thought it would be very small. The data showed him to be wrong. During the course of the experiment, the chance error grew in absolute terms but shrank in relative terms—just as the mathematics predicts. (See Figures 1 and 2 on pp. 241 and 242.) According to the square root law, the likely size of the chance error is $\sqrt{\text{number of tosses}} \times 1/2$. For instance, with 10,000 tosses the standard error is 50. When the number of tosses goes up to 1,000,000 the standard error goes up too, but only to 500. As the number of tosses goes up, the SE for the number of heads gets bigger and bigger in absolute terms, but smaller and smaller relative to the number of tosses, because of the square root. That is why the percentage of heads gets closer and closer to 50%. The square root law is the mathematical explanation for the law of averages.

EXERCISE SET D

1. A coin is tossed 16 times.
   (a) The number of heads is like the sum of 16 draws made at random with replacement from one of the following boxes. Which one and why?

   (i) | head | tail | (ii) | 0 | 1 | (iii) | 0 | 1 | 1 |

   (b) The number of heads will be around _____, give or take _____ or so.

2. True or false:
   (a) The SD of the box | 0 | 0 | 1 | 1 | 1 | is $\sqrt{2 \times 3}$.

   (b) The SD of the box | -1 | -1 | 1 | 1 | 1 | is $\sqrt{(2/5) \times (3/5)}$.

   (c) The SD of the box | 0 | 1 | 1 | is $\sqrt{(2/3) \times (1/3)}$.

   Explain your answers.

3. One hundred draws are made at random with replacement from the box | 1 | 2 | 3 | 4 | 5 |. What is the chance of getting between 8 and 32 tickets marked "5"?

4. According to the simplest genetic model, the sex of a child is determined at random—as if by drawing a ticket at random from the box

   | male | female |

   What is the chance that of the next 2,500 births (not counting twins or other multiple births) in a certain country, more than 1,275 will be females?

5. How many of the counts in Table 4 on p. 268 should be in the range 45 to 55? How many are? (Endpoints included.)

6. A coin is tossed 10,000 times. What is the chance that the number of heads will be in the range 4,850 to 5,150?

7. A coin is tossed 1,000,000 times. What is the chance that the number of heads will be in the range 498,500 to 501,500?

8. A computer program is written to do the following job. There is a box, with ten blank tickets. You tell the program what numbers to write on the tickets, and how many draws to make. Then, the computer will draw that many tickets at random with replacement from the box, add them up, and print out the sum—but not the draws. This program does not know anything about coin tossing. Still, you can use it to simulate the number of heads in 1,000 tosses of a coin. How?

$SE_{sum} = \sqrt{n} \times SD$

↑

# of draws

9. A die is rolled 100 times. Someone figures the expected number of aces as $100 \times 1/6 = 16.67$, with an SE of $\sqrt{100} \times \sqrt{1/6 \times 5/6} \approx 3.73$. What does statistical theory say? (An ace is $\boxed{\cdot}$.)

10. One draw will be made at random from one of the boxes below. You win $1 if you can guess the result exactly.
    (a) Which box is best? worst?
    (b) From which box is the draw least variable? Most variable?
    (c) Which box has the smallest SD? largest?

    (i) $\boxed{0}\ \boxed{0}\ \boxed{0}\ \boxed{1}\ \boxed{1}\ \boxed{1}$

    (ii) $\boxed{0}\ \boxed{0}\ \boxed{1}\ \boxed{1}\ \boxed{1}\ \boxed{1}$

    (iii) $\boxed{0}\ \boxed{0}\ \boxed{0}\ \boxed{0}\ \boxed{0}\ \boxed{1}$

The answers to these exercises are on p. A-50.

Technical note: The shortcut formula for the SD of a zero-one box can be proved by algebra, but the main idea can be understood by working out the SD for the box with a 1 and five 0's using the long method: in the end, this must give the same answer as the shortcut. Here is the calculation. The average of the numbers in the box is 1/6, so the deviations from average are

$$5/6 \qquad -1/6 \qquad -1/6 \qquad -1/6 \qquad -1/6 \qquad -1/6.$$

Their root-mean-square is

$$\sqrt{\frac{(5/6)^2 + 5 \times (1/6)^2}{6}} = \sqrt{1/6 \times (5/6)^2 + 5/6 \times (1/6)^2}$$
$$= \sqrt{1/6 \times 5/6 \times (5/6 + 1/6)} = \sqrt{1/6 \times 5/6}.$$

This is the same as the answer given by the shortcut method. In the equation, 1/6 and 5/6 were factored out, using the principle

$$a\,b\,c + a\,b\,d = a\,b\,(c + d).$$

5. REVIEW EXERCISES

1. When drawing at random with replacement from a box of numbered tickets, the SE for the sum is
    (i)   (number of draws) × (SD of box).
    (ii) (number of draws) × $\sqrt{\text{SD of box}}$.
    (iii) $\sqrt{\text{number of draws}}$ × (SD of box).

2. Take the sum of one hundred draws made at random with replacement from the box

    $\boxed{1}\ \boxed{2}\ \boxed{2}\ \boxed{4}\ \boxed{5}\ \boxed{10}$

    (a) How small can the sum be? How large?
    (b) The chance that the sum will be between 375 and 425 is closest to

    $\qquad$ 1%$\qquad$ 25%$\qquad$ 50%$\qquad$ 75%$\qquad$ 99%

3. True or false: the SD of the box $\boxed{-1}\ \boxed{-1}\ \boxed{1}$ is $\sqrt{(2/3)\times(1/3)}$. Explain.

4. A large group of people get together. Each one rolls a die 180 times, and counts the number of 6's. About what percentage of these people should get counts in the range 15 to 45?

5. A die will be thrown some number of times, and you have to guess the total number of spots that will be rolled. You lose a dollar for each spot that your guess is off. For instance, if you guess 200 and the total is 215, you lose $15. Which do you prefer: 50 throws, or 100? Explain.

6. The following two single-roll bets are available at craps

$$(A)\ \boxed{\bullet\ \ \bullet}\qquad\qquad (B)\ \boxed{\bullet\ \ \bullet\bullet}$$

If you bet on A, you win provided the shooter throws a 2 on his next turn. If you bet on B, you win provided the shooter throws a 3. Bet A pays thirty to one, bet B pays fifteen to one. To place a bet, you have to put a chip on the corresponding area of the table. The *Harrah's Club* booklet on craps advises: "And if you cannot reach these bets, the dealer will be happy to place them for you."
   (a) Why would the dealer be particularly happy to place these bets?
   (b) Why should the house pay thirty to one on A, and only fifteen to one on B?

7. You play roulette 1,000 times, staking $1 each time. There are two possibilities:
   (A) You bet on a split each time.
   (B) You bet on a number each time.
   (A split pays 17 to 1, and you have 2 chances in 38 to win; a number pays 35 to 1, and you have 1 chance in 38 to win.) True or false:
   (a) The chance of coming out ahead is the same with A and B.
   (b) The chance of winning more than $100 is bigger with B.
   (c) The chance of losing more than $100 is bigger with B.
   Explain your answers.

Box		Average	SD
1 ticket $\boxed{35}$	37 tickets $\boxed{-1}$	−0.05	5.8
2 tickets $\boxed{17}$	36 tickets $\boxed{-1}$	−0.05	4.0

8. A coin is tossed 10,000 times.
   (a) If there were 5,200 heads, what percentage of the time did the coin land heads?
   (b) If the coin landed heads 50.5% of the time, how many heads were there?
   (c) There are two options:
      (i) You win $1 if the number of heads is bigger than 5,100.
      (ii) You win $1 if the percentage of heads is bigger than 51%.
   Which option is better, or are they the same? Explain.

9. A number of draws will be made at random with replacement from the
box | -3 | -1 | 0 | 1 | 3 |

You win \$1 if the average of the draws is between −0.1 and 0.1. Which is
better, 100 draws or 1,000 draws? No calculations are necessary, but
explain your answer.

10. A coin is tossed 100 times.
(a) The difference ''number of heads − number of tails'' is like the
sum of 100 draws from one of the following boxes. Which one, and
why?

(i) | heads | tails |

(ii) | 0 | 1 |

(iii) | -1 | 1 |

(iv) | -1 | 1 | 1 |

(b) Find the expected value and standard error for the difference.

11. A true-false quiz has 16 questions. A correct answer is worth 2 points,
but a point is taken off for each incorrect answer. The passing score is set
at 20. A large number of students take the quiz, and each one guesses the
answers by tossing a coin (quietly). The percent who pass should be
closest to: 4%, 12%, 16%, 20%, 50%. Explain.

12. A box has seven tickets in it, numbered 1 through 7:

| 1 | 2 | 3 | 4 | 5 | 6 | 7 |

The average of this box is 4, with an SD of 2. Someone draws seven
tickets at random without replacement. True or false: The SE for the
sum of the draws is $\sqrt{7} \times 2$. Explain.

13. Imagine making a scatter diagram from Table 4 on p. 268 the following
way. Plot the point whose x-coordinate is the number of heads in tosses
#1–100, and whose y-coordinate is the number of heads in tosses #101–
200. This gives (44, 54). Then, plot the point whose x-coordinate is the
number of heads on tosses #201–300, and whose y-coordinate is the
number of heads in tosses #301–400. This gives (48, 53). And so on. Will
the scatter diagram look like (i) or (ii) or (iii) or (iv)? Explain your reasoning briefly. (To get a hint, fill in the blanks: the number of heads in one
hundred tosses should be around _____, give or take _____ or so.)

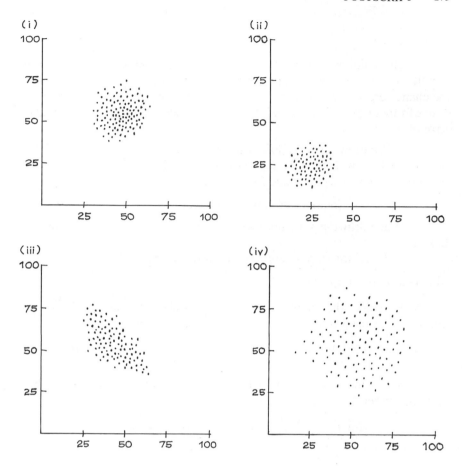

6. POSTSCRIPT

The exercises of this chapter teach a melancholy lesson: the more you gamble, the more money you lose. The basic reason is that all the bets are unfair, in the sense that your expected net gain is negative. So the law of averages works for the house, not for you. Of course, this chapter only discussed quite simple strategies, and gamblers have evolved quite complicated "systems" for betting on roulette, craps, and the like. But it is a theorem of mathematics that no system for compounding unfair bets can ever make your expected net gain positive. In proving this theorem, only two assumptions are needed:
- you aren't clairvoyant;
- your financial resources are finite.

The game of blackjack is an exception, in that under some circumstances there are bets with a positive expected net gain.[3] As a result, people have made a lot of money on blackjack.

7. SUMMARY

1. The number generated by a chance process should be somewhere around the *expected value,* but will be off by a chance error. The likely size of the chance error is given by the *standard error*. For instance, the sum of the draws from a box will be around the expected value, give or take a standard error or so.

2. When drawing at random with replacement from a box of numbered tickets, each draw adds to the sum an amount which is around the average of the box. So the expected value for the sum is

(number of draws) × (average of box).

3. When drawing at random with replacement from a box of numbered tickets,

SE for sum = $\sqrt{\text{number of draws}}$ × (SD of box).

This is a *square root law*.

4. Provided the number of draws is sufficiently large, the normal curve can be used to figure chances for the sum of the draws.

5. If you have to classify and count the draws, remember to put "1" on the tickets that count for you, "0" on the others.

6. When the box contains only 0's and 1's, its SD can be figured by a shortcut method as

$\sqrt{(\text{fraction of 1's in box}) \times (\text{fraction of 0's in box})}.$

18

The Normal Approximation for Probability Histograms

Everybody believes in the [normal approximation], the experimenters because they think it is a mathematical theorem, the mathematicians because they think it is an experimental fact.

—G. LIPPMANN (FRENCH PHYSICIST, 1845–1921)

1. INTRODUCTION

About 1700 a Swiss mathematician named James Bernoulli succeeded in giving a rigorous mathematical proof of the law of averages: if a coin is tossed a large number of times, the percentage of times it lands heads will be close to 50% with high probability. The matter rested there for about twenty years. Then, Abraham de Moivre made a substantial improvement on Bernoulli's work by showing how to compute the chance that the percentage of heads will fall within any given interval around the limiting value of 50%. The computation is not exact, but only approximate; however, the approximation gets better and better as the number of tosses goes up. So de Moivre could say, to a very good approximation, just how high Bernoulli's high probability really was.

Bernoulli and de Moivre both made the same assumptions about the coin: the tosses are independent, and on each toss the coin is as likely to land heads as tails. (The theory can be modified quite easily to handle an unsymmetric coin, as will be seen below.) From these assumptions, it follows that the coin is as likely to land in any specific pattern of heads and tails as in any other. What Bernoulli did was to show that for most patterns, about 50% of the entries are heads.

You can see this starting to happen even with five tosses. Imagine tossing the coin five times, and keeping a record of how it lands on each toss. One possible pattern is all heads:

$$H\ H\ H\ H\ H$$

This pattern has five heads. How many patterns are there with four heads? The answer is 5:

$$T\ H\ H\ H\ H \quad H\ T\ H\ H\ H \quad H\ H\ T\ H\ H \quad H\ H\ H\ T\ H \quad H\ H\ H\ H\ T$$

The pattern T H H H H, for instance, means that the coin landed tails on the first toss, then gave four straight heads. Table 1 shows how many patterns there are for any given number of heads.

Table 1. The number of patterns corresponding to a given number of heads, in five tosses of a coin.

Number of heads	Number of patterns
zero	1
one	5
two	10
three	10
four	5
five	1

With 5 tosses, there are altogether $2^5 = 32$ possible patterns in which the coin can land. And 20 patterns out of the 32 have nearly half heads, two or three out of five.

De Moivre managed to count, to within a small margin of error, the number of patterns having a given number of heads—for any number of tosses. To see what this involves, take 100 tosses. The number of patterns he had to think about is 2^{100}. This is quite a large number. If you tried to write all these patterns out, it might be possible to get a hundred of them on a page the size of this one. By the time you finished writing, you would have enough books to fill a shelf reaching from the earth to the farthest known star.

Still and all, mathematicians have a formula for the number of patterns with exactly fifty heads. It is given by the binomial coefficient (Chapter 15, but if you skipped this it won't matter here):

$$\frac{100!}{50! \times 50!} = \frac{100 \times 99 \times \cdots \times 51}{50 \times 49 \times \cdots \times 1}$$

Now this formula was of no immediate help to de Moivre, because it doesn't say how big the number is. The arithmetic looks very unpleasant, but by calculator[1] the number works out to 1.01×10^{29}, and 2^{100} is 1.27×10^{30}. So the chance of getting exactly 50 heads in 100 tosses of a coin is

$$\frac{1.01 \times 10^{29}}{1.27 \times 10^{30}} \approx .08 = 8\%.$$

Of course, de Moivre did not have anything like a modern calculator available—and maybe it's a good thing he didn't! He needed a mathematical way of estimating the binomial coefficients, without having to work the arithmetic out. And the right technique had just been invented by the English mathematician James Stirling. Using Stirling's formula, de Moivre was led to discover the normal curve. For example, he found that the chance of getting exactly 50 heads in 100 tosses of a coin was about equal to the area under the normal curve between -0.1 and $+0.1$, which is 8%. And in fact, he was able to prove mathematically that the whole *probability histogram* for the number of heads, when scaled in standard units, got closer and closer to the normal curve as the number of tosses went up. Modern researchers have extended this result to the sum of draws made at random from any box of tickets.

The mathematical details[2] of de Moivre's argument are too complicated to go into here, but we can retrace his steps graphically, using a computer to draw the pictures. First, however, probability histograms must be discussed.

2. PROBABILITY HISTOGRAMS

When a chance process generates a number, the expected value and standard error are a guide to where that number will be. But the *probability histogram* gives a complete picture.

> A probability histogram is a new kind of graph. It represents chance, not data.

Here is the idea, illustrated in a specific example. When playing craps, gamblers bet on the total number of spots shown when a pair of dice are rolled. The value of the total is determined by a chance process: rolling the dice. So the odds depend on the chance of rolling each possible total, from 2 through 12. To find these chances empirically, a casino might hire someone to roll a pair of dice over and over again, making a list of the totals observed on the different rolls, and from that a histogram (as in Chapter 3). This experiment was simulated on the computer. The results for the first hundred rolls are shown in Table 2 on the next page.

Table 2. Rolling a pair of dice. The computer simulated one hundred rolls of a pair of dice, and the table shows the total number of spots observed on each roll.

Roll	Total	Roll	Total	Roll	Total	Roll	Total	Roll	Total
1	8	21	10	41	8	61	8	81	11
2	9	22	4	42	10	62	5	82	9
3	7	23	8	43	6	63	3	83	7
4	10	24	7	44	3	64	11	84	4
5	9	25	7	45	4	65	9	85	7
6	5	26	3	46	8	66	4	86	4
7	5	27	8	47	4	67	12	87	7
8	4	28	8	48	4	68	7	88	6
9	4	29	12	49	5	69	10	89	7
10	4	30	2	50	4	70	4	90	11
11	10	31	11	51	11	71	7	91	6
12	8	32	12	52	8	72	4	92	11
13	3	33	12	53	10	73	7	93	8
14	11	34	7	54	9	74	9	94	8
15	7	35	7	55	10	75	9	95	7
16	8	36	6	56	12	76	11	96	9
17	9	37	6	57	7	77	6	97	10
18	8	38	2	58	6	78	9	98	5
19	6	39	6	59	7	79	9	99	7
20	8	40	3	60	7	80	7	100	7

The first panel in Figure 1 shows the histogram for the data in Table 2. The total of 12 came up 5 times, so the rectangle over 12 has an area of 5%, and similarly for the other possible totals. The next panel shows the empirical histogram for the first 1,000 rolls, and the third is for 10,000 rolls. (*Empirical* means "experimentally observed.") In the long run, these empirical histograms converge to the ideal probability histogram shown in the bottom panel of the figure.

Of course, this probability histogram could have been found by a theoretical argument. As shown in Chapter 14, there are 6 chances in 36 of rolling a seven. That's $16\frac{2}{3}\%$. So in the long run, the pair of dice will roll seven about $16\frac{2}{3}\%$ of the time. Consequently, the area of the rectangle over seven in the probability histogram equals $16\frac{2}{3}\%$. The width is 1, so the height must be $16\frac{2}{3}\%$. Similarly for the other rectangles. The probability histogram shows the chances for each possible number of spots: the area of the rectangle over a total tells how often that total will turn up, in the long run.

A probability histogram represents chance by area. The histogram is made up of rectangles. The base of each rectangle is centered at one of the numbers the chance process can generate; the area of the rectangle equals the probability of getting that number. The total area of the histogram is 100%.

Figure 1. Empirical histograms converging to a probability histogram. The computer simulated 100 rolls of a pair of dice. On each roll, it counted up the total number of spots. Then it made a histogram for these 100 numbers (top panel). This is an empirical histogram—based on observations. The second panel is for 1,000 rolls of the pair, the third panel for 10,000 rolls. The bottom panel is the ideal or probability histogram for the number of spots.

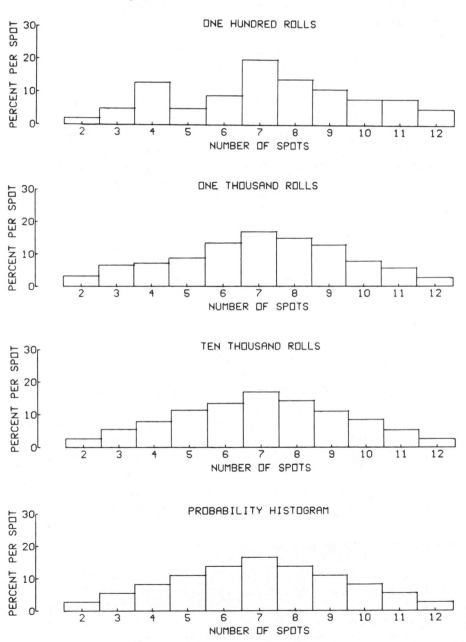

Rolling a pair of dice and counting the total number of spots is like taking the sum of two draws at random with replacement from the box

The same kind of reasoning can be used for any number of draws and any box. For instance, take the sum of twenty-five draws made at random with replacement from the box

The computer was programmed to simulate many repetitions of this chance experiment. The results of the first 100 repetitions were shown in Table 1 on p. 259. An empirical histogram for these 100 sums is shown at the top of Figure 2. For instance, 4 out of the 100 sums were equal to 75, so the rectangle over 75 has an area of 4%. An empirical histogram for 1,000 sums is shown next, and then for 10,000 sums. The probability histogram for the sum of twenty-five draws is shown at the bottom. Just as in the dice example, the empirical histograms get closer and closer to this theoretical probability histogram, as the number of repetitions of the experiment gets larger and larger. (Please distinguish between the number of draws and the number of repetitions. On each repetition of the experiment, the computer makes twenty-five draws at random from the box. This is so whether the number of repetitions is 100, or 10,000.)

Virtually all the area in the probability histogram (about 99.95%) is between 40 and 110. So the chance that the sum comes out between 40 and 110 is very close to 100%. In the first 1,000 repetitions of the chance experiment, all but one of the sums turned out to be between these two limits. In 10,000 repetitions, all but 6 of the sums turned out to be between 40 and 110. That is why the horizontal axis in Figure 2 only runs between these two values.

Now for some examples to show how probability histograms can be read. Suppose, for instance, that you are allowed to draw twenty-five times at random with replacement from one of box A or box B, and take the sum in dollars:

Which is better? Or are they the same?

Figure 2. Empirical histograms converging to a probability histogram. The computer simulated 100 repetitions of a chance experiment. On each repetition, it drew twenty-five tickets at random with replacement from the box

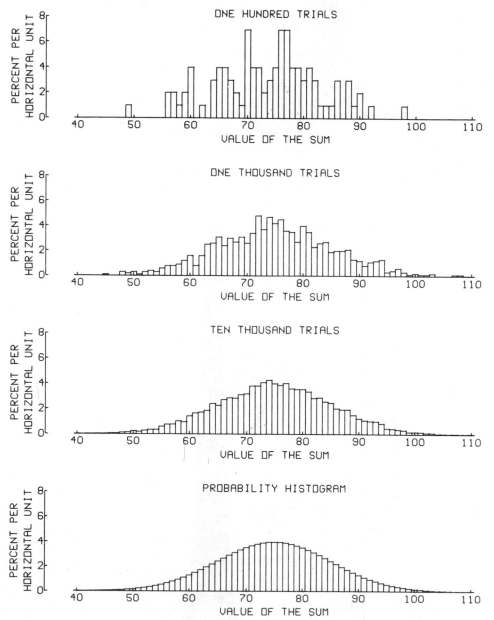

and added them up. It made a histogram for these 100 sums in the top panel. This is an empirical histogram. The second panel is for 1,000 repetitions, and the third for 10,000. The bottom panel shows the probability histogram for the sum. The empirical histograms converge to the probability histogram.

Figure 3. Comparing probability histograms. These histograms are for the sum of twenty-five draws made at random with replacement from box A and box B respectively:

The histogram for the sum from box A is more to the right, so this sum tends to be bigger. The histogram for the sum from B is taller and narrower, so this sum is easier to predict.

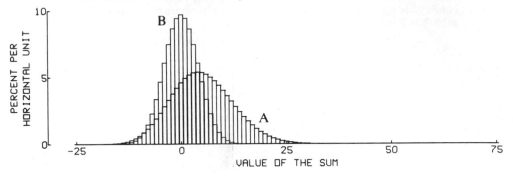

Figure 4. Probability histogram for the proportion of heads in one hundred tosses of a coin. Chance is represented by area, not height.

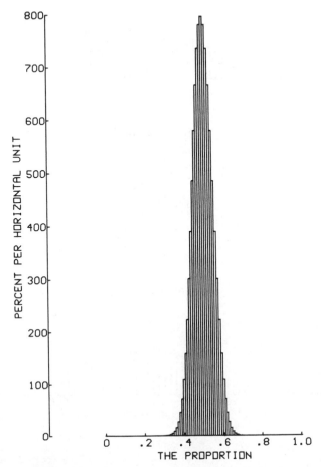

Figure 3 shows the probability histograms for the two sums. The one from A is on the whole to the right of the one from B, so the sum from A is on the whole bigger than the sum from B. Box A is better because it is likely to get you more dollars.

Suppose the rules of the game are changed. This time, you have to guess the value of the sum, and win ten dollars if you are right to within 5. Now which box is better? The histogram for the sum from B is taller and narrower, there is more area closer to the center. The sum from B is less variable, making it easier to predict. With the new rules, box B is the one.

As a final example, Figure 4 shows the probability histogram for the proportion of heads in one hundred tosses of a coin. The *proportion* is the number of heads, divided by the number of tosses. With 60 heads, that's $60/100 = 0.60$. With 47, it's $47/100 = 0.47$. The rectangle over 0.50 is 800% high. Do not be alarmed. This rectangle is only 0.01 wide, so its area is $800\% \times 0.01 = 8\%$. That is the chance that the proportion of heads will be exactly 0.50. Chance is represented by area, not by height.

EXERCISE SET A

1. The figure below is a probability histogram for the sum of 25 draws from the box ⎢ 1 2 3 4 5 ⎢. The shaded area represents the chance that the sum will be between _____ and _____ (inclusive).

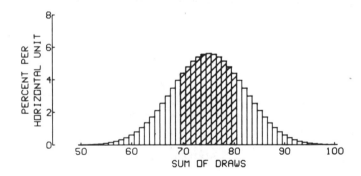

2. The figure below is the probability histogram for the sum of 25 draws made at random with replacement from a box. True or false: the shaded area represents the percentage of times you draw a number between 5 and 10 inclusive.

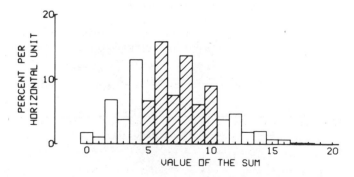

3. Using Figure 2: which value came up most often in 100 repetitions? in 1,000? in 10,000? Which value is likeliest?

4. Using Figure 3, say in each case whether the chance is closest to
1%   25%   50%   75%   99%.
   (a) The chance that the sum of 25 draws from box A will be positive.
   (b) The chance that the sum of 25 draws from box B will be positive.
   (c) The chance that the sum of 25 draws from box A will be more than 10.
   (d) The chance that the sum of 25 draws from box B will be more than 10.

5. Using Figure 4:
   (a) The chance that the proportion of heads in 100 tosses will be between 0.2 and 0.8 is closest to:
1%   10%   100%.
   (b) The chance that proportion of heads will be 0.40 exactly is closest to:
1%   10%   100%.

The answers to these exercises are on p. A-50.

3. PROBABILITY HISTOGRAMS AND THE NORMAL CURVE

The object of this section is to show, graphically, how the probability histograms for the number of heads gets close to the normal curve as the

Figure 5. Scaling. Probability histograms for the number of heads in four, sixteen or sixty-four tosses of a coin. As the number of tosses goes up, the histograms move off to the right and get flatter.

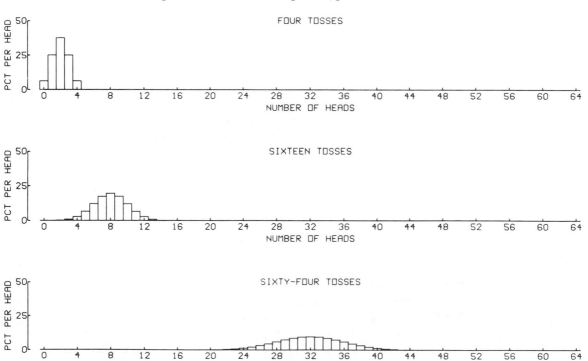

number of tosses goes up. One problem is the scaling, which is overcome by the use of standard units. In Figure 5, the computer drew the probability histogram for the number of heads in four tosses of a coin (top panel), sixteen tosses (middle panel), and sixty-four tosses (bottom panel). The thing to notice is that as the number of tosses increases, the histograms move off to the right, getting wider and flatter as they go. With more tosses, everything will disappear.

As a corrective measure, it will help to shift the histograms so that the expected values all line up—

- with 4 tosses, the expected number of heads is 2;
- with 16 tosses, the expected number of heads is 8;
- with 64 tosses, the expected number of heads is 32.

Figure 6 shows the results. The drift to the right has been stopped, but the histograms are still getting wider and flatter. (With 64 tosses, there was hardly any probability below 16 heads, or above 48, so only the central part of this histogram is shown. To bring out the shapes of the histograms, they have all been enlarged by the same factor.)

The histograms in Figure 6 are getting wider and flatter because they have been drawn so that one head covers an equal horizontal distance in all

Figure 6. Scaling. Probability histograms for the number of heads in four, sixteen, or sixty-four tosses of a coin, scaled so that the expected values line up.

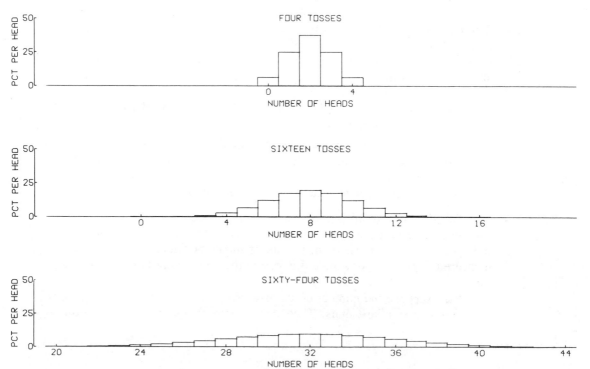

three pictures. But as far as the chances are concerned, one head means something different from picture to picture. The common unit for the chances is the standard error:

- with 4 tosses, the standard error for the number of heads is 1;
- with 16 tosses, the standard error for the number of heads is 2;
- with 64 tosses, the standard error for the number of heads is 4.

The trouble with Figure 6 is that a standard error in the bottom picture covers four times as much distance on the horizontal axis as a standard error in the top picture.

To keep the histograms from spreading out, they must be redrawn so that one standard error covers the same horizontal distance in all three cases. The left-hand panel of Figure 7 does this. The extra horizontal axes show standard units, or deviations from the expected value expressed as multiples of the standard error. With sixty-four tosses, for example, 40 heads is 8 heads more than the expected number, that's 2 SEs. So 2 on the standard-units axis falls directly below 40 on the number-of-heads axis.

As you look down the left-hand panel of Figure 7, the histograms are all about the same width: the spread has been stopped. But there is still something wrong. The histograms are still disappearing, because they are flattening out.

The bottom histogram in the left hand panel of Figure 7 has to be raised up, to make it match the top histogram. The best thing is to make the areas equal. Back in Figure 5, the two histograms did cover the same area on the page. But when the bottom histogram was made four times narrower, this decreased its area by a factor of four. So its height has to be jacked up by a factor of four, the SE. Similarly, the height of the middle histogram has to be doubled; its SE is 2. This final adjustment is made in the right-hand panel of Figure 7. Now the histograms all have just about the same shape—and that shape is the normal curve. The computer has drawn this curve over each histogram, for comparison.

The histograms on the right in Figure 7 are drawn in standard units. This involves three adjustments:

- shifting the histograms to make the expected values line up;
- changing the horizontal scale so the standard errors cover the same horizontal distance;
- changing the vertical scale so all the areas are equal.

Technically, multiplying the height of the histogram by its SE changes the units of the vertical scale from percent per head to percent per standard unit. As a result, the areas of the rectangles should be figured with the height measured in percent per standard unit, and the width measured in standard units.

Now suppose the number of tosses goes up, and you keep drawing the histograms in standard units. Then they will get closer and closer to the

Figure 7. Scaling. The left-hand panel shows probability histograms for the number of heads in four, sixteen, or sixty-four tosses of a coin, scaled so that the expected values line up, and the standard errors cover the same horizontal distance. The extra horizontal scale shows standard units. At the right, these histograms have been adjusted so they cover the same area. (When horizontal distances are measured on the "standard units" scale, and vertical distances on the "percent per standard unit" scale, this area is 100%.) The normal curve is shown for comparison.

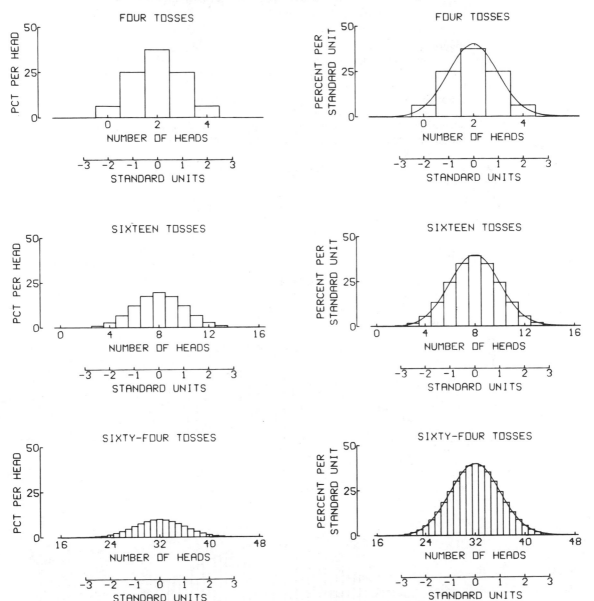

normal curve. You can see it happening in Figure 8. This shows probability histograms for the number of heads in 100, 400, and 900 tosses of a coin, all drawn in standard units. The figures only cover the range from −3 to +3 in standard units, because well over 99% of the probability is concentrated there. The histogram for 100 tosses is already very close to the normal curve. The one for 900 tosses is practically indistinguishable from the curve. In the early eighteenth century, de Moivre proved this convergence had to take place, by pure mathematical reasoning.

Figure 8. The normal approximation. Probability histograms are shown for the number of heads in 100, 400, and 900 tosses of a coin, scaled so the expected values line up, the standard errors cover the same horizontal distances, and the histograms cover the same area. (When horizontal distances are measured on the "standard units" scale and vertical distances on the "percent per standard unit" scale, this area is 100%.) The normal curve is shown for comparison. The histograms follow the normal curve better and better as the number of tosses goes up.

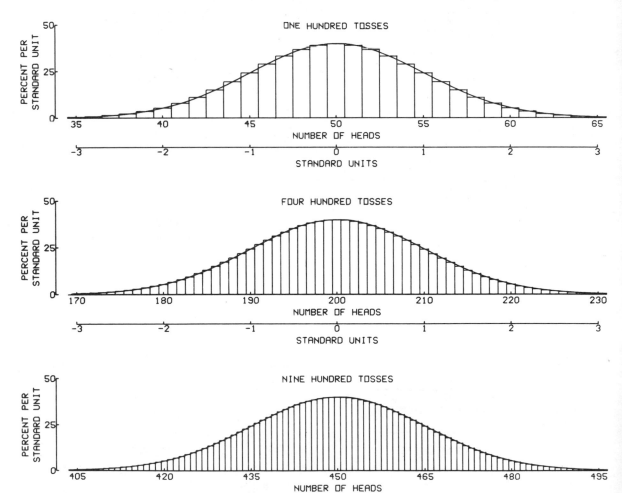

4. THE NORMAL APPROXIMATION

The normal curve has already been used in Chapter 17 to figure chances. The method is due to de Moivre, and this section will explain the logic and present a technique for taking care of endpoints. This technique should be used when the number of tosses is small or extra accuracy is wanted.

Example 1. A coin will be tossed 100 times. Estimate the chance of getting
    (a) between 45 and 55 heads inclusive.
    (b) between 45 and 55 heads exclusive.
    (c) exactly 50 heads.

Solution. The expected number of heads is 50, with a standard error of 5, as shown on p. 267 above.

Part (a). The chance of getting 45 to 55 heads inclusive is represented by the total area of the eleven rectangles over the values 45 through 55 in Figure 9. In other words, it's exactly the area under the histogram between 44.5 and 55.5 on the number-of-heads scale. Because the histogram follows the normal curve so closely, this is almost equal to the area under the curve between these two values expressed in standard units, that is, between -1.1 and $+1.1$. From the table, this normal area is about 72.87%. The exact chance is 72.87%, to two decimals.[3] The approximation is dead on.

Figure 9. The normal approximation. The chance of getting between 45 and 55 heads inclusive in 100 tosses of a coin is exactly equal to the area under the histogram between 44.5 and 55.5 on the number-of-heads scale, corresponding to -1.1 and $+1.1$ on the standard-units scale. Since the histogram follows the normal curve so closely, the chance is nearly equal to the area under the normal curve between -1.1 and $+1.1$.

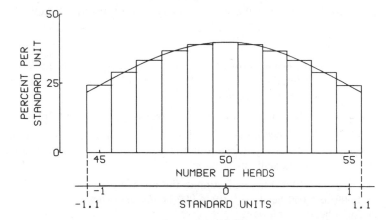

Part (b). The chance of getting 45 to 55 heads exclusive is represented by the total area of the nine rectangles over the values 46 through 54. That's the area under the histogram between 45.5 and 54.5 on the number-of-heads scale, which correspond to -0.9 and $+0.9$ on the standard-units scale. That's almost equal to the area between −0.9 and +0.9 under the normal curve, which is 63.19%. The exact chance is 63.18%, to two decimals.

Part (c). The chance of getting exactly 50 heads equals the area of the rectangle over 50, whose base goes from 49.5 to 50.5 on the number-of-heads scale. That's −0.1 to 0.1 in the standard units. So the chance is approximately equal to the area between −0.1 to 0.1 under the normal curve. From the normal table, the area is about 7.97%. The exact chance is 7.96%, to two decimals. This completes Example 1.

Usually, the problem will only ask for the chance that (for instance) the number of heads is between 45 and 55, without specifying the endpoints. Then, you can use the compromise procedure:

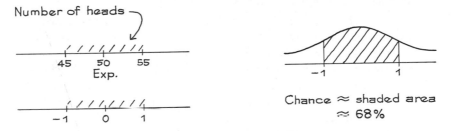

This amounts to replacing the area under the histogram between 45 and 55 by the area under the normal curve between these two values (in standard units). It splits the two end rectangles in half, and doesn't give quite as much precision as the method used in Example 1. Keeping track of the endpoints is worthwhile if the rectangles are big, or if a lot of precision is needed.

The normal approximation consists in replacing the actual probability histogram by the normal curve, before computing the area. This is legitimate when the probability histogram follows the normal curve. The point of the approximation is that probability histograms are often very hard to work out, while areas under the normal curve are easy to look up in the table.

EXERCISE SET B

1. In Figure 9, the chance of getting 52 heads is exactly equal to the area between 51.5 and 52.5 under the _____ . Fill in the blank, with one of the two options: normal curve, histogram.

2. A coin is tossed 100 times. Estimate the chance of getting 60 heads.

3. Kerrich's data on 10,000 tosses of a coin can be taken in successive groups of 100 tosses (Table 4 on p. 268). About how many groups should show exactly 60 heads? How many actually do?

4. A coin is tossed 10,000 times. Estimate the chance of getting
   (a) 4,900 to 5,050 heads.   (b) 4,900 heads or fewer.
   (c) 5,050 heads or more.

5. (a) Suppose you were going to estimate the chance of getting 50 heads or fewer in 100 tosses of a coin. Should you keep track of the edges of the rectangles?
    (b) Same, for the chance of getting 450 heads or fewer in 900 tosses.
    No calculations are needed, just look at Figure 8.

The answers to these exercises are on p. A-51.

5. THE SCOPE OF THE NORMAL APPROXIMATION

So far, the discussion has been about a fair coin, which lands heads or tails with chance 50%. Does the normal curve work for a biased coin? The answer is yes, as de Moivre showed, although more tosses are needed before the approximation takes hold. The worse the bias, the more tosses are needed. To illustrate the point, the computer was programmed to work out the probability histograms for a coin which is very biased—it lands heads with only one chance in ten. The histograms for 25, 100, or 400 tosses are shown in Figure 10, with a normal curve for comparison. The approximation

Figure 10. The normal approximation for a biased coin. The coin lands heads with chance 10%. The top panel shows the probability histogram for the number of heads in twenty-five tosses, the middle panel for one hundred tosses, the bottom panel for four hundred tosses. A normal curve is shown for comparison. The histograms are higher than the normal curve on the left, and lower on the right, because the coin is biased.[4] But with a hundred tosses, the normal approximation is quite satisfactory.

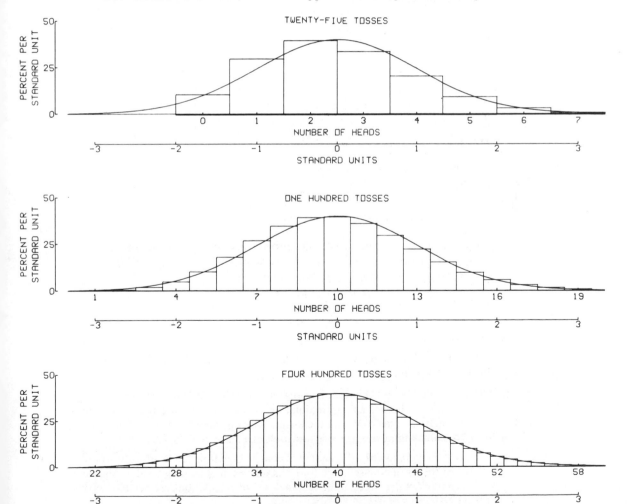

is not very good with twenty-five tosses, but is already quite good with one hundred. Things improve from there.

What about drawing from a box? Again, the normal approximation works perfectly well, with one caution. The further the histogram of the numbers in the box is from the normal curve, the more draws are needed before the approximation takes hold. For example, look at the box

The average of this box is 4, with an SD of about 3.6. A histogram for the tickets in the box is shown in Figure 11, with a normal curve for comparison. This histogram looks nothing like the curve.

Figure 11. Histogram for the contents of the box ⎣ 1 2 9 ⎦. This histogram is very different from the normal curve, but the probability histogram for the sum of one hundred draws from the box is already very close to the curve.

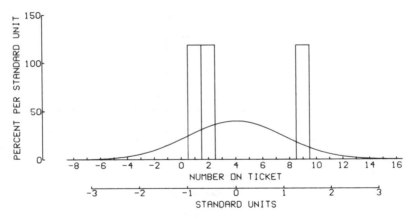

Even with twenty-five draws, the probability histogram for the sum is quite different from the curve—it shows waves (Figure 12).[5] With fifty draws, the waves are still there, but much smaller, and by one hundred draws the probability histogram is indistinguishable from the curve.

EXERCISE SET C

1. A biased coin has one chance in ten of landing heads. It is tossed four hundred times. Estimate the chance of getting exactly forty heads.

2. A biased coin has one chance in ten of landing heads. It is tossed twenty-five times. Suppose the normal approximation is used to estimate the chance of getting exactly one head. Would the estimate be just about right? too high? too low? No calculations are needed; just look at Figure 10.

3. Suppose you were asked to estimate the chance of getting ten heads or fewer in a hundred tosses of a biased coin (one chance in ten of landing heads). Should you keep track of the edges of the rectangles? No calculations are needed; look at Figure 10.

Figure 12. The normal approximation for a sum. Probability histograms are shown for the sum of draws from the box ⌊ 1 2 9 ⌋. The top panel is for twenty-five draws, and does not follow the normal curve especially well. (Note the waves.) The middle panel is for fifty draws, and the bottom panel is for one hundred draws. It already follows the normal curve very well.

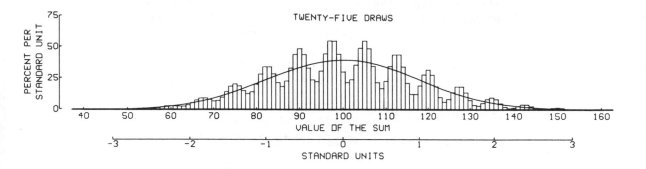

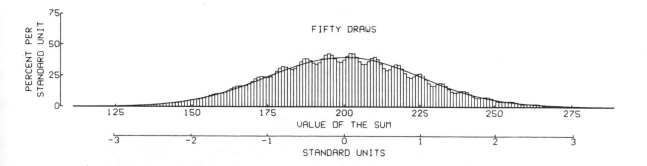

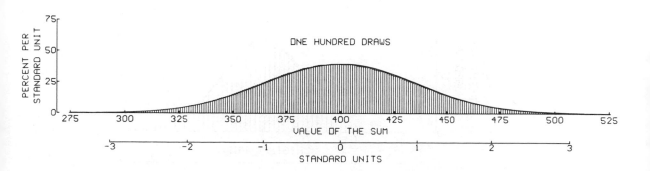

4. For the sum of twenty-five draws made at random with replacement from the box | 1 | 2 | 9 | :

   (a) Which of the following values is most likely? least likely?

             100   101   102   103   104   105.

   No calculations are necessary, just look at Figure 12.

   (b) What is the expected value for the sum?

5. Someone draws a hundred times at random from a box containing numbered tickets. Between successive draws, he always returns the ticket just drawn to the box, and also puts in ten new tickets, all identical with the one just drawn. The drawing begins with only two tickets in the box, one marked +1, the other marked −1. The sum of the draws will turn out to be somewhere between −100 and 100. Do you think the probability histogram will look like the normal curve? Explain your answer briefly.

The answers to these exercises are on pp. A-51–A-52.

6. CONCLUSION

The main point of the chapter can be stated as follows.

> Whatever is in the box, with a large enough number of draws the probability histogram for the sum (when put in standard units) follows the normal curve quite closely.[6]

This fact is the key to the statistical procedures in the rest of the book.

How many draws is enough? There is no set answer, because much depends on the contents of the box—remember the waves in Figure 12. However, for many boxes, the probability histogram for the sum of a hundred draws will be close enough to the normal curve for most purposes.

When the probability histogram does follow the normal curve, it can be summarized by the expected value and standard error. For instance, suppose you had to draw such a histogram—without any further information. In standard units you can do it, at least to a first approximation:

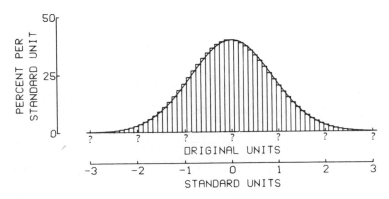

To finish the picture, you have to translate the standard units back into original units—you have to fill in the question marks. This is what the expected value and standard error do. So they say almost all there is to know about the histogram, because its shape is just like the normal curve.

> The expected value pins the center of the probability histogram to the horizontal axis, and the standard error fixes its spread.

Using the square root law, the expected value and standard error for a sum can be computed from

- the number of draws;
- the average of the box;
- the SD of the box.

So these three quantities just about determine the behavior of the sum. That is why the SD of the box is such an important measure of the spread shown by the numbers in the box.[7]

This chapter discussed two sorts of convergence for histograms, and it is important to separate them. In section 1, the number of draws from the box was fixed. The basic chance process of drawing from the box and taking the sum was repeated a larger and larger number of times. The empirical histogram from the observed values of the sum—a histogram for data—converged to the probability histogram for the sum. This probability histogram followed the normal curve, but was bumpier. In section 5, the number of draws from the box got larger and larger. Then the probability histogram for the sum—a histogram for chances—got smoother and smoother, and in the limit became the normal curve.

In Part II of the book, the normal curve was used for data. In some cases, this can be justified by a mathematical argument which uses the two types of convergence discussed in this chapter. When the number of repetitions is large, the empirical histogram for the sums of draws from a box will be close to the probability histogram for the sum. When the number of draws is large, the probability histogram for the sum will be close to the normal curve. Consequently, when the number of repetitions and the number of draws are both large, the empirical histogram for the sums will be close to the curve.[8] This is all a matter of pure logic: a mathematician can prove every step.

But there is still something missing in the explanation of why a histogram for data should look like the normal curve. It has to be shown that the process generating the data is like drawing numbers from a box and taking the sum. This sort of argument will be discussed in Part VII, and more than mathematics is involved. There will be questions of fact to settle as well.

7. REVIEW EXERCISES

1. The figure below shows the probability histogram for the total number of spots when a die is rolled thirty times. The shaded area represents the chance that the total will be between _____ and _____ (inclusive).

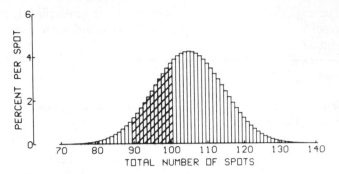

2. A hundred draws are made at random with replacement from the box $\boxed{\boxed{1}\;\boxed{2}\;\boxed{2}\;\boxed{5}}$. One of the graphs below is a histogram for the numbers drawn. Another is the probability histogram for the sum. And the third is irrelevant. Which is which?

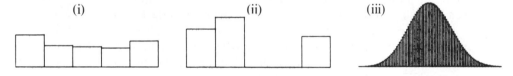

    (i)   (ii)   (iii)

3. A coin is tossed 100 times. The chance of getting exactly 55 heads is closest to: 1%, 5%, 10%, 25%, 55%.

4. A hundred draws are made at random with replacement from the box $\boxed{\boxed{3}\;\boxed{3}\;\boxed{3}\;\boxed{4}\;\boxed{5}}$. The chance of getting exactly twenty $\boxed{5}$'s is closest to: 10%, 20%, 50%, 90%.

5. Four hundred draws will be made at random with replacement from the box $\boxed{\boxed{0}\;\boxed{2}\;\boxed{3}\;\boxed{4}\;\boxed{6}}$.
    (a) Estimate the chance that the sum of the draws will be more than 1,250.
    (b) Estimate the chance that there will be fewer than a hundred $\boxed{6}$'s.

6. Say whether each statement is true or false, and explain why.
    (a) The SD of the list 0, 0, 1 is $\sqrt{2/3 \times 1/3}$.
    (b) The SD of the list $-1, 1, 1$ is $\sqrt{2/3 \times 1/3}$.

(c) When drawing at random with replacement from a box of numbered tickets, as the number of draws goes up the probability histogram for the sum (when put in standard units) follows the normal curve more and more closely.

(d) When drawing at random with replacement from a box of numbered tickets, as the number of draws goes up the histogram for the contents of the box (when put in standard units) follows the normal curve more and more closely.

(e) When drawing at random with replacement from a box of numbered tickets, as the number of draws goes up the histogram for the numbers drawn (when put in standard units) follows the normal curve more and more closely.

7. A pair of dice are thrown. The total number of spots is like

(i) one draw from the box $\boxed{2}\ \boxed{3}\ \boxed{4}\ \boxed{5}\ \boxed{6}\ \boxed{7}\ \boxed{8}\ \boxed{9}\ \boxed{10}\ \boxed{11}\ \boxed{12}$

(ii) the sum of two draws from the box $\boxed{1}\ \boxed{2}\ \boxed{3}\ \boxed{4}\ \boxed{5}\ \boxed{6}$

Explain.

8. Suppose you were to use the normal approximation to estimate the chance that the sum of 25 draws made at random with replacement from the box on p. 292 will equal 100 exactly. Would the estimate be somewhat too high? too low? or about right? (No calculations are necessary, just look at Figure 12 on p. 293.) Repeat this for the chance that the sum will be between 75 and 125.

9. A coin will be tossed 100 times. You get to pick 11 numbers. If the number of heads turns out to equal one of your 11 numbers, you win a dollar. Which 11 numbers should you pick, and what is your chance (approximately) of winning?

10. A sorcerer has hidden an M&M in one of an infinite row of boxes

You want to find it as quickly as possible. As a matter of fact, you only have time to look in eleven boxes. He says he will give you a hint. He will toss a coin 100 times and count the number of heads (all in a flash). He will not tell you this number, nor will he tell you the number of the box where he hid the M&M, but he will tell you the sum of these two numbers.

(a) If he tells you the sum is 75, which eleven boxes would you look in?

(b) As in (a), except replace "75" by "90."

(c) What is the general rule?

(d) Following this rule, how likely are you to locate the M&M?

11. A box contains ten tickets, five marked with a positive number and five with a negative number. All the numbers are between −10 and 10. A thousand draws will be made at random with replacement from the box. You are asked to estimate the chance that the sum will be positive.
    (a) Can you do it on the basis of the information already given?
    (b) Can you do it if you are also told the average and SD of the numbers in the box, but are not told the numbers themselves?
    Explain briefly.

12. Repeat Exercise 11, if you are asked to estimate the chance of getting 525 or more positive numbers.

13. Repeat Exercise 11, if you are asked to estimate the chance of getting 100 or more $\boxed{3}$'s.

14. A box contains ninety-nine tickets marked "0" and one ticket marked "1." Should the normal approximation be used for the sum of
    (a) 25 draws?   (b) 100 draws?   (c) 10,000 draws?
    Explain. (Hint: Compute the expected value and SE, and work out the normal approximation for the chance that the sum will be negative.)

8. SUMMARY

1. If the chance process for determining the value of a sum is repeated many times, the empirical histogram for the observed values tends to the *probability histogram*.

2. Area under a probability histogram represents chance.

3. Whatever is in the box, with a large enough number of draws, the probability histogram for the sum (when put in standard units) will follow the normal curve quite closely. This fact is used over and over again in statistics.

4. The normal approximation consists in replacing the actual probability histogram by the normal curve, before computing areas. Often, the accuracy of the approximation can be improved by keeping track of the edges of the rectangles.

5. Probability histograms which follow the normal curve can be summarized quite well by the expected value and SE. The expected value locates the center of the probability histogram, and the SE measures the spread.

PART VI

Sampling

19

Sample Surveys

"Data! data! data!" he cried impatiently. *"I can't make bricks without clay."*
— Sherlock Holmes

1. INTRODUCTION

An investigator usually wants to generalize about a whole class of individuals. This class is called the *population*. For example, in forecasting the results of a presidential election in the United States, one relevant population is the class of all eligible voters. However, studying the whole population is usually impractical. Only part of it can be examined, and this part is called the *sample*. The investigator will make generalizations from the part to the whole; in more technical language, he makes *inferences* from the sample to the population.[1]

Usually, there are some numerical facts about the population which the investigator wants to know. Such a numerical fact is called a *parameter*. In forecasting a presidential election in the United States, two relevant parameters are:

- the average age of all eligible voters;
- the percentage of all eligible voters who are currently registered to vote.

Ordinarily, parameters like these cannot be determined exactly, but can only be estimated from a sample. Then a major issue is accuracy: how close are the estimates going to be?

Parameters are estimated by *statistics,* or numbers which can be computed from a sample. For instance, with a sample of 10,000 Americans, an investigator could calculate the following two statistics in order to estimate the parameters mentioned above:

- the average age of the eligible voters in the sample;
- the percentage of the eligible voters in the sample who are currently registered to vote.

Statistics are what the investigator knows; parameters are what he wants to know.

Estimating the parameters of the population from the sample is justified when the sample represents the population. This is impossible to check just by looking at the sample, because it is impossible to say whether the sample resembles the population with respect to parameters which aren't known. What has to be done instead is to look at the way the sample was chosen. Some methods tend to do very badly; others are quite likely to give representative samples. The two main lessons of this chapter are

- the method of choosing the sample matters a lot
- the best methods involve the planned introduction of chance.

(Similar issues come up when assigning subjects to treatment or control in experiments: see Part I.)

2. THE *LITERARY DIGEST* POLL

In 1936, Franklin Delano Roosevelt was completing his first term of office as President of the United States. It was an election year, and the Republican candidate was Governor Alfred Landon of Kansas. The country was struggling to recover from the Great Depression. There were still nine million unemployed; real income had dropped by one third in the period 1929–1933, and was just beginning to turn upward. But Landon was campaigning on a program of economy in government, and Roosevelt was defensive about his deficit financing.[2]

Landon.   The spenders must go.
Roosevelt.   We had to balance the budget of the American people before we could balance the budget of the national government. That makes common sense, doesn't it?

The Nazis were rearming Germany, and the Civil War in Spain was moving to its hopeless climax. These issues dominated the headlines in the New York *Times,* but were ignored by both candidates.

Landon. We must mind our own business.

Most observers thought Roosevelt would be an easy winner. Not so the *Literary Digest* magazine, which predicted an overwhelming victory for Landon, by 57% to 43%. This prediction was based on the largest number of people ever replying to a poll—about 2.4 million individuals. It was backed by the enormous prestige of the *Digest,* which had called the winner in every presidential election since 1916. However, Roosevelt won the 1936 election by a landslide; 62% to 38%. (The *Literary Digest* went bankrupt soon after.)

The magnitude of the *Digest*'s error is staggering. It is the largest ever made by a major poll. Where did it come from? The number of replies was more than big enough. In fact, George Gallup was just setting up his survey organization.[3] Using a sample of only 3,000 people, he was able to predict what the *Digest* predictions were going to be, well in advance of their publication, with an error of only one percentage point. Using another sample of about 50,000 people, he correctly forecast the Roosevelt victory, although his prediction of Roosevelt's share of the vote was off by quite a bit: Gallup forecast 56% for Roosevelt; the actual percentage was 62%, so the error was 62% − 56% = 6 percentage points. (Survey organizations use "percentage points" as the units for the difference between actual and predicted percents.) The results are summarized in Table 1.

Table 1. The Election of 1936.

	Roosevelt's percentage
Gallup's prediction of the *Digest* prediction	44
The *Digest* prediction of the election result	43
Gallup's prediction of the election result	56
The election result	62

Note: Percentages are of the major-party vote. In the election, about 2% of the ballots went to minor-party candidates.
Source: George Gallup, *The Sophisticated Poll-Watcher's Guide*, 1972.

To find out where the *Digest* went wrong, you have to ask how they picked their sample. A sampling procedure should be fair, selecting people for inclusion in the sample in an impartial way, so as to get a representative cross section of the public. A systematic tendency on the part of the sampling procedure to exclude one kind of person or another from the sample is called selection bias. The *Digest*'s procedure was to mail questionnaires to ten million people; they got 2.4 million replies. The names and addresses of these ten million people came from sources like telephone books and club membership lists. That tended to screen out the poor, who were unlikely to have telephones or belong to clubs. (In 1936, there were 11 million residential telephones—and 9 million unemployed.) So there was a very strong selection bias in the *Digest*'s sampling procedure, against the poor. Prior to 1936, this bias against the poor may not have affected the predictions very much, because rich and poor voted along similar lines. But in 1936, the political split followed economic lines much more closely: the poor voted overwhelmingly for Roosevelt, the rich were for Landon. One reason for the magnitude of the *Digest*'s error was selection bias.

> When a selection procedure is biased, taking a large sample doesn't help. This just repeats the basic mistake on a larger scale.

So the *Digest* did very badly at the first step in sampling. But there is also a second step. After a survey organization has decided which people ought to be in the sample, it still has to get their opinions. This is harder than it looks. If a large number of those selected for the sample do not in fact respond to the questionnaire or the interview, this can create a serious distortion, called *nonresponse bias*. The nonrespondents differ from the respondents in one obvious way: they didn't respond. Experience shows they tend to differ in other important ways as well.[4] This can be seen in a special survey made by the *Digest* in 1936, where they mailed questionnaires to every third registered voter in Chicago. About 20% responded, and of those who responded over half favored Landon. But in the election Chicago went for Roosevelt, by a two-to-one margin.

> Nonrespondents can be very different from respondents. When there is a high nonresponse rate, look out for nonresponse bias.

The results of the main *Digest* poll were based on the responses of 2.4 million people, out of ten million selected for the sample. These 2.4 million respondents do not even represent the ten million people who were polled, let alone the population of all voters. The *Digest* poll was spoiled both by selection bias and nonresponse bias.[4a]

Special surveys have been carried out to measure the differences between respondents and nonrespondents. It turns out that lower-class and upper-class people tend not to respond to questionnaires, so that the middle class is overrepresented among respondents. For these reasons, modern survey organizations prefer to use personal interviews rather than mailed questionnaires. A typical response rate for personal interviews is 75%, compared with 25% for mailed questionnaires.[5] However, the problem of nonresponse bias still remains, even with personal interviews. Those who are not at home when the interviewer calls may be quite different from those who are at home, with respect to working hours, family ties, social background, and therefore with respect to attitudes. Good survey organizations keep this problem firmly in mind, and have evolved very ingenious methods for dealing with it: one of these methods will be discussed in section 6.

> Some samples are really bad. To find out whether a sample is any good, ask how it was chosen. Was there selection bias? nonresponse bias?

Going back to the 1936 election, how did Gallup predict the *Digest* predictions? He just chose 3,000 people at random from the same lists the *Digest* was going to use, and mailed them all a postcard asking how they planned to vote. He knew that a random sample was likely to be quite representative—as will be explained in the next two chapters.

3. THE YEAR THE POLLS ELECTED DEWEY

Thomas Dewey rose to fame as a crusading D.A. in New York City, and went on to capture the Governor's mansion in Albany. In 1948 he was the Republican candidate for president, challenging the incumbent Harry Truman. Truman began political life as a protégé of Boss Pendergast in Kansas City. After being elected to the Senate, Truman became FDR's vice-president, succeeding to the presidency when Roosevelt died. Truman was one of the most effective presidents of the century, as well as one of the most colorful. He kept a sign on his desk, "The buck stops here." Another of his favorite aphorisms became part of America's political vocabulary: "If you can't stand the heat, stay out of the kitchen." But Truman was the underdog in 1948, for it was a troubled time. World War II had barely ended, and the uneasy half-peace of the Cold War had just begun. There was disquiet at home, and complicated involvement abroad.

Three major polls covered the election campaign: Crossley, for the Hearst newspapers; Gallup, syndicated in about a hundred independent newspapers across the country; and Roper, for *Fortune* magazine. By fall, all three had declared Dewey the winner, with a lead of around five percentage points. Gallup's prediction was based on 50,000 interviews; and Roper's on 15,000. As the *Scranton Tribune* put it,

DEWEY AS GOOD AS ELECTED, STATISTICS CONVINCE ROPER

But on Election Day Truman scored an upset victory with just under 50% of the popular vote; Dewey got just over 45%. The results are shown in Table 2.

Table 2. The election of 1948.

| | *The Predictions* | | | |
The candidates	Crossley	Gallup	Roper	The results
Truman	45	44	38	50
Dewey	50	50	53	45
Thurmond	2	2	5	3
Wallace	3	4	4	2

Source: F. Mosteller and others, *The Pre-Election Polls of 1948*, New York: Social Science Research Council, 1949.

To find out what went wrong for the polls, it is necessary to find out how they chose their samples.[6] The method they all used is called *quota sampling.* With this procedure, each interviewer is assigned a fixed quota of subjects to interview; and the numbers falling into certain categories (usually by residence, sex, age, race, and economic status) are also fixed. In other respects, the interviewer was free to select anybody she liked. (In the books on survey work, nameless interviewers are always referred to as "she"; we are just following usage.) For instance, a Gallup poll interviewer in St. Louis was required to interview thirteen subjects, of whom[7]

- exactly six were to live in the suburbs, and seven in the central city;
- exactly seven were to be men, and six women.

Of the seven men (and there were similar quotas for women)

- exactly three were to be under forty years old, and four over forty;
- exactly one was to be a black, and six white.

The monthly rental to be paid by the six white men were specified also:

- one was to pay $44.01 or more;
- three were to pay $18.01 to $44.00;
- two were to pay $18.00 or less.

Remember, these are 1948 prices.

From a common-sense point of view, quota sampling looks good. It seems to guarantee that the sample will be like the voting population, with respect to all the important characteristics thought to affect voting behavior. (The distributions of residence, sex, age, race, and rent can be estimated quite closely from Census data.) But the 1948 experience shows this procedure worked very badly. We are now going to see why.[8]

The survey organizations want a sample which faithfully represents the nation's political opinions. However, and this is the first point to notice, no quotas can be set on Republican or Democratic votes. This is because the distribution of political opinion in the nation is precisely what the survey organizations do not know and are trying to find out. The quotas for the other variables are an indirect effort to make the sample reflect the nation's politics. But, and this is the second point to notice, there are many factors which influence voting behavior, besides the ones the survey organizations control for. There are rich white men in the suburbs who vote Democratic, and poor black women in the central cities who vote Republican. As a result, it is entirely possible for the survey organizations to hand-pick a sample which is a perfect cross section of the nation on all the demographic variables, and then to find the sample voting one way, while the nation goes the other. This possibility must have seemed theoretical—before 1948.

The next argument against quota sampling is the most important. And it involves a crucial feature of the method, which is easy to miss the first time through:

- Within the assigned quotas, the interviewers are free to choose anybody they like. This leaves a lot of room for human choice. And human choice is always subject to bias.

In 1948, the interviewers chose too many Republicans. On the whole, Republicans are wealthier and better educated than Democrats. They are more likely to own telephones, have permanent addresses, and they live on nicer blocks. Within each broad demographic group, the Republicans are marginally easier to interview. If you were an interviewer, you would probably end up with too many Republicans also.

In fact, the interviewers preferred Republicans in every presidential election from 1936 through 1948, as shown by the Gallup poll results in Table 3. Prior to 1948, the Democratic lead was so great that it overcame the

Republican bias of the polls, and Gallup was able to predict the winner. In 1948, the Democratic lead was much slimmer, and was overcome by the Republican bias in quota sampling.

Table 3. The Republican bias in the Gallup poll.

Year	Gallup's prediction of Republican vote	Actual Republican vote	Error in favor of the Republicans
1936	44	38	6
1940	48	45	3
1944	48	46	2
1948	50	45	5

Note: Percentages are of the majority-party vote, except in 1948.
Source: F. Mosteller and others, *The Pre-Election Polls of 1948,* New York: Social Science Research Council, 1949.

> In quota sampling, the sample is hand picked to resemble the population with respect to some key characteristics. The method seems sensible, but really does not work very well. The reason is unintentional bias.

The quotas in quota sampling are sensible enough, although they do not guarantee success—far from it. But the method of filling the quotas, free choice by the interviewers, was disastrous. The alternative is to use objective and impartial chance mechanisms to select the sample. That will be the topic of the next section.

4. USING CHANCE IN SURVEY WORK

Even in 1948 some survey organizations used chance, or *probability methods,* to select their samples; since 1948, virtually all survey organizations use chance methods. How is chance used to draw a sample? Just to get started, imagine carrying out a survey of 100 voters in a small town with a population of 1,000 eligible voters. Then, it is feasible to list all the eligible voters, to write the name of each one on a ticket, to put all the thousand tickets in a box, and to draw 100 tickets at random. Since there is no point interviewing the same person twice, the draws are made without replacement. In other words, the box is shaken well to mix up the tickets, and one is drawn out at random and set aside, leaving 999 in the box. The box is shaken again, a second ticket is drawn out and set aside. This is repeated until 100 tickets have been drawn. The people whose names have been drawn form the sample. This is called a *simple random sample.* The names have been simply drawn at random without replacement. At each draw, every name in the box has an equal chance to be chosen. The interviewers have no discretion at all in who they interview, and the procedure is impartial—everybody has the same chance to get into the sample. Consequently, the law of averages guarantees that the percentage of Democrats in the sample is very likely to be close to the percentage in the population.

Simple random sampling means drawing at random without replacement.

What happens in a more realistic setting, as when the Gallup poll tries to predict a presidential election? One idea would be just to take a nationwide simple random sample, of a few thousand eligible voters. This would give much better results than quota sampling. However, it isn't nearly as easy to do as it sounds. Drawing names at random, in the statistical sense, is very hard work. It is not at all the same as choosing people haphazardly. To begin drawing eligible voters at random, you would need a list of all 100 million of them. There is no such list. Even if there were, drawing a few thousand names at random from 100 million is a job in itself. (Remember, on each draw every name in the box has to have an equal chance of being selected.) And even if you could draw a simple random sample, the people would be scattered all over the map: it would be prohibitively expensive to send interviewers around to find them all.

It just isn't practical to take a simple random sample. Consequently, most survey organizations use some variant of *multistage cluster sampling*. The name is complicated, and so are the details.[9] But the idea is straightforward. For instance, the Gallup poll makes a separate study in each of the four geographical regions of the United States—Northeast, South, Midwest, and West. See Figure 1. Within each region, they group together all the population centers of similar sizes. One such grouping might be all towns in the Northeast with a population between 50 and 250 thousand. Then, a simple random sample of these towns is selected. This completes the first stage of sampling. Interviewers are stationed in the selected towns, and no interviews are conducted in the other towns of that group. For election purposes, each town is divided up into *wards,* and the wards are subdivided into *precincts.* At the second stage of sampling, some wards are selected—by drawing at random—from each sample town chosen in the stage before. At the third stage, some precincts are drawn at random from each of the previously selected wards. At the fourth stage, households are drawn at random from each selected precinct. Finally, some members of the selected households are interviewed. Even here, no discretion is allowed the interviewer. For instance, Gallup poll interviewers are instructed to "speak to the youngest man 18 or older at home, or if no man is at home, the oldest woman 18 or older."[10]

This design offers many of the advantages of quota sampling. For instance, it is set up so the distribution of the sample by residence is the same as the distribution for the nation. But each stage in the selection procedure uses an objective and impartial chance mechanism to select the sample units. This completely eliminates the worst feature of quota sampling: selection bias on the part of the interviewer.

Figure 1. Multistage cluster sampling.

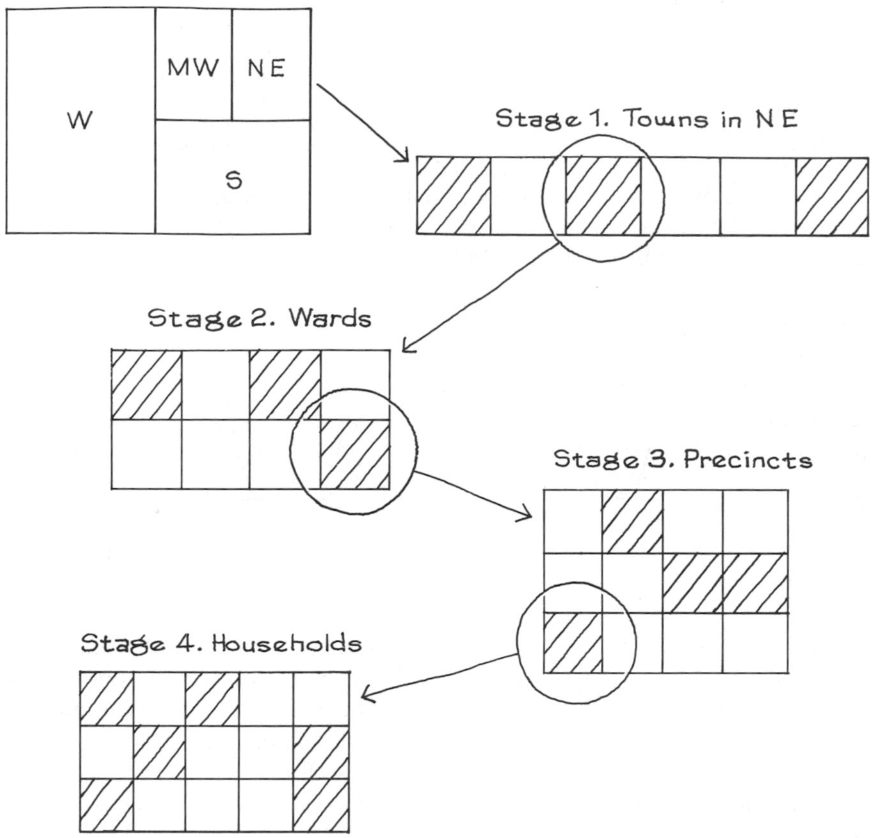

The details can be quite complicated, but all probability methods for sampling have two important properties in common with simple random sampling:

- the interviewers have no discretion at all as to whom they interview;
- there is a definite procedure for selecting the sample, and it involves the planned use of chance (for instance, drawing the names of the persons to be interviewed at random from a box).

As a result, with a probability method it is possible to compute the chance that any particular individuals in the population will get into the sample.

Quota sampling is not a probability method: it fails both tests. The interviewers have a lot of discretion in choosing subjects. And chance only enters in the most unplanned and haphazard way: what kinds of people does the interviewer like to approach? Who is going to be walking down a particular street at a particular time of day? No survey organization in the world can put numbers on these kinds of chances.

Usually, probability methods are designed so that each individual in the population will get into the sample with an equal chance. However, the Gallup procedure is biased in a minor way—against people who live in large households. See Figure 2. An adjustment is made to correct for this bias, as discussed in section 6.

Figure 2. Household bias. Selecting one person from each household produces a bias against people who live in large households. Imagine selecting one of the two households below at random; then select a person at random from the selected household. This produces a sample of size one. A person in the small household has a better chance of getting into the sample than a person in the large household.

5. HOW WELL DO PROBABILITY METHODS WORK?

Since 1948, the Gallup poll and all other major polls have used probability methods to choose their samples. The Gallup poll record in post-1948 presidential elections is shown in Table 4. There are three points to notice. The sample size has gone down sharply: the Gallup poll used a sample of size about 50,000 in 1948 and they now use samples of a fifth to a tenth that size. There is no longer any consistent trend favoring either Republicans or Democrats. And the accuracy has gone up appreciably. From 1936 to 1948, the errors were around 5%; since 1948, they vary around 2%. Using probability methods to select the sample, the Gallup poll has been able to predict the elections with startling accuracy, sampling less than five persons in a hundred thousand. This record proves the value of probability methods in sampling.

Table 4. The Gallup poll record in presidential elections after 1948.

Year	Sample size	Winning candidate	Gallup poll prediction	Election result	Error
1952	5,385	Eisenhower	51%	55.4%	+4.4%
1956	8,144	Eisenhower	59.5%	57.8%	−1.7%
1960	8,015	Kennedy	51%	50.1%	+0.9 of 1%
1964	6,625	Johnson	64%	61.3%	+2.7%
1968	4,414	Nixon	43%	43.5%	+0.5 of 1%
1972	3,689	Nixon	62%	61.8%	−0.2 of 1%
1976	3,439	Carter	49.5%	51.1%	−1.6%

Note: The percentages are of the major-party vote; in 1968, Wallace is counted with the major parties; the error is the difference "predicted − actual" Democratic vote.
Source: The Gallup poll (American Institute of Public Opinion).

Why do chance methods work so well? At first sight, it does seem that exercising judgment in the choice of the sample will lead to better results than leaving it all to chance. For instance, quota sampling absolutely guarantees that the percentage of men in the sample will be equal to the percentage of men in the population. With probability sampling, we can only say that the percentage of men in the sample is very likely to be very close to the percentage in the population. So absolute certainty is reduced to likelihood. But human judgment and choice usually show bias, while blind chance is impartial. That is why probability methods work better than human judgment.

> To minimize bias, an impartial and objective probability method should be used to choose the sample.

6. A CLOSER LOOK AT THE GALLUP POLL

Some degree of bias is almost inevitable even when probability methods are used to select the sample. This creates many practical difficulties for survey organizations. The discussion here is organized around the questionnaire used by the Gallup poll in the 1976 election. See Figure 3 (p. 312).

The nonvoters. In a typical presidential election in the United States, between one third and one half of the eligible voters fail to vote. The job of the Gallup poll is to guess how the voters will vote, so the nonvoters are irrelevant and should be screened out of the sample as far as possible. That isn't easy to do: there is a stigma attached to nonvoting, so many respondents tell the interviewer they will vote even if they know perfectly well that they won't. The problem of screening out the nonvoters is handled by questions 1 through 14 on the Gallup poll questionnaire. Question 3, for instance, asks where the respondent would go to vote; if he knows the answer, he is more likely to vote. Question 14 asks whether the respondent voted in the last election, and is phrased to make a negative answer easy to give (compensating for the stigma attached to non-voting). If he voted last time, he is more likely to vote this time.

This battery of questions is used to decide whether the respondent is likely to vote, and the final prediction of the election results is based only on that part of the sample judged likely to vote. It is a matter of record who actually votes in each election, and post-election studies by the Gallup organization have shown that their predictions as to who will vote are reasonably accurate. They have also shown that screening out the likely nonvoters materially increases the accuracy of the predictions—because the preferences of the likely voters are quite different from the preferences of the likely nonvoters.[11] Since the Gallup poll screens out most of the nonvoters, the ones left in the sample do not materially influence the estimates, so their presence can only lead to a small bias in the results.

The undecided. A further problem is that some percentage of the subjects being interviewed are undecided as to how they will vote. Question 7,

Figure 3. The Gallup poll questionnaire for the 1976 presidential election.

704-1
Form D
September 24, 1976

THE GALLUP POLL

SPONSORED BY LEADING REPUBLICAN, DEMOCRATIC AND INDEPENDENT NEWSPAPERS

SUGGESTED INTRODUCTION: I'm taking a GALLUP POLL. I'd like YOUR opinion on some topics of interest.

Time interview starts: ....

The first question I would like to ask you is . . .

1. How much thought have you given to the coming November elections — quite a lot, or only a little?
   1☐ Lot   2☐ Some   3☐ Little   ᵛ☐ None

2. Have you ever voted in this precinct or district?
   1☐ Yes   2☐ No   ᵛ☐ Don't Know

3. Where do people who live in this neighborhood go to vote?
   ....   ᵛ☐ Don't Know

4a. Are you NOW registered so that you can vote in the election this November?
   1☐ Yes   2☐ No   3☐ Don't have to reg.   ᵛ☐ DK

   ### IF NO or DON'T KNOW, ask 4 b:
   b. Do you plan to register so that you can vote in the November election?
   1☐ Yes   2☐ No   3☐ Other ....

5. Generally speaking, how much interest would you say you have in politics — a great deal, a fair amount, only a little, or no interest at all?
   1☐ Great deal   2☐ Fair   3☐ Little   ᵛ☐ None

6. How often would you say you vote — always, nearly always, part of the time, or seldom?
   1☐ Always   2☐ Nearly   3☐ Part time   4☐ Seldom
   5☐ Other .... ᵛ☐ Never vote

7. Suppose you were voting TODAY for President and Vice President of the United States. Here is a Gallup Poll Secret Ballot listing the candidates for these offices. (TEAR OFF ATTACHED BALLOT AND HAND TO RESPONDENT). Will you please MARK that secret ballot for the candidates you favor today — and then drop the folded ballot into the box.

   INTERVIEWER: If respondent hands back ballot and says he hasn't made up his mind, or refuses to mark it, say:
   "Well, would you please mark the ballot for the candidates toward whom you lean as of today?"
   If respondent still can't decide, or refuses to mark ballot, please write that on the ballot, and be sure to drop it into the box.

8. Right now, how strongly do you feel about your choice — very strongly, fairly strongly or not strongly at all?
   1☐ Very   2☐ Fairly   3☐ Not strongly at all
   4☐ Didn't make choice   ᵛ☐ Don't Know

9a. Do you ,yourself, plan to vote in the election this November, or not?
   1☐ Yes   2☐ No   ᵛ☐ Don't Know

   ### IF YES, ask 9b:
   b. How certain are you that you will vote — ABSOLUTELY certain, FAIRLY certain or NOT certain?
   3☐ Absolutely   4☐ Fairly   5☐ Not certain

10a. If the elections for Congress were being held TODAY, which party would you like to see win in this Congressional district, the Democratic party or the Republican party?
   1☐ Dem.   2☐ Rep.   3☐ Other   4☐ Undecided

   ### IF UNDECIDED or REFUSED, ask 10b:
   b. As of today, do you lean more to the Democratic party or more to the Republican party?
   5☐ Dem.   6☐ Rep.   7☐ Other
   8☐ Undecided   ᵛ☐ Refused to answer

11. Here is a picture of a ladder (HAND RESPONDENT LADDER). Suppose we say the top of the ladder (POINT) marked 10 represents a person who definitely will vote in the election this November, and the bottom of the ladder (POINT) marked zero represents a person who definitely will not vote in the election. How far up or down the ladder would you place yourself? (PLEASE CIRCLE THE APPROPRIATE NUMBER.)
   10   9   8   7   6   5   4   3   2   1   0   ᵛDK

12. Can you tell me who is the Director of the National "Get out the Vote" Committee?
   1☐ Yes, Who? .... 2☐ No

Now, here are a few questions so that my office can keep track of the cross-section of people I've talked to:

13. In politics as of TODAY, do you consider yourself a Republican, Democrat, or Independent?
   1☐ Rep.   2☐ Dem.   3☐ Ind.   4☐ Other

14. In the election in November 1972 - when Nixon ran against McGovern - did things come up which kept you from voting, or did you happen to vote? For whom?
   1☐ Nixon   4☐ Voted, don't remember whom
   2☐ McGovern   5☐ No, didn't vote
   3☐ Other   ᵛ☐ Don't remember if voted

15. Are you, or is your (husband/wife) a member of a labor union?
   1☐ Yes, respondent is   2☐ Yes, spouse is
   3☐ Yes, both are   ᵛ☐ No, neither is

16. Could you tell me the kind of business or industry the CHIEF WAGE EARNER (HEAD OF HOUSEHOLD) in your immediate family works in, and the kind of work he does there?
   Kind of business: ....
   Kind of work: ....

17. What was the last grade or class you COMPLETED in school?
   1☐ None or Grades 1-4   6☐ Tech., Trade, or Bus.
   2☐ Grades 5, 6, 7   7☐ College, Univ., inc.
   3☐ Grade 8   8☐ College, Univ., grad.
   4☐ H.S., inc. (Gd. 9-11)   ....
   5☐ H.S., grad. (Gd. 12)   (College, Univ. attended)

18. What is your religious preference — Protestant, Roman Catholic, or Jewish?
   1☐ Protestant   4☐ Other
   2☐ Roman Catholic   ᵛ☐ None
   3☐ Jewish

19. How many persons 18 years and over are there now living in this household, including yourself? Include lodgers, servants or other employees living in the household.
   Number ....

20. (HAND RESPONDENT CARD Y). From what nationality group (or groups) are you mainly descended? That is, from what country or countries did your family come before coming to the United States?
   1☐ A   2☐ B   3☐ C   4☐ D   5☐ E   6☐ F
   7☐ G   8☐ H   9☐ I   0☐ J   x☐ K   ᵛ☐ Don't know

21a. We are interested in finding out how often people are at home to watch TV or listen to the radio. Would you mind telling me whether or not you happened to be at home yesterday (last night/last Saturday) at this particular time?
   INTERVIEWER: See Interviewer Bulletin for handling this question.
   1☐ Yes, at home   2☐ No, not at home
   b. How about the day (night/Saturday) before at this time?
   1☐ Yes, at home   2☐ No, not at home
   c. And how about the day (night/Saturday) before at this time?
   That was ....
   1☐ Yes, at home   2☐ No, not at home

22. And what is your age?   Age ....

23. CHECK WHETHER:
   1☐ White Man   3☐ Black Man   5☐ Other Man
   2☐ White Woman   4☐ Black Woman   6☐ Other Woman

   So that my office can check my work in this interview if it wants to, may I have your name, address and telephone number please?

   NAME ....
   ADDRESS ....
   (House No. or RFD Route, St. or Rd., Apt. No.)
   CITY & STATE ....
   TEL. Area Code .... Phone No. .... ᵛ☐ No tel.

   PLACE INTERVIEWER BADGE NO. HERE

   I hereby attest that this is a true and honest interview.

   _____
   Interviewer

   Date of interview ....

   Time interview ended ....

Source: The American Institute of Public Opinion.

which asks for the preferences, is designed to keep this percentage as small as possible. To begin with, it asks how the respondent would vote the day of the interview, rather than Election Day. If the subject can't decide, he is asked to indicate "the candidates toward whom you lean as of today." A final device is the paper ballot, shown in Figure 4. Instead of naming his preferences out loud, the respondent just marks the ballot and drops it into a box carried by the interviewer.

Figure 4. The Gallup poll ballot, 1976. The interviewers use secret ballots, to minimize the number of undecided respondents.

DEMOCRATIC	REPUBLICAN	INDEPENDENT
☐ CARTER	☐ FORD	☐ McCARTHY

These techniques have been found to minimize the percentage of undecided; but there are still some left, and if they are thought likely to vote, the Gallup poll has to guess how. Some information about political attitudes is available from questions 13 and 14. This information can be used to predict how the undecided respondent will vote, but it is very difficult to say how well this works.

Response bias. The answers given by a respondent are influenced to some extent by the phrasing of the questions, and even the tone or attitude of the interviewer. This kind of distortion is called *response bias*. There was a striking example in the 1948 election survey: changing the order of the candidates' names was found to change the response by 5%, the advantage being with the candidate who was named first. To control response bias, all interviewers use the same questionnaire, and the interview procedure is standardized as far as possible. The ballot technique was found to reduce the effect of the political attitudes of the interviewer on the responses of the subjects.

Nonresponse bias. Even with personal interviews, many subjects are missed. Since they tend to be different from the subjects available for the interview, a nonresponse bias is created. To some extent, this bias can be adjusted out, by giving more weight to the subjects who were available but hard to get. This information is obtained from question 21, which asks whether the subject was at home on the previous days. This is done quite subtly, as you can tell by reading the question.

Household bias. The Gallup poll interviews only one person in each household selected for the survey. This discriminates against people who live in large households, not enough of them are represented in the sample. Consequently, more weight should be given to the ones who are represented. Household size is obtained from question 19.

"I'd say I'm about forty-two per cent for Nixon, thirty-nine per cent for Rockefeller, and nineteen per cent undecided."

Drawing by Dana Fradon; © 1959 The New Yorker Magazine, Inc.

Check data. In 1976, about 60% of the U.S. population aged 25 and over had completed high school or college. The Gallup sample usually includes too many of these better-educated people (who tend to be Republicans). In a detailed analysis, more weight is put on the responses of those subjects with less education (question 17). The other demographic data can be used in a similar way. This technique is called "ratio estimation." Do not confuse ratio estimation (a good technique) with quota sampling (a bad technique). Ratio estimation is an objective, arithmetic technique applied to the sample after it is chosen, to compensate for various small biases in the sampling procedure. Quota sampling is a method for choosing the sample. It has a large, subjective component—when the interviewer chooses the subjects—and this introduces large biases.

Interviewer control. In large-scale survey work, there is always the problem of making sure that the interviewers follow instructions. Some redundancy is built into the questionnaire, so the answers can be checked for consistency: inconsistencies suggest the interviewer may not be doing the job properly. A small percentage of the subjects are reinterviewed by administrative staff, as a further check on the quality of the work.

Talk is cheap. It is a very risky thing to predict what people will really do on Election Day from what they tell the interviewer they are going to do. It is a fact of experience that words and deeds often differ. People may be unwilling to reveal their true preferences. Even if they do, they may change their minds later.

7. CHANCE ERROR

The previous section discussed the practical difficulties faced by real survey organizations: people aren't at home, or even if they are at home they don't reveal their true preferences, or they go and change their minds. What happens if all these difficulties are assumed away? Imagine a box with a very large number of tickets, some marked "1" and some marked "0." This is the population. A survey organization is hired to estimate the percentage of 1's in the box. This is the parameter. The organization draws a thousand tickets at random without replacement from the box. This is the sample. There is no problem about response: the tickets are all there in the box. Drawing them at random eliminates selection bias. And the tickets don't change back and forth between 0 and 1. As a result, the percentage of 1's in the sample is going to be a good estimate for the percentage of 1's in the box. But it is still likely to be a bit off. The reason is that the sample is only part of the population. And since the sample is chosen at random, the amount off is governed by chance:

percentage of 1's in sample = percentage of 1's in box + chance error.

In more complicated situations, the equation has to take bias into account:

$$\boxed{\text{estimate} = \text{parameter} + \text{bias} + \text{chance error.}}$$

Now there are some questions to ask about chance errors:

- How big are they likely to be?
- How do they depend on the size of the sample? The size of the population?
- How big does the sample have to be in order to keep the chance errors under control?

These questions will all be anwered in the next two chapters. Unfortunately, there are no formulas for dealing with bias.

EXERCISE SET A

1. The California poll sometimes conducts pre-election surveys by telephone. Could this bias the results? How?

2. About 1930, a survey was conducted in New York on the attitude of former black slaves towards their owners and conditions of servitude.[12] Some of the interviewers were black, some white. Would you expect the two groups of interviewers to get similar results? Give your reasons.

3. One study on slavery estimated that "11.9% of slaves were skilled craftsmen." This estimate turns out to be based on the records of thirty plantations in Plaquemines Parish, Louisiana.[13] Is it trustworthy?

4. A study was conducted in Holland in 1968, to relate the intelligence of eighteen-year-old men to the number of their brothers and sisters.[14] In Holland, all men take a military preinduction exam at age 18. The exam includes an intelligence test known as "Raven's progressive matrices," and includes questions about demographic variables like family size. The study related scores on the intelligence test with family size. The records of all the exams taken in 1968 were used.
   (a) What is the population? the sample?
   (b) Is there any chance error?

5. In one study, the Educational Testing Service needed a representative sample of college students.[15] To draw the sample, they first divided up the population of all colleges and universities into relatively homogeneous groups (one group consisted of all public universities with 25,000 or more students; another group consisted of all private four-year colleges with 1,000 or fewer students; and so on). Then, they used their judgment to choose one representative school from each group. That created a sample of schools. Each school in the sample was then asked to pick a sample of students. Was this a good way to get a representative sample of students?

6. The monthly Gallup poll opinion survey is based on a sample of about 1,500 persons, "scientifically chosen as a representative cross section of the American public." The Gallup poll thinks the sample is representative mainly because
   (i) it resembles the population with respect to such characteristics as race, sex, age, income, and education
or
   (ii) it was chosen using a probability method.

7. A survey is carried out at a university to estimate the percentage of undergraduates living at home during the current quarter.
   (a) What is the population? The parameter?
   The registrar keeps an alphabetical list of all undergraduates, with their current addresses. Suppose there are 10,000 undergraduates in the current quarter. Someone proposes to choose a number at random from one to a hundred, count that far down the list, taking that name and every hundredth name after it for the sample.
   (b) Is this a probability method? Is it the same as simple random sampling? Someone else proposes to go out and take the first hundred undergraduates she sees as the sample.
   (c) Repeat (b) for this method.

The answers to these exercises are on p. A-52.

8. REVIEW EXERCISES

Review exercises may cover material from previous chapters.

1. Two surveys are conducted to measure the effect of an advertising campaign for a certain brand of detergent.[16] In the first survey, interviewers

Rose Kennedy, center, receives a bi... front
rig... ...er, F...

Hong Kong in suspense

ty,
ill
id-
a
t is
will

colony.

Despite China's pledges to maintain the colony's capitalist system, however, many Hong Kong leaders have told China that capitalism and communism are incompatible.

Among the concrete issues that the two sides are just beginning to discuss are the form and structure of a Peking-run Hong Kong government, elections, the future role of Britain in Hong Kong, and the authority of existing Hong Kong law, which is based largely on British law.

There are also unresolved questions about the future of Hong Kong's free press, the right of residents to travel freely, and the future of Hong Kong's membership in the General Agreement on Tariffs and Trade (GATT) and its observer status in the European Community.

There is a gnawing feeling among many Hong Kong ▢▢▢ that some Peking leader▢

ss-
is-
at
r-

to
a
r
e

rant and ill-prepared for the responsibility of integrating Hong Kong into China. One delegate, who spent a half an hour explaining to a senior Chinese official the significance of the battered Hong Kong dollar, said he was dumbfounded when the official asked what the strength of the U.S. dollar had to do with the Hong Kong dollar. The answer is "plenty," particularly for China, which earns more than 30 percent of its hard currency through Hong Kong. Thus, each pronouncement from Peking on Hong Kong's future sets off financial tremors, eroding Peking's profits.

The Bank of China's Hong Kong director, Jiang Wengui, recently confirmed that China's exports to Hong Kong, which exceed $5 billion a year, had lost 20 to 30 percent of their value.

There are signs, however, that Peking is absorbing the lessons of the investment crisis and the Hong Kong dollar. Chin▢▢ ▢▢▢wn diversifying Delhi.

its economic presence in the colony, stepping up investments in everything from department stores and cement works to textiles anu high-technology firms to help stabilize the jittery climate, Hong Kong analysts note.

Policeman kills 11 in Hindu temple

NEW DELHI, India (AP) — A po▢ liceman went berserk at a Hindu ten▢ ple in the town of Mandsaur Saturda▢ shooting 11 people to death ar▢ wounding nine before a fellow offic▢ shot and killed him, United News India reported.

The agency said among the peop▢ slain was the priest at the temple the town 340 miles southwest of Ne ask housewives whether they use that brand of detergent. In the second, the interviewers ask to see what detergent is being used. Would you expect the two surveys to reach similar conclusions? Give your reasons.

2. In a study on peach-tree farming, questionnaires were mailed to a sample of farmers.[17] A first wave of replies was received. If a farmer did not reply in the first wave, a follow-up letter was sent to him; this generated a second wave of replies. Finally, agents were sent to interview those who replied neither in the first nor in the second wave. This generated a third wave of replies. One question asked for the size of the farm. Would you expect the sizes of the farms in the three waves to be similar? Give your reasons.

3. One study on slavery estimated that a slave had only a 2% chance of being sold into the interstate trade each year. This estimate turns out to be based on auction records in Ann Arundel County, Maryland.[18] Is it trustworthy?

4. In one study, it was necessary to draw a representative sample of Japanese-Americans resident in San Francisco.[19] The procedure was as follows. After consultation with representative figures in the Japanese community, the four most representative blocks in the Japanese area of the city were chosen; all persons resident in those four blocks were taken for the sample. However, a comparison with Census data shows that the sample did not include a high-enough proportion of Japanese with college degrees. How can this be explained?

5. (Hypothetical.) A small undergraduate college has 1,000 students, evenly distributed among the four classes: freshman, sophomore, junior, senior. A study is being designed to estimate the percentage of marijuana smokers in the college. It is decided to interview a sample of 100 students. One person proposes the following method for taking the sample: Write the name of each student on a ticket, put the 1,000 tickets in a box, stir them up for an hour so they are thoroughly mixed, and then take 100 tickets at random from the box. Another person objects, on the grounds that this sample is unlikely to have exactly 25 students in each of the four classes. So, a second design is proposed: Hire four interviewers, one from each class, and ask each interviewer to select 25 representative students from her class. Which method is better?

6. The Nielsen organization used to rate TV shows by the following method. They chose a panel of homes with TV sets, and attached a meter to each set. The meter recorded the times at which the set was on, and which channel it was tuned to. At the end of the month, the meters were read, and the organization computed the total number of hours spent by panel members watching each TV show. "The Nielsen Rating" was based on this total.[20] The panel was seldom changed. What kind of bias would you suspect?

7. A hypothetical polling organization tries to predict the results of a state election as follows. It takes a simple random sample of one thousand voters and interviews them. Whatever the results of the interviews, the

organization ignores them and predicts a 60% Republican vote. (This isn't very sensible, and the organization is bound to stay hypothetical.) Is there any chance variability in this estimate? What about bias?

8. (Hypothetical.) A survey is carried out by the planning department to determine the distribution of household size in a certain city. They draw a simple random sample of 1,000 households; but after several visits, the interviewers find people at home in only 853 of the sample households. Rather than face such a high nonresponse rate, the planners draw a second batch of households, and use the first 147 completed interviews in the second batch to bring the sample up to its planned strength of 1,000 households. They count 3,087 people in these 1,000 households, and estimate the average household size in the city to be about 3.1 persons. Is this estimate likely to be too low, too high, or just about right? Explain.

9. One hospital has 218 live births during the month of January.[21] Another has 536. Which is likelier to have 55% or more male births? Or is it equally likely? Explain. (There is about a 52% chance for a live-born infant to be male.)

9. SUMMARY

1. A *sample* is part of a *population*.

2. A *parameter* is a numerical fact about a population. Usually, a parameter can't be determined exactly, but can only be estimated.

3. A *statistic* can be computed from a sample, and used to estimate a parameter. A statistic is what the investigator knows; a parameter is what he wants to know.

4. When estimating a parameter, one major issue is accuracy: how close is the estimate going to be?

5. Some methods for choosing samples are likely to produce accurate estimates. Others are spoiled by *selection bias* or *nonresponse bias*. Therefore, when thinking about a sample survey, ask yourself:
   • What is the population? The parameter?
   • How was the sample chosen?
   • What was the response rate?

6. Large samples offer no protection against bias.

7. In *quota sampling,* the sample is hand-picked by the interviewers to resemble the population in some key ways. This method seems logical, but often gives bad results. The reason: unintentional bias on the part of the interviewers.

8. *Probability methods* for sampling use an objective chance process to pick the sample, and leave no discretion to the interviewer. The hallmark of a probability method is that the investigator can compute the chance that any particular individuals in the population will be selected for the sample.

9. One probability method is *simple random sampling*. This means drawing names at random without replacement.

10. With a probability method, the sample is quite likely to be representative: blind chance is impartial.

11. But even when using probability methods, the estimate may differ a bit from the parameter due to bias and chance error:

$$\text{estimate} = \text{parameter} + \text{bias} + \text{chance error}.$$

20

Chance Errors in Sampling

To all the ladies present and some of those absent.
—THE TOAST USUALLY PROPOSED BY JERZY NEYMAN

1. INTRODUCTION

All sample surveys involve chance errors. This chapter will explain how to find the likely size of the chance error in a percentage, for simple random samples from a population whose composition is known. It turns out that this mainly depends on the size of the sample, and not on the size of the population. But first, an example.

Part II discussed some findings from Cycle I of the Health Examination Survey of 1961, based on a representative cross section of 6,672 Americans aged 18 to 79. A sociologist now wishes to do a follow-up study on these people. He does not have the resources to do them all, in fact he only has enough money to study a sample of 100 of them. To avoid bias, he is going to draw the sample at random. In the imaginary dialogue which follows, he is discussing the problem with his statistician.

Soc. I guess I have to write all the 6,672 names on separate tickets, put them in a box, and draw out 100 tickets at random. It sounds like a lot of work.

Stat. We have the files on the computer, code-numbered from 1 to 6,672. So you could just draw 100 numbers at random, in that range. Your sample would be the people with those code numbers.

Soc.   Yes, but then I still have to write the numbers from 1 to 6,672 on the tickets. You haven't saved me much time.

Stat.   That isn't really what I had in mind. In fact, with such a large box, it would be very hard for you to mix the tickets properly. If you didn't, most of your draws would probably be from the tickets you put in last. That could be a real bias.

Soc.   So what do you suggest?

Stat.   The computer has a random-number generator. It will pick a number at random from 1 to 6,672 for us. The person with that code-number goes into the sample. Then it will pick a second code-number at random, different from the first. That's the second person to go into the sample. It keeps on going that way until it gets a hundred people. Instead of you trying to mix all the tickets, let the random numbers do the mixing for you. Besides, they save all that writing.

Soc.   OK. But if we use the computer, do you guarantee that my sample will be representative?

Stat.   What do you have in mind?

Soc.   Well, for instance there were 3,091 men and 3,581 women in the original survey: 46% were men. I want my sample to have 46% men. Besides that, I want them to have the right age distribution. Then there's income and educational level to think about. Of course, what I really want is a group whose attitudes to health care are typical.

Stat.   Let's not get into attitudes right now. First things first. I drew a sample to show you. Look at Table 1. The first person chosen by the computer was female, so was the second. But the third was male. And so on. Altogether, you got 54 men. That's pretty close.

Table 1. The sex of a hundred sample people: 46% of the subjects in Cycle I of the Health Examination Survey were men; a hundred of these people were chosen at random, and 54% of the sample people turned out to be men.

F	F	M	F	F	F	F	F	M	F	F	M	M	M	M	F	M	M	M	F
F	F	M	M	M	F	F	F	F	M	M	F	F	F	M	M	F	M	M	F
F	M	F	F	M	M	F	F	F	M	M	F	M	M	F	M	M	F	F	M
F	M	F	F	M	M	M	F	F	M	M	M	F	M	M	M	F	M	M	M
F	M	M	M	F	M	M	M	F	F	M	M	F	M	F	M	M	M	F	M

Soc.   But there should only be 46 men. This sample has the proportions reversed. There must be something wrong with the computer.

Stat.   No, not really. Remember, the people in the sample are drawn at random. Just by the luck of the draw, you could get too many men—or too few. I had the computer take some more samples for you, 250 in all. [Table 2 on the next page.] The number of men ranged from a low of 32 to a high of 59. Only 13 samples out of the lot had exactly 46 men. There's a histogram. [Figure 1 on the next page.]

Soc.   What stops the numbers from being 46?

Stat.   Chance variability. Remember the Kerrich experiment I told you about the other day?

Table 2. Two hundred fifty random samples (each of a hundred people) were drawn from the respondents to Cycle I of the Health Examination Survey (of whom 46% were men). The number of men in each sample is shown below.

```
54 50 47 53 53 53 47 42 44 42 33 37 50 42 41 42 47 39 52 52 43 42 42 38 50
36 38 48 43 49 55 38 46 44 42 48 53 47 49 44 48 49 43 36 57 41 46 47 45 40
41 48 47 44 52 40 48 50 43 55 52 47 45 44 52 48 46 48 45 49 45 45 52 48 50
50 51 47 45 45 43 49 48 46 43 49 45 55 44 50 38 45 49 37 50 50 45 49 47 42
47 45 39 52 52 44 47 37 47 36 51 51 45 39 45 49 43 47 52 44 47 43 50 35 47
39 44 54 52 41 47 49 45 45 51 55 40 54 53 42 34 45 45 46 54 45 46 45 52 45
48 49 46 40 53 47 48 41 38 41 42 48 48 49 43 46 46 38 35 40 40 45 50 50 44
41 45 41 48 43 48 48 42 44 40 38 54 48 44 47 51 55 42 51 42 40 43 47 42 50
55 48 40 42 43 43 32 43 52 49 49 50 54 52 46 42 43 49 46 37 43 49 45 48 48
46 49 46 43 41 37 59 37 49 41 41 42 48 47 42 47 45 50 49 47 51 52 50 48 42
```

Soc. Yes, but that was about coin-tossing, not sampling.

Stat. Well, there isn't much difference between coin-tossing and sampling. Each time you toss the coin, you either get a head or a tail, and if you're counting heads the number either goes up by one or stays the same. The chances are 50–50 each time. It's the same with sampling. Each time the computer chooses a person for the sample, it either gets a man or a woman, and if it's counting men the number either goes up by one or stays the same. The chances are just about 46 to 54 each time—taking a hundred tickets out of the box can't change the proportions in the box very much.

Soc. What's the point?

Stat. The chance variability in sampling is just like the chance variability in coin-tossing.

Soc. Hmmm. What happens if we increase the size of the sample? Won't it come out more like the population?

Stat. Right. For instance, suppose we increase the sample size by a factor of four, to 400. I got the computer to draw another 250 samples, this time

Figure 1. Histogram for the numbers of men as reported in Table 2.

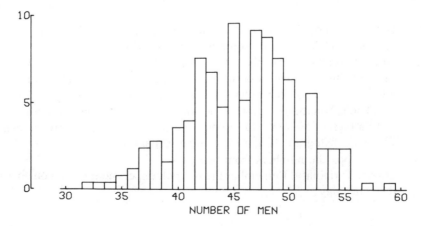

with 400 people in each sample. With some of these samples, the percentage of men is below 46%, with others it is above. The low is 38%, the high is 54%. There's a histogram in Figure 2. You can compare it with the histogram for samples of size 100. Multiplying the sample size by four cuts the likely size of the chance error in the percentage down to half of what it was.

Figure 2. Histogram for the percentages of men in samples of size 400. There are 250 samples, drawn at random from the respondents to Cycle I of the Health Examination Survey.

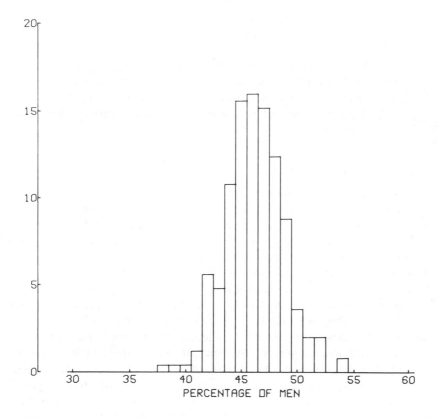

Soc. Can you get more specific about this chance error?

Stat. Let me write an equation:

> percentage of men in sample
>   = percentage of men in population + chance error.

Of course, the chance error will be different from sample to sample— remember the variability in Table 2.

Soc. So if I let you draw one sample for me, with this random-number business, can you say how big my chance error will be?

Stat. Not exactly, but I can tell you its likely size. This is all very much like coin tossing. The likely size of your chance error is given by the standard error, and to compute that we need a box model.

Soc. Look, there's one point I missed earlier. How can you have 250 different samples with 100 people each? I mean, $250 \times 100 = 25,000$, and you only started with 6,672 people.

Stat. The samples are all different, but they have some people in common. Look at the sketch. The inside of the circle is like the 6,672 people, and each shaded strip is like a sample:

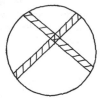

The strips are different, but they overlap. Actually, we only scratched the surface with our sampling. The number of different samples of size 100 is over 10^{200}. That's 1 followed by two hundred 0's. Some physicists don't even think there are that many elementary particles in the whole universe.

2. THE STANDARD ERROR

The sociologist of the previous section was thinking about taking a sample of size 100 from the population of 6,672 subjects in the HES study. He realized that the percentage of men in the sample would not equal the percentage in the whole population, but would be off by a chance error. How large is this chance error likely to be? The answer is given by the standard error. For the sociologist's problem, the standard error is five percentage points. In other words, the sociologist should expect the percentage of men in his sample to be off the percentage in his population by five percentage points or so. The method for calculating such standard errors will now be presented. The idea is to find the SE for the number of men in the sample, and then to convert this to a percent, relative to the size of the sample. The *size* of the sample just means the total number of people in it—100, in this case.

The key step in calculating the SE is making up a box model. Take the sociologist's sample of size 100. It is taken from a population of 6,672 persons. This is like drawing 100 tickets at random from a box of 6,672 tickets. There is one ticket in the box for each person in the population, and one draw for each person in the sample. The sample is taken without replacement, so the tickets must be drawn without replacement too. The sociologist is classifying people by sex and counting the men, so there should only be 0's and 1's in the box (p. 267). Since it is men that are being counted, the 3,091 tickets corresponding to the men should be marked "1," and the 3,581 tickets corresponding to the women should be marked "0." Now the number of men in the sample is like the sum of 100 draws from the box. The reason is that each person coming into the sample either pushes the number of men up by one or leaves it alone, just as each draw from the box either pushes the sum up by one or leaves it alone—and the chances in the two situations are the same. This completes the box model.

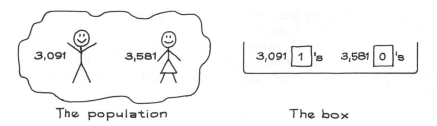

The population   The box

The point of setting up the box model is that the SE for the number of men is given by the SE for the sum of the draws, and the SE for the sum can be figured using the square root law. The fraction of 1's in the box is 0.46, so the SD of the box is $\sqrt{0.46 \times 0.54} \approx 0.50$. The SE for the sum of 100 draws is $\sqrt{100} \times 0.5 = 5$. That is, the sum of 100 draws from the box will be around 46, give or take 5 or so. Coming back to the sociologist, the number of men in his sample of 100 is expected to be around 46, give or take 5 or so. The SE for the number of men is 5. The box model has delivered the standard error for the number of men.

Now 5 out of 100 is 5%, so the percentage of men in the sample is expected to be around 46%, give or take 5% or so. This give-or-take number of 5% is the SE for the percentage of men in the sample. The sociologist just has to face up to chance errors of this magnitude.

> To compute the SE for a percentage, first get the SE for the corresponding number; then convert to a percent, relative to the size of the sample.

What happens with a larger sample? For instance, if the sociologist took a sample of size 400, the SE for the number of men in his sample would be $\sqrt{400} \times 0.5 = 10$. Now 10 represents $2\frac{1}{2}\%$ of 400, the size of the sample, so the SE for the percentage of men in a sample of 400 would be $2\frac{1}{2}\%$. In this instance, multiplying the size of the sample by 4 divided the SE for the percentage by $\sqrt{4} = 2$. The basic reason is that the SE for the number of men goes up much more slowly than the size of the sample—because of the square root in the formula for the SE for the sum. So, as a percentage of the sample size, the SE for the number of men goes down. Large samples are more accurate than small ones.

The general pattern can be stated as follows.

> Multiplying the size of a sample by some factor divides the SE for a percent by the square root of that factor. (For instance, multiplying the size of the sample by 4 divides the SE for the percent by $\sqrt{4} = 2$.) This is exact when drawing with replacement. It is a good approximation even when drawing without replacement, provided the number of draws is small by comparison with the number of tickets in the box.

The calculation for the sociologist's SE slurred over the distinction between drawing with or without replacement, because the number of draws was small by comparison with the number of tickets in the box. No matter which hundred tickets he pulls out, the percentage of 1's in the box stays very close to 46%. So he might just as well draw with replacement from the box

$$\boxed{46 \; \boxed{1}\text{'s} \quad 54 \; \boxed{0}\text{'s}}$$

Example 1. In a certain town, the telephone company has 100,000 subscribers. It plans to take a simple random sample of 400 of them, as part of a market research study. About 20% of the company's subscribers earn over $25,000 a year. Find the SE for the percentage of persons in the sample with incomes over that level.

Solution. The first step is to make up a box model. To draw a sample of 400 from this population is like drawing 400 tickets at random from a box of 100,000 tickets. There is one ticket in the box for each person in the population, and one draw for each person in the sample. The drawing is done at random without replacement. The problem involves classifying the people in the sample according to whether their incomes are more than $25,000 a year or not, and then counting the ones whose incomes are over that level. So each ticket in the box should be marked "1" or "0." Those people earning more than $25,000 a year get 1's, and the others get 0's. It is given that 20% of the people in this population earn more than $25,000 a year, so 20,000 of the tickets in the box are marked "1," and 80,000 are marked "0." The number of people in the sample who earn more than $25,000 a year is like the sum of 400 draws from this box.

Five people   Their tickets

This completes the first step, setting up the box model. There was the possibility for a fatal error here, putting each person's income on the ticket. That model would answer questions about the sum of the incomes of the sample people, but it does not answer questions about the number earning over $25,000 a year.

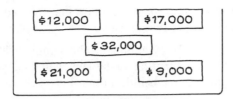

Fatal error

The next step is to calculate the standard error. First, the SD of the box is needed. This is $\sqrt{0.2 \times 0.8} = 0.4$. There are 400 draws, so the SE for the sum is $\sqrt{400} \times 0.4 = 8$. The sum will be around $400 \times 0.20 = 80$, give or take 8 or so. In other words, the number of people in the sample earning more than $25,000 a year is expected to be around 80, give or take 8 or so. Now 8 people out of 400 (the size of the sample) is 2%. Therefore, the percentage of people in the sample earning more than $25,000 a year is expected to be around 20%, give or take 2% or so. This give-or-take number of 2% is the SE for the percentage, completing the example.

For reasonably large samples, the normal approximation can be used just as in Part V:
 • the chance error is smaller than 1 SE with probability about 68%;
 • the chance error is smaller than 2 SEs with probability about 95%;
 • the chance error is smaller than 3 SEs with probability about 99.7%.

Example 2. Continuing the previous example, estimate the chance that between 18% and 22% of the persons in the sample earn more than $25,000 a year.

Solution. The expected value for the sample percent is 20%, with an SE of 2%. Now convert to standard units:

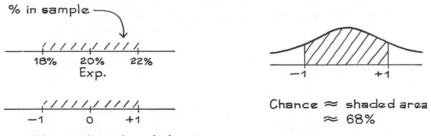

This completes the solution.

Changing the box came up before (p. 267). The incomes of the people in the sample start out as quantitative data—numbers. Examples 1 and 2, however, involved classifying and counting: each person is classified as earning more than $25,000 a year, or less. Then the high earners are counted. In other words, the data are treated as being qualitative: each income either has or doesn't have the quality of being more than $25,000 a year.

Typically, sampling problems are about some number computed from the sample. The problem asks for the chance that this number will be in a given interval. Now there are two possibilities:

• the number is computed by adding up the sample values (for instance, incomes would be treated as quantitative data when finding the chance that the total income of the sample persons in Example 1 is more than $6,000,000);

• the number is computed by classifying the sample values and then counting or taking percents (for instance, counting incomes over $25,000 a year—making them qualitative).

In the first case, you need lots of different numbers in the box. In the second case, you only need 0's and 1's.

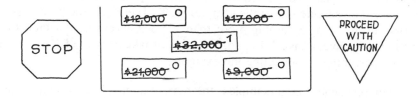

EXERCISE SET A

1. A box contains 4 red marbles and 1 blue one. Four hundred draws will be made at random with replacement from this box. The percentage of red marbles among the draws will be around _____%, give or take _____% or so.

2. According to the Census, a certain town has a population of 100,000 people aged 18 and over. Of them, 80% are married, 20% have incomes over $25,000 a year, and 10% have college degrees.[1] As part of a pre-election survey, a simple random sample of 1,600 people will be drawn from this population.
   (a) To find the chance that 78% or less of the people in the sample are married, a box model is needed. Should the number of tickets in the box be 1,600, or 100,000? Explain. Then find the chance.
   (b) To find the chance that 21% or more of the people in the sample have incomes over $25,000 a year, a box model is needed. Should each ticket in the box show the person's income? Explain. Then find the chance.
   (c) Find the chance that between 9% and 11% of the people in the sample have a college degree.

3. (a) What is the chance that the sociologist in section 1 will get between 41% and 51% men in his sample of 100?
   (b) With 25 samples of size 100, about how many should have been between 41% and 51% men?
   (c) How many actually were, in the first row of Table 2 on p. 322.

4. Repeat Exercise 3(a) for samples of size 400.

5. (a) What is the chance that the sociologist in section 1 will get exactly 46 men in a simple random sample of size 100?
   (b) With 50 samples of size 100, about how many should have exactly 46 men?
   (c) How many do, in the first two rows of Table 2?

6. Someone is going to toss a coin 10,000 times, and wants to compute the SE for the percentage of heads as "$\sqrt{10,000} \times 0.5 = 50\%$." Comment.

7. The box $\boxed{0}\ \boxed{0}\ \boxed{0}\ \boxed{1}\ \boxed{2}$ has an average of 0.6, with an SD of 0.8. True or false: The SE for the percentage of 1's in 400 draws can be found as follows:

$$\text{SE for number of 1's} = \sqrt{400} \times 0.8 = 16$$
$$\text{SE for percent of 1's} = \frac{16}{400} \times 100\% = 4\%.$$

Explain.

8. The age distribution of all students registered at a certain large university is unknown. However, in a simple random sample of 400 students, it turned out that 200 were over the age of 21. Choose one option and explain why.
  (i)   Exactly 50% of all the registered students are over the age of 21.
  (ii) About 50% of all the registered students are over the age of 21, but this may be off by a few percentage points.
  (iii) About 50% of all the registered students are over the age of 21, but this may be off by ten or twenty percentage points.

9. Someone plays a dice game 100 times. On each play, he rolls a pair of dice, and then advances his token along the line by a number of squares equal to the total number of spots thrown. (See the diagram.) About how far will he get? Give or take how much?

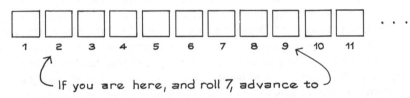

10. (Hard.) In a certain population, the distribution of height follows the normal curve, with an average of 5 feet 8 inches and an SD of 3 inches. A simple random sample of size 500 is drawn from this population. To find the chance that between 7% and 11% of the people in the sample are over 6 feet tall, a box model is needed. Should each ticket in the box show the person's height? Explain. Then find the chance. (The arithmetic does not work out evenly.)

The answers to these exercises are on pp. A-53–A-54.

Technical note: When drawing at random with replacement from a zero-one box, the SE for the number of 1's among the draws is $\sqrt{\#\text{draws}} \times$ SD of box. So the SE for the percentage of 1's among the draws is

$$(\sqrt{\#\text{draws}} \times \text{SD of box}/\#\text{draws}) \times 100\%.$$

By algebra, this simplifies to (SD of box/$\sqrt{\#\text{draws}}$) $\times$ 100%.

3. THE CORRECTION FACTOR

In 1976, there were about 0.5 million eligible voters in Delaware, and about 12.5 million in the state of New York. Suppose one polling organization takes a simple random sample of 2,500 voters in Delaware, in order to estimate the percentage of the voters in that state who are Democratic. Another polling organization takes a simple random sample of 2,500 voters from New York, in order to estimate the percentage of Democratic voters in that state. Both polls use exactly the same techniques. Each estimate is likely to be a bit off, by chance error. For which poll is the chance error likely to be smaller? The Delaware poll is sampling one voter out of 200, while the New York poll is only sampling one voter out of 5,000. So it certainly seems that the Delaware poll should be much more accurate than the New York poll. However, this is one of the places where intuition comes into head-on conflict with statistical theory—and it is intuition which has to give way. In fact, the accuracy to be expected from the New York poll is just about the same as the accuracy to be expected from the Delaware poll!

> When estimating percents, it is the absolute size of the sample which determines accuracy, not the size relative to the population. This is true for ordinary cases, where the sample is only a small part of the population.

A box model will help in focusing the issue. We are going to make up two boxes, D and N. Box D represents Delaware, box N represents New York. To begin with box D, it has 500,000 tickets, one for each voter. The tickets corresponding to Democrats are marked "1," the others are marked "0," and to keep it simple, we make the percentage of 1's equal to 50%. Now we hire a polling organization to take a simple random sample from box D—without telling them what is in the box. (Remember, taking a simple random sample means drawing at random without replacement.) Their job is to estimate the percentage of 1's in the box, and naturally they use the percentage of 1's in their sample.

They cannot expect to be exactly right. For instance, the percentage of 1's in their sample might be 51%, which would mean a chance error of +1 percentage point. Or the sample percentage might be 48%, corresponding to a chance error of −2 percentage points. In general,

$$\text{percentage of 1's in sample of 2,500 tickets from box D} = \text{percentage of 1's in box D} + \text{chance error.}$$

Now for Box N. This represents New York, so it has 12,500,000 tickets; again, we make the percentage of 1's in the box equal to 50%, the others being 0's. Another polling organization is hired to take a simple random sample of 2,500 tickets from box N, without knowing the composition of that box. They too estimate the percentage of 1's in the box by the percentage in the sample, and are off by a chance error:

$$\text{percentage of 1's in sample of 2,500 tickets from box N} = \text{percentage of 1's in box N} + \text{chance error.}$$

Box D and box N have been set up with the same percentage composition, and the two samples are the same size. Intuition would insist that the organization sampling from box D will have a much smaller chance error—because box D is so much smaller. But statistical theory shows that the likely size of the chance error is just about the same for the two polls.

The issue has now been stated sharply. How does statistical theory justify itself? To begin with, suppose the samples were drawn with replacement. Then it really wouldn't matter at all which box was used. There would be a 50–50 chance to get a 0 or a 1 on each draw, and the size of the box would be completely irrelevant. Box D and box N have the same SD of 0.5, so if they drew with replacement, both polling organizations would have the same SE for the number of 1's among the draws: $\sqrt{2,500} \times 0.5 = 25$. As a result, they would both have the same SE for the percentage of 1's among the draws:

$$\frac{25}{2,500} \times 100\% = 1\%.$$

If they drew at random with replacement, both organizations would be off by about the same amount—1 percentage point or so—and the size of the box would be completely irrelevant.

So far, so good. But the assumption behind this argument is that the draws are made with replacement. In fact, the draws are made without replacement. How does statistical theory get over that one? The answer is that the number of draws is just a tiny fraction of the number of tickets in the box. So taking the draws without replacement barely changes the composition of the box. On each draw, the chance of getting a "1" must still be very close to 50%, and similarly for "0." As far as the chances are concerned, there is almost no difference between drawing with or without replacement. Therefore, even drawing without replacement, box D and box N are about the same.

In essence, that is why the size of the population has almost nothing to do with the accuracy of estimates. Still, there is a shade of difference between drawing with and without replacement. How does the theory take that into account? Here is the answer. When drawing without replacement, the box does get a bit smaller, reducing the variability slightly. So the SE for drawing without replacement is a bit smaller than the SE for drawing with replacement. There is a mathematical formula to deal with this:

$$\begin{array}{c}\text{SE when drawing} \\ \text{WITHOUT replacement}\end{array} = \begin{array}{c}\text{correction} \\ \text{factor}\end{array} \times \begin{array}{c}\text{SE when drawing} \\ \text{WITH replacement}\end{array}$$

The formula for the correction factor is complicated:

$$\sqrt{\frac{\text{number of tickets in box} - \text{number of draws}}{\text{number of tickets in box} - \text{one}}}$$

But the point is that when the number of tickets in the box is very large by comparison with the number of draws, the correction factor is nearly 1, and can be ignored. Table 3 shows this when the number of draws is fixed at 2,500, and the number of tickets in the box ranges from 5,000 to 12,500,000.

Table 3. The correction factor. The number of draws is fixed at 2,500.

Number of tickets in the box	Correction factor (to five decimals)
5,000	0.70718
10,000	0.86607
100,000	0.98743
500,000	0.99750
1,000,000	0.99875
12,500,000	0.99990

Now let us return to the key issue: Why it is that the survey organizations sampling from box D and box N get chance errors of the same size. Using the correction factor, we can really pin this down. The SE for drawing with replacement depends only on the number of draws and the SD of the box: the number of tickets in the box does not come in at all. So the size of the population—the number of tickets in the box—only comes into the SE through the correction factor. When the sample is a small part of the population, the correction factor is nearly one, so the size of the population doesn't matter.

We will work this out for 2,500 draws made at random without replacement from box D or box N. On p. 331, the SE for the percentage of 1's was seen to be 1%—when the draws are made with replacement. There are 500,000 tickets in box D, so the correction factor is 0.99750, from Table 3. With 2,500 draws made at random WITHOUT replacement from box D, the formula on p. 331 gives the SE for the percentage of 1's among the draws as

$$0.99750 \times 1\%.$$

Similar reasoning for box N shows that with 2,500 draws made at random WITHOUT replacement from this much larger box, the SE for the percentage of 1's among the draws is

$$0.99990 \times 1\%.$$

At this point, intuition just has to change. There is no place left to go.

When estimating percents, the size of the population has very little to do with accuracy: the absolute size of the sample is what counts. This begins to explain how the Gallup poll can make accurate predictions of presidential elections using samples of only a few thousand voters out of a hundred million: the enormous size of the population doesn't affect the accuracy of the poll.

Another part of the mathematical explanation for Gallup's success is the square root law. As the size of the sample goes up, the SE for a percentage goes down. With 2,500 draws, it is already quite small, as Table 4 shows. For instance, with a sample of size 2,500 drawn from a population which is evenly split, the sample percentage will be off the population percentage by 1 percentage point or so. With uneven splits, the standard errors are smaller.

This section has been a bit technical. At this point, a nonmathematical analogy may be quite helpful. Suppose you took a drop of liquid from a bottle, for chemical analysis. If the liquid is well mixed, the chemical composition of the drop would reflect quite faithfully the composition of the whole bottle—and it really wouldn't matter if the bottle was a test tube or a gallon jug. The chemist won't care whether the drop is 1% or 1/100 of 1% of the solution. The analogy is precise. There is one ticket in the box for each molecule in the bottle. If the liquid is well mixed, the drop is drawn at random. The number of molecules in the drop corresponds to the number of tickets drawn. This number—the sample size—is so large that the chance error in the percentages is negligible.

Table 4. Standard error for the percentage of 1's when drawing at random from a very large 0–1 box. The percentage of 1's in the box is shown along the top, the number of draws along the left, and the SE is in the body of the table.

| Number of draws | Percentage of 1's in the box | | | | |
	10%	20%	50%	80%	90%
100	3%	4%	5%	4%	3%
2,500	0.6 of 1%	0.8 of 1%	1%	0.8 of 1%	0.6 of 1%
10,000	0.3 of 1%	0.4 of 1%	0.5 of 1%	0.4 of 1%	0.3 of 1%

This section focused on simple random sampling, but the conclusions hold for most probability methods. The likely size of the chance error in the sample percentages depends mainly on the absolute size of the sample, hardly at all on the size of the population, and the chance error is already small for samples of size several thousand. This completes the explanation for the success of the Gallup poll in predicting elections. Gallup's sample of a few thousand voters really is big enough to predict election results to within a percentage point or so. And the huge size of the population he is sampling from does not affect the accuracy of the estimates. That is what statistical theory shows, and the theory is confirmed by experience: look at Table 4 on p. 310.

The main thing which governs the accuracy of the Gallup poll is the absolute, numerical size of the sample. Since 1948, Gallup poll press releases almost always give the size of the sample on which their estimates are based. And newspaper editors almost always edit this information out. Newspaper readers may not be ready for the concept of chance error.

EXERCISE SET B

1. One public opinion poll uses a simple random sample of size 1,500 drawn from a town with a population of 25,000. Another poll uses a simple random sample of size 1,500 from a town with a population of 250,000. Other things being equal:
    (i)   the first poll is likely to be quite a bit more accurate than the second.
    (ii) the second poll is likely to be quite a bit more accurate than the first.
    (iii) there is not likely to be much difference in accuracy between the two polls.

2. You have hired a polling organization to take a simple random sample from a box of 100,000 tickets, and estimate the percentage of 1's in the box. Unknown

to them, the box contains 50% 0's and 50% 1's. How far off should you expect them to be:

    (a) if they draw 2,500 tickets?

    (b) if they draw 25,000 tickets?

    (c) if they draw 100,000 tickets?

3. A survey organization wants to take a simple random sample in order to estimate the percentage of people who have seen a certain television program. To keep the costs down, they want to take as small a sample as possible. But their client will only tolerate chance errors of 1 percentage point or so in the estimate. Should they use a sample of size 100, 2,500, or 10,000? (See Table 4 on p. 333; you may assume the population to be very large.)

4. A box contains 2 red marbles and 8 blue ones. Four marbles are drawn at random. Find the SE for the percentage of red marbles drawn, when the draws are made

    (a) with replacement   (b) without replacement.

(The arithmetic does not work out evenly.)

5. The following news item is reproduced, with no changes or omissions, from the *San Francisco Chronicle* of May 29, 1975.

    (a) One key fact is missing. What is it?

    (b) How else can the findings be interpreted, apart from "sagging confidence"?

CALIFORNIA POLL

SAGGING CONFIDENCE
By Mervin D. Field

As the U. S. approaches its bi-centennial, the California public's confidence in many of the nation's basic governmental and social institutions continues to slump.

In a statewide survey just completed, the California Poll finds that the level of public confidence in a score of different groups and institutions has declined from the already low levels found in a similar survey two years ago.

The findings in this survey show the social and political malaise that has been observed not just in California, but throughout the country.

For only one group, research scientists, does more than one-half of the public says it has "a lot of confidence." Even with this group, however, the proportion saying it has "a lot of confidence" is down from 1973.

The institutions and organizations tested covered a spectrum of our society: medical and legal professions, Congress and the President, universities, public schools, organized religion, manufacturing and food companies and others.

A yardstick for seeing how each institution looks to the public can be obtained by comparing the percentage of public support for each institution with the percent of the public that expresses a lack of confidence. When this is done, the institutions fall into groups as follows:

Most Favorable "lot of confidence" rating exceeds 50 percent and "not much confidence" rating is below 5 percent: research scientists.

Somewhat Favorable "lot of confidence" rating exceeds 40 percent; "not much confidence" rating is below 16 percent: local police department.

Slightly Favorable "lot of confidence" rating exceeds 33 percent; "not much confidence" rating is below 22 percent: the medical profession, consumer protection groups, FBI—Supreme Court.

Stand-off Situation "lot of confidence" rating about the same as the "not much confidence" rating: news media, environmental protection groups and organizations, universities and colleges, organized religion and churches, Legislature, Congress and financial institutions.

Somewhat Unfavorable "not much confidence" rating exceeds "lot of confidence" rating: government regulatory agencies, legal profession and President of the United States.

Most Unfavorable "not much confidence" rating far exceeds "lot of confidence" rating: food companies, manufacturing corporations, labor unions, public school system and public utilities.

Some other interesting sidelights to the two survey findings:

Two professions which not too long ago were held in high public esteem—law and medicine—have been receiving unfavorable publicity recently. The current medical malpractice insurance debate has featured lawyers attacking doctors and vice versa. Other medical controversies involving overcharging have been in the headlines, and the image of lawyers was tarnished by Watergate events in which many lawyers were implicated.

While there is diminishing confidence in all professions, the medical profession seems to rank higher in overall public esteem than the legal profession today.

The sharpest decline in public confidence observed is that toward public utilities. In 1973, the ratio of "lot of confidence" to "not much confidence" was 33 percent to 17 percent on the favorable side. That ratio now has reversed—17 percent to 36 percent unfavorable.

Since 1973 most public utilities have had to raise rates sharply because of increases in fuel and other costs and as a result of this they become the target of considerable consumer protest. Other ramifications of the energy crisis issue has also focused unprecedented attention on the utilities.

Here is the comparison of the level of confidence in terms of percentage of the public in the institutions rated between the 1973 and 1975 surveys.

		Lot of confidence	Some confidence	Not much confidence
Research scientists	1975	54	36	5
	1973	58	34	5
Local police	1975	46	35	16
	1973	51	40	7
Medical profession	1975	35	43	20
	1973	43	44	13
Consumer groups	1975	33	39	21
	1973	37	42	14
FBI	1975	35	39	22
	1973	43	40	13
Supreme Court	1975	35	38	21
	1973	31	45	21
Environmental groups	1975	27	44	23
	1973	30	48	17
Universities, colleges	1975	28	43	24
	1973	25	62	11
Organized religion	1975	27	43	26
	1973	24	46	28

[Table continues on next page.]

		Lot of confidence	Some confidence	Not much confidence
Financial institutions	1975	22	48	26
	1973	22	51	25
Legislature	1975	17	61	18
	1973	12	67	16
Congress	1975	19	58	21
	1973	30	53	15
* Regulatory agencies	1975	17	51	23
* Legal profession	1975	18	53	24
News media	1975	27	49	22
	1973	27	55	18
President of U.S.	1975	22	46	30
	1973	34	34	31
Food companies	1975	10	47	40
	1973	9	52	35
Manufacturers	1975	13	47	32
	1973	9	55	30
Labor unions	1975	16	39	40
	1973	13	50	34
Public schools	1975	15	45	38
	1973	23	51	25
Public utilities	1975	17	43	36
	1973	33	48	17

* Not rated in 1973 survey.
The "no opinion" group accounts for differences between the sum of these figures and 100 per cent.
Copyright 1975, Field Research Corp. Reproduced by permission of Mr. Mervin Field.

The answers to these exercises are on p. A-54.

4. REVIEW EXERCISES

Review exercises may also cover material from previous chapters.

1. Complete the following table for the coin-tossing game.

	Number of heads		Percent of heads	
Number of tosses	Expected value	SE	Expected value	SE
100	50	5	50%	5%
2,500				1%
10,000				
1,000,000				

2. A coin is tossed 400 times. Someone wishes to compute the SE for the percentage of heads among the tosses as "$\sqrt{400} \times 0.5 = 10\%$." Comment.

3. A box contains 1 red marble and 9 blue ones. Nine hundred draws are made at random with replacement from this box. The percentage of red marbles among the draws should be around _____%, give or take _____ percentage points or so. Fill in the blanks.

4. A group of 10,000 tax forms shows an average gross income of $17,000, with an SD of $10,000. Furthermore, 10% of the forms show a gross income over $30,000. A group of 900 forms is chosen at random for audit. To estimate the chance that between 9% and 11% of the forms chosen for audit show gross incomes over $30,000, a box model is needed.

(a) Should the number of tickets in the box be 900, or 10,000?

(b) Each ticket in the box shows

   a zero or a one   a gross income.

(c) True or false: The SD of the box is $10,000.

(d) True or false: The number of draws is 900.

(e) Find the chance (approximately) that between 9% and 11% of the forms chosen for audit show gross incomes over $30,000.

(f) With the information given, can you find the chance (approximately) that between 19% and 21% of the forms chosen for audit show gross incomes over $20,000?

5. As in Exercise 4, except it is desired to find the chance (approximately) that the total gross income of the audited forms is over $15,000,000. Work parts (a) through (d), then find the chance.

6. On the average, hotel guests weigh about 150 pounds with an SD of 25 pounds. An engineer is designing a large elevator for a convention hotel, to lift 100 people. If he designs it to lift 15,500 pounds, the chance it will be overloaded by a random group of 100 people is closest to
   0.1 of 1%   2%   5%   50%   95%   98%   99.9%.

7. The Census Bureau is planning to take a sample amounting to 1/10 of 1% of the population in each state in order to estimate the percentage of the population in that state earning over $10,000 a year. Other things being equal:

   (i)   The accuracy to be expected in California (population twenty million) is about the same as the accuracy to be expected in Nevada (population half a million).

   (ii) The accuracy to be expected in California is quite a bit higher than in Nevada.

   (iii) The accuracy to be expected in California is quite a bit lower than in Nevada.

   Explain.

8. In 1965, the U.S. Supreme Court decided the case of *Swain* v. *Alabama*.[2] Swain, a black man, was convicted in Talladega County, Alabama, of raping a white woman. He was sentenced to death. The case was appealed to the Supreme Court on the grounds that there were no blacks on the jury; even more, no black "within the memory of persons now living has ever served on any petit jury in any civil or criminal case tried in Talladega County, Alabama."

   The Supreme Court denied the appeal, on the following grounds. As provided by Alabama law, the jury was selected from a panel of about 100 persons. There were 8 blacks on the panel. (They did not serve on the jury because they were "struck," through peremptory challenges by the prosecution. Such challenges are constitutionally protected.) The presence of 8 blacks on the panel showed "The overall percentage disparity has been small and reflects no studied attempt to include or exclude a specified number of Negroes."

At that time in Alabama, only men over the age of 21 were eligible for jury duty. There were 16,000 men over the age of 21 in Talladega County, of whom about 26% were black. If 100 people were chosen at random from this population, what is the chance that 8 or fewer would be black?

9. A certain university has a student population of 25,000. A survey organization wants to take a sample of 100 students. They propose to station an interviewer at a central point on the campus at 10:00 A.M. the following Tuesday morning, and to choose the first hundred students the interviewer sees for the sample. Since the interviewer is not choosing purposefully, the survey organization maintains it will get a random sample, and every student will have an equal chance to get into the sample. What does statistical theory say?

10. The following quote is from "The Grab Bag" by L. M. Boyd in the *San Francisco Chronicle*.

"The Law of Averages says that if you throw a pair of dice 100 times, the number tossed will add up to just about 683."

(a) As far as the chances are concerned, the total of the numbers thrown is just like

(i) the sum of 100 draws from the box

$$\boxed{2}\ \boxed{3}\ \boxed{4}\ \boxed{5}\ \boxed{6}\ \boxed{7}\ \boxed{8}\ \boxed{9}\ \boxed{10}\ \boxed{11}\ \boxed{12}$$

(ii) the sum of 200 draws from the box

(b) Is Boyd right?
(The average of box (i) is 7, with an SD of 3.2; the average of box (ii) is 3.5, with an SD of 1.7.)

11. There are 50,000 households in a certain city. The average number of persons aged 16 and over living in each household is known to be 2.28, with an SD of 1.78. A survey organization plans to take a simple random sample of 400 households, and interview all persons aged 16 and over living in the sample households. The total number of interviews will be around _____ , give or take _____ or so.

12. A certain town has 25,000 families. These families own 2.1 cars, on the average, with an SD of 0.80. And 20% of them have no cars at all. As part of an opinion survey, a simple random sample of 1,600 families is chosen. What is the chance that between 19% and 21% of the sample families will not own cars?

5. SUMMARY

1. The sample is only part of the population, so the percentage composition of the sample usually differs a bit from the percentage composition of the whole population, by chance error.

2. The likely size of the chance error is given by the standard error.

3. To figure the SE for a percentage, first figure the SE for the corresponding number, then convert to percent.

4. To figure the SE, a box model is needed. When the problem involves classifying and counting the draws, or taking percents, there should only be 0's and 1's in the box. Change the box, if necessary.

5. When the sample is only a small part of the population, the number of individuals in the population has almost no influence on the accuracy of the estimates. It is the absolute size of the sample (that is, the number of individuals in the sample) which matters, not the size relative to the population.

6. The square root law is exact when the draws are made with replacement. When the draws are made without replacement, the formula gives a good approximation—provided the number of draws is small by comparison with the number of tickets in the box.

7. When drawing without replacement, to get the exact SE it is necessary to multiply by the correction factor

$$\sqrt{\frac{\text{number of tickets in box } - \text{ number of draws}}{\text{number of tickets in box } - \text{ one}}}$$

8. When the number of tickets in the box is much larger than the number of draws, the correction factor is nearly one.

21

The Accuracy of Percentages

In solving a problem of this sort, the grand thing is to be able to reason backward. That is a very useful accomplishment, and a very easy one, but people do not practise it much. . . . Most people, if you describe a train of events to them, will tell you what the result would be. They can put those events together in their minds, and argue from them that something will come to pass. There are few people, however, who, if you told them a result, would be able to evolve from their own inner consciousness what the steps were which led up to that result. This power is what I mean when I talk of reasoning backward. . . .

—*Sherlock Holmes*[1]

1. INTRODUCTION

The previous chapter reasoned from the box to the draws. Draws were made at random from a box whose composition was known, and a typical problem was finding the chance that the percentage of 1's among the draws would be in a given interval. As Sherlock Holmes points out, it is often very useful to turn this reasoning around, and go from the draws to the box. A statistician would call this *inference* from the sample to the population, and inference is the topic of this chapter.

For instance, suppose a survey organization wants to know the percentage of Democrats in a certain district. They might estimate it by taking a simple random sample. Naturally, the percentage of Democrats in the sample would be used to estimate the percentage of Democrats in the district—an example of reasoning from the draws to the box. And because the sample was chosen at random, it is even possible to say how accurate this estimate is likely to be—just from the size and composition of the sample. This chapter will explain how to evaluate the accuracy of percentages estimated from a simple random sample. The technique is one of the key ideas in statistical theory.

We begin with an example. A political candiate wants to enter a primary in a district with 100,000 eligible voters, but only if he has a good chance of winning. He hires a survey organization, which takes a simple random sample of 2,500 voters. In the sample, 1,328 favor the candidate, so his percentage is

$$\frac{1,328}{2,500} \times 100\% \approx 53\%.$$

The candidate is discussing this result with his pollster.

Politician. So I win.

Pollster. Not so fast. You want to know the percentage you'd get among all the voters in the district. We only have it in the sample.

Politician. But with a good sample, it's bound to be the same.

Pollster. Not true. It's what I said before. The percentage you get in the sample is different from what you'd get in the whole district. The difference is what we call chance error.

Politician. Could the sample be off by as much as three percentage points? If so, I could lose.

Pollster. Actually, you can be about 95% confident that we're right to within two percentage points. It looks good.

Politician. What tells you the size of the chance error?

Pollster. The standard error. Remember, we talked about that the other day. As I was telling you. . .

Politician. Sorry, I'm expecting a phone call now.

The politician has arrived at the crucial question to ask when considering survey data: how far wrong are the estimates likely to be? As the pollster said, the likely size of the chance error is given by the standard error, and to figure that a box model is needed.

There should be one ticket in the box for each voter, making 100,000 tickets in all. Each ticket should be marked "1" or "0," where "1" means a vote for the candidate, "0" a vote against him. To get the SE, the survey organization needs the SD of the box. This is

$$\sqrt{(\text{fraction of 1's}) \times (\text{fraction of 0's})}.$$

At this point, they seem to be stuck. They certainly don't know how each ticket in the box should be marked. And they don't even know the fraction of 1's in the box. That parameter represents the fraction of voters in the district who favor their candidate, which is exactly what they were hired to find out!

$$\boxed{?? \boxed{0}\text{'s} \quad ?? \boxed{1}\text{'s}}$$

100,000 tickets

Survey organizations lift themselves over this kind of obstacle by their own bootstraps. They substitute the fractions observed in the sample for the unknown fractions in the box. In the example, 1,328 people out of the sample of 2,500 favored the candidate. So $1,328/2,500 \approx 0.53$ of the sample favored him, and the other 0.47 were opposed. The estimate is that about 0.53 of the 100,000 tickets in the box are marked "1," the other 0.47 being marked "0." On this basis, the SD of the box is estimated as $\sqrt{0.53 \times 0.47} \approx 0.50$. The SE for the number of voters in the sample who favor the candidate is then estimated as $\sqrt{2,500} \times 0.50 = 25$. Now 25 people out of 2,500 (the size of the sample) is 1%. So the SE for the percentage of voters in the sample favoring the candidate is estimated as 1%. This completes the bootstrap procedure for estimating the standard error.

As far as the candidate is concerned, this calculation shows that his pollster's estimate of 53% is only likely to be off by one percentage point or so. It is very unlikely to be off by as much as three percentage points—that's three SEs. So he is well on the safe side of 50%. He should enter the primary.

Now for some terminology. Usually, the object of a survey is to estimate the percentage of individuals in a certain population who have a specified property, such as favoring a particular candidate. This parameter is called the *population percentage*. It is unknown. It can be estimated by drawing a simple random sample from the population, and taking the percentage of individuals in the sample who have the quality. This statistic is called the *sample percentage*. The survey organization is off by the difference between the sample percentage and population percentage. This difference is our old friend, the chance error in the sample percentage. Its likely size is given by the standard error for the sample percentage, which can be estimated by the bootstrap procedure.

When sampling from a zero-one box whose composition is unknown, the SD of the box can be estimated by substituting the known fractions of 0's and 1's in the sample for the unknown fractions in the box. The estimate is good when the sample is reasonably large.

This bootstrap procedure may seem crude. But even with moderate-sized samples, the fraction of 1's among the draws is likely to be quite close to the fraction in the box (pp. 332–333). Similarly for the 0's. So, if the survey organization uses the sample fractions in the formula for the SD of the box, they are not likely to be far wrong in estimating their SE.

Example. In fall quarter, 1977, a certain university had 25,000 registered students. A survey was made that quarter to estimate the percentage whose parents had both completed college. A simple random sample of 400 students was drawn, and it turned out that for 317 of them, both parents had college degrees. Estimate the percentage of students at the university that quarter whose parents were both college graduates, and attach a standard error to the estimate.

Solution. The sample percentage is $317/400 \times 100\% \approx 79\%$, and this is the estimate for the population percentage. For the standard error, a box model is needed. There are 25,000 tickets in the box, one for each student in the population. There are 400 draws from the box, one for each student in the sample. This problem involves classifying and counting, so each ticket in the box should be marked "1" or "0." We are counting students whose parents were both college graduates. The tickets corresponding to these students should be marked "1," the others, "0." The number of students in the sample whose parents both have college degrees is like the sum of 400 draws from the box. This completes the model.

$$\boxed{??\boxed{0}\text{'s} \quad ??\boxed{1}\text{'s}}$$

25,000 tickets

The fraction of 1's in the box is a parameter. It represents the fraction of all students at this university whose parents have college degrees. It is unknown, but can be estimated as 0.79—the fraction of such students in the sample. Similarly, the fraction of 0's in the box is estimated as 0.21. So the SD of the box is estimated by the bootstrap method as $\sqrt{0.79 \times 0.21} \approx 0.41$. The SE for the number of students in the sample whose parents are both college graduates is then estimated as $\sqrt{400} \times 0.41 \approx 8$. This is 2% of 400, the size of the sample. So the SE for the sample percentage is estimated as 2%. Therefore, the population percentage is around 79%, give or take 2 percentage points or so. This is the answer. About 79% of the students had parents with college degrees—give or take 2 percentage points or so.

The discussion in this section focussed on simple random sampling, where the mathematics is easiest. In practice, survey organizations use much more complicated designs. Even so, with probability methods it is always possible to say how far wrong the estimates are likely to be. This is one of the great advantages of probability methods for drawing samples (pp. 307 ff.). If probability methods are not used to draw the sample, there is no sensible way to calculate standard errors for the estimates.

EXERCISE SET A

1. A box contains 100,000 marbles, some of which are red and others blue. The percentage of red marbles in the box is unknown. To estimate it, a simple random sample of 1,600 marbles is drawn, and 322 of the marbles in the sample turn out to be red. The percentage of red marbles in the box is estimated as _____, give or take _____ or so.

2. In a certain city, there are 100,000 persons aged 20 to 24. A simple random sample of 1,600 such persons was drawn, of whom 322 turned out to be currently enrolled in school. Estimate the percentage of all persons aged 20 to 24 in that city who were currently enrolled in school.[2] Put a give-or-take number on the estimate.

3. In a simple random sample of 100 graduates from a certain college, 48 were earning $15,000 a year or more. Estimate the percentage of all graduates of that college earning $15,000 a year or more. Attach a standard error to the estimate.

4. A simple random sample of size 400 is taken from the population of all manufacturing establishments in a certain state; 16 firms in the sample had 250 employees or more. Estimate the percentage of all manufacturing establishments in the state which had 250 employees or more. Attach a standard error to the estimate.[3]

5. In the same state, a simple random sample of size 400 is taken from the population of all persons employed by manufacturing establishments; 260 people in the sample worked for firms with 250 employees or more. Estimate the percentage of people in the population (of all manufacturing employees) who work for firms with 250 employees or more. Attach a standard error to the estimate.

6. Is the difference between the percentages in Exercises 4 and 5 due to chance error?

The next two exercises are designed to illustrate the bootstrap method for estimating the SD of the box.

7. Suppose there is a box of 100,000 tickets, each marked "0" or "1." Suppose that in fact, 20% of the tickets in the box are 1's. Calculate the standard error for the percentage of 1's in 400 draws from the box.

8. Three different people take simple random samples of size 400 from the box in Exercise 7, without knowing its contents. The number of 1's in the first sample is 72; in the second, it is 84; in the third, it is 98. Each person estimates the SE by the bootstrap method.
    (a) The first person estimates the percentage of 1's in the box as _____ give or take _____ or so.
    (b) The second person estimates the percentage of 1's in the box as _____ give or take _____ or so.
    (c) The third person estimates the percentage of 1's in the box as _____ give or take _____ or so.

The answers to these exercises are on pp. A-54–A-55.

2. CONFIDENCE INTERVALS

In the example of the previous section, 79% of the students in a sample came from families where both parents had finished college. In more technical language, the sample percentage was 79%. How far can the population percentage be from 79%? (Here, "population percentage" means the percentage of all students at that university whose parents both had college degrees.) The standard error was estimated as 2%, suggesting a chance error of around 2% in size. So the population percentage could easily be 77%. This would mean a chance error of exactly 2%:

$$\text{sample percentage} = \text{population percentage} + \text{chance error}$$
$$79\% = 77\% + 2\%$$

The population percentage could also be 76%, corresponding to an error of 3%. This is getting unlikely, because 3% represents 1.5 SEs. The population percentage could even be as small as 75%, but this is still more unlikely; the error of 4% represents two SEs. Of course, the population percentage could be on the other side of the sample percentage, corresponding to negative chance errors. For instance, the population percentage could be 83%. Then the estimate is low by 4%, and the chance error is −4%, that is, −2 SEs.

With chance errors, there is no sharp dividing line between the possible and the impossible. Errors larger in size than two SEs do occur—infrequently. It is natural to ask what happens if you try to cut off at two SEs, by looking only at the interval from two SEs below the sample percentage to two SEs above—from 75% to 83%:

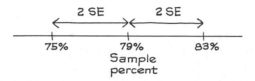

This is a *confidence interval* for the population percentage, with a *confidence level* of about 95%. You can be about 95% confident that the population percentage is caught inside the interval from 75% to 83%.

What happens if you want a different confidence level? Anything except 100% is possible, by going the right number of SEs in either direction from the sample percentage. For instance:

• the interval "sample percentage ± 1 SE" is an approximate 68%-confidence interval for the population percentage;
• the interval "sample percentage ± 2 SEs" is an approximate 95%-confidence interval for the population percentage;
• the interval "sample percentage ± 3 SEs" is an approximate 99.7%-confidence interval for the population percentage.

No multiple of an SE gives 100% confidence, because there is always the remote possibility of a very large chance error. Mathematically, this is reflected in the fact that there are no definite limits to the normal curve. No matter how large a finite interval you choose, the normal curve has some area outside that interval.

Example. A simple random sample of 1,600 persons is taken, to estimate the percentage of Democrats among the 25,000 eligible voters in a certain town. It turns out that 917 people in the sample are Democrats. Find an approximate 95%-confidence interval for the percentage of Democrats among all 25,000 eligible voters.

Solution. The percentage of Democrats in the sample is

$$\frac{917}{1,600} \times 100\% \approx 57\%.$$

The estimate is that about 57% of the eligible voters are Democrats—and 43% aren't. For the standard error, a box model is needed.

There is one ticket in the box for each eligible voter in the town, making 25,000 tickets in all. There are 1,600 draws, corresponding to the sample size of 1,600. This problem involves classifying people (Democrat or not) and counting, so each ticket is marked "1" or "0." It is Democrats that are being counted, so the tickets corresponding to the Democrats are marked "1," the others are marked "0." The number of Democrats in a sample of size 1,600 is like the sum of 1,600 draws from the box. This completes the model.

The fraction of 1's in the box (translation: the fraction of Democrats among the 25,000 eligible voters) is unknown, but can be estimated by 0.57, the fraction of Democrats in the sample. Similarly, the fraction of 0's in the box is estimated as 0.43. So the SD of the box is estimated by the bootstrap method as $\sqrt{0.57 \times 0.43} \approx 0.50$. The SE for the number of Democrats in the sample is then estimated as $\sqrt{1,600} \times 0.50 = 20$. As a percentage of the size of the sample, this is

$$\frac{20}{1,600} \times 100\% = 1.25\%.$$

So the SE for the percentage of Democrats in the sample is 1.25%. In other words, the percentage of Democrats in the sample is likely to be off the percentage of Democrats among all 25,000 eligible voters by 1.25 percentage points or so. An approximate 95%-confidence interval for the percentage of Democrats among all the eligible voters is

$$57\% \pm 2 \times 1.25\%.$$

That is the answer. We can be about 95% confident that between 54.5% and 59.5% of the 25,000 eligible voters in this town are Democrats.

Confidence levels are quoted as being "about" so much. One reason is that the standard errors have been estimated from the data. Another is that the normal approximation has been used. If the normal approximation does not apply, neither do the methods of this chapter. There is no hard-and-fast rule to follow in deciding. The best way to proceed is to imagine that the percentage composition of the population is reflected in the sample. Then try to decide whether the normal approximation would work for the sum of the draws from the corresponding box. In general, a sample percentage near 0% or 100% suggests that a very large number of draws is needed before the normal approximation takes over. On the other hand, if the sample percentage is near 50%, the normal approximation should be satisfactory when there are only a hundred draws or so. (It is like coin-tossing: when the coin is biased, more tosses are needed before the probability histogram looks like the normal curve, as shown on p. 291.)

There is a final caution to observe. The mathematics depends on certain assumptions, met by ideal sample surveys. Real surveys do not obey the assumptions exactly, and this too can throw the calculations off. For instance, in Cycle I of the Health Examination Survey (Chapter 4), the nonresponse rate was about 15%. If the nonrespondents were different from the respondents, the nonresponse bias could be quite a bit larger than the chance error being described by confidence intervals. Mathematics can only do so much.

EXERCISE SET B

1. Refer back to Exercise 1 on p. 343.
   (a) Find an approximate 95%-confidence interval for the percentage of red marbles in the box.
   (b) Repeat, if possible, for a confidence level of 99.7%.
   (c) Repeat, if possible, for a confidence level of 100%.

2. Refer back to Exercise 2 on p. 343. Find an approximate 95%-confidence interval for the percentage of all persons aged 20 to 24 in that city who were currently enrolled in school.

3. As in Exercise 2, but supposing the size of the sample was 2,500, of whom 503 were currently enrolled in school.

4. A box contains 10,000 marbles, of which some are red and the others blue. To estimate the percentage of red marbles in the box, 100 are drawn at random without replacement. Among the draws, 1 turns out to be red. The percentage of red marbles in the box is estimated as 1%, with an SE of 1%. True or false: an approximate 95%-confidence interval for the percentage of red marbles in the box is 1% ± 2%. Explain.

5. You are advising a political candidate on opinion polls. Two organizations have submitted bids to do a survey to estimate the extent of his support. Both will use nationwide probability samples of 2,500 people, and both charge the same price. One organization only states that it has a 95% chance to get the estimate right to within 3%. The other absolutely guarantees that its estimate will be right to within 3%. Other things being equal, which bid should be accepted?

The answers to these exercises are on p. A-55.

3. INTERPRETING A CONFIDENCE INTERVAL

Going back to the example on pp. 342–343, a simple random sample was taken to estimate the percentage of students registered at a certain university in fall, 1977, whose parents were both college graduates. An approximate 95%-confidence interval for this percentage ran from 75% to 83%, because

$$\text{sample percentage} \pm 2 \text{ SE} = 75\% \text{ to } 83\%.$$

It seems more natural to say "There is a 95% chance that the population percentage is between 75% and 83%." But there is a problem here. In the frequency theory of chance, a chance represents the percentage of the time that something will happen. No matter how many times you take stock of all the students registered at that university in the fall of 1977, the percentage with parents who were both college graduates won't change. Either this percentage is between 75% and 83%, or it isn't. So there really isn't any way to define the chance that the parameter will be in this interval—or any other.[4] That is why statisticians have to turn the problem around slightly. They realize that the chances are in the sampling procedure, not in the parameter, and they use the new word "confidence" to remind you of this.

So the confidence level of 95% has to say something about the sampling procedure, and we are going to see what this is. The first point to notice is that the confidence interval depends on the sample. If the sample had come out differently, the confidence interval would have been different, because the percentages in the sample would have been different. With some samples, the interval "sample percentage ± 2 SE" does trap the population percentage. (The word statisticians use is *cover*.) But with other samples, the interval fails to cover. It's like buying a used car. Sometimes you get a lemon—a confidence interval which doesn't cover the parameter.

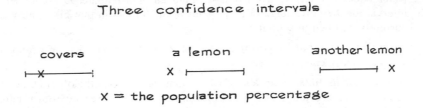

The interval "sample percentage ± 2 SE" has a confidence level of 95%. This can now be interpreted: For about 95% of all samples, the interval "sample percentage ± 2 SE" covers the population percentage, and for the other 5% it fails.

Of course, an investigator usually can't tell whether or not his particular interval covers the population percentage: he doesn't know this parameter, it's what he is trying to estimate. But he followed a procedure which works 95% of the time—taking a simple random sample, and then going two SEs either way from the sample percentage. It is as if his interval was drawn at random from a box full of intervals, of which 95% cover the parameter, and only 5% are lemons. It beats buying a used car.

This interpretation of confidence levels is a bit difficult, because it involves thinking not only about the actual sample, but about other samples that could have been drawn. The interpretation is illustrated in Figure 1.[5] A hundred survey organizations are hired to estimate the percentage of red marbles in a large box. Unknown to the organizations, this percentage is 80%. Each organization takes a simple random sample of 2,500 marbles, and computes a 95%-confidence interval for the percentage of reds in the box, using the formula "percentage of reds in sample ± 2 SE." The percentage of reds is different from sample to sample, and so is the estimated standard error. As a result, the confidence intervals have different centers and lengths. Some of the organizations get intervals covering the percentage of red marbles in the box, others fail. In the figure, these intervals are drawn at different heights, so you can tell them apart. About 95% of them should cover the percentage of red marbles in the box, marked by a vertical line. And in fact, 96 out of 100 do.

> The chances are in the sampling procedure, not in the parameter.

Figure 1. Interpreting confidence intervals. The 95%-confidence interval is shown for a hundred different samples. The interval changes from sample to sample. For about 95% of the samples, the interval covers the population percentage, marked by a vertical line.

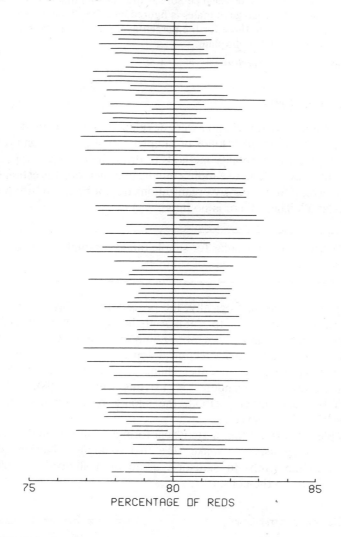

PERCENTAGE OF REDS

EXERCISE SET C

1. Refer to Exercises 7 and 8 on p. 344. Compute a 95%-confidence interval for the percentage of 1's in the box, using the data obtained by the person in Exercise 8(a). Repeat for the other two people. Which of the three intervals cover the population percentage—that is, the percentage of 1's in the box? Which do not?

2. A simple random sample of 1,000 persons is taken to estimate the percentage of Democrats in a large population. It turns out that 643 of the people in the sample are Democrats. The sample percentage is $(643/1,000) \times 100\% \approx 64\%$. The SE for the sample percentage of Democrats is figured as 1.5%. True or

false: There are about two chances in three for the percentage of Democrats in the population to be in the range 64% ± 1.5%. Explain.

3. At a large university, 64% of the students are male and 36% are female. A simple random sample of 1,000 persons is drawn from this population. The SE for the sample percentage of males is figured as 1.5%. True or false: There are about two chances in three for the percentage of males in the sample to be in the range 64% ± 1.5%. Explain.

The answers to these exercises are on p. A-56.

4. CAVEAT EMPTOR

The methods of this chapter were developed for simple random samples. They do not apply to other kinds of samples. Most survey organizations use fairly complicated probability methods to draw their samples (pp. 307–310). As a result, they have to use more complicated methods for estimating their standard errors. Some survey organizations do not bother to use probability methods at all. Then, there is no hope.

> Warning. The formulas for simple random samples should not be used for other kinds of samples.

Here is the reason. Logically, the procedures in this chapter all come out of the square root law (p. 258). And the square root law only applies when the numbers are drawn at random from a box. The phrase "at random" is used here in its technical sense: at each stage, every ticket in the box has to have an equal chance to be chosen. Simple random samples are drawn "at random" in just this sense (p. 308). If the sample is taken by some other method, the square root law does not apply, and may give silly answers.

People often think that a statistical formula will somehow check itself while it is being used, to make sure that it applies. Nothing could be further from the truth. In statistics, as in old-fashioned capitalism, the responsibility is on the consumer.

Caveat emptor   Let the buyer beware

$$\bar{x} \ \pm \ z_\alpha \times s/\sqrt{n}$$

EXERCISE SET D

1. A psychologist administers a test of passivity to 100 students in his class and finds that 20 of them score over 50. He concludes that approximately 20% of all

students would score over 50 on this test. He recognizes that this estimate may be off a bit, and he estimates the likely size of the error as follows:

$$\text{SE for number scoring over } 50 \approx \sqrt{100} \times \sqrt{0.2 \times 0.8} = 4$$
$$\text{SE for percent} \approx (4/100) \times 100\% = 4\%$$

What does statistical theory say?

2. A small undergraduate college has 1,000 students, evenly distributed among the four classes: freshman, sophomore, junior, and senior. In order to estimate the percentage of students who have ever smoked marijuana, a sample is taken by the following procedure: 25 students are selected at random without replacement from each of the four classes. As it turns out, 35 out of the 100 sample students admit to having smoked. So, it is estimated that 35% out of the 1,000 students at the college would admit to having smoked. A standard error is attached to this estimate, by the following procedure:

$$\text{SE for number} \approx \sqrt{100} \times \sqrt{0.35 \times 0.65} \approx 5$$
$$\text{SE for percent} \approx (5/100) \times 100\% = 5\%.$$

What does statistical theory say?

The answers to these exercises are on p. A-56.

5. THE GALLUP POLL

The Gallup poll does not use a simple random sample (p. 308). As a result, they do not estimate their standard errors using the method explained in section 1. However, it is interesting to compare their results with the standard errors estimated by that method. For instance, in 1952 they predicted a 51% vote for Eisenhower, based on a sample of 5,385 people. Going through the arithmetic,

$$\text{SE for number} \approx \sqrt{5,385} \times \sqrt{0.51 \times 0.49} \approx 37$$
$$\text{SE for percent} \approx \frac{37}{5,385} \times 100\% \approx 0.7 \text{ of } 1\%$$

In fact, Eisenhower got 55.4% of the vote in that election. The Gallup poll estimate was off by 4.4 percentage points, and this is more than 6 times the SE computed for a simple random sample. Table 1 shows the comparison for every presidential election from 1952 to 1976.

Table 1. Comparing the Gallup poll with a simple random sample. The errors of prediction are on the whole quite a bit bigger than those to be expected from a simple random sample of the same size.

Year	Sample size	SE—for simple random sample	Actual error
1952	5,385	0.7 of 1%	4.4%
1956	8,144	0.6 of 1%	1.7%
1960	8,015	0.6 of 1%	0.9 of 1%
1964	6,625	0.6 of 1%	2.7%
1968	4,414	0.8 of 1%	0.5 of 1%
1972	3,689	0.8 of 1%	0.2 of 1%
1976	3,439	0.9 of 1%	1.6%

Source: See Table 4 on p. 310.

Except in 1968 and 1972, the errors in the Gallup poll predictions were considerably larger than the SEs computed for simple random samples. One reason is that the Gallup poll predictions are based only on part of the sample—those people judged likely to vote (p. 311). This by itself eliminates about half the sample, so the effective size of the sample must be reduced by half. Table 2 compares the errors made by the Gallup poll with SEs computed for simple random samples whose size equals the number of likely voters interviewed by the poll. The simple random sample formula is still not doing too good a job at predicting the size of the errors. This is because the Gallup poll is not taking simple random samples.

Table 2. The accuracy of the Gallup poll compared to that of a simple random sample whose size equals the number of likely voters in the Gallup poll sample.

Year	Number of likely voters	SE—for simple random sample	Actual error
1952	3,350	0.9 of 1%	4.4%
1956	4,950	0.7 of 1%	1.7%
1960	5,100	0.7 of 1%	0.9 of 1%
1964	4,100	0.8 of 1%	2.7%
1968	2,700	1.0%	0.5 of 1%
1972	2,100	1.1%	0.2 of 1%
1976	2,000	1.1%	1.6%

Source: The Gallup poll (American Institute of Public Opinion).

EXERCISE SET E

1. A Gallup poll pre-election survey based on a sample of 1,000 people estimates a 65% vote for the Democratic candidate in a certain election. True or false: The likely size of the chance error in this estimate can be figured as follows—

$$\sqrt{1,000} \times \sqrt{0.65 \times 0.35} \approx 15, \quad \frac{15}{1,000} \times 100\% = 1.5\%$$

Explain.

The answer to this exercise is on p. A-56.

6. REVIEW EXERCISES

Review exercises may cover material from previous chapters.

1. A simple random sample of 3,600 persons is taken, to estimate the percentage of smokers in a certain large population.[6] It turns out that 1,217 people in the sample are smokers. The percentage of smokers in the population is estimated as

$$\frac{1,217}{3,600} \times 100\% \approx 34\%.$$

The standard error is estimated as 0.8 of 1%, because

$$\sqrt{3,600} \times \sqrt{0.34 \times 0.66} \approx 28, \quad \frac{28}{3,600} \times 100\% \approx 0.8 \text{ of } 1\%.$$

What does statistical theory say?

2. The National Assessment of Educational Progress (NAEP) administers tests each year in a variety of different subjects, to a nationwide probability sample of high school students aged 17. In 1975, one of the tests covered American political institutions.[7] Two of the findings were:

- Only 53% of the students in the sample knew that each state has two United States senators.
- Almost 50% of the students in the sample believed that the President can appoint members of Congress.

Assume these percentages are based on a simple random sample of 1,000 students, drawn from the population of all seventeen year olds enrolled in school.

(a) Find an approximate 95%-confidence interval for the percentage of all persons in this population who know that each state has two United States senators.

(b) Find an approximate 99.7%-confidence interval for the percentage of all persons in this population who believe that the President can appoint members of Congress.

3. Sherlock Holmes says,

While the individual man is an insoluble puzzle, in the aggregate he becomes a mathematical certainty. You can, for example, never foretell what any one man will be up to, but you can say with precision what an average number will be up to. Individuals vary, but percentages remain constant. So says the statistician.[8]

The statistician doesn't quite say that. What is Sherlock Holmes forgetting?

4. A monthly opinion survey is based on a sample of 1,500 persons, "scientifically chosen as a representative cross section of the American public." The press release warns that the estimates are subject to chance error, but guarantees that they are "reliable to within two percentage points." The word "reliable" is ambiguous. According to statistical theory, the guarantee should be interpreted as follows:

(i) In virtually all these surveys, the estimates will be within two percentage points of the parameters.

(ii) In most such surveys, the estimates will be within two percentage points of the parameters, but in some definite percentage of the time larger errors are expected.

Explain.

5. When using probability methods to select a sample for an opinion survey, is the main thing which affects accuracy the size of the population, or the size of the sample?

6. True or false: With a well-designed sample survey, the sample percentage is very likely to equal the population percentage. Explain.

7. In a sample of about 400 students at University of California, Berkeley (the Statistics 2 class of fall, 1976), about 45% reported family incomes over $25,000 a year. Census figures show that only about 15% of American families have incomes over that level. If a simple random sample of 400 American families is taken, what is the chance (approximately) that over 45% of families in the sample will have incomes over $25,000 a year?

8. A coin is tossed 400 times. Estimate the chance of getting 200 heads.

9. One year at the University of California, Berkeley, there were 2,500 freshmen, of whom 55% had grade point averages of 3.0 or better. Of course, this percentage changes from year to year. A statistician estimates the likely size of these changes as follows:

$$\text{SE for number scoring over } 3.0 \approx \sqrt{2{,}500} \times \sqrt{0.55 \times 0.45} \approx 25$$

$$\text{SE for percent} \approx \frac{25}{2{,}500} \times 100\% = 1\%.$$

Comment.

10. (Hypothetical.) A bank wants to estimate the amount of change people carry, in a certain city. They take a simple random sample of 100 people, and find that on the average, people in the sample carry 60¢ in change. They figure the standard error as 5¢, because

$$\sqrt{100} \times \sqrt{0.60 \times 0.40} \approx 5, \quad 5/100 = 0.05$$

Comment.

11. The National Longitudinal Survey was a sample survey on the career patterns of women between the ages of 30 and 44. One interesting finding was that very few women worked steadily at their careers. For instance, of all the sample women who were currently employed, only 1% had worked at least six months in every year since leaving school.[9]

Suppose that a simple random sample of 400 people is drawn from the population of all women aged 30 to 44 who are currently employed. And suppose it turns out that only 3 women in the sample have worked at least six months in every year since leaving school. Can you find an approximate 95%-confidence interval for the percentage of all such women in the population from which the sample is drawn? Explain briefly.

12. In the National Longitudinal Survey, of the sample women who were currently employed, only 17% had worked at least six months in all but three of the years since leaving school.[10]

Suppose that a simple random sample of 400 people is drawn from the population of all women aged 30 to 44 who are currently employed. And suppose it turns out that only 69 women in the sample have worked at least six months in all but three of the years since leaving school. Can you find an approximate 95%-confidence interval for the percentage of all such women in the population from which the sample is drawn? Explain briefly.

7. SUMMARY

1. A large population is given. It is desired to estimate the percentage of people in the whole population who have a given quality. This parameter is called the *population percentage*. It is unknown. A simple random sample is taken. The percentage of people in the sample who have the given quality is a statistic, called the *sample percentage*. The sample percentage is used to estimate the population percentage.

2. The accuracy of this estimate depends mainly on the absolute size of the sample, larger samples being more accurate. The standard error uses the size of the sample and its composition to judge the accuracy.

3. The standard error for the sample percentage can be estimated by making a box model. The fraction of 0's and 1's in the box is unknown, so the fractions observed in the sample are substituted in the formula for the SD of the box. The resulting estimate is good when the sample is large.

4. A *confidence interval* for the population percentage is obtained by taking the right number of standard errors either way from the sample percentage. The confidence level is read off the normal curve. This method should only be used with large samples.

5. In the frequency theory of probability, parameters aren't subject to chance variation. That is why confidence statements are made, instead of probability statements.

6. When a sample is taken by a probability method it is possible not only to estimate the parameter, but also to figure the likely size of the chance error in the estimate.

7. The formulas for simple random samples shouldn't be used for other kinds of samples. If the sample was not chosen by a probability method, there is no sensible way to calculate standard errors for it.

22

Measuring Employment
and Unemployment

The country is hungry for information; everything of a statistical character, or even a statistical appearance, is taken up with an eagerness that is almost pathetic; the community have not yet learned to be half skeptical and critical enough in respect to such statements.
—GENERAL FRANCIS A. WALKER, SUPERINTENDENT OF THE 1870 CENSUS

1. INTRODUCTION

The unemployment rate is one of the most important economic indicators published by the government. Its graph is shown in Figure 1. The unemployment rate was only 3% in 1929, before the stock market crash. It reached 25% in the depths of the depression, and remained fairly high until the United States entered World War II. In the early 1970s, the unemployment rate hovered around 5%, rising to 9% in the 1975 recession.

The government agency in charge of measuring employment and unemployment is the Bureau of Labor Statistics. But how does the Bureau know who is employed or unemployed? As it turns out, employment statistics are estimated from a sample survey—the Current Population Survey. This massive and beautifully organized sample survey is conducted monthly for the Bureau of Labor Statistics by the Census Bureau.[1] During the week containing the 19th day of the month, a field staff of about a thousand interviewers canvasses a nationwide probability sample of about a hundred thousand

Figure 1. The unemployment rate from 1929 to 1976.

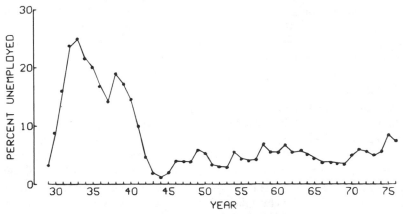

Source: Table A-1 of *Employment and Earnings*, October, 1976.

people. The size of the labor force, the unemployment rate, and a host of other economic and demographic statistics (like the distribution of income and educational level) are estimated from survey results, at a cost of about five million dollars a year. The results are published in

* *Monthly Labor Review;*
* *Employment and Earnings* (monthly);
* *Current Population Reports* (irregular);
* *Special Labor Force Reports* (irregular);
* *Statistical Abstract of the United States* (annual);
* *Economic Report of the President* (annual).

The object of this chapter is to present the Current Population Survey in detail, from the ground up. This will illustrate and consolidate the ideas introduced in previous chapters. It will also make other large-scale surveys much easier to understand.

Two main conclusions should be drawn from this case study:

* In practice, fairly complicated probability methods must be used to draw samples. Simple random sampling is only a building-block in these designs.
* The standard error formulas for simple random samples do not apply to these complicated designs, and other methods must be used for estimating the standard errors.

2. THE DESIGN OF THE CURRENT POPULATION SURVEY

The Current Population Survey is redesigned by the Census Bureau every ten years, to take advantage of population counts developed by the decennial Census. The current form of the Survey was designed in 1970. At that time, the United States was divided into 3,146 counties and independent cities, which were grouped together to form 1,931 *Primary Sampling Units* (or PSUs, for short). Each PSU consists of either a city, or a county, or a group of contiguous counties.[2] These PSUs were then grouped further into 376 *strata,* chosen so the PSUs in each stratum would resemble each other in certain demographic and economic characteristics (like population density, rate of growth, type of principal industry or agriculture), and would be as close together geographically as possible. Some of the largest PSUs, like New York or Los Angeles, were thought to be unique, and put in strata all by themselves.

The sample was then chosen in three stages. To begin with, one or sometimes two PSUs were chosen from each stratum, using a probability method which ensured that within each stratum, the chance of a PSU getting into the sample was proportional to its population. In all, 461 PSUs found their way into the sample. This completed the first stage. During the 1970s all interviewing for the Survey takes place in these PSUs and in no others.[2a] A map showing these PSUs isn't available, but the ones used in 1961 are shown in Figure 2.

Ultimately, the Bureau isn't interested in the PSUs but in the people who inhabit them. How are they chosen? Each PSU is made up of Census *Enumeration Districts,* which are compact geographical areas containing about 300 households. As the second stage in the sampling procedure, a

Figure 2. Primary Sampling Units, 1961 Current Population Survey.

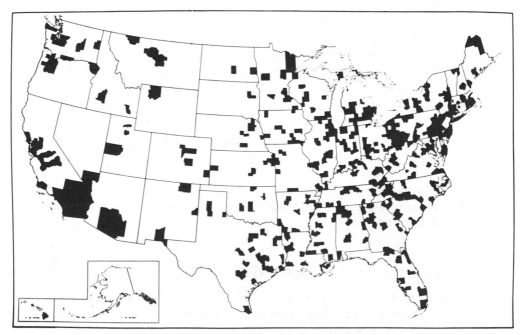

Source: *Measuring Employment and Unemployment,* Bureau of the
Census, Department of Commerce, 1962.

chance mechanism was used to choose several Enumeration Districts from each of the previously selected PSUs. The selected Enumeration Districts were divided up into *Ultimate Sampling Units* (or USUs), and at the third and final stage, some of them were selected for the sample by a chance mechanism. Within each PSU, all USUs had the same chance of being selected for the sample. Each USU consists of about four households, and every person aged sixteen and over living in a household in a selected USU in a selected Enumeration District in a selected PSU gets into the sample.[3]

One upshot of this complicated multistage sampling procedure is that every person in the United States, aged sixteen and over, has an equal chance of getting into the sample.[4] This chance is about one in 1,600. However, the procedure is different from simple random sampling. The difference may be hard to see, because the design is so complicated. To bring it out more clearly, in a simple hypothetical situation, imagine sampling from a minipopulation of four individuals, named A, B, C, and D. They are grouped into two USUs:

A B		C D
USU #1		USU #2

To select a sample of two individuals, a mini-Bureau of Labor Statistics might flip a coin. If it lands heads, they take everybody in USU #1. If it lands tails, they take everybody in USU #2. Everybody has half a chance to get into the sample. But A and C have no chance whatsoever to get into the sample together, because they live in different USUs. With simple random sampling, however, A and C would have some chance (one in six) to get into the sample together.

Simple random sampling treats all individuals alike. It also treats all pairs of individuals (and in fact all finite sets) alike. In this way, simple random sampling is fairer than multistage sampling. Why does the Bureau use its complicated procedure instead of simple random sampling? One reason is that there is no list showing all the people aged 16 and over in the United States, with their current addresses. Even if there were such a list, taking a simple random sample from it would produce people scattered thinly all over the country, and the cost of finding them would be prohibitive. With the Bureau's procedure, the sample is bound to come out in clumps in relatively small and well-defined areas, so the interviewing cost is quite manageable—about $10 a household. (These issues were discussed in the context of the Gallup poll on p. 308 above.)

The Bureau's choices for the sample in 1970–1980 were all made back in 1970: they even made provisions for sampling people who were going to live in housing yet to be constructed. And in fact, the Bureau chose not just one sample, but about fifteen different ones—because part of the sample is changed every month. After it gets into the sample, a household stays there for four months, drops out for eight months, and then returns for a final four months, after which the Bureau loses interest forever. There are two things to ask about this pattern of rotation. Why change the sample at all? For one reason, the interviewers probably wear out their welcome after a while.

Besides that, people's responses probably change as a result of being inter-viewed, progressively biasing the sample (this is called *panel bias*). For instance, there is some evidence to show that people are more likely to say they are looking for a job the first time they are interviewed than the second time. Then why not change the sample completely every month? It's quite a bit cheaper to keep part of it the same. Besides that, having some overlap in the sample makes it easier to estimate the monthly changes in employment and unemployment.

3. CARRYING OUT THE SURVEY

The Survey design, as determined in 1970, produces about 55,000 households to be canvassed each month. Of these, about 8,000 are ineligible for the sample (being vacant, or even demolished since 1970). Another 2,000 or so are unavailable, because no one is at home, or because those at home refuse to cooperate. That leaves about 45,000 households in the Survey. All persons aged 16 or over in these households are asked about their work experience in the previous week. On the basis of their answers, they are classified as:

   • employed (those who did any paid work in the previous week, or worked at least 15 hours in a family business, or were on leave from a regular job);

   • or unemployed (those who were not employed the previous week but were available for work and looking for work during the past four weeks);

   • or outside the labor force (defying Aristotle, the Bureau defines this as the state of being neither employed nor unemployed).

The employed are asked about the hours they work and the kind of job they have. The unemployed are asked about their last job, when and why they left it, and how they are looking for work. Those outside the labor force are asked whether they are keeping house, or going to school, or unable to work, or other (in which case, they are asked to specify what). In September, 1976, this came out as shown in Table 1.

Table 1. Bureau of Labor Statistics Estimates, September, 1976.

The noninstitutional population aged 16 and over

	Total (millions)
Civilian	
Employed	88.0
Unemployed	7.0
Outside the labor force	59.5
Total civilian	154.5
Military	2.1
Total	156.6

Source: *Employment and Earnings,* October, 1976, Table A-1.

By definition, the *civilian labor force* consists of those civilians who are either employed or unemployed, and in September, 1976, that amounted to $88.0 + 7.0 = 95.0$ million people.[5] The *unemployment rate* is the percentage of the civilian labor force which is unemployed, and in September, 1976, that amounted to

$$\frac{7.0}{95.0} \times 100\% \approx 7.4\%$$

This 7.4% is an average rate of unemployment—for all the subgroups of the population. Like many averages, it conceals some striking differences. These differences are brought out by a process of cross-tabulation. Unemployment falls more heavily on teenagers, on women, and on blacks, as shown by Table 2. (In September, 1976, there were about 16.8 million people aged 16 to 19 in the United States; only 6.5 million were going to school.)

Table 2. Bureau of Labor Statistics unemployment rates by race, age, and sex, September, 1976.

Race	Sex	Age group		
		16–19	*20–64*	*65 and over*
White	Male	16.2	4.7	3.9
White	Female	17.2	7.4	5.2
Black and other	Male	39.7	8.5	6.9
Black and other	Female	42.7	12.6	0.8

Source: Table A-3, *Employment and Earnings*, October, 1976.

The overall unemployment rate is quite variable, as shown in Figure 1. But the pattern of rates in Table 2 is quite stable. For instance, the unemployment rate for blacks has been just about double the unemployment rate for whites over the twenty-year period from 1956 to 1976: see Figure 3.

Figure 3. The ratio of the unemployment rate for blacks to the unemployment rate for whites.

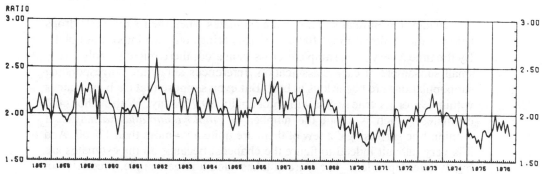

Source: Chart 11 of *Employment and Earnings*, October, 1976.

Unemployment numbers are published for much finer classifications than the ones shown in Table 2, including: marital status, type of last job, reason for unemployment (fired or quit), and "job search" method. See Figure 4. The Bureau starts with about 100,000 people in the sample. But by the time it comes down to the white men aged 20–24, who are widowed, divorced, or separated, whose last job was salesman for fabricated metal products, which they quit within the last five weeks, and who are looking for work by reading the want ads, there might be only a few cases left.

Figure 4. Page 33 from *Employment and Earnings,* October 1976.

HOUSEHOLD DATA

A-19. Unemployed persons by duration, occupation, and industry of last job

Occupation and industry	Thousands of persons					Average (mean) duration, in weeks	Less than 5 weeks as a percent of unemployed in group		15 weeks and over as a percent of unemployed in group	
	Total	Less than 5 weeks	5 to 14 weeks	15 to 26 weeks	27 weeks and over					
	September 1976						Sept. 1975	Sept. 1976	Sept. 1975	Sept. 1976
OCCUPATION										
White-collar workers	2,235	1,035	560	301	339	13.8	42.9	46.3	31.5	28.6
Professional and managerial	803	357	182	128	137	14.8	38.7	44.5	33.4	33.0
Sales workers	299	141	66	33	59	15.7	45.8	47.2	33.1	30.7
Clerical workers	1,132	537	312	141	142	12.5	45.0	47.4	29.6	25.0
Blue-collar workers	2,594	1,065	729	282	519	16.4	35.2	41.0	40.6	30.8
Craft and kindred workers	696	240	221	75	160	17.4	35.9	34.4	38.7	33.7
Operatives, except transport	1,056	447	298	121	191	16.3	34.3	42.3	44.7	29.5
Transport equipment operatives	215	75	61	28	52	18.7	35.8	34.7	38.2	36.9
Nonfarm laborers	627	304	149	59	116	14.7	36.2	48.4	35.3	27.8
Service workers	1,192	573	347	117	155	13.0	49.5	48.1	23.4	22.9
INDUSTRY[1]										
Agriculture	151	83	44	18	7	8.3	57.5	54.8	14.8	16.0
Construction	572	257	172	37	106	14.8	40.1	44.9	33.8	25.0
Manufacturing	1,472	614	367	192	299	17.0	31.7	41.7	46.4	33.4
Durable goods	827	317	204	97	209	19.0	27.5	38.4	51.0	37.0
Nondurable goods	645	297	163	95	90	14.4	39.0	46.0	38.3	28.7
Transportation and public utilities	268	95	85	34	53	16.6	40.5	35.5	35.3	32.6
Wholesale and retail trade	1,500	635	430	198	238	14.7	43.3	42.3	29.1	29.0
Finance and service industries	1,808	908	467	190	243	11.6	46.8	50.2	26.4	23.9
Public administration	224	88	65	28	43	16.7	44.6	39.2	25.6	31.6
No previous work experience	897	438	275	101	83	11.1	51.9	48.8	17.9	20.5

[1] Includes wage and salary workers only.

In fact, the Bureau will not make estimates when their subsample drops below a threshold of about 100 cases. This brings out two important points. By the time a very large sample is cross-tabulated, there might be only a very small subsample in each classification. Inferences about the corresponding subpopulations are hazardous at best. But even suppose that each estimate is within 1% of its true value with probability 99.7%, say. With a thousand estimates (which is about the number in *Employment and Earnings*), it wouldn't be surprising if a few of them are quite a bit more than 1% off. And it is very difficult indeed to figure the chances, because all the estimates are highly interdependent. The Bureau takes such a large sample, precisely because it has to make many estimates about many subpopulations. It wants to be reasonably confident that they are all fairly accurate.

4. ANALYZING THE DATA

Suppose that one month, in the Bureau's sample of 100,000 persons, there were 4,300 unemployed.[6] The Bureau is sampling 1 person in 1,600 (from the civilian noninstitutional population aged 16 and over). So, it is natural to think that each person in the sample represents 1,600 people in the nation. Then the way to estimate the total number of unemployed people in the nation is to weight up the sample number of 4,300 by the factor 1,600:

$$1,600 \times 4,300 = 6,880,000.$$

However, the Bureau does not give everybody in their sample the same weight. Instead, they divide the sample up into groups (by age, sex, and race), and then they weight up each group separately.

One group, for example, consists of white men aged 16 to 19. The Bureau uses Census data to estimate the total number of these people in the whole country, correcting for known trends in population growth. In September, 1976, there were about 7.2 million people in this group. About 0.3 million were in the Army and other institutions, leaving a civilian noninstitutional population of about 6.9 million. Suppose that in September, 1976, 4,400 people in this group turned up in the sample, and 418 were unemployed. Now the Bureau weights each white man aged 16 to 19 in its sample by the factor 6.9 million/4,400 $\approx$ 1,568. Each of these men represents not 1,600 people in general but 1,568 white men aged 16 to 19. So the Bureau estimates the total number of unemployed white men aged 16 to 19 in the whole civilian noninstitutional population as

$$1,568 \times 418 \approx 656,000.$$

The same procedure is followed for each of the groups considered by the Bureau, and each group gets its own weight—depending on the composition of the sample. The total number of unemployed persons is estimated then by adding up the separate estimates for all the different groups. This technique is called *ratio estimation*. Table 1 was derived using it.[7]

The key idea is to put different weights on the different groups. The advantage of reweighting the sample this way is that some chance variability in the estimates can be eliminated. For instance, suppose that by the luck of the draw, the sample contains proportionately too many teenagers. Weighting everybody up by the same factor would tend to make the estimated number of unemployed too high, because unemployment falls more heavily on the young (Table 2). Using different weights for the different groups compensates for differences between the demographic composition of the sample and the population, by putting more weight on the groups who are underrepresented in the sample, and less weight on those who are overrepresented. In the example, there were a few too many young white men in the sample, and this group was only weighted up by the factor 1,568, rather than 1,600.

5. STANDARD ERRORS

The unemployment rate must be measured with precision. To say, for example, that it is 7% give or take 0.1 of 1% gives a definite picture of the economy. To say that it is 7% give or take 2% covers everything from boom to bust. So, it is important to know how good the Bureau's estimates really are. The trouble is that they are not using simple random sampling—so the procedures we have developed so far just do not apply.

With simple random sampling, each person drawn into the sample provides additional information, almost independent of the persons drawn previously. With samples of the kind used by the Bureau, this is not so. To be more specific, at the last stage of its sampling procedure, the Bureau chooses some ultimate sampling units (USUs); a USU is a cluster of about four adjacent households. Every person aged 16 and over living in a chosen USU gets into the sample. With a cluster, it's all or nothing: either everybody in the cluster gets into the sample, or nobody does. This kind of sample is called a *cluster sample* and it is really different from a simple random sample. With a simple random sample, if one person gets into the sample, the neighbors still only have 1 chance in 1,600 to get in; with a cluster sample, the neighbors are bound to get in.

Now people living in a cluster tend to be similar to one another in some ways, and information about each one says something about all the others. The family backgrounds, educational histories, and employment records of such people are all interrelated, rather than independent. So the Bureau's cluster sample of 100,000 persons contains less information than a simple random sample of the same size.

> Cluster samples are less informative than simple random samples of the same size. So the simple random sample formulas for the standard error do not apply.

Clustering tends to reduce the precision of the Bureau's estimates. On the other hand, the use of ratio estimates improves the precision. All in all, judging the precision of the Bureau's estimates is a very delicate business.

As it turns out, with a cluster sample the standard errors can themselves be estimated very closely from the data, using the *half-sample method*. Although the details are quite complicated (it would take twenty hours of computer time each month to do the necessary computations), the idea is simple and will now be explained. If the Bureau wanted to see how accurate the Current Population Survey was, one thing to do would be make another independent survey, following exactly the same procedures. The difference between the results of the two surveys would give some idea of how reliable each set of results was.

Of course, nobody would seriously propose to replicate the Current Population Survey, at a cost of another $5 million a year, just to see how reliable it is. But the Bureau could get just about the same effect, by splitting the Survey into two independent pieces which have the same chance behavior. (Hence the name "half-sample method.") Suppose for instance that one piece of the survey estimates the civilian labor force at 95.2 million, and the other estimates it at 94.8 million. This difference is due to chance error. The pooled estimate of the civilian labor force is

$$\frac{95.2 + 94.8}{2} = 95.0 \text{ million.}$$

The two individual estimates are 0.2 million away from their average, and the standard error for the pooled estimate is estimated by this difference of 0.2 million.

Of course, an estimated standard error based on only one split may not be too reliable by itself. But there are many different ways to split the sample. It would be possible to look at a number of them and combine the estimated standard errors by taking their root-mean-square. This completes the outline of the half-sample method. The Bureau used to estimate the standard errors this way; recently, it switched to a procedure which is much easier to do on the computer—but harder to explain.[7a] Some of the estimated standard errors for September, 1976, are shown in Table 3.

Table 3. Estimated standard errors, September, 1976.

	Estimate	*Standard error*
Civilian labor force	95.0 million	229,000
Employment	88.0 million	241,000
Unemployment	7.0 million	114,000
Unemployment rate	7.4%	0.1 of 1%

Source: Basic Surveys Section, Bureau of the Census.

These estimated standard errors can be interpreted as before. They measure the uncertainty in the estimating procedure caused by the chance selection of the sample back in 1970. For example, the Bureau's estimate of the civilian labor force is apt to miss by a bit due to the chance variation introduced by the sampling procedure, and by a bit due to bias. The equation is

estimated size of civilian labor force
= actual size of civilian labor force + bias + chance error.

The likely size of the chance error is measured by the standard error. For instance, we can be about 95% confident that the chance error is smaller in size than twice the standard error. This tells us by how much chance error is likely to throw the estimate off. But it does not measure the effect of bias.

> Bias cannot be detected by comparing results from different parts of a survey, when it affects them all in a similar way.

How do the estimated standard errors in Table 3 compare to those for a simple random sample of the same size and composition? Calculations show that for estimating the size of the labor force, the Bureau's standard error is about 5% smaller than that for a simple random sample: the ratio estimates are doing a good job. For estimating the number of unemployed, however, the Bureau's sample is about 15% worse than a simple random sample: the clustering hurts.[8]

These calculations bring out a real issue. Just from looking at data, it is impossible to compute the SE—because the right calculation depends on how the sample was picked. With a simple random sample, you would get one SE. With a cluster sample, you would get another. And if you don't know how the sample was picked, there is no sensible way to get an SE. (This issue came up before, in the context of the Gallup poll, p. 351.)

> The formulas for the standard error have to take into account the details of the probability method used to draw the sample. If you don't know how the sample was picked, there is no way to calculate the standard error.

A *sample of convenience* is a sample that isn't chosen by a probability method. (An example would be some instructor's first-year psychology class.) Some people use the simple random sample formulas on samples of convenience. That is a real blunder. With samples of convenience, there is no way to define the chances, and therefore no way to get standard errors.

EXERCISE SET A

1. (Hypothetical.) The Health Department in a certain city takes a simple random sample of 100 households. In 80 of these households, all the occupants have been vaccinated against polio. So the Department estimates that for 80% of the households in that city, all the occupants have been vaccinated against polio—give or take 4% or so. What does statistical theory say?

2. Continuing Exercise 1, the Department interviews every person aged 25 and over in these sample households. They find 144 such persons, of whom 15 have college degrees. They estimate that 10% of the people aged 25 and over in that city have college degrees—give or take 2.5% or so. What does statistical theory say?

3. One month, the Current Population Survey sample amounted to 100,000 people. Of them, 55,000 were employed, and 5,000 were unemployed. True or false: the Bureau would estimate the unemployment rate in the nation as

$$\frac{5,000}{55,000 + 5,000} \times 100\% \approx 8.3\%.$$

Explain.

4. In Table 2, which estimate is more trustworthy: for white males aged 20–64, or black males aged 20–64? Explain.

5. The Current Population Survey sample is split into two independent halves. From one half, the number of unemployed persons is estimated as 7.1 million; from the other, it is estimated as 6.9 million. Combine these two estimates, and attach a standard error to the result.

6. In election years, the Bureau makes a special report on voting, using the Current Population Survey sample. In 1972, about 63% of all the people of voting age in the sample said they voted; but only 56% of the total population of voting age did in fact vote.[9] Can the difference be explained as a chance error? If not, how else can it be explained? (You may assume that the Bureau's sample is the equivalent of a simple random sample of 75,000 people.)

The answers to these exercises are on p. A-56.

6. THE QUALITY OF THE DATA

The data collected by the Survey are of very high quality. For instance, Survey data are considered to be more accurate than Census data. In any large-scale field operation, mistakes are inevitable; since the Survey operates on a much smaller scale than the Census, it can afford much better quality control. The key is careful selection, training, and supervision of the field staff. Interviewers, for instance, are given about four days of training in survey procedures before they start work, and several hours a month of retraining while they are on the job. At least once a year, their work is observed by their supervisors. In addition, about 5% of the monthly sample (chosen by a separate probability sampling procedure) is reinterviewed by supervisors. All discrepancies are discussed with the interviewers. The interviewers' reports are *edited,* that is, checked for incomplete or inconsistent entries. Error rates of 0.5 of 1% are typical, and each error is reviewed with the person who made it.

7. BIAS

As in any survey, there are some problems with bias. To begin with, it is believed that the Census usually misses between 2% and 5% of the population. This percentage is hard to pin down. And even if the Bureau knew it, they would still have a hard time adjusting the estimated number of unemployed (say) to compensate for the undercount, because the people missed by the Census could be quite different from the ones the Census finds. So it isn't just a matter of increasing all the estimated numbers by some factor, the pattern of unemployment might be a bit off too. Furthermore, the Current Population Survey misses about 3% of the people counted by the Census. To some extent, this is corrected by the ratio estimates. But there remains a problem of nonresponse bias: These people are probably a bit different from the ones the Survey finds, while the ratio estimates guess they are the same.

Next, there are real conceptual problems in defining what it means to be employed or unemployed. For example, people who have a part-time job but would like full-time work are classified as employed, but they really are partially unemployed. There were about 3.3 million such people in September, 1976, as shown in Table A-8 of *Employment and Earnings*. Another problem is that people who really want to work but have given up looking are classified as outside the labor force, although they probably should be classified as unemployed. The criterion used by the Bureau for unemployment, namely being without work, available for work, and looking for work, is necessarily subjective. In practice, it is a bit slippery. In consequence, it is believed that the number of unemployed is substantially higher than the Bureau's estimate—by several hundred thousand people. In this case, the bias is several times larger than the sampling error.

The main evidence concerning this bias comes from the *reinterview program,* in which about 5% of the sample is reinterviewed by supervisors. In over 10% of the cases, there is a disagreement about who is unemployed; and there is a similar rate of disagreement about who is working part-time and about who is "with a job but not at work."[10] By comparison, the rate of disagreement about who has a full-time job is under 2%. The evidence suggests that the Bureau can find out reasonably well who has a full-time job, and who is outside the labor force. The problem is with a third group, the marginal workers who are classified either as part-time workers, or with a job but not at work, or unemployed.[11] The total size of this third group is known quite accurately. But the classification into three subgroups is not very accurate, as shown by the reinterview program. The estimated number of employed depends on this classification, explaining the magnitude of the bias in the unemployment rate.

In theory, cluster samples and ratio estimates can create small biases. In practice, however, with reasonably large samples the bias from these two sources has proved to be negligible.

8. REVIEW EXERCISES

Review exercises may cover previous chapters as well.

1. One month, there were 100,000 people in the Current Population Survey sample, of whom 60,000 were in the labor force. True or false: The Bureau would estimate the percentage of the population in the labor force as 60%. Explain.

2. One month, there were 100,000 people in the Current Population Survey sample, and the Bureau estimated that 60% of the population was in the labor force. True or false: The standard error for this percentage should be estimated as follows

$$\text{SE for number} = \sqrt{100{,}000} \times \sqrt{0.6 \times 0.4} \approx 155$$

$$\text{SE for percent} = \frac{155}{100{,}000} \times 100\% = 0.155 \text{ of } 1\%.$$

3. One month, the Current Population Survey sample was split into two independent replicates. Using one replicate, the size of the civilian labor force is estimated as 96 million. The other replicate produced an estimate of 95 million. Using this information, estimate the size of the civilian labor force, and attach a standard error to the estimate.

4. Using the data in Exercise 3, what can be said about the bias in the estimate?

5. (Hypothetical.) The American Medical Association estimates that 80% of its members oppose compulsory medical insurance. This estimate is based on 2,500 questionnaires filled out by AMA members attending a convention. True or false: the SE for this estimate is 0.8 of 1%, because

$$\sqrt{2{,}500} \times \sqrt{0.8 \times 0.2} = 20, \quad \frac{20}{2{,}500} \times 100\% = 0.8 \text{ of } 1\%.$$

6. As part of a study on drinking, the attitudes of a sample of alcoholics are assessed by interview.[12] Cases are assigned to interviewers at random. Some of the interviewers are teetotalers, others drink. Would you expect the two groups of interviewers to reach similar conclusions? If not, give your reasons.

7. At a large university, a simple random sample of 400 students is chosen; 240 students in the sample turn out to wear glasses.

   (a) If possible, find an approximate 95%-confidence interval for the percentage of all students at that university wearing glasses.
   (b) If possible, find an approximate 95%-confidence interval for the percentage of all university students in the United States wearing glasses.

8. A survey organization takes a simple random sample to estimate the percentage of Democrats in a certain town. They find that a 95%-confidence interval for this percentage is 55% ± 4%. A 99.7%-confidence interval for this percentage is then

   (i) impossible to compute because the sample size isn't given

   (ii) $55\% \pm \frac{3}{2} \times 4\%$ (iii) $55\% \pm 2 \times 4\%$

   (iv) $55\% \pm \frac{99.7}{95} \times 4\%$ (v) $55\% \pm \frac{100 - 95}{100 - 99.7} \times 4\%.$

9. A university has 5,000 registered students, with the names listed in al-phabetical order in a card file. Someone draws a sample of 100 of these students, by going through the list and taking every fiftieth name. Of the students in the sample, 53 are over the age of 21. It is estimated that 53% of all the 5,000 registered students are over the age of 21, give or take 5% or so. What does statistical theory say?

10. "Toss a hundred pennies in the air and record the number of heads that come up when they fall. Do this several thousand times and plot a histogram for the number of heads that come up on each of the thousands of throws of the group of one hundred pennies. You will have a histogram that closely approximates the normal curve, and the more times you toss the hundred pennies the closer you will get to the curve."[13] If you keep on tossing this group of a hundred pennies, will your histograms converge to the normal curve? Explain briefly.

11. In Keno, there are 80 balls, numbered from 1 through 80. If you play a single number, you win if that number is chosen. This bet pays 2 to 1, and you have 1 chance in 4 of winning. If you play a double-number, you win if both numbers are chosen. This bet pays 11 to 1, and you have very close to a 6% chance of winning.[14]

    (a) Find the expected value and SE for your net gain, if you play 100 times and stake $1 on a single number each time.

    (b) Find the expected value and SE for your net gain, if you play 100 times and stake $1 on a double-number each time.

9. SUMMARY

1. The unemployment rate in the United States is estimated by a monthly sample survey, called the Current Population Survey.

2. This survey uses a nationwide probability sample of about 100,000 persons. The design is much more complicated than simple random sampling.

3. The Survey uses ratio estimates, which reweight the sample so it agrees with Census data on age, sex, race, and certain other characteristics influencing employment status.

4. The methods used to calculate standard errors must take into account the details of the method used to draw the sample. The formulas which apply to simple random samples will usually underestimate the standard errors in *cluster samples* of the kind used in the Current Population Survey.

5. If the sample wasn't drawn by probability methods, or the details of the method are unknown, a standard error cannot be computed. The simple random sample formulas should not be used to deal with other kinds of samples.

6. The standard errors for cluster samples can be obtained by the *half-sample method*, splitting the sample into two halves and seeing how well they agree.

7. When bias is relatively constant, it cannot be detected by the half-sample method, because it affects all parts of the survey in the same way.

8. The Current Population Survey, like all surveys, is subject to a number of small biases. The bias in the estimate of the unemployment rate is thought to be several times larger than its standard error.

win   +2

lose   −1

1 +2 3 −1   chances (1 in 4)

↓ 100 draws

Avg $= \dfrac{2-1-1-1}{4} = .25$

SD $= \sqrt{\dfrac{2.25^2 + 3 \times .75^2}{4}} = 1.3$

23

The Accuracy of Averages

God bless our sample persons.

—SIGN AT HEADQUARTERS OF THE
HEALTH EXAMINATION SURVEY

1. INTRODUCTION

The object of this chapter is to estimate the accuracy of an average computed from a simple random sample. This section deals with a preliminary question: How much chance variability is there in the average of numbers drawn at random from a box?

Take the box

$$\boxed{1}\ \boxed{2}\ \boxed{3}\ \boxed{4}\ \boxed{5}\ \boxed{6}\ \boxed{7}$$

The computer was programmed to make 25 draws at random with replacement from this box:

2 4 3 2 5 7 5 6 4 5 4 4 1 2 4 4 6 4 7 2 7 2 5 7 3.

The sum of these numbers is 105, so their average is 105/25 = 4.2. Of course, when this little experiment was repeated, it came out differently:

5 1 4 3 4 5 2 1 7 7 1 2 3 2 4 7 1 6 5 3 6 6 3 3 4.

The sum of these numbers is 95, so their average is 95/25 = 3.8. The average of the draws, like the sum, is subject to chance variability. The problem is to calculate the expected value and standard error for the average of the draws. The method will be indicated by example.

Example 1. Twenty-five draws will be made at random with replacement from the box

$$\boxed{1}\ \boxed{2}\ \boxed{3}\ \boxed{4}\ \boxed{5}\ \boxed{6}\ \boxed{7}$$

The average of the draws will be around _____ , give or take _____ or so.

Solution. The average of the box is 4, with an SD of 2. So the sum of 25 draws will be around $25 \times 4 = 100$, give or take $\sqrt{25} \times 2 = 10$ or so. Now the average of these draws is just their sum, divided by 25. So the average of the 25 draws will be around 100/25 = 4, the average of the box. And it will be off by 10/25 = 0.4 or so. This completes the calculation. The average of the draws will be around 4 give or take 0.4 or so. The 4 is the expected value for the average of the draws, and the 0.4 is the SE for the average of the draws.

The procedure outlined in Example 1 is perfectly general.

> When drawing at random from a box, the expected value for the average of the draws is the average of the box. And the SE for the average of the draws equals
> $$\frac{\text{SE for their sum}}{\text{the number of draws}}.$$

If the number of draws is large enough, the SE for the average can be used with the normal curve to figure chances, in the usual way. For instance, in Example 1 there is about a 68% chance that the average of the draws will be in the range 4 ± 0.4, there is about a 95% chance that the average of the draws will be in the range 4 ± 0.8, and so on.

Example 2. One hundred draws will be made at random with replacement from the box in Example 1.

(a) The average of the draws will be around _____ , give or take _____ or so.

(b) Estimate the chance that the average of the draws will be more than 4.2.

Solution. As in Example 1, the sum of the draws will be around $100 \times 4 = 400$, give or take $\sqrt{100} \times 2 = 20$ or so. Dividing through by 100, the average of the draws will be around $400/100 = 4$, give or take $20/100 = 0.2$ or so. The SE for the average of 400 draws is 0.2.

Part (b) is handled by the normal approximation.

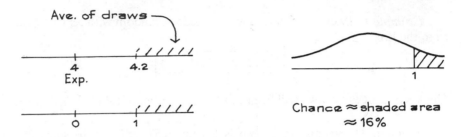

The chance is around 16%. This completes the solution.

Comparing Examples 1 and 2, when the number of draws went up by a factor of 4, from 25 to 100, the SE for the average of the draws went down by a factor of $\sqrt{4} = 2$, from 0.4 to 0.2. This is so in general.

> When drawing at random with replacement from a box of tickets, multiplying the number of draws by a factor (like 4) divides the SE for the average of the draws by the square root of that factor ($\sqrt{4} = 2$).

The SE for the sum and the SE for the average behave quite differently as the number of draws goes up. The SE for the sum gets bigger and the SE for the average gets smaller. Here is the reason. The SE for the sum goes up, but only by the square root of the number of draws. As a result, while the SE for the sum gets bigger in absolute terms, relative to the number of draws it gets smaller. Then the division by the number of draws makes the SE for the average go down. Keep this difference between the two SEs in mind.

When drawing without replacement, the exact SE for the average of the draws can be found using the correction factor (p. 331):

$$\text{SE without} = (\text{correction factor}) \times (\text{SE with}).$$

Usually, the number of draws is small by comparison with the number of tickets in the box, and the correction factor will be so close to 1 that it can be ignored.

EXERCISE SET A

1. A hundred draws are made at random with replacement from a box.
    (a) If the sum of the draws is 110, what is their average?
    (b) If the average of the draws is 0.9, what is their sum?

2. Fill in the table below, for draws made at random with replacement from the box

$$\boxed{0}\ \boxed{2}\ \boxed{3}\ \boxed{4}\ \boxed{6}$$

Number of draws	SE for sum of draws	SE for average of draws
25		
100		
400		

3. A box contains 10,000 tickets. The numbers on these tickets average out to 50, with an SD of 20.
   (a) A hundred tickets are drawn at random with replacement. The average of these draws will be around _____ give or take _____ or so.
   (b) What if the draws are made without replacement?

4. A box of tickets averages out to 50, with an SD of 20. One hundred draws are made at random with replacement from this box.
   (a) Find the chance (approximately) that the average of the draws will be in the range 30 to 70.
   (b) Repeat, for the range 48 to 52.

5. A hundred draws will be made at random with replacement from a box of tickets. The average of the numbers in the box is 200. The SE for the average of the draws is computed, and turns out to be 10. True or false:
   (a) About 68% of the tickets in the box are in the range 190 to 210.
   (b) There is about a 68% chance for the average of the hundred draws to be in the range 190 to 210.

6. There are three boxes of numbered tickets; the average of the numbers in each box is 200. However, the SD of box A is 10, the SD of box B is 20, and the SD of box C is 40. Now
   • 100 draws are made from box A
   • 200 draws are made from box B
   • 400 draws are made from box C.
   (The draws are made with replacement.) The average of each set of draws is computed. Here they are, in scrambled order:

$$203.6 \qquad 198.1 \qquad 200.4$$

   (a) Which average comes from which box?
   (b) Could it possibly be otherwise?
   Explain your answers.

The answers to these exercises are on pp. A-56–A-57.

Technical note: When drawing at random with replacement from a box, the SE for the sum of the draws is $\sqrt{\text{no. of draws}} \times$ SD of box. So the SE for the average of the draws is ($\sqrt{\text{no. of draws}} \times$ SD of box)/no. of draws. This simplifies to SD of box/$\sqrt{\text{no. of draws}}$.

2. THE SAMPLE AVERAGE

In section 1, the numbers in the box were known, and the problem was to say something about the average of the draws. This section reasons in the opposite—and more practical—direction: a simple random sample is taken from a box of unknown composition, and the problem is to estimate the average of the box. Naturally, the average of the draws is used as the estimate. How accurate is this estimate? Because the draws were made at random, it is possible to answer this question, just from the spread in the data and the size of the sample. This is one of the major achievements of statistical theory.

The method to be presented in this section only applies to simple random samples which are reasonably large: the small-sample theory is discussed in Chapter 24. The SE for the sample average, and the normal curve, will be used to judge the accuracy of an estimate, just as in Chapter 21. Along the way, there will be two important technical questions to answer:

- What is the distinction between the SD of the sample and the SE for the sample average?
- Why is it legitimate to use the normal curve in figuring confidence levels?

We begin with an example. Suppose that a city manager wants to know the average income of the 25,000 families living in his town. He hires a survey organization to take a simple random sample of 900 families. The average of the 900 incomes of the sample families turns out to be $14,400 with an SD of $9,000. On this basis, the average income for all 25,000 families is estimated to be around $14,400. Of course, this is off by a chance error. So the problem is to put a give-or-take number on the estimate:

$$\$14,400 \pm \$\underline{\hspace{1cm}} \ ?$$

To solve this problem, a box model is needed. There should be one ticket in the box for each family in the town. The data are quantitative, so each ticket should show a number—the income of the corresponding family (p. 328). Then, 900 draws are made at random without replacement from the box, to get the sample. The survey organization is estimating the average of the box from the average of the draws, and trying to find out how far wrong this estimate is likely to be. It is the SE for the average which does the job for them.

The SE for the average of the draws can be found by the method of section 1. The first step is to find the SE for the sum of the draws. Since 900 is such a small fraction of 25,000, there is no point in bothering with the correction factor. The SE for the sum is approximately

$$\sqrt{900} \times \text{SD of box.}$$

Of course, the survey organization does not know the SD of the box, but they can estimate it by the SD of the sample. This is another example of the bootstrap method discussed on p. 342.

> With a simple random sample, the SD of the sample can be used to estimate the SD of the box. The estimate is good when the sample is large enough.

In the example, the SD of the box would be estimated by the SD of the sample as $9,000. So the SE for the sum is estimated as

$$\sqrt{900} \times \$9,000 = \$270,000.$$

Dividing through by the number of families in the sample, the SE for their average income is estimated as $270,000/900 = $300. This is the answer. The average of the draws is something like $300 off the average of the box. So the average of the incomes of all 25,000 families in the town can be estimated as

$$\$14,400 \pm \$300.$$

Keep the interpretation of the $300 in mind: it is the margin of error for the estimate. This completes the example.

> The SE for the average of the draws says how far the average of the draws is likely to be from the average of the box.

Confidence intervals for percents (qualitative data) were discussed on p. 344. The same idea can be used to get confidence intervals for the average of the box (quantitative data). For example, an approximate 95%-confidence interval for the average of the incomes of all 25,000 families in the town is obtained by going two SEs either way from the sample average:

$$\$14,400 \pm 2 \times \$300 = \$13,800 \text{ to } \$15,000.$$

Two different numbers came up in this calculation. The SD of the sample was $9,000, and the SE for the sample average was $300. These two numbers do different things.

- The SD measures the spread in the 900 incomes of the sample families.
- The SE says how far the average of all 900 sample incomes is from the average income for the whole town.

People who confuse the SD with the SE might think that somehow, 95% of the families in the town had incomes in the range $14,400 ± $600. This would be ridiculous: the SD of income is about $9,000. And the confidence interval is saying something very different. What the 95% means is that for about 95% of all samples, if you go two SEs either way from the average of the incomes of the 900 sample families, this confidence interval will cover the average of the incomes of all 25,000 families in the town. For the other 5% of the samples, the interval will miss. The word "confidence" is to remind you that the chances are in the sampling procedure; the average of the box isn't moving around. (These issues were discussed before, on pp. 347–348.)

Example 1. As part of an opinion survey, a simple random sample of 400 persons aged 16 and over is taken in a certain town. The average educational level (years of schooling completed) of the sample persons is 12.7 years, with an SD of about 4 years. Find an approximate 95%-confidence interval for the average educational level of all persons aged 16 and over in this town.

Solution. A box model is needed. There should be one ticket in the box for each person aged 16 and over in the town, showing the number of years of schooling completed by that person. Then 400 draws are made at random without replacement to get the sample. The SE for their average is needed. Now the SE for the sum of the draws is $\sqrt{400} \times$ SD of box. The SD of the box is unknown, but can be estimated by the SD of the sample, as 4 years. So the SE for the sum of the draws is estimated as $\sqrt{400} \times 4 = 80$ years. Now the SE for their average is 80 years/400 = 0.2 years. The average educational level of the persons in the sample will be off the average for the town by 0.2 years or so. An approximate 95%-confidence interval for the average educational level of all persons aged 16 and over in the town is

$$12.7 \pm 0.4 \text{ years.}$$

This is the answer.

At this point, it is natural to object that the histograms for income and education look nothing like the normal curve (pp. 31 and 33). Why is it legitimate to use the curve to figure confidence levels? The main point is that the curve is not being used to approximate the histogram for the data. Instead, it is being used to approximate the probability histogram for the sample average. The distinction between these two kinds of histograms is crucial, and it will be reviewed here in the context of the income example.

The box model will help in focusing the issues. Remember that there is one ticket in the box for each of the families in the town, showing their income. To have a definite example, we made up a box: a histogram for its contents is shown at the top of Figure 1. This histogram represents the incomes of all 25,000 families in the town. It has the same shape as real income histograms, and it is nothing like the normal curve. Now 900 draws must be made at random without replacement from the box, to get the sample. The computer was programmed to do this. A histogram for the 900 draws is shown at the bottom of the figure. This represents the distribution of income for the 900 sample families. It is very similar to the first histogram, although there are a few too many families with incomes in the range $35,000 to $40,000—a chance variation. (Figure 1 indicates why the SD of the sample is a good estimate for the SD of the box—the two histograms show just about the same amount of spread.)

So far, we have seen two histograms, both for data. Now a probability histogram comes in, for the average of the draws. This is the limit of empirical histograms (p. 278). The computer was programmed to draw a thousand samples from a box, each sample consisting of 900 draws. (The 900 tickets were replaced in the box after each sample was taken.) The average of the first 900 draws was $13,939. The next average was $14,128. The next was $13,843. And so on. A histogram for the thousand sample averages is shown at the top of Figure 2 on the next page. This is quite ragged—chance variation. Of course, there is no reason to stop with a thousand samples. If the computer kept on going, eventually it would get the smooth probability histogram shown at the bottom of Figure 2.

This probability histogram represents the chances for the sample aver-

Figure 1. Income histograms. The top panel shows the distribution of income in the whole town. The bottom panel shows the distribution of income in the sample. These are histograms for data.

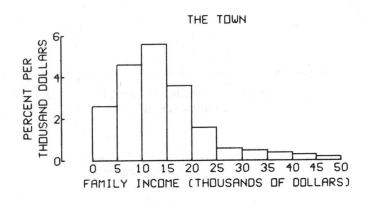

THE TOWN

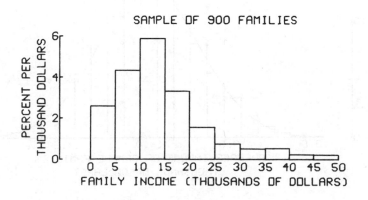

SAMPLE OF 900 FAMILIES

age, not the distribution of incomes in the sample. For instance, take the area under the probability histogram between $13,500 and $14,500. This area represents the chance that the average of 900 draws from the box will be between $13,500 and $14,500. The area works out to 90%. So, for 90% of the

Figure 2. The normal curve appears. The computer drew a thousand samples, each of 900 families, and computed the average income for each sample. A histogram for these thousand sample averages is shown in the top panel. If the computer kept on going, in the limit it would get the probability histogram shown in the bottom panel. This probability histogram follows the normal curve very well indeed.

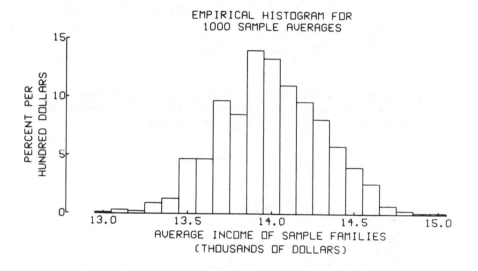

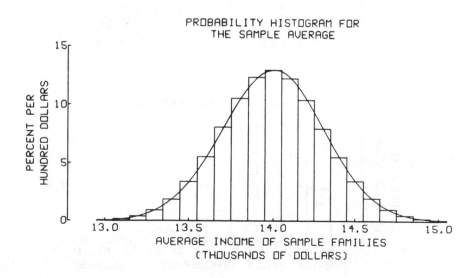

samples, the average of the incomes of the 900 sample families will be in the range $13,500 to $14,500. The other 10% of the time, the sample average will be outside this range. Any area under the probability histogram can be interpreted in a similar way.

Now you can see where the normal curve comes in. As the figure shows, the curve is a very good approximation to the probability histogram for the average of the draws, even though the contents of the box were nothing like the curve. This follows from the mathematics of Chapter 18. The probability histogram for the sum of the draws will be close to the normal curve. The average of the draws equals their sum, divided by 900. This division by 900 is just a change of scale, and washes out in standard units.

> The probability histogram for the average of many draws from a box will follow the normal curve, even if the tickets in the box don't.[1]

And that is why the normal curve can be used to figure confidence levels. (With a small number of draws, however, you can't use the normal curve—see Chapter 18.)

EXERCISE SET B

1. A university has 30,000 registered students; as part of a survey, 900 of these students are chosen at random. The average age of the sample students turns out to be 22.3 years, with an SD of 4.5 years. Find an approximate 95%-confidence interval for the average age of all 30,000 registered students.[2]

2. The Census Bureau makes its Annual Housing Survey for the Department of Housing and Urban Development.[3] In 1974, for instance, the Bureau estimated that there were 70.8 million housing units in the United States, of which 25.0 million were rental units. The average rent paid on these units was about $150, with an SD of $75.

   A certain town has 10,000 occupied rental units. A local real estate office commissions a survey of these units: 400 are chosen at random, and the occupants are interviewed. Among other things, the rent paid in the previous month is determined. The 400 sample rents average out to $184, with an SD of $80. A histogram is plotted for the sample rents, and it does not follow the normal curve.

   (a) If possible, find an approximate 68%-confidence interval for the average rent paid in the previous month on all 10,000 occupied rental units in this town. If this is not possible, explain why not.

   (b) True or false: For about 68% of all the occupied rental units in this town, the rent paid in the previous month was between $180 and $188.

3. The figure below is the probability histogram for the average of 900 draws from the box in Figure 1. What does the shaded area represent?

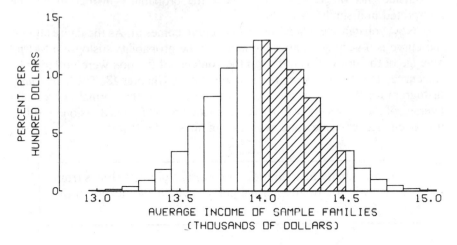

The answers to these exercises are on p. A-57.

3. A REMINDER

The previous section explained how to evaluate the accuracy of an average computed from a simple random sample. The arithmetic can be carried out on any list of numbers: find the SD, multiply by the square root of the number of entries, then divide by the number of entries. Sometimes, this will give a sensible SE—for instance, when the numbers come from a simple random sample. In other cases, the result is nonsense. It all depends on where that list of numbers comes from. This issue was discussed before, on pp. 350 and 366.

> A reminder. The formulas for simple random samples should not be used for other kinds of samples. And if the sample was not taken by probability methods, there is no way to evaluate the accuracy of the estimates.

EXERCISE SET C

This exercise set also covers material from previous chapters.

1. A utility company serves 50,000 households. As part of a survey of customer attitudes, they take a simple random sample of 400 of these households. The average number of television sets in the sample households turns out to be 1.70, with an SD of 0.90. If possible, find an approximate 95%-confidence interval for the average number of television sets in all 50,000 households. If this isn't possible, explain why not.

2. Out of the 400 households in the survey of the previous exercise, 399 have at least one television set. If possible, find an approximate 68%-confidence interval for the percentage of all the 50,000 households having at least one television set.[4] If this isn't possible, explain why not.

3. Continuing Exercises 1 and 2, out of the 400 sample households, 150 have dishwashers. If possible, find an approximate 68%-confidence interval for the percentage of all the 50,000 households having dishwashers.[5] If this isn't possible, explain why not.

4. As part of the survey described in Exercise 1, all persons aged 16 and over in the 400 sample households are interviewed. This makes 900 people. On the average, the sample people watched 3.20 hours of television the Sunday before the survey, with an SD of 1.50 hours. If possible, find an approximate 95%-confidence interval for the average number of hours spent watching television by all persons aged 16 and over in the 50,000 households on that Sunday. If this isn't possible, explain why not.

The answers to these exercises are on p. A-57.

4. REVIEW EXERCISES

Review exercises may cover material from previous chapters.

1. A box of tickets has an average of 100, with an SD of 20. Four hundred draws will be made at random with replacement from this box.
    (a) Estimate the chance that the average of the draws will be in the range 80 to 120.
    (b) Estimate the chance that the average of the draws will be in the range 99 to 101.

2. Four hundred draws are made at random with replacement from a box with 10,000 tickets. The average of the box is unknown, however, the average of the draws was 51.3, with an SD of about 22. True or false:
    (a) About 68% of the tickets in the box are in the range 51.3 ± 1.1.
    (b) An approximate 68%-confidence interval for the average of the tickets in the box is 51.3 ± 1.1.

3. As part of the Annual Housing Survey, the Census Bureau determines how far the head of a household has to commute to work.[6] In 1974, this averaged 13 miles, with an SD of 13 miles too! (Distances are one-way, not round-trip, throughout this exercise and the next one.)

    A real estate office wants to make a similar survey in a certain town, which has about 20,000 occupied households. A simple random sample of 400 households is chosen, the occupants are interviewed, and it is determined that on the average, the heads of the sample households commuted 12.7 miles to work, with an SD of 12.0 miles. (If someone isn't working, the commute distance is defined to be 0.)

    If possible, find an approximate 95%-confidence interval for the average distance that all heads of households in that town commute to work. If this isn't possible, explain why not.

4. Continuing Exercise 3, the real estate office interviewed all persons aged 16 and over in the 400 sample households; there were 900 such persons. On the average, these 900 people commuted 8.7 miles to work, with an SD of 13.8 miles. (Again, if someone isn't working, the commute distance is defined to be 0; and all distances are one-way.) If possible, find an approximate 95%-confidence interval for the average commute distance for all people aged 16 and over in this town. If this isn't possible, explain why not.

5. Continuing Exercise 4, in 321 out of the 400 sample households, the head of the household commuted by car. If possible, find an approximate 95%-confidence interval for the percentage of all households in the town where the head of the household commutes by car.[7] If this isn't possible, explain why not.

6. As part of a survey on physical fitness, a large university takes a simple random sample of 900 male students. The average height of these sample students turns out to be 5 feet 10.2 inches, with an SD of about 3 inches. A histogram for their heights is plotted, and follows the normal curve. The average height of all the male students at this university is estimated as 5 feet 10.2 inches, give or take 0.1 inch or so.

   Say whether each of the following statements is true or false, and explain why.
   (a) It is estimated that 68% of the male students in this university are between 5 feet 10.1 inches and 5 feet 10.3 inches in height.
   (b) An approximate 68%-confidence interval for the average height of all male students at this university runs from 5 feet 10.1 inches to 5 feet 10.3 inches.
   (c) If someone takes a simple random sample of 900 male students at this university, and goes one SE either way from the average height of the 900 sample students, there is about a 68% chance that this interval will cover the average height of all male students at this university.
   (d) About 68% of the men in the sample had heights in the range 5 feet 7.2 inches to 6 feet 1.2 inches.

7. There are about 2,700 institutions of higher learning in the United States (including junior colleges and community colleges). In 1976, as part of a continuing study of higher education, the Carnegie Commission took a simple random sample of 400 of these institutions.[8] The average enrollment in the 400 sample schools was 3,700, with an SD of 6,500. They estimate the average enrollment at all 2,700 institutions to be around 3,700, give or take 325 or so. Say whether each of the following statements is true or false, and explain why.
   (a) It is estimated that 68% of the institutions of higher learning in the United States enroll between $3,700 - 325 = 3,375$ and $3,700 + 325 = 4,025$ students.
   (b) An approximate 68%-confidence interval for the average enrollment of all 2,700 institutions runs from 3,375 to 4,025.

   (c) If someone takes a simple random sample of 400 institutions, and goes one SE either way from the average enrollment of the 400 sample schools, there is about a 68% chance that this interval will cover the average enrollment of all 2,700 schools.

   (d) The normal curve can't be used to figure confidence levels here at all, because the data doesn't follow the normal curve.

   (e) About 68% of the schools in the sample had enrollments in the range 3,700 ± 6,500.

8. Continuing Exercise 7, there are about 600,000 faculty members at institutions of higher learning in the United States. As part of its study, the Carnegie Commission took a simple random sample of 2,500 of these faculty persons.[9] On the average, these 2,500 sample persons had published 1.7 research papers in the two years prior to the survey, with an SD of 2.3 papers. If possible, find a 95%-confidence interval for the average number of research papers published by all 600,000 faculty members in the two years prior to the survey. If this isn't possible, explain why not.

9. In fall, 1976, at the University of California, Berkeley, 400 students took the final in Statistics 2. Their scores averaged 60 out of 100, with an SD of 25. Now

$$\sqrt{400} \times 25 = 500, \quad 500/400 = 1.25$$

Is 60 ± 2.5 a 95%-confidence interval? If so, for what? If not, why not?

10. What's the worst thing about a sample of size one?

11. (Hard.) Draws are made at random with replacement from a box. Match the six quantities on the left with their values on the right.

The expected value for the sum of the draws	812
The SE for the sum of the draws	800
The average of the box	400
The SE for the average of the draws	20
The number of draws	2
The sum of the draws	0.05

Now find the SD of the box.

5. SUMMARY

   1. When drawing at random from a box, the expected value for the average of the draws equals the average of the box. And the SE for the average of the draws equals the SE for their sum, divided by the number of draws.

   2. The SE for the average of the draws says how far the average of the draws is likely to be from the average of the box.

   3. When drawing with replacement, multiplying the number of draws by some factor divides the SE for their average by the square root of that factor.

4. With enough draws, the probability histogram for the average of the draws will follow the normal curve, even if the contents of the box do not.

5. With a simple random sample, the SD of the sample can be used to estimate the SD of the box. The estimate is good when the sample is large enough.

6. With a simple random sample, a confidence interval for the average of the box can be found by going the right number of SEs either way from the average of the draws. The confidence level is read off the normal curve. This method should only be used with large samples.

7. The formulas for simple random samples should not be used for other kinds of samples (reminder).

PART VII

Chance Models

24

A Model for Measurement Error

Upon the whole of which it appears, that the taking of the Mean of a number of observations, greatly diminishes the chance for all the smaller errors, and cuts off almost all possibility of any great ones: which last consideration, alone, seems sufficient to recommend the use of the method, not only to astronomers, but to all others concerned in making the experiments of any kind (to which the above reasoning is equally applicable). And the more observations or experiments there are made, the less will the conclusions be liable to error, provided they admit of being repeated under the same circumstances.
—THOMAS SIMPSON (ENGLISH MATHEMATICIAN, 1710–1761)

1. INTRODUCTION

In this part of the book, the frequency theory of chance will be used to study measurement error and genetics. Historically, the frequency theory was developed to handle problems of a very special kind—figuring the odds in games of chance. Some effort is needed to apply the theory to situations outside the gambling context. In each case, it is necessary to show that the situation being studied resembles a process—like drawing from a box—to which the theory applies. In other words, a box model is needed. (Such models are often referred to as *chance models* or *stochastic models*.) This section will take up a preliminary question: When is it appropriate to use a box model? There is no easy answer, but here is one rule.

> If the data show any trend or pattern, it isn't like the result of drawing at random from a box. So a box model would not apply.

The reason is that a sequence of draws from a box is very unlikely to show any trend or pattern. The next two examples indicate how this rule can be used.

Table 1 shows the population of the United States as determined by the Census from 1790 to 1970. Does this data look like the results of drawing at random from a box? The answer is no. The population of the United States has been going up quite steadily. Numbers drawn at random from a box don't do that: sometimes they go up and other times they go down.

Table 1. Population of the United States, 1790 to 1970.

1790	3,929,214
1800	5,308,483
1810	7,239,881
1820	9,638,453
1830	12,866,020
1840	17,069,453
1850	23,191,876
1860	31,443,321
1870	39,818,449
1880	50,155,783
1890	62,947,714
1900	75,994,575
1910	91,972,266
1920	105,710,620
1930	122,775,046
1940	131,669,275
1950	150,697,361
1960	178,464,236
1970	203,211,926

Source: *Statistical Abstract*, 1976, Table 1.

For another example, take the daily maximum temperature at San Francisco airport. Are these data like draws from a box? Again, the answer is no. There is a definite seasonal pattern to these data: it is warmer in the summer, colder in the winter. There even are local patterns to the data—because the temperature on one day tends to be like the temperature on the day before. These patterns are shown in Figure 1, where the temperature data are graphed in the top panel. There is a dot above each day of the year for 1975, showing the maximum temperature on that day. The seasonal pattern shows up clearly: on the whole, the dots for the summer months are higher than the ones in the winter months. Also, there is an irregular wavy pattern within each season. The crest of a wave represents a stretch of warm days—a warm spell. The cold spells are in the troughs.

By comparison, the second panel in Figure 1 is for a mythical airport where the climate is on average like that of San Francisco, but the daily maximum temperatures are like draws from a box. There is no pattern to these hypothetical data.

Figure 1. Temperatures and box models. The first panel shows the daily maximum temperature at San Francisco airport in 1975. There is a seasonal pattern to the data, warmer in summer than winter. Also, there are local patterns: warm spells and cold spells. A box model would not apply. The second panel shows what the temperatures would look like, if they were generated by drawing from a box.

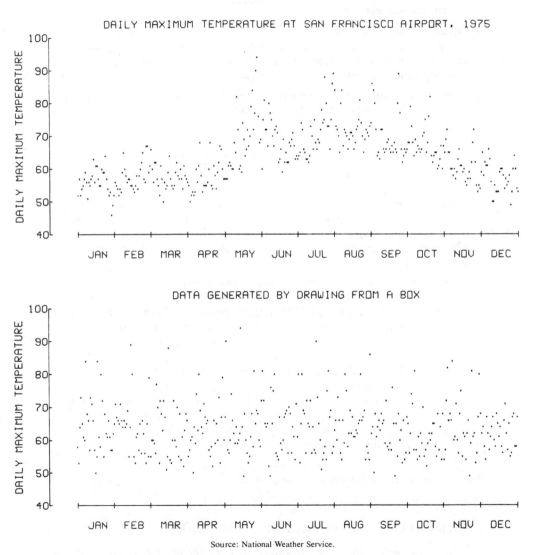

Source: National Weather Service.

EXERCISE SET A

1. A thumbtack is thrown in the air. It lands either point up or point down.

   Someone proposes the following box model: drawing with replacement from the box  , where U means "point up" and D means "point down." Someone else suggests using the box ⎪U⎥ D⎥ D⎪ . How could you decide which box was better?

2. In San Francisco, it rains on about 17% of the days in an average year. Someone proposes the following chance model for the sequence of dry and rainy days: draw with replacement from a box containing one card marked "rainy" and five cards marked "dry." Is this a good model?

3. Someone goes through the phone book, and makes a list showing the last digit of each phone number. Can this be modeled by a sequence of draws (with replacement) from the box

   ⎪0⎥1⎥2⎥3⎥4⎥5⎥6⎥7⎥8⎥9⎪

   What about a list of first digits?

4. Someone makes a list showing the first letter of each family name in the phone book, going name by name through the book in order. Is it sensible to model this sequence of letters by drawing at random with replacement from a box? (There would be 26 tickets in the box, each ticket marked with one letter of the alphabet.) Explain.

5. "The smart professional gambler, when heads comes up four times in a row, will bet that it comes up again. A team that's won six in a row will win seven. *He believes in the percentages.* The amateur bettor will figure that heads can't come up again, that tails is 'due.' He'll bet that a team on a losing streak is 'due' to win. *The amateur believes in the law of averages.*"

   —Jimmy the Greek, *San Francisco Chronicle,* July 23, 1975

   (a) What is the main difference between the two situations considered by Jimmy the Greek: tossing a coin and playing a competitive team sport?

   (b) Kerrich's coin (Chapter 16) will be tossed until it lands heads four times in a row. Suppose Jimmy the Greek offers five to four that the coin will land tails on the next toss. (On tails, he pays you $5; on heads, you pay him $4.) Do you take the bet?

The answers to these exercises are on p. A-58.

2. ESTIMATING THE ACCURACY OF AN AVERAGE

Measurement error was introduced in Chapter 6: any measurement is subject to chance error, and if repeated it would come out a bit differently. To get at the size of the chance error, the best thing to do is to repeat the measurement process a number of times. Then the spread in the measurements, as shown by their SD, will give an estimate of the likely size of the chance error in a single measurement. Chapter 6 stopped there. In brief, that chapter showed how to use a series of repeated measurements to get an estimate of the likely size of the chance error in a single measurement. This chapter continues the discussion, but the focus is on the average of the measurements in the series, rather than a single measurement. More explicitly, a method will be given for estimating the likely size of the chance error in this average. The method will be presented cold—without saying when it works, or why. These questions will be answered in the next section.

The context for the discussion in this section will be the hundred measurements on NB 10, reported in Table 1 on p. 91. This checkweight weighs a bit less than ten grams. The numbers in the table are the amounts, in micrograms, by which the individual measurements fall short of ten grams. The measurements were all made by the same procedure, but they varied a bit. In fact, the SD of the 100 numbers in the table is about 6 micrograms. So a single measurement made by this procedure is only accurate up to 6 micrograms or so. The right give-or-take number to put on a single measurement is 6 micrograms.

The best guess for the weight of NB 10 is given by the average of all 100 measurements. But since each measurement is thrown off by chance error, the average can't be exactly right either. Of course, the average is going to be more accurate than any single measurement, so it is going to be off by less than 6 micrograms. The question is, what is the right give-or-take number to put on the average:

$$\text{average} \pm \underline{\quad\quad}?$$

The answer is suggested by the square root law. The SE for the sum of 100 measurements can be estimated as

$$\sqrt{100} \times 6 \text{ micrograms} = 60 \text{ micrograms}.$$

Then the SE for their average is

$$60 \text{ micrograms}/100 = 0.6 \text{ micrograms}.$$

This completes the calculation. The likely size of the chance error in the average of all 100 measurements is estimated to be 0.6 micrograms. The average of all 100 numbers in the table is 404.6 micrograms. So NB 10 really weighs about 404.6 micrograms below ten grams, plus or minus 0.6 micrograms or so. (A microgram is the weight of a speck of dust.)

Two numbers come up in the calculation: 6 micrograms and 0.6 micrograms. The first is the SD of the 100 measurements, the second is the SE for their average. What is the difference between the two?

- The SD says that a single measurement is accurate up to 6 micrograms or so.
- The SE says that the average of all 100 measurements is accurate up to 0.6 micrograms or so.

The distinction between the SD of the measurements and the SE for the average is crucial.

Example 1. A hundred measurements are made on the weight of a one kilogram checkweight. Their SD is 80 micrograms.

    (a) Is a single measurement likely to be off the exact weight by around 8 micrograms, or 80 micrograms?

    (b) Is the average of all 100 measurements likely to be off the exact weight by around 8 micrograms, or 80 micrograms?

Solution. A single measurement is off by a chance error similar in size to the SD of the measurements. This is 80 micrograms. The answer to (a) is 80 micrograms. For (b), the SE for the sum of the measurements is estimated as $\sqrt{100} \times 80 = 800$ micrograms. So the SE for the average is $800/100 = 8$ micrograms. That is the answer to (b).

The method of this section should be used only where there is a fairly large number of measurements. With fewer than 25 measurements, most statisticians would use a slightly different procedure, based on what is called the *t-distribution* (p. 409).

EXERCISE SET B

1. Some modern scales use electrical *load cells*. The weight is distributed over a number of cells. Each cell converts the weight it carries to an electrical current, which is fed to a central scanner. This scanner adds up all the currents, and computes the corresponding total weight, which it prints out. This process is repeated several dozen times a second. As a result, a loaded boxcar (weighing about 100,000 pounds) can be weighed as it crosses a special track, with chance errors of only several hundred pounds in size.[1] Suppose 25 readings on the weight of a boxcar show an average of 82,670 pounds, with an SD of 500 pounds.
       (a) Is a single reading off by around 100 pounds, or 500 pounds?
       (b) Is the average of the 25 readings off by around 100 pounds, or 500 pounds?

2. (Hypothetical.) The British Imperial Yard is sent to Paris for calibration against The Meter. Its length is determined 100 times. This sequence of

measurements averages out to 91.4402 cm., with an SD of 800 microns. (A micron is the millionth part of a meter.)

    (a) Is a single reading off by around 80 microns, or 800 microns?

    (b) Is the average of all 100 readings off by around 80 microns, or 800 microns?

The answers to these exercises are on p. A-58.

3. THE MODEL

The previous section gave the standard method for estimating the likely size of the chance error in the average of a series of repeated measurements. The method is easy enough to carry out, but it doesn't apply to every set of data. The purpose of this section is to explain the reasoning behind the method.

The explanation begins with a model proposed around 1800 by C. F. Gauss (Germany, 1777–1855), in connection with astronomical data he was studying. Gauss realized that variability in repeated measurements was like variability in numbers drawn at random with replacement from a box. The reason is that a measurement, like a draw from a box, can be repeated independently and under the same conditions. In our terms, Gauss proposed a box model for chance error.

The model will now be described in more detail. The basic situation is that a series of repeated measurements are made on some quantity, with each measurement differing from the exact value by a chance error. What the model says is that this error is like a draw made at random from a box of

Carl Friedrich Gauss (Germany, 1777–1855)

tickets, called the *error box*. To capture the idea that the chance errors aren't systematically positive, or systematically negative, it is assumed that the average of the numbers in the error box equals 0. The draws from the error box are made with replacement, so the process can be repeated independently and under the same conditions.

> In the Gauss model, each time a measurement is made, a ticket is drawn at random with replacement from the error box. The number on the ticket is the chance error. It is added to the exact value to give the actual measurement. The average of the error box is equal to 0.

This completes the description. In the model, it is the SD of the error box which gives the likely size of the chance errors. Usually, this SD is unknown, but it can be estimated by the SD of the measurements. The reason is that according to the model, the spread in the measurements reflects the spread of the numbers in the error box.

> When the Gauss model applies, the SD of a series of repeated measurements is an estimate for the SD of the error box. The estimate is good when there are enough measurements.

As section 5 will show, the NB 10 data fits the model: the chance errors in the measurements look like draws from a box. The number of measurements is large, so the SD of the error box can be estimated by the SD of the data, 6 micrograms. A single measurement is accurate up to 6 micrograms or so. Then, by the method of the previous section, the average of the 100 measurements is accurate up to 0.6 micrograms or so. This represents an improvement by the factor of 10—the square root of the number of measurements. The question is, why this factor? The object of what follows is to explain how the square root comes out of the model. Without the box model, this square root wouldn't make sense.

To begin with, according to the model, each measurement on NB 10 is around the exact weight, but is off by a draw from the error box:

> 1st measurement = exact weight + 1st draw from error box,
> 2nd measurement = exact weight + 2nd draw from error box,
>
> .
>
> .
>
> .
>
> 100th measurement = exact weight + 100th draw from error box.

Therefore, the sum of the 100 measurements will be around 100 times the exact weight, being off by the sum of the 100 draws from the box. So the average of all 100 measurements will be off the exact weight by the average of the 100 draws from the error box. Since the average of the box is 0, the average of the draws will be around 0 too—give or take an SE or so. In other words, it is the SE for the average of the draws which is the give-or-take number to put on the average of the measurements. Since we are dealing with draws from a box, the method of p. 373 can be used to calculate the SE for the average, justifying the arithmetic of the previous section.

In essence, then, the square root of the number of measurements is the right factor because of the square root law for the sum of draws from a box. The square root law applies because the chance errors in the measurements on NB 10 are like draws from a box. If this weren't so, the SE of 0.6 micrograms wouldn't make any sense. Without a chance model, there is no reasonable way to estimate the likely size of the error in an average.

This completes the reasoning behind the method of the previous section. The Gauss model is what makes it possible to judge the likely size of the chance error in the average of a series of repeated measurements, starting only from the spread in these measurements. That is why the model is so useful.

One feature of the model hasn't been mentioned yet: the error box belongs to the measuring process, rather than the thing being measured. When there is a lot of experience with the procedure, it is usually better to estimate the SD of the error box from a large amount of past data rather than a small amount of current data.

Example 1. (Hypothetical.) After making several hundred measurements on NB 10, and finding their SD to be about 7 micrograms, the Bureau misplaces this checkweight. They go out and buy a new one. They measure its weight by exactly the same procedure as for NB 10, and on the same scale. After a week the Bureau accumulates 25 measurements. These average out to 605 micrograms above ten grams, with an SD of 8 micrograms. Assuming the Gauss model, the new weight is 605 micrograms above ten grams, give or take about

7 microgramms   8 micrograms   1.4 micrograms   1.6 micrograms.

Solution. First, the model. The chance error in each measurement is like a draw from the error box. The error box belongs to the scales, not the weight. So the SD of the error box should be estimated by the SD of the large amount of past data on NB 10, not the small amount of current data on the new weight. The SD of the error box is therefore estimated as 7 micrograms. This tells the likely size of the chance error in a single measurement. But the likely size of the chance error in the average of 25 measurements is smaller. The SE for the sum of the 25 measurements is $\sqrt{25} \times 7 = 35$ micrograms. The SE for their average is $35/25 = 1.4$ micrograms. This is the answer.

The increase in SD, from 7 micrograms for NB 10 to 8 micrograms for the new weight, just seems to reflect chance variability. However, if the increase in SD had been much larger, or if it persisted over many more measurements, then the Bureau would be forced to consider that the error box had indeed changed. Also, if the Bureau started weighing hundred-gram weights on the same scale as they used for NB 10, the historical data would not be so relevant.

A final point. The version of the Gauss model presented here makes the tacit assumption that there is no bias in the measuring procedure. When bias is present, each measurement is the sum of three terms:

exact value + bias + chance error.

Then the SE for the average no longer says how the average of the measurements is from the exact value, but only how far it is from

exact value + bias.

The methods of this chapter are no help in judging bias. We did not take it into account for the measurements on NB 10, because other lines of reasoning suggest that the bias in precision weighing at the National Bureau of Standards is negligible.

EXERCISE SET C

1. A good scale for weighing people makes chance errors of several ounces in size. Which of the following is the best error box? Explain briefly.

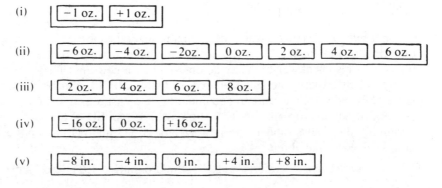

2. A carpenter who works carefully can measure wood to about 1/64 of an inch. Which of the following is the best error box? Explain briefly.

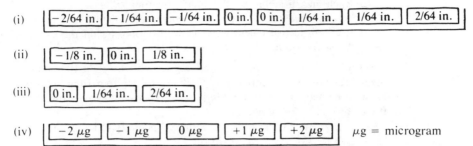

(i) −2/64 in.   −1/64 in.   −1/64 in.   0 in.   0 in.   1/64 in.   1/64 in.   2/64 in.

(ii) −1/8 in.   0 in.   1/8 in.

(iii) 0 in.   1/64 in.   2/64 in.

(iv) −2 µg   −1 µg   0 µg   +1 µg   +2 µg   µg = microgram

Note: The error box need not contain equal numbers of the different tickets. Box (i), for instance, models the idea that an error of 1/64 inch is twice as likely as an error of 2/64 inch.

3. Would taking the average of 25 measurements divide the likely size of the chance error by a factor of 5, or 10, or 25?

4. (a) A ten-gram checkweight is being weighed. Assume the Gauss model with no bias. If the exact weight is 501 micrograms above ten grams, and the number drawn from the error box is +3 micrograms, what would the measurement be?
   (b) Repeat, if the exact weight is 510 micrograms above ten grams, and the number drawn from the error box is −6 micrograms.

5. The first measurement on NB 10 was 409 micrograms below ten grams. Assuming the Gauss model with no bias, this measurement equals

$$\text{exact weight} + \text{chance error}.$$

Can you figure out the numerical value for each of the two terms? Explain briefly.

6. Suppose you sent a nominal ten gram weight off to the Bureau, and asked them to weigh it 25 times, and tell you the average weight. They will use the same procedure as on NB 10, where the SD of several hundred measurements was about 7 micrograms. The 25 measurements average out to 307 micrograms above ten grams, with an SD of about 5 micrograms. Your weight is 307 micrograms above ten grams, give or take around:

   5 micrograms   7 micrograms   1 microgram   1.4 micrograms.

(You may assume the Gauss model, with no bias.)

7. A scientist wishes to weigh an ore sample with high precision. He plans to take several measurements and average them. He has two procedures to choose from. Procedure A has an SD of only one milligram, but each weighing by A costs $3. Procedure B has an SD of two milligrams, but each weighing by B costs only $1. Other things being equal, is he better off spending $9 on A, or $9 on B? (You may assume the Gauss model, with no bias.)

8. Twenty-five measurements are made on the speed of light. These average out to 300,007 with an SD of 10, the units being kilometers per second. Say

whether each statement is true or false, and explain why. (You may assume the Gauss model, with no bias.)
   (a) Each measurement is off 300,007 by 10 or so.
   (b) The average of all 25 measurements is off 300,007 by 2 or so.
   (c) If a 26th measurement were made, it would be off the exact value for the speed of light by 2 or so.

9. The Bureau is about to weigh a one-kilogram checkweight 100 times, and take the average of the measurements. They are willing to assume the Gauss model, with no bias, and on the basis of past experience they estimate the SD of the error box to be 50 micrograms.
   (a) The average of all 100 measurements is likely to be off the exact weight by _____ or so.
   (b) The SD of all 100 measurements is likely to be around _____
   (c) Estimate the probability that the average of all 100 measurements will be within 10 micrograms of the exact weight.

10. A surveyor is measuring the distance between five points A, B, C, D, E. They are all on a straight line. He finds that each of the four distances AB, BC, CD, and DE measures one mile, give or take an inch or so. These four measurements are made independently, by the same procedure.

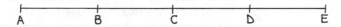

The distance from A to E is about four miles, give or take around
        4 inches   2 inches   1 inch   ½ inch   ¼ inch.
Explain briefly. (You may assume the Gauss model, with no bias.)

The answers to these exercises are on pp. A-58–A-59.

Historical note: There is a hidden connection between the theory of measurement error and neon signs. In 1890, the atmosphere was believed to consist of nitrogen (about 80%), oxygen (a little under 20%), carbon dioxide, water vapor—and nothing else. Chemists were able to remove the oxygen, carbon dioxide, and water vapor. The residual gas should have been pure nitrogen. Lord Rayleigh undertook to compare the weight of this residual gas with the weight of an equal volume of chemically pure nitrogen. One measurement on the weight of the residual gas gave 2.31001 grams. And one measurement of the pure nitrogen gave a bit less, 2.29849 grams. However, the difference of 0.01152 grams was rather small, and in fact was comparable to the chance errors made by the weighing procedure.

So this difference could have resulted from chance error. But if it didn't, the residual gas had to contain something heavier than nitrogen. The way Rayleigh solved the problem was by replicating the experiment, until he had enough measurements to prove to everyone's satisfaction that the residual gas from the atmosphere really was heavier than pure nitrogen. He went on to isolate the rare gas called *argon*, which is heavier than nitrogen and present in the atmosphere in minute quantities. Other researchers later discovered the similar gases neon, krypton and xenon, all occurring naturally (in minute amounts) in the atmosphere. These gases are what make "neon" signs glow in different colors.[2]

4. CONFIDENCE INTERVALS

A measurement is always surrounded by fuzz—chance error. Averaging some measurements reduces this error, but does not eliminate it. The standard error for the average gives the likely size of the error that is left. With the NB 10 data, the SE for the average was estimated as 0.6 micrograms. On this basis, the exact weight of NB 10 was estimated as 404.6 micrograms below ten grams, give or take 0.6 micrograms or so.

To go at this more carefully, statisticians use confidence intervals. Provided there are a reasonable number of measurements, the method is just the same as for sampling. A 95%-confidence interval for the exact weight of NB 10, for example, can be obtained by going two SEs in either direction from the average. In this case, the average is 404.6 micrograms below ten grams, and two SEs is $2 \times 0.6 = 1.2$ micrograms, so NB 10 weighs somewhere between 403.4 and 405.8 micrograms below ten grams—with confidence about 95%.

Again, the word "confidence" is there as a reminder that the chances are in the measurement process and not in NB 10. The exact weight isn't subject to chance variability (at least the Bureau hopes it doesn't keep changing between measurements). So, in the frequency theory, there is no way to talk about the chance that the exact weight is in any particular interval. Either it is, or it isn't. (For a similar discussion in the sampling context, see p. 347.)

As always, interpreting a confidence level takes an act of imagination. You have to think not only about the actual data, but also about all the other ways the data could have come out. In fact, you have to think about the whole procedure that was followed to get the interval. The 95% is the chance that this procedure gives an interval which covers the exact weight.

To be more specific, imagine that 100 other investigators set out to determine how much NB 10 really weighs, using the same equipment as the Bureau. Suppose that each one went through the same procedure as the Bureau followed: taking a hundred measurements, computing their average and estimating its SE, and then going two SEs either way from the average. About 95 of these people should get intervals which cover the exact weight of NB 10, and the other 5 should miss.

This experiment was simulated on the computer, with the results shown in Figure 2 (next page). To get started, the program was told that NB 10 weighed 405 micrograms below ten grams; a vertical line is drawn over this value. The program then generated a hundred measurements for each investigator—by drawing the chance errors at random from an error box.[3] Finally, the program calculated the interval "average ± 2 SE" for each investigator. In the figure, these intervals are placed at different heights, so you can tell one from the other. Notice that these intervals have different centers: this is because the investigators got different averages. Notice too that the intervals have different lengths: this is because the investigators got different SDs. The confidence level says that about 95 of these 100 intervals should cover the vertical line. In fact, 96 do.

Figure 2. A computer simulation of 100 confidence intervals, each with confidence level 95%. The intervals have different centers and lengths, due to chance variation; 96 out of 100 cover the exact weight, represented by the vertical line.

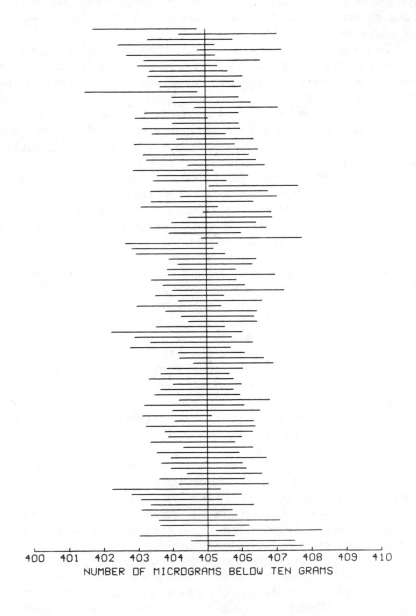

> The chances are in the measuring procedure, not the thing being measured.

The histogram for the NB 10 data is quite different from the normal curve (p. 94). Why is it legitimate to use the normal curve in figuring confidence levels? The first point is that the confidence level relates not to any single measurement, but to their average. Now the model comes into play. As shown in Chapter 18, the probability histogram for the sum of 100 draws from the error box will follow the normal curve, even though the tickets in the box don't. As a result, the probability histogram for the average of the 100 draws will also follow the curve—the division by 100 just represents a change of scale, and washes out in standard units. But each measurement differs from the exact weight by a draw from the error box. So the average of the measurements differs from the exact weight by the average of 100 draws from the error box. And that is why the normal curve can be used to approximate the probability histogram for the average of the measurements, even though the histogram for the data is far from the curve. (This point came up before in the sampling context, on p. 381.)

There is no hard-and-fast rule as to how many measurements are enough to justify using the curve. However, if the histogram for the error box isn't wildly different from the curve, the approximations are quite good even with only 25 measurements.[4] (See Chapter 18 for more discussion of this point.) With fewer than 25 measurements, statisticians would prefer to use a different method, based on the *t*-distribution (section 6).

So far in this section, bias has been assumed to be negligible. When bias is suspected, all that can be said about going 2 SEs either way from the average of the measurements is that with 95% confidence, this interval covers

exact value + bias.

This may not be too helpful. The point is that the SE tells you about chance error, not bias. And usually, there is no way to find out just from internal evidence whether a measuring process is biased. The results have to be compared to some external standard.

EXERCISE SET D

1. A laboratory has a method for measuring lengths, which they consider to be unbiased. Their next job is to calibrate a yardstick. They measure it 25 times. The average of their measurements is 0.910835 meters, with an SD of 45 microns. Find an approximate 95%-confidence interval for the exact length of this yardstick. (A *micron* is one millionth of a meter. With modern laser technology, this kind of length can be measured to within one wave length of visible light, which is about half a micron.)

2. The 95%-confidence interval for the exact weight of NB 10 is the range from 403.4 to 405.8 micrograms below ten grams. Say whether each of the following

statements is true or false, and explain why. (You may assume the Gauss model, with no bias.)
  (a) About 95% of the measurements are in this range.
  (b) There is about a 95% chance that the next measurement will be in this range.
  (c) About 95% of the time that the Bureau goes 2 SEs either way from the average of the measurements, they succeed in covering the exact weight.

The answers to these exercises are on p. A-59.

5. THE ROLE OF THE MODEL

NB 10 is just a chunk of metal. It is weighed on a contraption of platforms, gears, and levers, like the one shown in Figure 3 below. The results of these weighings have been subjected to an elaborate statistical analysis, involving the standard error, the normal curve, and the confidence intervals. What connects NB 10 to all this mathematics? Only an idea: the Gauss model. The chance errors are like draws from a box, so their average is like the average of the draws. The number of draws is so large that the probability histogram for the average will follow the normal curve very closely, as indicated on p. 403. Without the model there would be no box, no standard error, and no confidence levels.

How well do the measurements on NB 10 fit the Gauss model? One thing to check is that the data show no trend or pattern. The top panel in Figure 4 is a graph of the data. There is one point for each measurement. The *x*-coordinate says which measurement it was: number 1, or number 2, or number 3, and so on. The *y*-coordinate says how many micrograms below ten grams the measurement was. The points do not show any trend or pattern;

Figure 3. Precision weighing machine built at the National Bureau of Standards. This machine is for weights of about one kilogram, and was given to the International Bureau of Weights and Measures in Paris. It is one of the most precise weighing machines ever built.

they look as random as draws from a box. In fact, the second panel shows hypothetical data generated on the computer using the Gauss model.[5] If you didn't know which was which, it would be very hard to tell the difference between these two panels. By comparison, the third panel (also for computer-generated data) shows a very strong pattern, and the Gauss model would not apply.

Figure 4. The Gauss model and NB 10. The top panel graphs the repeated measurements on NB 10 (p. 91). The middle panel graphs hypothetical data, generated by a computer simulation of the Gauss model. These two panels are very similar, showing how well the Gauss model represents the real data. The bottom panel graphs data showing a very strong pattern: the Gauss model would not apply.

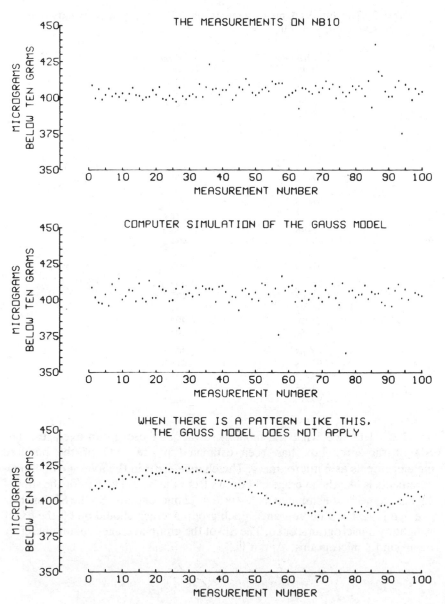

The model has passed the first test. Here is a harder one. The most striking prediction of the model is the square root law. For example, averaging 4 measurements together should reduce the likely size of the chance error by the factor $\sqrt{4}$. One way to test this theory on the NB 10 data is to break the hundred measurements down into successive groups of 4, and examine the group averages—as in Table 2. The first group average is figured by adding up measurements numbers 1 through 4, and dividing by 4; the result is 403.5 micrograms below ten grams. The second group average is figured by adding up measurements numbers 5 through 8, and dividing by 4; the result is 403.0 micrograms below ten grams. And so on. Altogether, there are $100/4 = 25$ group averages, as shown in the table.

Table 2. The hundred measurements on NB 10, taken in twenty-five successive groups of four.

Measurement numbers	Average of the group in micrograms below ten grams
1–4	403.5
5–8	403.0
9–12	401.2
13–16	402.2
17–20	402.0
21–24	402.0
25–28	401.0
29–32	400.8
33–36	410.2
37–40	404.8
41–44	403.8
45–48	408.8
49–52	404.0
53–56	408.2
57–60	405.8
61–74	402.0
65–68	405.2
69–72	407.2
73–76	405.2
77–80	404.8
81–84	406.8
85–88	415.8
89–92	403.2
93–96	400.5
97–100	402.8

Now let's see what the model says about these group averages. The SD of the error box has been estimated by the SD of the hundred measurements as 6 micrograms. The chance error in the average of 4 measurements is like the average of 4 draws from the box. The SE for the sum of 4 measurements is estimated as $\sqrt{4} \times 6 = 12$ micrograms. So the SE for their average is $12/4 = 3$ micrograms. Each group average should be off the exact weight by 3 micrograms or so. The SD of the group averages should therefore be around 3 micrograms. And it is.[6]

Statistical inference means the use of chance methods to draw conclusions from data. Attaching a standard error to an average would be an example. Now it is always possible to go through the procedure mechanically. Many desk calculators will do the work for you. It is even possible to label the output as a ''standard error.'' But it is important not to get hypnotized by the arithmetic or the terminology. The procedure only makes sense because of the theory of chance—the square root law. The arithmetic makes the implicit assumption that the data are like the results of drawing from a box. If this assumption doesn't hold, calculating the ''standard error'' is silly. The sad fact is that many investigators ignore this kind of point. As a result, the standard errors they report are often meaningless. (This point was discussed before, in the sampling context: pp. 350, 366, 382.)

> Statistical inference can be justified only in terms of an explicit chance model for the data. No box, no inference.

Parts II and III focused on *descriptive statistics:* drawing diagrams or calculating numbers which summarize data and bring out their salient features. Such techniques can be used very generally, because they do not involve any hidden assumptions about where the data came from.

Example. Table 1 on p. 390 reported the population of the United States every ten years from 1790 to 1970. These 19 numbers average out to 70 million, with an SD of 61 million. An investigator attaches a standard error to the average, by the following procedure:

$$\text{SE for sum} \approx \sqrt{19} \times 61 \text{ million} \approx 266 \text{ million},$$
$$\text{SE for average} \approx 266/19 \approx 14 \text{ million}.$$

Is this sensible?

Solution. The average and SD make sense, as descriptive statistics. They summarize part of the information in Table 1, although they miss quite a bit—for instance, the fact that the numbers increase steadily. The SE of 14 million, however, is silly. If the investigator wants to know the average of the 19 numbers in the table, that's been computed, and there is no need to worry about chance error.

Of course, he may want to know something else, like the average of a list showing the population of the United States in every year from 1790 to 1970. (Every tenth number on that list is shown in Table 1; the numbers in between are unknown, because the Census is only taken every ten years.) The investigator would then be making an inference—using the average from Table 1 to estimate this other average. And the estimate would be off by some amount. But the square root law can't possibly tell him anything about this amount off. The reason is that the errors—the differences between the numbers in Table 1 and the average that the investigator wants to know—aren't like draws from a box. In this example, there is no way to make sense out of the standard error.

> The square root law only applies to draws from a box.

EXERCISE SET E

1. An investigator is trying to determine the effects of a new drug on the blood pressure of experimental animals. At this stage, he is using five monkeys. Under standard conditions, before the drug is administered, each monkey has an equilibrium blood pressure (which differs from monkey to monkey). The procedure for measuring this blood pressure is unbiased, but subject to large chance errors.
    (a) Over a period of time, the investigator takes ten blood pressure readings on one monkey. He wants to estimate the equilibrium blood pressure for that monkey by the average of the ten readings. Can he use the Gauss model to judge the accuracy of his estimate?
    (b) Next, he takes ten readings on each of the other monkeys, so he has a total of 50 measurements. They average out to 117, with an SD of 32. He concludes that with a confidence of about 95%, the average equilibrium blood pressure of the five monkeys is in the interval from 117 mm − 2 SE to 117 mm + 2 SE. He figures the SE for the average as follows. The SD of the data is 32 mm, so

    SE for the sum of the blood pressures = $\sqrt{50} \times 32 \approx 226$ mm.
    SE for their average = $226/50 \approx 4.5$ mm.

    What does statistical theory say?

2. The concept of measurement error is often applied to the results of psychological tests. The equation is

    actual test score = true test score + chance error.

    The chance error term reflects accidental factors, like the mood of the subject, or luck. Do you think that the Gauss model applies?

3. A psychologist has a theory about what situations people find pleasant or unpleasant. To test it, he makes up 25 items, of the following kind:

    Imagine that you like John, you like cats, and John likes cats. How pleasant is this situation?

    In each item, then, the subject is asked to imagine that he likes some other person, and some object, and the other person likes the object as well. This is called a situation of *cognitive consonance*. The psychologist now presents the 25 items to a volunteer, one at a time. After each presentation, the volunteer rates the pleasantness of the situation on a scale:

    1   2   3   4   5

    unpleasant   neutral   pleasant

    The average of the 25 ratings is 4.2, with an SD of 1.0; the psychologist infers that this subject's average response to a situation of cognitive consonance is in the interval 4.2 ± 2 SE. He figures the SE for the average as follows:

    SE for the sum of the scores = $\sqrt{25} \times 1.0 = 5.0$
    SE for their average = $5.0/25 = 0.2$

    What does statistical theory say?

4. Say whether each statement is true or false.
   (a) The square root law only applies to draws from a box.
   (b) If the tickets in the box don't follow the normal curve, you can't calculate the SE for the sum of the draws from the box.
   (c) If the Bureau made more and more measurements on NB 10, the histogram for the measurements would get closer and closer to the normal curve.

5. An investigator gives a psychological test to 100 students. Their scores average 53, with an SD of 10. The likely size of the chance error in the average is:

$$1 \qquad 10 \qquad 100 \qquad \text{can't say.}$$

The answers to these exercises are on p. A-59.

6. STUDENT'S *t*-CURVE

This section will explain how to find confidence intervals based on a small number of repeated measurements, assuming the histogram for the error box is not too different from the normal curve, and its SD is being estimated from the current measurements. This section is somewhat technical, and the material will not be used in the rest of the book.[7]

We begin with an example. The first four measurements on NB 10 were below ten grams by the following amounts (in micrograms):

$$409 \qquad 400 \qquad 406 \qquad 399.$$

Suppose the Bureau had only made these four measurements, did not know the SD of the measurement procedure, and wanted a 95%-confidence interval for the exact weight of NB 10. It would be natural to go 2 SEs either way from the average. In this case, the average of the four measurements is 403.5 micrograms. Their SD is 4.2 micrograms, and it seems right to estimate the SD of the error box by this amount. So the SE for the sum of the four measurements is $\sqrt{4} \times 4.2 = 8.4$ micrograms, and the SE for their average is $8.4/4 = 2.1$ micrograms. On this basis, the confidence interval is the range from $403.5 - 2 \times 2.1$ to $403.5 + 2 \times 2.1$ micrograms below ten grams.

However, this procedure is overlooking something. In the model, the SD of the measurements is only an estimate for the SD of the error box. And in this case, the number of measurements is so small that their SD is not likely to be a good estimate for the SD of the error box. To take this extra uncertainty into account, the interval has to be made longer, and the adjustment depends on the number of measurements. Usually, it is made in two steps.

Step 1. With a large number of measurements, the SD of the error box is estimated by the SD of the measurements. But with a small number of measurements, it is estimated by the larger quantity[8]

$$\sqrt{\frac{\text{number of measurements}}{\text{number of measurements} - \text{one}}} \times \text{SD of measurements.}$$

This quantity is abbreviated as SD⁺.

In the example, the number of measurements is 4 and their SD is 4.2 micrograms. So SD⁺ is $\sqrt{4/3} \times 4.2 \approx 1.15 \times 4.2 \approx 4.8$ micrograms. Then, the SE is figured in the usual way: the SE for the sum is $\sqrt{4} \times 4.8 = 9.6$ micrograms; the SE for the average is $9.6/4 = 2.4$ micrograms.

Step 2. With a large number of measurements, a 95%-confidence interval is obtained by going 2 SEs either way from the average. But with a small number of measurements, this multiplier of 2 has to be replaced by a larger number. This is the important step.

Originally, the multiplier 2 was found by solving an equation:

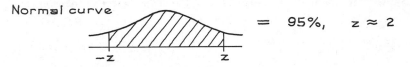

Mathematically, it turns out that the normal curve is appropriate only when the SD of the error box is known, or when it can be estimated fairly closely from the data. When it can be estimated only loosely from the data, the normal curve must be replaced by what statisticians call *Student's t-curve*.[9] We will explain how to work with Student's curve in a moment, but in the example the right multiplier turns out to be 3.18. So an approximate 95%-confidence interval for the exact weight of NB 10, based on the first four measurements, goes from

average $- 3.18 \times$ SE $\approx 403.5 - 3.18 \times 2.4 \approx 395.9$ micrograms

below ten grams to

average $+ 3.18 \times$ SE $\approx 403.5 + 3.18 \times 2.4 \approx 411.1$ micrograms

below ten grams. The interpretation of the confidence level is as before: about 95% of the time, this procedure will give an interval covering the exact value, and about 5% of the time, it will fail.

The interval "average ± 2 SE" computed using the normal curve runs from 399.3 to 407.7 micrograms below ten grams. With only 4 measurements, it is much too short to be a 95%-confidence interval. (In fact, its confidence level is only about 80%.)

To see where the multiplier of 3.18 came from, we must now look at Student's curve. The name is somewhat misleading, because there is a whole family of these curves—one for each number of *degrees of freedom*. In the present context,

degrees of freedom = number of measurements − one.

The term "degrees of freedom" does have a slightly baroque quality, but the image was introduced for the following reason. The length of a confidence interval depends on the SD of the measurements, and this in turn depends on the deviations from the average. But the sum of these deviations has to be 0, so they cannot all vary freely. The constraint that the sum be 0 eliminates one degree of freedom.

Student's curves for three and nine degrees of freedom are shown in Figure 5, with normal curves for comparison. Student's curves look quite a lot like the normal curve, but they are less piled up in the middle and more

Figure 5. Student's curves. The dashed line is Student's curve for three degrees of freedom (top) or nine degrees of freedom (bottom). The solid line is a normal curve, for comparison.

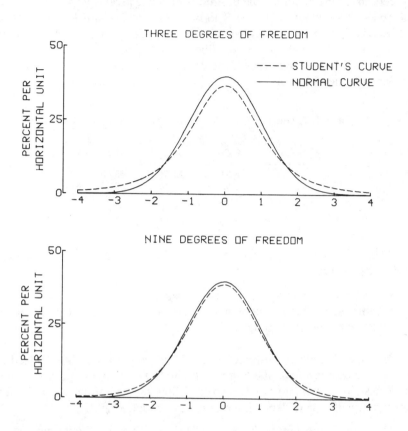

spread out. As the number of degrees of freedom goes up, his curves get closer and closer to the normal, reflecting the fact that the SD of the measurements is getting closer and closer to the SD of the error box. His curves are all symmetric around 0, and the total area under each equals 100%.[10]

Coming back to the example, with four measurements, there are $4 - 1 = 3$ degrees of freedom. Let *t* stand for the multiplier which gives a 95%-confidence interval, of the form

$$\text{average} \pm t \times \text{SE}.$$

This *t* has to be found by solving an equation:

Solving this equation looks like a formidable job, but in fact it is easy to do with the help of a special table on p. A-71, part of which is reproduced here as Table 3. The columns are headed by confidence levels, the rows by degrees of freedom. In the example, the desired confidence level is 95% and there are 3 degrees of freedom. In Table 3 look down the column headed "95%" and across the row for 3 degrees of freedom. They intersect at the entry "3.18," so $t = 3.18$ solves the equation. This means that the area between -3.18 and 3.18 under Student's curve with 3 degrees of freedom is 95%. That is where the multiplier of 3.18 came from.

Table 3. A short t-table. (The full table is on p. A-71.)

Degrees of freedom	90%	95%	99%
1	6.31	12.71	63.66
2	2.92	4.30	9.92
3	2.35	3.18	5.84
4	2.13	2.78	4.60
5	2.02	2.57	4.02

There are two key assumptions behind the use of Student's curve to get confidence intervals:

- The Gauss model applies, with no bias.
- A histogram for the contents of the error box should not look too different from the normal curve.

Example 1. Five measurements on a hundred-gram weight averaged out to 358 micrograms above 100 grams, with an SD of 27 micrograms. Assume the Gauss model, with no bias and errors following the normal curve. Find an approximate 95%-confidence interval for the exact weight.

Solution. To begin with, $SD^+ = \sqrt{5/4} \times SD \approx 30.2$ micrograms, estimating the SD of the error box on the basis of this small number of measurements. The SE for the sum of the 5 measurements is then estimated as $\sqrt{5} \times 30.2$ micrograms ≈ 67.5 micrograms; so the SE for their average is $67.5/5 = 13.5$ micrograms. The number of degrees of freedom is $5 - 1 = 4$. In Table 3, look for the "95%" column and the "4 degrees of freedom" row. They intersect at 2.78.

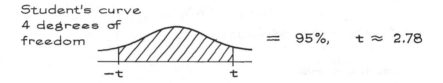

The approximate 95%-confidence interval is

$$\text{average} \pm t \times SE \approx 358 \pm 2.78 \times 13.5 \approx 358 \pm 37.5$$

micrograms above a hundred grams, completing the example.

Example 2. A calibration laboratory has been measuring the lengths of meter sticks for several years using the same procedure; the Gauss model is known to apply, and bias is known to be negligible. Several hundred repeated measurements of the length of one of their own standard meters produced a histogram which had an SD of 18 microns, and followed the normal curve very well. Now they make nine measurements of the length of someone else's meter-stick. These average out to 298 microns above a meter, with an SD of 21 microns. They are about 95% confident that the length of this new meter-stick is 298 microns above a meter, give or take _____ .

Solution. First, the model. The chance error in each measurement is like a draw from the error box. This box belongs to the measurement procedure, not the meter-stick. So its SD should be estimated by the SD of the past data, as 18 microns. Now the SE for the sum of 9 measurements is estimated as $\sqrt{9} \times 18 = 54$ microns. The SE for their average is 54/9 = 6 microns.

Student's method does not apply, because the SD of the error box is not estimated from the current data. However, the numbers in the error box follow the normal curve, and when this is so, the probability histogram for their average will follow the curve too, even with a small number of draws. Therefore the 95%-confidence interval involves a give-or-take number of 2 SE = 12 microns, completing the solution.

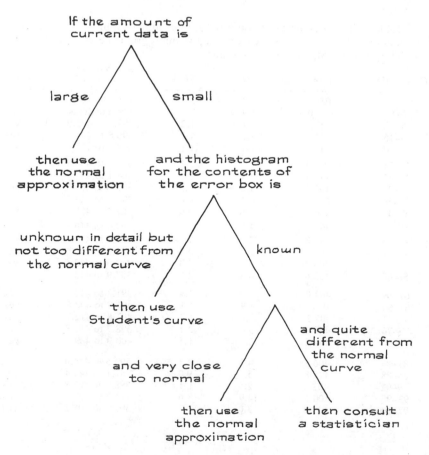

If you only have a small amount of current data to work with, but the histogram for the error box is known from past data and does not follow the normal curve, then neither the normal curve nor Student's curve should be used. Consult a statistician.[11]

The theory behind Student's procedure can be tested using the data on NB 10 in Table 1 on p. 91. For example, suppose an investigator only had access to the first four measurements in the table. As on p. 410, he could calculate a 95%-confidence interval for the exact weight as

$$\text{average} \pm 3.18 \times \text{SE} \approx 403.5 \pm 7.6 \text{ micrograms}$$

below ten grams. Another investigator might only have access to measurements 5 through 8. His 95%-confidence interval would be

$$\text{average} \pm 3.18 \times \text{SE} \approx 403.0 \pm 3.4 \text{ micrograms}$$

below ten grams. And so on. There are twenty-five groups of four successive measurements to be considered, as shown in Table 4. For each group of four measurements, the table gives the average, the SD, the SE, and the 95%-confidence interval computed by Student's method as "average $\pm$ 3.18 SE." (The SE is computed using SD^+.) The table indicates how hard it would be to estimate the SD of the error box from four measurements: the SDs range from 1.4 to 15.7. They can't all be right.

Table 4. The measurements on NB 10, in groups of four. The averages and confidence intervals are expressed in micrograms below ten grams. The SDs and SEs are in micrograms.

Observations	Average	SD	SE	95%-confidence interval	Covers
1–4	403.5	4.2	2.4	395.9 to 411.1	yes
5–8	403.0	1.9	1.1	399.6 to 406.4	yes
9–12	401.2	2.1	1.2	397.5 to 405.0	yes
13–16	402.2	3.0	1.7	396.8 to 407.7	yes
17–20	402.0	1.9	1.1	398.6 to 405.4	yes
21–24	402.0	3.7	2.1	395.2 to 408.8	yes
25–28	401.0	3.8	2.2	394.1 to 407.9	yes
29–32	400.8	1.5	0.9	398.0 to 403.5	no
33–36	410.2	8.1	4.6	395.5 to 425.0	yes
37–40	404.8	1.6	0.9	401.7 to 407.8	yes
41–44	403.8	3.7	2.1	397.0 to 410.5	yes
45–48	408.8	2.7	1.5	403.8 to 413.7	yes
49–52	404.0	1.4	0.8	401.4 to 406.6	yes
53–56	408.2	2.4	1.4	403.9 to 412.6	yes
57–60	405.8	4.3	2.5	397.9 to 413.6	yes
61–64	402.0	5.9	3.4	391.2 to 412.8	yes
65–68	405.2	1.9	1.1	401.7 to 408.8	yes
69–72	407.2	3.0	1.7	401.8 to 412.7	yes
73–76	405.2	3.6	2.1	398.7 to 411.8	yes
77–80	404.8	2.6	1.5	400.0 to 409.5	yes
81–84	406.8	4.0	2.3	399.5 to 414.0	yes
85–88	415.8	15.7	9.0	387.1 to 444.4	yes
89–92	403.2	2.5	1.4	398.7 to 407.8	yes
93–96	400.5	14.9	8.6	373.2 to 427.8	yes
97–100	402.8	3.0	1.7	397.3 to 408.2	yes

About 95% of the confidence intervals in Table 4 should succeed in covering the exact weight of NB 10. This prediction can't be checked directly, because we don't know the exact weight. However, the average of all one hundred measurements—404.6 micrograms below ten grams—is a very good estimate for the exact weight. So, about 95% of the intervals should cover this value. The last column in the table says whether each interval succeeds at this job or not: 95% of 25 = 23.75, and 24 intervals do it.

When using Student's method in practice, an investigator will base his confidence interval on all his data, not just part of it. So he will not be able to check the method as done here. However, he can rely on the confidence level—because of empirical checks like this one, and because of the mathematical theory.[12]

EXERCISE SET F

1. (a) Find the multiplier t such that the area between $-t$ and t under Student's curve with three degrees of freedom equals 70%.
   (b) Find the multiplier t such that the area between $-t$ and $+t$ under Student's curve with nine degrees of freedom equals 70%.
   (c) Find the multiplier t such that the area between $-t$ and t under Student's curve with nine degrees of freedom equals 95%.

2. The first ten measurements on NB 10 averaged out to 403.0 micrograms below ten grams, with an SD of 3.0 micrograms. On the basis of these measurements, find a 95%-confidence interval for the exact weight
   (a) using the normal curve.
   (b) using Student's curve.

3. Continuing Exercise 2,
   (a) which interval is longer?
   (b) which interval would you use?

4. An investigator makes repeated measurements of the length of a metric standard. He is willing to assume the Gauss model for measurement error with negligible bias, and he believes that a histogram for the error box would follow the normal curve. However, he does not know the SD of the error box. He wants an approximate 95%-confidence interval for the exact length. Compute it for him, assuming he made
   (a) two measurements and got the result 1.0001 meters, 1.0003 meters.
   (b) ten measurements and got an average of 1.0002 meters, with an SD of 0.0001 meters.
   (c) one hundred measurements and got an average of 1.0002 meters, with an SD of 0.0001 meters.

5. Which of the following areas is largest, and which is smallest? No computations are needed, look at Figure 5.
   (i) between -1 and 1 under Student's curve for three degrees of freedom.
   (ii) between -1 and 1 under Student's curve for nine degrees of freedom.
   (iii) between -1 and 1 under the normal curve.

6. A weighing procedure makes chance errors which follow this histogram

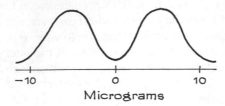

Micrograms

The average is 0, and the SD is 6 micrograms. Four measurements are made on a weight, and a confidence interval is required. Can you use the normal curve? Student's curve? In each case, say briefly why or why not. Repeat with 100 measurements. (You may assume the Gauss model, with no bias.)

7. Repeat Exercise 6, for chance errors which follow the normal curve.

The answers to these exercises are on pp. A-59–A-60.

7. REVIEW EXERCISES

Only Exercises 11 and 12 use ideas from section 6.

1. The speed of light was measured 2,500 times. The average was 299,774 kilometers per second, with an SD of 14 kilometers per second.[13] Assume the Gauss model, with no bias. Find an approximate 95%-confidence interval for the speed of light.

2. In Exercise 1, light was timed as it covered a certain distance. The distance was measured 57 times, and the average of these measurements was 1,594,265 millimeters. What else do you need to know to decide how accurate this value is?

3. Exercise 2 points to one possible source of bias in the measurements described in Exercise 1. What is it?

4. New laser altimeters can measure elevation to within a few inches, without bias, and with no trend or pattern to the measurements. As part of an experiment, 25 readings were made on the elevation of a mountain peak. These averaged out to 73,631 inches, with an SD of 10 inches. True or false:
    (a) 73,631 ± 4 inches is a 95%-confidence interval for the elevation of the mountain peak.
    (b) About 95% of the readings were in the range 73,631 ± 4 inches.
    (c) There is about a 95% chance that the next reading will be in the range 73,631 ± 4 inches.

5. True or false: "If the data don't follow the normal curve, you can't use the curve to get confidence levels." Explain your answer.

6. You may assume the Gauss model for both parts of this question. Say whether each assertion is true or false, and give a reason.
    (a) If all you have is one measurement, you can't estimate the likely size of the chance error in it—you'd have to take another measurement, and see how much it changes.

(b) If all you have is one hundred measurements, you can't estimate the likely size of the chance error in their average—you'd have to take another hundred measurements, and see how much the average changes.

7. In 1975, the average of the daily maximum temperature at San Francisco airport was 63.7 degrees, with an SD of 8 degrees (Figure 1 on p. 391). Now

$$\sqrt{365} \times 8 \text{ degrees} \approx 153 \text{ degrees}, \quad 153 \text{ degrees}/365 \approx 0.4 \text{ degrees}.$$

True or false: A 95%-confidence interval for the average daily maximum temperature at San Francisco airport is 63.7 ± 0.8 degrees. Explain.

8. A butter machine makes sticks of butter that average 4.0 ounces in weight, with an SD of 0.1 ounces. There is no trend or pattern in the weights. There are 4 sticks to a package.
   (a) A package weighs _____ give or take _____ or so.
   (b) A store buys 100 packages. Estimate the chance that they get 100 pounds of butter, to within 2 ounces.

9. In a long series of trials, a computer program is found to take on average 32 seconds of CPU time to execute, with an SD of 1 second. There is no trend or pattern in the times. It will take about _____ seconds of CPU time to execute the program 100 times, give or take _____ seconds or so. (The CPU is the "central processing unit," where the machine does logic and arithmetic.)

10. "All measurements were made twice. If two staff members were present, the duplicate measurements were made by different people. In order to minimize gross errors, discrepancies greater than certain arbitrary limits were measured a third time, and if necessary a fourth, until two measurements were obtained which agreed within the set limits. In cases of discrepancy the measurers decided which of the three or four results was most 'representative' and designated it for inclusion in the statistical record. In cases of satisfactory agreement, the statistical record was based routinely on the first measurement recorded." Comment briefly.[14]

The following two exercises depend on the ideas in section 6.

11. A calibration laboratory has been measuring a one-kilogram checkweight by the same procedure for several years. The SD of this sequence of measurements is 55 micrograms, and their histogram follows the normal curve. Someone now sends them a one-kilogram weight to be calibrated by the same procedure. If possible, find a 95%-confidence interval:
    (a) if they make one measurement, which turns out to be 183 micrograms above a kilogram.
    (b) if they make four measurements, which average 183 micrograms above a kilogram, with an SD of 47 micrograms.
    (You may assume the Gauss model, with no bias.)

12. A weighing procedure is known to make chance errors following the normal curve, but the SD is unknown. Four weighings of the same object give an average of 522 micrograms, with an SD of 12 micrograms. The following confidence interval for the exact weight is proposed:

$$522 \pm 12 \text{ micrograms.}$$

Which of the following is true? Explain briefly.
   (i)   The confidence level is just about 95%.
   (ii) The confidence level is quite a bit lower than 95%.
   (iii) The confidence level is somewhat higher than 95%.
(You may assume the Gauss model, with no bias.)

8. SUMMARY

1. According to the *Gauss model* for measurement error, each time a measurement is made, a ticket is drawn at random with replacement from the *error box*. The number on the ticket is the chance error. It is added to the exact value of the thing being measured, to give the actual measurement. The average of the error box is equal to 0. This model assumes that bias is negligible.

2. When the Gauss model applies, the SD of many repeated measurements is an estimate for the SD of the error box. This tells the likely size of the chance error in an individual measurement.

3. The average of the series is more precise than any individual measurement, by a factor equal to the square root of the number of measurements. This assumes that the data follow the model.

4. An approximate confidence interval for the exact value of the thing being measured can be found by going the right number of SEs either way from the average of the measurements; the confidence level is taken from the normal curve. The approximation is good provided there are enough measurements. The Gauss model is assumed, with no bias.

5. This method for getting confidence intervals applies even with a small number of measurements, provided the SD of the error box is known, and the numbers in the error box follow the normal curve.

6. In the frequency theory, the chance variability is in the measuring process, not the thing being measured. The word "confidence" is to remind you of this.

7. If the model does not apply, neither do these methods for getting confidence intervals. In particular, if there is any trend or pattern in the data, do not use these methods.

8. *Statistical inference* can only be justified in terms of an explicit *chance model*. Without a model, no inference is legitimate.

9. When the Gauss model applies, with no bias, and the numbers in the error box follow the normal curve reasonably well, but the SD of the error box is estimated from a small amount of current data, the method for computing confidence intervals has to be modified as follows:

- use SD^+ of the data, instead of the SD, to estimate the SD of the error box;
- use *Student's t-curve,* instead of the normal, to figure the confidence level.

The *degrees of freedom* for this procedure are one less than the number of measurements.

25

Chance Models in Genetics

I shall never believe that God plays dice with the world.

—ALBERT EINSTEIN (1879–1955)

1. HOW MENDEL DISCOVERED GENES

This chapter is hard, and it can be skipped without losing the thread of the argument in the book. It is included for two reasons:

- Mendel's theory of genetics is great science;
- the theory shows the power of simple chance models in action.

In 1865, Gregor Mendel published an article which provided a scientific explanation for heredity, and eventually caused a revolution in man's understanding of life.[1] By a curious twist of fortune, this paper was ignored for about thirty years, until the theory was simultaneously rediscovered by three men, Correns in Germany, de Vries in Holland, and Tschermak in Austria. De Vries and Tschermak are now thought to have seen Mendel's paper before they published, but Correns apparently found the idea by himself.

Mendel's experiments were all carried out on garden peas; here is a brief and simplified account of one of these experiments. Pea seeds are either yellow or green,[2] and a plant can bear seeds of both colors. Mendel bred a pure yellow strain, that is, a strain in which every plant in every generation had only yellow seeds; and separately he bred a pure green strain. He then crossed plants of the pure yellow strain with plants of the pure green strain: for instance, he used pollen from the yellows to fertilize ovules on plants of the green strain. (The alternative method, using pollen from the greens to fertilize plants of the yellow strain, gave exactly the same results.) The seeds resulting from a yellow-green cross, and the plants into which they grow, are called *first-generation hybrids*. First-generation hybrid seeds are all yellow, indistinguishable from seeds of the pure yellow strain. So the green seems to have disappeared completely.

These first-generation hybrid seeds grew into first-generation hybrid plants which Mendel crossed with themselves, producing *second-generation hybrid* seeds. Some of these second-generation seeds were yellow, but some were green. So the green disappeared for one generation, but reappeared in the second. Even more surprising, the green reappeared in a definite simple

Gregor Mendel (Austria, 1822–1884).
From the collection of the Moravian Museum, Brno.

proportion: of the second-generation hybrid seeds, about 75% were yellow and 25% were green.

What is behind this regularity? To explain it, Mendel postulated the existence of the entities now called *genes*.[3] According to Mendel's theory, there were two different variants of a gene which paired up to control seed color. They will be denoted here by *y* (for yellow) and *g* (for green). So there are four different gene-pairs: *y/y, y/g, g/y,* and *g/g.* The gene-pair controls seed color by the rule:

- *y/y, y/g,* and *g/y* make yellow
- *g/g* makes green.

As is usually said, *y* is *dominant* and *g* is *recessive*. It is the gene-pair in the seed which determines what color the seed will be, and all the cells making up a seed contain the same color gene-pair. This completes the first part of the model.

Now the seed grows up and becomes a plant; all cells in this plant also carry the seed's color gene-pair—with one exception. Sex cells, either sperm or eggs, contain only one gene of the pair.[4] For instance, a plant whose ordinary cells contain the gene-pair *y/y* will produce sperm cells each containing the gene *y;* similarly, it will produce egg cells each containing the gene *y.* On the other hand, a plant whose ordinary cells contain the gene-pair *y/g* will produce some sperm cells containing the gene *y,* and some sperm cells containing the gene *g.* In fact, half its sperm cells will contain *y,* and the other half will contain *g;* similarly, half its eggs will contain *y,* and the other half will contain *g.*

This model accounts for the experimental results. Plants of the pure yellow strain have the color gene-pair *y/y,* so the sperm and eggs all just contain the gene *y.* Similarly, plants of the pure green strain have the gene-pair *g/g,* so their pollen and ovules just contain the gene *g.* Crossing a pure yellow with a pure green amounts for instance to fertilizing a *g*-egg by a *y*-sperm, producing a fertilized cell bearing the gene-pair *y/g.* This cell reproduces itself and eventually becomes a seed, in which all the cells bear the gene-pair *y/g* and are yellow in color. The model has explained why all first-generation hybrid seeds are yellow, and none is green.

What about the second generation? Well, a first-generation hybrid seed grows into a first-generation hybrid plant, bearing the gene-pair *y/g.* In due course, this plant produces sperm cells, of which half will contain the gene *y* and the other half will contain the gene *g;* it also produces eggs, of which half will contain *y* and the other half will contain *g.* When two first-generation hybrids are crossed, each resulting second-generation hybrid seed gets one gene at random from each parent—because it is formed by the random combination of a sperm cell and an egg. From the point of view of the seed, it's as if one ticket (or gene) was chosen at random from each of two boxes (or parents): in each box, half the tickets are marked *y,* and the other half are marked *g.*

Figure 1. Mendel's chance model for the genetic determination of seed-color: one gene is chosen at random from each parent. The chance of each combination is shown. The sperm gene is listed first, but the combinations *y/g* and *g/y* are not distinguishable after fertilization.

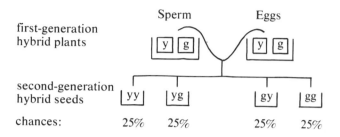

As shown in Figure 1, the seed has a 25% chance to get a gene-pair with two *g*'s and be green; it has a 75% chance to get a gene-pair with one or two *y*'s and be yellow. The number of seeds is small by comparison with the numbers of pollen grains, so the selections for the various seeds are essentially independent. The conclusion is, the color of second-generation hybrid seeds will be determined as if by a sequence of draws with replacement from the box

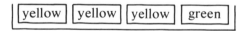

And that is how the model accounts for the reappearance of green in the second generation, for about 25% of the seeds.

Mendel made a very bold leap indeed from his experimental evidence to his theoretical conclusions. His reconstruction of the chain of heredity was based entirely on statistical evidence, of the kind discussed here. But as it turns out he was right. Most of the process whose existence he guessed at can now be photographed. Modern research in genetics and molecular biology is uncovering the chemical basis of heredity, and has provided ample direct proof for the existence of Mendel's hypothetical entities. As we know today, genes are segments of DNA on chromosomes—the dark patches in Figure 2 on the next page.

Essentially the same mechanism of heredity operates in all forms of life, from dolphins to fruit flies. So the genetic model proposed by Mendel unlocks one of the great mysteries of life. How is it that a pea-seed always produces a pea, and never a tomato, or a whale? Furthermore, the answer turns out to involve chance in a crucial way, despite Einstein's quote at the opening of the chapter.

Figure 2. Photomicrograph. These cells are from the root tip of a pea plant, and are magnified about 2,000 times. The cell shown in the center is about to divide. At this stage, each individual chromosome consists of two identical pieces, lying side by side. There are fourteen chromosomes arranged in seven homologous pairs, indicated by the Roman numerals from I to VII. The gene-pair controlling seed-color is located on chromosome pair I, one of the genes being on each chromosome.[5]

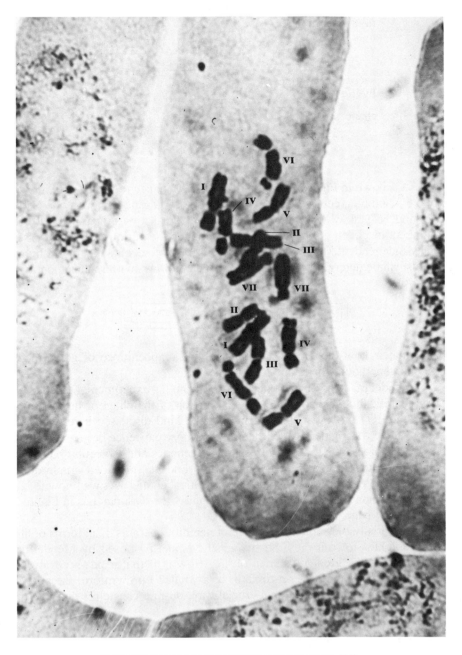

Source: New York State Agriculture Experiment Station, Geneva, N.Y.

EXERCISE SET A

1. In some experiments, a first-generation hybrid pea is back-crossed with one parent. If a *y/g* plant is crossed with a *g/g,* about what percentage of the seeds will be yellow? Of 1,600 such seeds, what is the chance that over 850 will be yellow?

2. Flower color in snapdragons is controlled by one gene-pair. There are two variants of the gene, *r* (for red) and *w* (for white). The rules are:
   > *r/r* makes red flowers,
   > *r/w* and *w/r* make pink flowers,
   > *w/w* makes white flowers.

   So neither *r* nor *w* is dominant. Their effects are *additive,* like mixing red paint with white paint.
   (a) Work out the expected percentages of red-, pink-, and white-flowered plants resulting from the following crosses: white × red, white × pink, pink × pink.
   (b) With four hundred plants from pink × pink crosses, what is the chance that between 190 and 210 will be pink-flowered?

3. Snapdragon leaves come in three widths: wide, medium, and narrow. In breeding trials, the following results are obtained:
   > wide × wide → 100% wide
   > wide × medium → 50% wide, 50% medium
   > wide × narrow → 100% medium
   > medium × medium → 25% narrow, 50% medium, 25% wide.

   (a) Can you work out a genetic model to explain these results?
   (b) What results would you expect from each of the following crosses: narrow × narrow, narrow × medium?

4. Eye color in humans is determined by one gene-pair, with brown dominant and blue recessive. In a certain family, the husband had a blue-eyed father; he himself has brown eyes. The wife has blue eyes. They plan on having three children. What is the chance that all three will have brown eyes? (It is better to work this out exactly rather than using the normal approximation.)

The answers to these exercises are on p. A-60.

2. DID MENDEL'S FACTS FIT HIS MODEL?

Mendel's discovery ranks as one of the greatest in science. Today, his theory is both amply proved and extremely powerful. But how good was his own experimental proof? Did Mendel's data prove his theory? Only too well, answered R. A. Fisher:

> . . . the general level of agreement between Mendel's expectations and his reported results shows that it is closer than would be expected in the best of several thousand repetitions. The data have evidently been sophisticated systematically, and after examining various possibilities, I have no doubt that Mendel was deceived by a gardening assistant, who knew only too well what his principal expected from each trial made.[6]

In other words, leaving the gardener aside, Fisher is saying that Mendel's data were fudged. The basis for this claim is that Mendel's observed frequencies were uncomfortably close to his expected frequencies, much closer than ordinary chance variability would permit. In one experiment, for instance, Mendel obtained 8,023 second-generation hybrid seeds. He expected ¼ × 8,023 ≈ 2,006 of them to be green, and observed 2,001, for a discrepancy of 5. According to his own chance model, that's like drawing 8,023 times with replacement from the box

$$\boxed{\;\boxed{\text{yellow}}\;\boxed{\text{yellow}}\;\boxed{\text{yellow}}\;\boxed{\text{green}}\;}$$

In this model, what is the chance of observing a discrepancy of 5 or less between the number of greens and the expected number? In other words, what is the probability that the number of greens will be

between ¼ × 8,023 − 5 ≈ 2,001 and ¼ × 8,023 + 5 ≈ 2,011?

This is like drawing 8,023 times with replacement from the box

$$\boxed{\;\boxed{0}\;\boxed{0}\;\boxed{0}\;\boxed{1}\;}$$

and asking for the chance that the sum will be between 2,001 and 2,011 inclusive. The chance can be estimated using the normal approximation, but keeping track of the edges of the rectangles, as on p. 289.

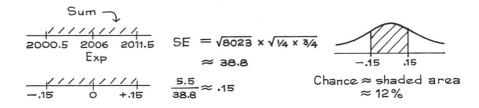

$$SE = \sqrt{8023} \times \sqrt{¼ \times ¾}$$
$$\approx 38.8$$
$$\frac{5.5}{38.8} \approx .15$$

Chance ≈ shaded area ≈ 12%

That is, about 88% of the time, chance variation would cause a discrepancy between Mendel's expectations and his observations greater than the one he reported.

By itself, this evidence is not overwhelming. The trouble is, every one of Mendel's experiments (with an exception to be discussed below) shows this kind of unusually close agreement between expectations and observations. Using the χ^2-test to pool the results (Part VIII), Fisher showed that the chance of agreement as close as that reported by Mendel is about four in a hundred thousand. To put this another way, suppose millions of scientists were busily repeating Mendel's experiments. For each scientist, imagine measuring the discrepancy between his observed frequencies and the expected frequencies by the χ^2-statistic. And suppose that Mendel's chance model is absolutely right. Then by the laws of chance, about 99,996 out of every 100,000 of these imaginary scientists would report a discrepancy between observations and expectations greater than the one reported by Mendel. So there are two possibilities:

- either Mendel's data were massaged
- or he was pretty lucky.

The first possibility is easier to believe.

One aspect of Fisher's argument deserves closer attention; however, the discussion is technical, and readers can skip to the beginning of the next section. Mendel worked with six characteristics other than seed color; one of them, for instance, was the shape of the pod, which was either inflated (the dominant form) or constricted (the recessive form). The hereditary mechanism is very similar to that for seed color: pod shape is controlled by one gene-pair. There are two variants of the shape-gene, denoted by i (inflated) and c (constricted). The gene i is dominant, so i/i or i/c or c/i make inflated pods, and c/c makes constricted pods. The gene-pair controlling seed color acts independently of the pair controlling pod shape.

There is one difference between seed color and pod shape. Pod shape is a characteristic of the parent plant, and is utterly unaffected by the fertilizing pollen. Thus, if a plant of a pure strain showing the recessive constricted form of seed pods is fertilized with pollen from a plant of a pure strain showing the dominant inflated form, all the resulting seed pods will have the recessive constricted form. But when the seeds of this cross grow up into mature first-generation hybrid plants, and make their own seed pods, they will all exhibit the dominant inflated form.

If first-generation hybrids are crossed with each other, of the second-generation hybrid plants, about 3/4 will exhibit the dominant form and 1/4 the recessive form. As Figure 1 shows, of the second-generation hybrid plants with the dominant inflated form, about

$$\frac{25\%}{25\% + 25\% + 25\%} = \frac{1}{3}$$

should be i/i's and the other 2/3 should be i/c or c/i. Mendel checked this out on 600 plants, finding 201 i/i's, a result too close to the expected 200 for comfort.[7] (The chance of such close agreement is only 10%.)

But worse is yet to come. You can't tell the i/i's from the others just by looking, the appearances are identical. So how did Mendel classify them? Well, if undisturbed by naturalists, a pea plant will pollinate itself. So Mendel took his second-generation hybrid plants showing the dominant inflated form, and selected 600 at random. He then raised ten offsprings from each of his selected plants. If the plant bred true and all ten offspring showed the dominant inflated form, he classified it as i/i; but if the plant produced any offspring showing the recessive constricted form, he classified it as i/c or c/i.

There is one difficulty with this scheme, which Mendel seems to have overlooked. As Figure 1 shows, the chance that the offspring of a self-fertilized i/c will contain at least one dominant gene i, and hence show the dominant inflated form, is 3/4. So the chance that ten offsprings of an i/c crossed with itself will all show the dominant form is $(3/4)^{10}$. And that's a fairly substantial chance, about 6%. Similarly for c/i's. With this in mind, the expected frequency of plants classified as i/i is a bit higher than 200, because about 6% of the 400 i/c's and c/i's will be incorrectly classified as i/i. Taking this chance of misclassification into account, the expected frequency of plants classified as i/i, correctly or incorrectly, is

$$200 + 0.06 \times 400 = 224.$$

Mendel's observed frequency (201 classified as i/i) is rather too far from this corrected expectation. The chance of such a large discrepancy is about 5%. As Fisher concludes: "There is no easy way out of the difficulty."

3. THE LAW OF REGRESSION

Part III discussed Galton's work on heredity, and presented his finding that on the average a child is halfway between the parent and the average. In 1918, Fisher proposed a chance model[8] based on Mendel's ideas, which explained Galton's finding on regression, as well as the approximate normality of many biometrical characteristics like height (Chapter 5). The model can be made quite realistic, at the expense of introducing many complications. This section begins with a stripped-down version, necessarily unrealistic, but easier to understand. Later, some of the possible refinements will be mentioned. The model will focus on heights, although exactly the same argument could be made for other characteristics.

The first assumption in the model is

(1)   height is controlled by one gene-pair.

The second assumption is

(2)   the genes controlling height act in a *purely additive* way.

The symbols h^*, h^{**}, h', h'' will be used to denote four typical variants of the height-gene. Then assumption (2) means for instance that h^* always contributes a fixed amount to an individual's height, whether it is combined with another h^*, or with an h', or whatever. These genes act very differently from the y's and g's controlling seed color in Mendel's peas: g contributes green to the seed color when it is combined with another g, but when it is combined with a y it has no effect. The height genes act much more like the snapdragon genes in Exercises 2 and 3 on p. 425 above.

On assumption (2), each gene contributes a fixed amount to an individual's height. This contribution (say in inches) will be denoted by the same letter as used to denote the gene, but in capitals. Thus, an individual with the gene-pair h/h' will have height equal to the sum $H + H'$. In the first instance, the letters refer to the genes; in the second, to their contributions to height.

Fisher assumed, with Mendel, that a child gets one gene of the pair controlling height at random from each parent, as depicted in Figure 3. To be more precise, the father has a gene-pair controlling height, and so does the mother. Then one gene is drawn at random from the father's pair and one from the mother's pair to make up the child's pair.

For the sake of argument, suppose the father has the gene-pair h^*/h^{**}, and the mother has the gene-pair h'/h''. The child has chance 1/2 to get h^* and chance 1/2 to get h^{**} from the father. Therefore, the father's expected contribution to the child's height is $\frac{1}{2}H^* + \frac{1}{2}H^{**} = \frac{1}{2}(H^* + H^{**})$,

Figure 3. The simplified Mendel-Fisher model for the genetic determination of height. Height is controlled by one gene-pair, with purely additive genetic effects. One gene is drawn at random from the gene-pair of each parent to make up the child's pair.

namely one-half the father's height. And this is where Galton's regression coefficient of one-half comes from. Similarly, the mother's expected contribution equals one-half her height. In other words, taking a large number of children of parents where the father's height is fixed at one level, and the mother's height is fixed at another level, then the average height of these children must be about equal to

(3)   ½(father's height + mother's height).

The expression (3) is called the *mid-parent height*. For instance, with many families where the father is 72 inches tall and the mother is 68 inches tall, the mid-parent height is $\frac{1}{2}(72'' + 68'') = 70''$, and on the average the children will be about 70 inches tall at maturity, give or take a small chance error. And this is the biological explanation for the statistical regularity that Galton found—the law of regression to mediocrity.

The assumption (1), that height is controlled by one gene-pair, isn't really needed in the argument; it was made to avoid complicated sums. If for instance three gene-pairs are involved, it is only necessary to assume additivity of the genetic effects and randomness in drawing one gene from each pair for the child, as depicted in Figure 4.

Figure 4. The simplified Mendel-Fisher model for the genetic determination of height, assuming three gene-pairs with purely additive effects. One gene is drawn at random from each gene-pair of each parent to make up the corresponding gene-pair of the child.

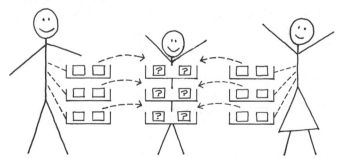

So far, the model has not taken into account sex-differences in height. One way to get around this is by "adjusting" women's heights, increasing them by around 8% so they are just as tall as men. More elegant (and more complicated) methods are available too.

How well does the model fit? For the Pearson-Lee study discussed in Part III, the regression of son's height on parents' heights was approximately[9]

(4)   estimated son's ht. $= 15'' + 0.8 \times \dfrac{\text{father's ht.} + 1.08 \times \text{mother's ht.}}{2}$

The regression coefficient of 0.8 is noticeably lower than the 1.0 predicted by a purely additive genetic model. Some of the discrepancy may be due to environmental effects, and some to nonadditive genetic effects. It should also be remembered that the sons averaged 1 inch taller than the fathers. This too cannot be explained by a purely additive genetic model.[10] (There were 1,078 families in the study, so a chance variation of 1 inch in the average is very unlikely.)

The correlation between the heights of the fathers and mothers was about 0.25; so the regression of son's height on father's height was very nearly

(5)   estimated son's height $= 35'' + 0.5 \times$ father's height.

Now equation (5) can also be derived from equation (3) in the additive model, by assuming that there is no correlation between the heights of the parents.[11] Basically, however, this is a case of two mistakes cancelling. The additive model is a bit off, and the heights of parents are somewhat correlated; but these two facts work in opposite directions, and balance out in equation (5).

Technical note: To derive equation (3) from the model, no assumptions are necessary about the independence of draws from different gene-pairs; all that mattered was each gene having a 50% chance to get drawn. No assumptions are necessary about statistical relationships between the genes in the different parents (such as independence). And no assumptions are necessary about the distribution of the genes in the population (like equilibrium).

4. AN APPRECIATION OF THE MODEL

Genetics represents one of the most satisfying applications of statistical methods. To review the development, Mendel found some striking empirical regularities—like the reappearance of a recessive trait in one-fourth of the second-generation hybrids. He made up a chance model involving what are now called genes to explain his results. He invented these genes by pure reasoning—he never saw any. Independently, Galton and Pearson found another striking empirical regularity: on the average, a son is halfway between his father and the overall average for sons.

At first sight, the Galton-Pearson results look very different from Mendel's, and it is hard to see how they can both be explained by the same biological mechanism. But Fisher was able to do it. He explained why the average height of children equaled mid-parent height, and even why there are deviations from average. These deviations are caused by chance variation, when genes are chosen at random to pass from the parents to the children.

Chance models are now used in many fields. But usually, they only assert that certain entities behave as if they were determined by drawing tickets at random from a box, and little effort is spent establishing a physical basis for this claim of randomness. These models seldom say explicitly what is like the box, or what is like the tickets. The genetic model is quite unusual, in that it answers these questions.

Briefly, there are two main sources of randomness in the model: one being the random allotment of chromosomes (one from each pair) to sex cells; and the other being the random pairing of sex cells to produce the fertilized egg. The two sources of randomness will now be discussed in more detail.

Figure 5. Production of sex-cells and body-cells by splitting. Chromosomes are denoted here by capital letters, like C. Chromosomes come in homologous pairs, as indicated by primes. Thus, C and C' form a homologous pair; they are chemically similar, but not identical. As a preliminary to splitting, a cell doubles all its chromosomes. When doubled, C will be denoted by C-C. The two pieces are chemically identical, and loosely attached. Similarly, C' doubles to C'-C'.

(a) Splitting to make body cells   (b) Splitting to make sex cells

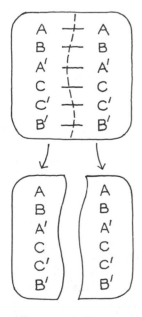

Chromosomes naturally come in *homologous* pairs. The one matching C is denoted C'; the chromosomes C and C' are chemically similar, but not identical. A gene-pair has one gene located on each chromosome of a homologous chromosome-pair. A body cell can divide to form other cells. As a preliminary step, each chromosome in the parent cell doubles itself, as shown in Figure 2 and (schematically) in Figure 5. When chromosome C is in this doubled condition, it will be denoted C-C. The two pieces are chemically identical and loosely joined together. The production of ordinary body cells is shown in Figure 5a. The parent cell splits in two. Each fragment becomes a separate cell with one-half of each doubled chromosome, winding up with exactly the same complement of chromosomes as the parent cell (before doubling). There is nothing random about the resulting chromosomes—it is a matter of copying the whole set. Homologous chromosomes are not treated in any special way.

The production of sex cells is shown in Figure 5b. The doubled chromosomes move into position with one doubled chromosome from each homologous pair on opposite sides of the line along which the cell will split. Which side of the line? This seems to be random, like coin-tossing. Sometimes one side, sometimes the other, just as a coin sometimes lands heads, sometimes tails. In the model, it is assumed to be random. The result of this first step is shown at the top of Figure 5b. Next, the cell splits as shown in the middle of Figure 5b. Each fragment contains doubled chromosomes—but only one chromosome of each homologous pair is represented. Finally, each of these fragments splits again, as shown at the bottom of Figure 5b, and the results of the second split are the sex cells.[12] The lining-up of the homologous pairs at the top of Figure 5b is a critical step. The result is that a sex cell contains ordinary, undoubled chromosomes—but only one chromosome out of each homologous pair. Which one? One chosen at random. This is one physical source of randomness in Mendelian genetics.

A fertilized egg results from the union of one male sex cell and one female, out of the many which are produced. Which ones? This seems to be random, like drawing tickets at random from a box. In the model, it is assumed to be random. This is the second main physical source of randomness in Mendelian genetics.

When thinking about any other chance model, it is good to ask two questions:

- What are the physical entities which are supposed to act like the tickets and the box?
- Do they really act like that?

5. REVIEW EXERCISES

1. Mendel discovered that for peas, the unripe pods are green or yellow. Their color is controlled by one gene-pair, with variants g for green and y for yellow, g being dominant. In a set of breeding trials, plants with known pod color but unknown genetic makeup were crossed. The results are tabulated below. For each line of the table, guess the genetic makeup of each parent—g/g, g/y or y/g, y/y.[13]

Pod color of parents	Number of progeny with	
	green pods	yellow pods
green × yellow	82	78
green × green	118	39
yellow × yellow	0	50
green × yellow	74	0
green × green	90	0

2. Mendel found that pea seeds were either smooth or wrinkled. He bred a pure smooth strain and a pure wrinkled strain. Interbreeding these two strains gave first-generation hybrids, which all turned out to be smooth. Mendel crossed the first-generation hybrids with themselves to get second-generation hybrids; of 7,324 second-generation hybrid plants, 5,474 turned out to be smooth, and 1,850 were wrinkled. Make up a genetic model to account for these results. In the model, what is the chance of agreement between the expected frequency of smoothies and the observed frequency as close as that reported by Mendel?

3. Peas flower at three different times: early, intermediate, and late.[14] Breeding trials gave the following results:

early × early → early
early × late → intermediate
late × late → late.

What results would you expect from the cross

intermediate × intermediate?

If you had 2,500 plants resulting from the intermediate × intermediate cross, what is the chance that 1,300 or more would be intermediate-flowering?

4. In humans, there is a special chromosome-pair which determines sex. Males have the pair X-Y, while females have the pair X-X. A child gets one X-chromosome automatically, from the mother; from the father, it has half a chance to get an X-chromosome and be female, half a chance to get Y and be male. Some genes are carried only on the X-chromosome; these are said to be *sex-linked*. An example is the gene for male-pattern baldness. (Color blindness and hemophilia are other sex-linked characteristics.)

   (a) If a man has a bald father, is he more likely to go bald?
   (b) If a man's maternal grandfather was bald, is he more likely to go bald?

Explain briefly.

5. Sickle cell anemia is a genetically controlled disease in humans. In the United States, it is especially prevalent among blacks, where one person in four hundred suffers from it. The disease is controlled by one gene-pair, with variants A and a, where a causes the disease but is recessive:

A/A, A/a, a/A—healthy person
a/a—sickle-cell anemia.

   (a) Suppose one parent has the gene-pair A/A. Can the child have sickle-cell anemia? How?

(b) Suppose neither parent has sickle-cell anemia. Can the child have it? How?

(c) Suppose both parents have sickle-cell anemia. Can the child avoid having it? How?

6. SUMMARY

1. Whenever reproduction is sexual, the mechanism of heredity is based on gene-pairs. The offspring gets one gene of each pair drawn at random from the corresponding pair in the maternal organism, and one at random from the corresponding pair in the paternal organism. The two genes in a pair are very similar, but not identical.

2. Gene-pairs can control biological characteristics in several ways. One way is *dominance*. In this case, there may be only two varieties (alleles) of the gene, say *d* and *r*. The gene-pairs *d/d, d/r, r/d* all produce the dominant characteristic, while *r/r* produces the recessive characteristic. (Seed color in peas is an example.) Another way is *additivity*. In this case, each variety of the gene has an effect, and the effect of the gene-pair is the sum of the two effects. (Flower color in snapdragons is an example.)

3. Fisher showed that Galton's law of regression was a mathematical consequence of this genetic mechanism, assuming only additive effects.

4. The genetic model explains (at least part of the reason) why children resemble their parents, and also why they differ.

PART VIII

Tests of Significance

26

Tests of Significance

Who would not say that the glosses [commentaries on the law] increase doubt and ignorance? It is more of a business to interpret the interpretations than to interpret the things.

—MICHEL DE MONTAIGNE (FRANCE, 1533–1592)[1]

1. INTRODUCTION

Was it just due to chance, or something else? Statisticians have invented *tests of significance* to deal with this sort of question. These tests are among the statistical techniques most widely used in the social sciences. In fact, it is almost impossible to read a research article in the social sciences without running across them. Therefore, it is a good idea to find out what they mean. As Montaigne suggests, the main difficulty is in the language. Here is a passage which exemplifies the problem:

With a one-tailed z-test, the null hypothesis was rejected at the 5% level. The difference must be regarded as statistically significant.

The object in Chapters 26 through 28 is to explain the ideas behind tests of significance, and some of the language. Some of the limitations will be pointed out in Chapter 29. This section presents an example, to bring out the main idea.

Many companies are experimenting with "flex-time," allowing employees to choose their schedules within broad limits set by management.

Among other benefits, flex-time is supposed to reduce absenteeism. Of course, a firm with many thousands of employees might have a hard time studying them, except on a sample basis. Suppose one firm knows that in each of the past few years, employees have averaged 6.3 days off from work (apart from vacations). This year, the firm introduces flex-time. Management chooses a simple random sample of 100 employees to follow in detail over the year. At the end of the year, these sample employees are found to have averaged only 5.4 days off work. The personnel committee is now meeting to assess this result.

Investigator.   Among other benefits, flex-time reduced absenteeism. On the average, days off work dropped from 6.3 to 5.4 this year.

Chance skeptic.   Now look, your sample only amounted to 100 employees out of the 5,000 at the plant. By the luck of the draw, you just happened to get too many hard workers in your sample. All you're showing us is a chance variation.

Investigator.   Not true.

Chance skeptic.   What was the SD in your sample?

Investigator.   Days off averaged 5.4, with an SD of about 3 days.

Chance skeptic.   Well, the difference between 6.3 and 5.4 days is a lot less than one SD. To me, that looks like a chance variation.

Investigator.   OK. I want to make a calculation, and for that I need a box model. There are 5,000 tickets in the box, one for each employee, showing the number of days taken off work this year. We drew 100 tickets at random from the box, for our sample. What we really know for sure is these 100 tickets. Agreed?

Chance skeptic.   Agreed.

Investigator.   And what we're arguing about is the average of the 5,000 numbers in the box. You say that's still 6.3 days, and I say it's gone down.

Chance skeptic.   Right. What I say is that by the luck of the draw, you got too many small numbers in your sample.

Investigator.   I understand. One more thing. We don't know the SD of the box, but we can agree to estimate it by the SD of the sample—3 days. OK?

Chance skeptic.   That seems reasonable to me.

Investigator.   Fine. Now, the SE for the sum of 100 draws from the box is $\sqrt{100} \times 3 = 30$ days, so the SE for their average is $30/100 = 0.3$ days.

Chance skeptic.   So?

Investigator.   So, suppose you're right and the average of the box is still 6.3 days. Then, you have to expect the average of the sample to be around 6.3 days—and you're off by 3 SEs:

$$\frac{5.4 - 6.3}{0.3} = -3.$$

Chance skeptic. Hmmm.

Investigator.   We have enough draws here so that we can use the normal approximation on the probability histogram for the average of the draws. The area to the left of −3 under the normal curve is about 0.1 of 1%.

Chance skeptic. That's true.

Investigator.   Now you have a hard choice to make. Either you can say that the average of the box is still 6.3 days, or you can agree with me that the average of the box went down. But if you stick with the 6.3 days, you need a minor miracle to explain our data—there's only one chance in a thousand of being that far below your expected value.

Chance skeptic. So what do you think the average of the box is?

Investigator.   I estimate it as 5.4 ± 0.3 days. We have a small reduction in days off work here, but a real one. I mean you can't just dismiss it as a chance variation.

Chance skeptic. Does this prove that flex-time reduces absenteeism?

Investigator.   No, but we've proved that absenteeism really went down. It wasn't just the luck of the draw.

Our first pass at the example is now complete. The issue in the example comes up over and over again: the investigator thinks he has a real difference, but a skeptic might say it's all chance variation. The investigator tries to forestall the skeptic by making a chance calculation, as in the dialog: this calculation is called a *test of significance*. The next two sections will pick out the key ingredients in this calculation, and say what their technical names are.

EXERCISE SET A

1. In the example of this section, suppose the average number of days off work in the sample had turned out to be 5.9 with the same SD of 3 days; would you be willing to believe that the average for all the employees was still 6.3 days?

2. A die is rolled 100 times. The total number of spots is 368, instead of the expected 350. Can this be explained as a chance variation, or is the die loaded?

3. A die is rolled 1,000 times. The total number of spots is 3,680, instead of the expected 3,500. Can this be explained as a chance variation, or is the die loaded?

The answers to these exercises are on p. A-61.

2. THE NULL AND THE ALTERNATIVE

This section will introduce the *null hypothesis* and the *alternative hypothesis,* and will discuss the role of models in testing. In the example of the previous section, the investigator and the chance skeptic had some data, for 100 employees. Both saw the difference between the sample average of 5.4 days and the old average of 6.3 days. What they were arguing about was the interpretation of this difference. In statistical shorthand, the investigator claimed the difference between the sample average and the old average was "real." This may sound odd: of course there's a difference between 5.4 and 6.3 days. But the question was whether this difference just reflected chance variation, as the skeptic said, or whether it indicated a real change—that on the average, absenteeism for all 5,000 employees had gone down.

The argument between the chance skeptic (who says it's only chance variation) and the investigator (who says it's a real change) is so common that the two positions have been given official names:

•   The *null hypothesis* is that an observed difference (like the one between the sample average and the old average) just reflects chance variation. This is the position of the chance skeptic in the dialog.

•   The *alternative hypothesis* is that the observed difference is real. This is the position of the investigator in the dialog.

The names are unsettling, because the "alternative hypothesis" is usually the main one: what the investigator wants to prove. The "null" hypothesis is what most people would consider an alternative, and dull, explanation for the findings, in terms of chance variation. However, there is no help for it, the terminology is completely standard.

In order to persuade the chance skeptic to abandon the null hypothesis, the investigator set up a box model for the problem. He translated the null hypothesis and the alternative into numerical statements about the box:

•   null hypothesis—the average of the box equals 6.3 days;
•   alternative hypothesis—the average of the box is less than 6.3 days.

> To make a test of significance, the null hypothesis has to be formulated as a statement about a box model. Usually, the alternative does too.

In the dialog, the important thing was the average of the box, not the average of the sample. So the argument was about the box, not the sample. A test of significance only makes sense in a debate about the numbers in the box. (This point will be discussed again in Chapter 29, section 5.)

> Every legitimate test of significance involves a box model. The test gets at the question of whether an observed difference is real, or just a chance variation. A real difference is one that says something about the box, and doesn't just reflect a fluke in the sampling.

EXERCISE SET B

1.   True or false: The alternative hypothesis is that nothing is going on, besides chance variation.

2.   True or false: In order to make a statistical test of a null hypothesis, this hypothesis has to be translated into a statement about a box model.

3. A hundred draws are made at random with replacement from a box. The average of the draws is 102.7, with an SD of 10. Someone claims that the average of the box equals 100. Is this plausible?

The answers to these exercises are on p. A-61.

3. TEST STATISTICS AND SIGNIFICANCE LEVELS

The object of this section is to introduce *test statistics* and *significance levels*. In the dialog of section 1, the investigator made a box model for the data. He formulated the chance skeptic's position as the null hypothesis about the model—that the average of the box was still 6.3 days. Then, he wanted to show that this null hypothesis would lead to a ridiculous conclusion. For the sake of the argument, he temporarily assumed the null hypothesis to be right. On this basis, he calculated how many SEs away the observed value of the sample average was from its expected value:

$$\frac{5.4 - 6.3}{0.3} = -3.$$

This is an example of a *test statistic*. Notice that the expected value for the sample average (6.3 days) was calculated assuming the null hypothesis to be right.

> A test statistic is used to measure the difference between the data and what is expected on the null hypothesis.

The test statistic used in section 1 can be put in shorthand:

$$\frac{\text{observed} - \text{expected}}{\text{SE}}.$$

This ratio is usually abbreviated as "z," and tests using it are called *z-tests*. Remember the interpretation: z says how many SEs away an observed value is from its expected value. The expected value is always calculated using the null hypothesis—that is, assuming the chance skeptic to be right.

> The z-statistic converts the observed value to standard units, on the basis of the null hypothesis.[1a]

With flex-time, the test statistic worked out to -3: the observed value of the sample average (5.4 days) was 3 SEs below what was expected on the null hypothesis (6.3 days). This -3 stopped the chance skeptic in his tracks. Why was it so intimidating? After all, three is not a very big number. The answer, of course, is that the area to the left of -3 under the normal curve is ridiculously small. The skeptic's chance of getting a sample average 3 SEs or more below his expected value was 1 in 1,000—just about nil. It is the smallness of this chance which overwhelmed the skeptic, and forced him to concede that the average of the box had dropped below 6.3 days. This sort of chance is called an *observed significance level*, and it should be computed as part of any test of significance.

> The observed significance level is the chance of getting a test statistic as extreme as or more extreme than the observed one. The chance is computed on the basis that the null hypothesis is right. The smaller this chance is, the stronger the evidence against the null.

The observed significance level is usually denoted by P, for probability, and it is often called the *P-value* of the test. In the flex-time example, P was 1 in 1,000.

At this point, the logic of the test can be seen more clearly. It is an argument by contradiction, designed to show that the null hypothesis will lead to an absurd conclusion and must therefore be rejected. You look at the data, compute the test statistic, and then the observed significance level. Take, for instance, a P of 1 in 1,000. To interpret this number, you begin by assuming that the null hypothesis is right, and you imagine many other investigators repeating your experiment. What the 1 in 1,000 says is that your test statistic is really far out: only one investigator in a thousand would get a test statistic as extreme as, or more extreme than, the one you got. So the difference between your data and what is expected on the null hypothesis would be very difficult to explain as a chance variation. This is almost a contradiction, and it makes you want to abandon the null hypothesis. In general, the smaller the observed significance level is, the more you want to reject the null hypothesis—it is involving you in absurdities. Statisticians use the phrase "reject the null" to emphasize the point that with a test of significance, the argument is by contradiction.[2]

An argument by contradiction is complicated, and it is tempting to think that P gives the probability of the null hypothesis being right. Do not be misled. No matter how often you do the draws, the null hypothesis about the average of the box is either always right, or always wrong. According to the frequency theory, there is no question of chance here. (A similar point for confidence intervals is discussed on p. 347.)

> The observed significance level gives the chance of getting evidence against the null hypothesis as strong as the evidence at hand—or stronger. It does not give the chance of the null hypothesis being right. In fact, P is computed on the basis of the null hypothesis.

The calculation of P in the flex-time example will now be reviewed very briefly, to illustrate the terminology. The observed value of the test statistic was -3. Of course, if the data had turned out differently, the test statistic would have also. For instance, if the sample average had been 5.3 days, with an SD of 2.9 days,

$$z = \frac{5.3 - 6.3}{0.29} \approx -3.4$$

This is even stronger evidence against the chance skeptic. On the other hand, if the sample average had been 5.5 days, with an SD of 3.1 days,

$$z = \frac{5.5 - 6.3}{0.31} \approx -2.6$$

This is weaker evidence. In other words, it is the values to the left of -3 which are more extreme than the observed value of the test statistic.

To compute P, we assume that the null hypothesis is right—the average of the box is still 6.3 days. We consider the experiment of drawing 100 tickets at random from the box, and computing the test statistic

$$z = \frac{\text{average of 100 draws} - 6.3 \text{ days}}{\text{SE for average of 100 draws}}$$

Then P is the chance that z will turn out to be -3 or less. This chance is about equal to the area to the left of -3 under the normal curve, because there are enough draws to use the normal approximation on the probability histogram for the average of the draws, and because z has already converted the average of the draws to standard units.

$$P \approx \text{} \approx 1 \text{ in } 1{,}000$$

(From the table, this area is 0.135 of 1%. Rounding off, the P-value is about 0.1 of 1%. This is 0.1 of 0.01 = 0.001 = 1/1,000.)

It is the null hypothesis which told us to subtract off 6.3 days, and not some other number, in the numerator of z. This is the exact point where the null hypothesis was used in figuring P.

The procedure for calculating z and P can be summarized as follows:

$$z = \frac{\text{observed} - \text{expected}}{\text{SE}}, \quad P \approx$$

This procedure is used for reasonably large samples. With small samples, statisticians prefer a slightly different technique (the t-test, p. 463).

The z-test assumes that the data are like the result of drawing from a box. If there is any trend or pattern in the data, or any dependence, the formulas do not apply and will give misleading answers.

EXERCISE SET C

1. In the flex-time example, suppose the sample average were 5.5 days, with an SD of 2.9 days. Compute the test statistic, and the observed significance level.

2. Repeat Exercise 1 for a sample average of 5.5 days and an SD of 2.7 days.

3. True or False: The observed significance level depends on the data.

4. True or false: If the observed significance level is 1%, there is only 1 chance in 100 for the null hypothesis to be right.

5. According to one investigator's model, his data are like 400 draws made at random from a large box. The null hypothesis is that the average of the box equals 50; the alternative is that the average of the box is more than 50. In fact, his data averaged out to 52.7, with an SD of about 25. Compute z and P. What should he conclude?

The answers to these exercises are on p. A-61.

4. MAKING A TEST OF SIGNIFICANCE

Making a test of significance is a fairly complicated job. To begin with, there is some data. Then, an investigator has to go through the following steps:

- making a box model for the data;
- formulating the null hypothesis (and usually the alternative hypothesis) as statements about the box model;
- defining a test statistic to measure the difference between the data and what is expected on the null hypothesis;
- computing the observed significance level, P.

This completes the test.

The choice of test statistic depends on the model and the hypotheses being considered. The tests discussed in this chapter are "one-tailed z-tests," and are based on the z-statistic. (Two-tailed tests will be discussed on p. 496.) There are also "t-tests," based on the t-statistic (p. 464), χ^2-tests based on the χ^2-statistic (p. 472), and many others not even mentioned in this book. However, all tests follow the steps outlined above, and their P-values can be interpreted in the same way as with the z-test.

After making the test, the investigator has to decide whether or not to reject the null hypothesis—that is, to reject chance as the explanation for an observed difference. Finally, if he concludes that the difference is real, he has to decide what it means in terms of the problem he is studying. (These questions will be discussed again in Chapter 29, sections 1 and 2.)

At this point, it is natural to ask how small the observed significance level has to be before an investigator should reject the null hypothesis. Many statisticians draw a line at 5% or 1%.

- If P is less than 5%, the result is called *statistically significant*.

There is another line at 1%:

- If P is less than 1%, the result is called *highly statistically significant*.

Section 3 of Chapter 29 will take up the question of whether these lines make sense.

EXERCISE SET D

1. True or false: If the observed significance level is 4%, then the result is "statistically significant."

2. True or false: If the *P*-value of a test is 1.1%, the result is "highly statistically significant."

3. True or false: A "statistically significant" result cannot possibly be explained by chance.

4. True or false: If a difference is "highly statistically significant," there is less than a 1% chance for the null hypothesis to be right.

5. True or false: If a difference is "highly statistically significant." then *P* is less than 1%.

6. True or false: If *P* is 43%, the null hypothesis looks plausible.

The answers to these exercises are on p. A-61.

5. ZERO-ONE BOXES

The *z*-test can also be used when the situation involves classifying and counting. It is a matter of putting 0's and 1's in the box (p. 267). This section will give two examples. These examples will also have a new feature: the alternative hypothesis can't be translated into a statement about the box model, as will be explained below.

In 1973, Charles Tart ran an experiment at the University of California, Davis, to demonstrate ESP.[3] Tart used a machine called the "Aquarius." The Aquarius has an electronic random number generator, and four "targets." Using its random number generator, the machine picks one of the four targets at random: it does not indicate which one. Then, the subject guesses which target was chosen, by pushing a button. Finally, the machine lights up the target it chose, ringing a bell if the subject guessed right. The machine keeps track of the number of trials, and the number of correct guesses.

Tart selected 15 subjects who had previously shown clairvoyant abilities. Each of the subjects made 500 guesses on the Aquarius, for a total of $15 \times 500 = 7{,}500$ guesses. Out of this total, 2,006 were correct. Now even if the subjects had no clairvoyant abilities whatsoever, they would still be right about one fourth of the time. In other words, about $1/4 \times 7{,}500 = 1{,}875$ correct guesses are expected, just by chance. True, there is a surplus of $2{,}006 - 1{,}875 = 131$ correct guesses, but can't this be explained as a chance variation?

Tart could (and did) defend himself against this charge by making a test of significance. The null hypothesis is that the surplus of correct guesses is just a chance variation—no ESP is involved. To translate this into a box model, we assume that the Aquarius generates numbers by a perfectly random process, so that on each trial each of the four targets has exactly 1 chance in 4 to be chosen, and we assume that there is no ESP. Then each guess has exactly 1 chance in 4 to be correct. The data consist of a record of the 7,500 guesses, showing whether each is correct or not. On the null hypothesis, this is like 7,500 draws from the box

$$\boxed{0}\;\boxed{0}\;\boxed{0}\;\boxed{1}$$ 0 = incorrect, 1 = correct

On the null hypothesis, the number of correct guesses is like the sum of 7,500 draws from the box, completing the model. The machine is classifying each guess as correct or incorrect, and counting the number of correct guesses. That is why a zero-one box is needed.

There is no sensible way to translate the alternative hypothesis into a box model. This has nothing to do with zero-one boxes. The reason is that if the subjects do have ESP, the chances for each guess to be correct may well depend on the outcomes of previous trials, and may change from trial to trial. Then the data will not be like draws from any box.[4]

Once the null hypothesis has been translated into a box model, it is easy to test. The average of the box is 0.25, with an SD of $\sqrt{0.25 \times 0.75} \approx 0.43$. The expected value for the sum of 7,500 draws is $7{,}500 \times 0.25 = 1{,}875$, with an SE of $\sqrt{7{,}500} \times 0.43 \approx 37$. The observed value of 2,006 is 3.5 SEs above the expected value:

$$z = \frac{\text{observed} - \text{expected}}{\text{SE}} = \frac{2{,}006 - 1{,}875}{37} \approx 3.5$$

So P is rather small:

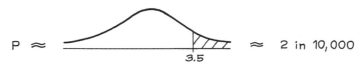

$$P \approx \qquad \approx \quad 2 \text{ in } 10{,}000$$

This is decisive evidence against the null hypothesis. The surplus of correct guesses is very hard to explain away as a chance variation. Of course, this doesn't prove that ESP exists. For example, the Aquarius random number generator may not be very good.[5] Or the machine may be giving the subject some subtle clues as to which target it picked. There may be many reasonable explanations for the results, besides ESP. But chance variation isn't one of them. That is what the test of significance shows, finishing the ESP example.

This test used the same z-statistic as before:

$$z = \frac{\text{observed} - \text{expected}}{\text{SE}}$$

To work z out correctly in a problem, however, you have to pay attention to the details. In some cases, it is hard to decide how to compute the SE. To get started, first decide whether the data are quantitative or not. Then set up the box model. In the quantitative case, you have to decide whether to work with the sum of the draws, or their average. You also have to decide whether the SD of the box is given, or must be estimated from the data.

If the problem involves classifying and counting, remember to put 0's

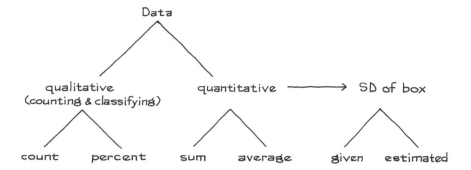

and 1's in the box. Then, you have to decide whether to work with numbers or percents. However, the SD of a zero-one box is often given implicitly—by the null hypothesis. The reason is that the null hypothesis usually specifies the fraction of 1's in the box. For instance, in the ESP example, this fraction was specified as 1/4. The fraction of 0's is then automatically determined:

$$\text{fraction of 0's} = 1 - \text{fraction of 1's.}$$

Now, the SD is

$$\sqrt{(\text{fraction of 1's}) \times (\text{fraction of 0's}).}$$

The balance of this section will be spent on the Salk vaccine field trial (p. 3). To review briefly, there were about 400,000 children involved in the randomized controlled experiment. Each child was assigned to treatment (the vaccine) or control (a placebo) by the toss of a coin.[6] As it turned out, 57 children in the treatment group got polio, compared to 142 in the control group. This was interpreted as decisive evidence that the vaccine worked. But couldn't it be explained otherwise—as a chance variation? After all, 57 and 142 are tiny compared to 400,000.

Some thought is needed to fit this problem into the framework of this chapter. Suppose that assignment to treatment or control had absolutely no effect on the outcome—the devil's advocate position. Altogether there were 57 + 142 = 199 polio cases among the 400,000 children in the experiment. The devil's advocate position is that these 199 children were doomed to get polio, whichever group they were assigned to. On this basis, the smaller number of polio cases in the treatment group just reflects chance variation in the coin-tossing.

This null hypothesis will now be put to the test, and the first step is making a box model. Focus on the 199 children who were doomed to get polio. Each time the coin was tossed for one of them, there was a 50% chance for the child to be assigned to treatment, and therefore a 50% chance for the count of polio cases in the treatment group to go up by 1. Likewise, there was a 50% chance for this count to stay the same. On the null hypothesis, then, the count of polio cases in the treatment group is like the sum of 199 draws from the box

0 = assignment to control   1 = assignment to treatment

Now it is just like the ESP example. Again, there is no reasonable way to translate the alternative hypothesis into a statement about the box. For instance, if you put more 0's in the box, the chance of assignment to control will rise above 50%; that isn't right.

Once the null hypothesis has been set up as a box model, it is easy to test. On the null hypothesis, the number of cases in the treatment group is like the sum of 199 draws from the box, and is expected to be around $199 \times 0.5 = 99.5$, with an SE of $\sqrt{199} \times 0.5 \approx 7$. The test statistic is

$$z = \frac{\text{count of polio cases in treatment} - \text{expected count}}{\text{SE}}$$

$$= \frac{57 - 99.5}{7} \approx -6.$$

Then

$$P \approx \text{[curve figure with shaded region at } -6] \approx 1 \text{ in } 1,000,000,000$$

(This P can't be read off the table; it was worked out by a more advanced method.)[6a]

The null hypothesis was that the difference between cases in the treatment and the expected 99.5 just represented chance variation in the coin-tossing. This hypothesis did not survive the test. The Salk vaccine field trial was so carefully designed that only one other explanation seems possible—the vaccine worked.

EXERCISE SET E

1. As part of a statistics project, one day Mr. Frank Alpert approached the first 100 students he saw on Sproul Plaza at the University of California, Berkeley, and found out the school or college in which they were enrolled. His sample included 53 men and 47 women. From Registrar's data, it was known that 67% of the students at Berkeley that quarter were male. Is his sampling procedure like taking a simple random sample?
   (a) Formulate the null hypothesis as a box model (there is no alternative about the box).
   (b) Compute z and P.
   (c) What do you conclude?

2. A coin is tossed 10,000 times, and it lands heads 5,167 times. Is the chance of getting heads on any one toss equal to 50%? Or are there too many heads for that?
   (a) Formulate the null and alternative hypotheses on terms of a box model.
   (b) Compute z and P.
   (c) What do you conclude?

3. Repeat Exercise 2 if the coin lands heads 5,067 times, as it did for Kerrich (p. 240).

4. A vocabulary test for six-year-old children is standardized on a large nationwide sample to have an average score of 50 out of 100, with an SD of 10. School

authorities in Maine choose a statewide simple random sample of 400 six-year-olds. These children average 51.3 on the test, with an SD of 12. Is it safe to infer that if the test had been administered to all six-year-olds in Maine, they would have averaged above 50? Or can the 1.3 point difference be explained as a chance variation?

(a) Formulate the null and alternative hypotheses as statements about a box model.

(b) Compute z and P.

(c) Was it a chance variation?

5. A colony of laboratory mice consisted of several hundred animals. Their average weight was about 30 grams, with an SD of about 5 grams. As part of an experiment, graduate students were instructed to choose 25 animals haphazardly, without any definite method.[7] The average weight of these animals turned out to be around 33 grams. Is choosing animals haphazardly the same as drawing them at random? Discuss briefly, formulating the null hypothesis in terms of a box model, and computing z and P. (There is no need to formulate an alternative hypothesis about the box.)

6. (This exercise involves the measurement error model.) Several thousand measurements on a checkweight average out to 512 micrograms above a kilogram, with an SD of 50 micrograms. Then, the weight is cleaned. The next 100 measurements average out to 508 micrograms above one kilogram, with an SD of 52 micrograms. Apparently, the weight got 4 micrograms lighter. Or is this a chance variation? (You may assume the Gauss model with no bias.)

(a) Formulate the null and alternative hypotheses as statements about a box model.

(b) Would you estimate the SD of the box as 50 or 52 micrograms?

(c) Compute z and P.

(d) Did the weight get lighter? If so, by how much?

The answers to these exercises are on pp. A-61–A-62.

6. REVIEW EXERCISES

1. With a perfectly balanced roulette wheel, in the long run red numbers should turn up 18 times in 38. To test its wheel, one casino records the results of 3,800 plays, finding 1,868 red numbers. Is that too many reds? Or a chance variation?

(a) Formulate the null and alternative hypotheses as statements about a box model.

(b) Compute z and P.

(c) Were there too many reds?

2. One kind of plant has only red flowers and white flowers. According to a genetic model, the offspring of a certain cross have a 75% chance to be red-flowering, and a 25% chance to be white-flowering, independently of one another. Two hundred seeds of such a cross are raised, and 142 turn out to be red-flowering. Is this consistent with the model? (Hint: There is no alternative hypothesis about the box. In this case, if the investigator believes the model, he will be happy not to reject the null hypothesis: large P-values will be good for him.)

3. One large course has 800 students, broken down into section meetings with 25 students each. The section meetings are led by teaching assistants. On the final, the class average is 65, with an SD of 15. However, in one section the average is only 61. The TA is accused of bad teaching. He argues this way:

> If you took 25 students at random from the class, there is a pretty good chance they would average below 61 on the final. That's what happened to me—chance variation.

Evaluate this argument. (Hint: Formulate the null hypothesis in terms of a box model. The TA is anxious not to have to reject the null. No alternative hypothesis about the box is needed.)

4. There are about 2,500 colleges and universities in the United States. The National Center for Educational Statistics calculated the average undergraduate enrollment in 1974 to have been around 3,700 students per institution.[8] Current data is hard to get, so one organization takes a simple random sample of 225 institutions. Among the sample institutions, the average enrollment in the current year is 3,500 students per institution, with an SD of 6,000. Does this show that the average enrollment among all colleges and universities has gone down? Or can the difference be explained as a chance variation? Formulate the null and alternative hypotheses in terms of a box model before deciding.

5. On November 9, 1965, the power went out in New York City, and it stayed out for a day—The Great Blackout. Nine months later, the newspapers suggested that New York was experiencing a baby boom. The table below shows the number of babies born every day during a twenty-five day period, centered nine months and ten days after The Great Blackout.[9] These numbers average out to 436. This turns out not to be unusually high for New York. But there is an interesting twist to the data: the three Sundays only average 357. How likely is it that the average of three days chosen at random from the table will be 357 or less? What do you infer?

Number of births in New York, August 1–25, 1966.

August	1 Mon.	451	August	14 Sun.	377
''	2 Tues.	468	''	15 Mon.	451
''	3 Wed.	429	''	16 Tues.	497
''	4 Thur.	448	''	17 Wed.	458
''	5 Fri.	466	''	18 Thur.	429
''	6 Sat.	377	''	19 Fri.	434
''	7 Sun.	344	''	20 Sat.	410
''	8 Mon.	448	''	21 Sun.	351
''	9 Tues.	438	''	22 Mon.	467
''	10 Wed.	455	''	23 Tues.	508
''	11 Thur.	468	''	24 Wed.	432
''	12 Fri.	462	''	25 Thur.	426
''	13 Sat.	405			

(Hint: The SD of the 25 numbers in the table is about 40. Formulate the null hypothesis in terms of a box model; no alternative hypothesis about the box is necessary. In this case, the normal approximation can be used even with 3 draws.)

6. The National Assessment of Educational Progress administers nation-wide tests on academic subjects to students and young adults.[10] One item on a recent mathematics test asked

    Do the following addition: $1/2 + 1/3 =$ ———— .

    About 66% of the seventeen-year-olds in school in the United States knew the correct response, 5/6. (The other popular response was 2/5.)

    Suppose that a school board in California wants to compare the performance of their seventeen-year-old students with the national standard on this item. They take 400 of these students at random, and ask them the question; 243 answer correctly. Can the difference between this result and the national standard of 66% be explained as a chance variation? Formulate the null and alternative hypothesis in terms of a box model before deciding.

7. Say whether each of the following statements is true or false.
    (a) The P-value of a test equals its observed significance level.
    (b) The alternative hypothesis is another way of explaining the results, in terms of chance variation.

8. Discount stores often introduce new merchandise at a special low price in order to induce people to try it. However, a psychologist predicted that in the long run this practice would reduce sales. With the coopera-tion of a discount chain, an experiment was performed in 1968 to test this.[11] Twenty-five pairs of stores were selected, matched according to such characteristics as location and sales volume. These stores did not advertise, and displayed their merchandise in very similar ways. A new kind of cookie was introduced in all fifty stores. Within each pair of stores, one was chosen at random to introduce the cookies at the special low price of 25¢ a box, the price increasing to 29¢ a box after two weeks; the other store in the pair introduced the cookies at the regular price of 29¢ a box. Total sales of the cookies were computed for each store for six weeks from the time they were introduced. In 18 of the 25 pairs, the store which introduced the cookies at the regular price turned out to have sold more of them than the other store. Can this be explained as a chance variation? Or does this result support the prediction that introducing merchandise at a low price reduces long-run sales? (Formulate the null hypothesis in terms of a box model; there is no alternative hypothesis about the box.)

9. Does the psychological environment affect the anatomy of the brain? This question was studied experimentally by Mark Rosenzweig and others at the University of California, Berkeley.[12] The subjects for the experiment came from a genetically pure strain of rats. From each litter, one rat was selected at random for the treatment group, and one for the

control group. Both groups got exactly the same kind of food and drink—as much as they desired. But each animal in the treatment group lived with eleven others in a large cage, furnished with playthings which were changed daily. Animals in the control group lived in isolation, with no toys. After a month, the experimental animals were killed and dissected. On the average, the control animals were heavier, and had heavier brains, perhaps because they ate more and got less exercise. However, the treatment group had consistently heavier cortexes (the "gray matter," or thinking part of the brain). This experiment was repeated many times; results from the first five trials are shown below. "T" means treatment, and "C" is for control. Each line refers to one pair of animals. In the first pair, the animals in treatment had a cortex weighing 689 milligrams; the one in control had a lighter cortex, weighing only 657 milligrams. And so on. Two methods of analyzing the data will be presented in the form of exercises. Both methods take into account the pairing, which is a crucial feature of the data. The pairing results from the fact that the two animals come from the same litter.

Cortex weights (in milligrams) for experimental animals. The treatment group (T) had an enriched environment. The control group (C) had a deprived environment.

| Expt. #1 | | Expt. #2 | | Expt. #3 | | Expt. #4 | | Expt. #5 | |
T	C	T	C	T	C	T	C	T	C
689	657	707	669	690	668	700	662	640	641
656	623	740	650	701	667	718	705	655	589
668	652	745	651	685	647	679	656	624	603
660	654	652	627	751	693	742	652	682	642
679	658	649	656	647	635	728	578	687	612
663	646	676	642	647	644	677	678	653	603
664	600	699	698	720	665	696	670	653	593
647	640	696	648	718	689	711	647	660	672
694	605	712	676	718	642	670	632	668	612
633	635	708	657	696	673	651	661	679	678
653	642	749	692	658	675	711	670	638	593
		690	621	680	641	710	694	649	602

First analysis

(a) How many pairs are there in all? In how many of these pairs did the treatment animal have a heavier cortex?

(b) Suppose treatment had no effect, so each animal of the pair had a 50–50 chance to have the heavier cortex. Suppose also that this is independent from pair to pair. Under these assumptions, how likely is it that an investigator would get as many pairs as Rosenzweig did, or more, with the treatment animal having the heavier cortex? What do you infer? (The calculation in (a) and (b) is called the *sign test*.)

Second analysis

(c) For each pair of animals, compute the difference in cortex weights "treatment–control." Find the average and SD of all these differences. The null hypothesis is that these differences are like draws made at random with replacement from a box, whose average is 0—the treatment makes no difference. Make a z-test of this hypothesis. What do you infer?

Another point

(d) To ensure the validity of the analysis, the following precaution was taken: "The brain dissection and analysis of each set of littermates was done in immediate succession but in a random order and identified only by code number so that the person doing the dissection does not know which cage the rat comes from." Comment briefly on the following: What is the point of this precaution? Was it a good idea? Should they have done more? If so, what?

10. (This exercise involves the measurement error model.) Several hundred measurements on NB 10 averaged out to 406 micrograms below ten grams, with an SD of 7 micrograms. Then, this checkweight was cleaned. The next 25 measurements averaged out to 409 micrograms below ten grams, with an SD of 8 micrograms. Is this a chance variation? Or did the weight get lighter? (You may assume the Gauss model, with no bias.)

7. SUMMARY

1. A *test of significance* gets at the question of whether an observed difference is real—the *alternative hypothesis*—or just a chance variation—the *null hypothesis*.

2. A legitimate test of significance requires a box model for the data. The null hypothesis, and usually the alternative, have to be translated into statements about the box model.

3. A *test statistic* is used to measure the difference between the data and what is expected on the null hypothesis. The *z-test* uses the statistic

$$z = \frac{\text{observed} - \text{expected}}{\text{SE}}.$$

The expected value is computed on the null hypothesis.

4. The *observed significance level* (also called "P," or the "P-value") is the chance of getting a test statistic as extreme as or more extreme than the observed one. The chance is computed on the basis that the null hypothesis is right. Therefore, P does not give the chance of the null hypothesis being right.

5. Small values of P are evidence against the null hypothesis; they indicate that something besides chance was operating to make the difference.

27

More Tests for Averages

Vive la différence!

1. THE STANDARD ERROR FOR A DIFFERENCE

In the previous chapter, the average of a sample was compared to an external standard. The z-test can also be used to compare the averages of two samples, but it is a little more complicated, because both averages are affected by chance error. The first problem is to measure the chance variability in the difference between two sample averages. That is the topic of this section.

For example, suppose there are two boxes, A and B, with averages and SDs as shown.

Box A
Average = 110
SD = 60

Box B
Average = 90
SD = 40

Four hundred draws are made at random with replacement from box A, and independently one hundred draws are made at random with replacement from box B. Then the average of the draws from box B is subtracted from the average of the draws from box A.

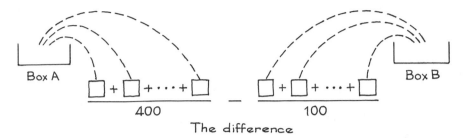

The difference

The problem is to find the expected value and standard error for this difference. The first step is to compute the expected value and SE for each average separately (p. 373):

average of 400 draws from box A = 110 ± 3 or so
average of 100 draws from box B = 90 ± 4 or so.

The expected value for the difference is just 110 − 90 = 20, but the new problem is how to put the two SEs together:

$$(110 \pm 3) - (90 \pm 4) = 20 \pm \underline{\quad} ?$$

A natural first guess is just to add the SEs: 3 + 4 = 7. However, this ignores the possibility of cancellation. A square root law is needed, and one can be stated as follows.

> The standard error for the difference of two independent quantities is $\sqrt{a^2 + b^2}$, where
> - a is the SE for the first quantity;
> - b is the SE for the second quantity.

In the example, the draws from the two boxes are made independently, so the two averages are independent, and the square root law applies. Now a is 3, and b is 4, so the SE for the difference between the two averages is

$$\sqrt{3^2 + 4^2} = \sqrt{25} = 5.$$

Example 1. A hundred draws are made at random with replacement from box C, shown below. Independently, one hundred draws are made at random with replacement from box D. Find the expected value and SE for the difference between the number of 1's drawn from box C and the number of 5's drawn from box D.

C | 1 2 | D | 4 5 |

Solution. The number of 1's will be around 50 give or take 5 or so. The number of 5's will also be around 50 give or take 5 or so. The expected value for the difference is 50 − 50 = 0. The draws are made independently, so the two numbers are independent, and the square root law applies. The SE for the difference is $\sqrt{5^2 + 5^2} \approx 7$.

Example 2. A hundred draws are made at random with replacement from the box

$$\boxed{\;\boxed{1}\;\boxed{2}\;\boxed{3}\;\boxed{4}\;\boxed{5}\;}$$

The expected number of 1's is 20, with an SE of 4. The expected number of 5's is also 20 with an SE of 4. True or false: the SE for the difference between the number of 1's and the number of 5's is $\sqrt{4^2 + 4^2}$.

Solution. This is false. The two numbers are dependent, because when one is large the other must be small. The square root law does not apply.

EXERCISE SET A

1. Box A has an average of 100 with an SD of 10. Box B has an average of 50 with an SD of 18. Now 25 draws are made at random with replacement from box A, and independently 36 draws are made at random with replacement from box B. Find the expected value and standard error for the difference between the average of the draws from box A and the average of the draws from box B.

2. A coin is tossed 500 times. Find the expected value and SE for the difference between the percentage of heads in the first 400 tosses and the percentage of heads in the last 100 tosses.

3. A box contains 5,000 numbered tickets, which average out to 50 with an SD of 30. Then, 200 tickets are drawn at random without replacement. True or false: the SE for the difference between the average of the first 100 draws and the average of the second 100 draws is approximately $\sqrt{3^2 + 3^2}$. (Hint: What if the draws were made with replacement?)

4. Repeat Exercise 3 if the box only contains 200 tickets.

5. One hundred draws are made at random with replacement from box X: the average of these draws is 51 with an SD of 3. Independently, four hundred draws are made at random with replacement from box Y: the average of these draws is 48 with an SD of 8. Someone claims that both boxes have the same average. What do you think?

The answers to these exercises are on pp. A-62–A-63.

2. COMPARING TWO SAMPLES

An example will indicate how to test the difference between two averages, using the formula of the previous section. In the 1960's, a "negative income tax" was proposed to supplement low incomes instead of taxing them. The Federal Government gave serious thought to scrapping the existing welfare programs, and switching to a negative income tax plan. One crucial question was the extent to which income subsidies would cause the recipients to stop working. In 1967, the Wisconsin Institute for Research on

Poverty, in conjunction with Mathematica (a research firm based in Princeton), began the New Jersey Negative Income Tax Experiment.[1] The point of the experiment was to measure the effect of income subsidies on the time spent working by the recipients. The subjects in the experiment were chosen (by probability methods) in three New Jersey cities: Trenton, Passaic, and Patterson. As it turned out, negative income tax only brought about a slight reduction in the hours worked by the men, with a somewhat larger reduction for married women. However, for black men the negative income tax led to a small increase in hours worked—contrary to predictions.

The details of the experiment were quite complicated, but a simplified version of it can be presented here. Suppose that an investigator is planning a study on the effects of a negative income tax on the number of hours worked by low-income families in a certain town. There are 10,000 such families in the town, and he can only afford to do the study on a sample basis. So he takes two simple random samples:

- 400 families are chosen at random for the control group;
- 225 families are chosen at random for the treatment group.

Both groups are followed for three years. The control group is taxed in the usual way, while the treatment group is put on the negative income tax. At the end of the experiment, it turns out that the control families averaged 7,100 hours of paid work during the three-year period, with an SD of 4,000 hours. (This includes all work done for pay by all family members.) The treatment families, on the other hand, only averaged 6,200 hours of paid work, with an SD of 3,400 hours. Is this difference between 7,100 and 6,200 hours real, or just a chance variation?

To settle this question, a box model is needed. And this time, two boxes are involved—a control box and a treatment box. Each box has 10,000 tickets, one for each low-income family in the town. The number of hours that would be worked by a family during the experiment, if they were taxed in the usual way, is marked on the corresponding ticket in the control box. The data for the control group consist of the total number of hours worked by each of the 400 families, and these data are like 400 draws from the control box.

The treatment box is harder to visualize: each ticket represents the number of hours the corresponding family would have worked during the experiment, if it had been put on the negative income tax. The data on the treatment group consist of the total number of hours worked by each of the 225 treatment families, and these data are like 225 draws from the treatment box.

To understand the treatment box better, consider the following hypothetical experiment. Put every low-income family in the town on the negative income tax, but choose only 225 at random for study. Drawing 225 times at random from the treatment box is a perfect model for this hypothetical study. The point is that drawing 225 families at random and putting only them on the negative income tax gives exactly the same information as the hypothetical study. In either case, the investigator just gets to see what happens to the 225 sample families. The real design is much cheaper!

Implicit in the model is the assumption that a family's response to the negative income tax does not depend on the tax treatment given to the other families. If this assumption is seriously wrong, the effect of the negative income tax cannot be studied on a sample basis.

This completes the box model. The null hypothesis is that the difference between the averages for the treatment group and the control group is just due to chance. The alternative hypothesis is that this difference is real: putting people on the negative income tax reduces the number of hours they will work. These two hypotheses can be formulated in terms of the model:

- null hypothesis—the average of the treatment box equals the average of the control box;
- alternative hypothesis—the average of the treatment box is smaller than the average of the control box.

The next thing to look at is the difference between the two averages:

$$\text{treatment average} - \text{control average} = 6,200 - 7,100$$
$$= -900 \text{ hours.}$$

If the null hypothesis is right, the difference is expected to be around 0. The test statistic z takes the form

$$z = \text{observed difference/SE.}$$

The denominator is the SE for the difference between the two averages, and it can be worked out following the method of section 1.

The SD for the treatment box is estimated by the SD of the treatment group as 3,400 hours—another instance of the bootstrap method (p. 342). The SE for the total of the hours worked by all 225 treatment families is $\sqrt{225} \times 3,400 = 51,000$ hours. The SE for the treatment average is

$$51,000/225 \approx 230 \text{ hours.}$$

Similarly, the SE for the control average is around 200 hours. The two averages are practically independent, so the SE for their difference can be figured using the square root law:

$$\sqrt{230^2 + 200^2} \approx 300 \text{ hours.}$$

Now $z = -900/300 = -3$. On the null hypothesis, the difference between the two averages is 3 SEs below its expected value. The more extreme values of z are those below -3: they would be even stronger evidence against the null hypothesis.

The number of draws is so large that the probability histograms for the two sample averages both follow the normal curve. Then it is a mathematical fact that the probability histogram for the ratio z will follow the curve too:

$$z = \frac{\text{average of 225 draws from treatment box} - \text{average of 400 draws from control box}}{\text{SE for difference}}$$

And on the null hypothesis, this ratio is in standard units. So, if the null hypothesis is right, and the experiment is repeated, the chance of getting a z-value of -3 or less can be read directly off the normal curve, completing the test:

This procedure is called a *two-sample z-test* and it can be summarized as follows:

$$z = \frac{\text{1st average} - \text{2nd average}}{\text{SE for difference}}, \quad P \approx$$

This test is appropriate for two independent and reasonably large simple random samples from two separate boxes, the null hypothesis being that the averages of the two boxes are equal.

In the example, the z-test shows that the difference between the treatment group and the control group is hard to explain as a chance variation. One effect of the negative income tax is a reduction in the number of hours worked for pay. How big is this reduction? The estimate is that if the negative income tax were applied to all families in this town, the average number of hours worked for pay over a three-year period would go down 900 hours give or take 300 hours or so. This is a small effect, but a real one. If a more precise estimate of it is wanted, larger samples would be needed.

The two-sample z-test can also be used when the problem involves classifying and counting. For instance, suppose that 88% of the heads of households in the control group were employed, compared to 82% in the treatment group. Is this difference real, or a chance variation? The investigator classifies each family according to whether the head of household was employed, and then counts those where the head was employed. So zero-one boxes are needed.

The control box should have 10,000 tickets, one for each low-income family in the town. Imagine that all the families are taxed in the usual way. If the head of household would be employed, the corresponding ticket is marked "1"; on the other hand, if the head would be unemployed, the ticket is marked "0." The data on the control group show, for the 400 families, whether the head was employed ("1") or not ("0"). These data are like 400 draws from the control box.

The treatment box is similar. Imagine that all the familes are put on the negative income tax. If the head of household would be employed in this situation, the corresponding ticket is marked "1"; on the other hand, if the head would be unemployed, the ticket is marked "0." The data on the treatment group are like 225 draws from the treatment box.

The percentages of 1's in the two samples are different: 88% compared to 82%. The null hypothesis is that the difference just reflects chance variation. The alternative is that the difference is real—the negative income tax tends to reduce participation in the labor force. These hypotheses can now be formulated in terms of the model:

- null hypothesis—the percentage of 1's in the treatment box is equal to the percentage of 1's in the control box;
- alternative hypothesis—the percentage of 1's in the treatment box is smaller than the percentage of 1's in the control box.

The test statistic is

$$z = \frac{\text{percentage of 1's in treatment sample} - \text{percentage of 1's in control sample}}{\text{SE for difference}}$$

Next, the SE for the difference must be worked out, and for this the SEs for the two percents are needed. The fraction of 1's in the control box is estimated by the fraction in the sample, 0.88. So the fraction of 0's must be 0.12, and the SD of the control box is estimated as $\sqrt{0.88 \times 0.12} \approx 0.32$. The SE for the number of 1's in the sample is $\sqrt{400} \times 0.32 \approx 6.4$, so the SE for the percentage is

$$\frac{6.4}{400} \times 100 \approx 1.6 \text{ percentage points.}$$

Similarly, the SE for the percentage in the treatment group is estimated as 2.6 percentage points. The two percentages are practically independent. So the SE for the difference is

$$\sqrt{1.6^2 + 2.6^2} \approx 3 \text{ percentage points.}$$

Now

$$z = \frac{82 - 88}{3} = -2.$$

Again, P can be read off the normal curve, completing the test:

$$P \approx \qquad \approx 2\%$$

The difference between the percents who work in the two groups is hard to explain as a chance variation.

As before, if the negative income tax plan were applied to all low-income families in the town, the estimate is that the percent of heads of household who work would be reduced slightly, by 6 ± 3 percentage points.

One point has been slurred over: the independence. To review, 400 families are chosen at random to form the control group, so 400 tickets are drawn at random from the control box. Then, 225 families are chosen at random for the treatment group, and 225 tickets are drawn at random from the treatment box. A family cannot be both in treatment and in control. Thus, the 400 tickets corresponding to the control group should in principle be removed from the treatment box, before the 225 tickets corresponding to the treatment group are drawn. This creates a slight dependence between the two averages. But it makes almost no difference whether 400 tickets out of 10,000 are in or out of the box. The two averages (or percentages) are practically independent.

Of course, the numbers used in this example were hypothetical, but they are close to the ones obtained in the New Jersey Negative Income Tax Experiment. In that experiment, for instance, the negative income tax reduced the average number of hours worked per week by the white families from 46 to 40. The reduction for black and Hispanic families was somewhat smaller.

EXERCISE SET B

1. Cycle III of the Health Examination Survey, completed in 1970, used a nation-wide probability sample of youths aged 12 to 17. One object of the survey was to estimate the percentage of youths who were illiterate.[2] A test was developed to measure literacy. It consisted of seven brief passages, with three questions about each, like the following:

> There were footsteps and a knock at the door. Everyone inside stood up quickly. The only sound was that of the pot boiling on the stove. There was another knock. No one moved. The footsteps on the other side of the door could be heard moving away.
> —The people inside the room
>   (a) Hid behind the stove
>   (b) Stood up quickly
>   (c) Ran to the door
>   (d) Laughed out loud
>   (e) Began to cry
> —What was the only sound in the room?
>   (a) People talking
>   (b) Birds singing
>   (c) A pot boiling
>   (d) A dog barking
>   (e) A man shouting
> —The person who knocked at the door finally
>   (a) Walked into the room
>   (b) Sat down outside the door
>   (c) Shouted for help
>   (d) Walked away
>   (e) Broke down the door.

This test was designed to be at the fourth-grade level of reading, and subjects were defined to be literate if they could answer more than half the questions correctly.

There turned out to be some difference between the performance of males and females on this test: 7% of the males were illiterate, compared to 3% of the females. Is this difference real, or the result of chance variation? You may assume that the investigators took a simple random sample of 1,600 male youths, and an independent simple random sample of 1,600 female youths. Formulate the null and alternative hypotheses in terms of a box model before answering the question.

2. Cycle II of the Health Examination Survey, completed in 1962, used a nation-wide probability sample of children aged 6 to 11. One object of the survey was to study the relationship between the children's scores on intelligence tests and their family backgrounds.[3] One intelligence test used was the WISC vocabulary scale. This consists of 40 words which the child has to define; 2 points are given for a correct answer, and 1 point for a partially correct answer. The scores turned out to be strongly related to the income and educational level of the parents. There was also some relationship between test scores and the type of community in which the parents lived. For example, big-city children averaged 26 points on the test, with an SD of 10 points; but rural children only averaged 25 points with the same SD of 10 points. Can this difference be explained as a chance variation? You may assume that the investigators took a simple random sample of 400 big-city children, and an independent simple random sample of 400 rural children. Formulate the null and alternative hypotheses in terms of a box model before answering the question.

3. Repeat Exercise 2, if both samples were of size 1,000 instead of 400.

4. Exercise 9 on p. 451 described an experiment in which 59 animals were put in treatment (enriched environment), and 59 were in control. The cortex weights for the treatment group averaged 683 milligrams, with an SD of 31 milligrams. The cortex weights for the control group averaged 647 milligrams with an SD of 29 milligrams. Someone proposes to make a two-sample z-test:

$$\text{SE for sum of treatment weights} \approx \sqrt{59} \times 31 \approx 238 \text{ milligrams}$$

$$\text{SE for average of treatment weights} \approx 238/59 \approx 4.0 \text{ milligrams}$$

$$\text{SE for sum of control weights} \approx \sqrt{59} \times 29 \approx 223 \text{ milligrams}$$

$$\text{SE for average of treatment weights} \approx 223/59 \approx 3.8 \text{ milligrams}$$

$$\text{SE for difference} \approx \sqrt{4.0^2 + 3.8^2} \approx 5.5 \text{ milligrams}$$

$z = (683 - 647)/5.5 \approx 6.5,$ P $\approx$ $\approx$ 0

What does statistical theory say?

5. An investigator wants to show that first-born children score higher on IQ tests than second-borns. In one school district, he finds 400 two-child families with both children enrolled in elementary school. He gives these children the WISC vocabulary test (described in the previous exercise), with the following results:
   • The 400 first-borns average 30 with an SD of 10.
   • The 400 second-borns average 28 with an SD of 10.
(Scores have been corrected to compensate for age differences.) He then makes a two-sample z-test:

$$\text{SE for first-born average} \approx 0.5$$

$$\text{SE for second-born average} \approx 0.5$$

$$\text{SE for difference} = \sqrt{(0.5)^2 + (0.5)^2} \approx 0.7$$

$z = (30 - 28)/0.7 \approx 2.9,$ P $\approx$ $\approx$ 2 in 1,000

What does statistical theory say?

The answers to these exercises are on p. A-63.

3. THE *t*-TEST

With small samples, the *z*-test has to be modified. If the tickets in the box do not follow the normal curve, and the number of draws is small, the normal approximation cannot be used on the probability histogram for the sum or average of the draws (Chapter 18). However, even if the tickets in the box do follow the normal curve, there is a new problem when their SD is unknown. The reason is that the SD of a small sample need not be close to the SD of the box. This ruins the calculation of the observed significance level from the *z*-test. To fix things up, statisticians use SD^+ instead of SD in figuring the test statistic, and Student's curve instead of the normal in figuring *P*. This section will develop the technique, by example. However, the discussion is a bit technical, and it can be skipped. The material in section 6 of Chapter 24, on the *t*-distribution for confidence intervals, will be used here. Only the one-sample *t*-test will be presented.[4]

In Los Angeles, many studies have been conducted to determine the concentration of CO—carbon monoxide—near freeways, under various conditions of traffic flow. The basic technique involves capturing air samples in special bags, and then determining the CO concentrations in the bag samples by using a machine called a *spectrophotometer*. These machines can measure concentrations up to about 100 ppm. (parts per million, by volume) with errors on the order of 10 ppm. However, they are quite delicate and have to be calibrated every day. This involves using the machine to measure CO concentration in a manufactured gas sample, called *span gas,* where the CO concentration is very precisely controlled at 70 ppm. If the machine reads close to 70 ppm. on the span gas, it's ready for use; if not, it has to be adjusted. A complicating factor is that the size of the measurement errors varies from day to day. On any particular day, however, it is reasonable to assume that the errors are independent and follow the normal curve, with some unknown SD which depends on the day.

One day, a technician makes five readings on span gas, and gets

$$78 \quad 83 \quad 68 \quad 72 \quad 88.$$

Four out of five of these numbers are higher than 70, and some of them by quite a bit. Can this be explained on the basis of chance variation? Or does it show bias, perhaps from improper adjustment of the machine?

A test of significance is called for, and to make one a box model is needed. The one to use is the Gauss model (p. 396). According to this model, each measurement equals the true value of 70 ppm., plus bias, plus a draw with replacement from the error box. (The tickets in the error box average out to 0, and their SD is unknown.)

$$\text{Measurement} = 70 \, \text{ppm.} + \text{bias} + \square$$

The key parameter here is the bias. The null hypothesis is that the bias equals 0. On this hypothesis, the average of the five measurements has an expected value of 70 ppm.; the difference between the observed and expected values is explained as a chance variation. The alternative hypothesis is that the bias differs from 0, so the difference between the average of the sample and 70 ppm. is real.

As before, the appropriate test statistic to use is

$$\frac{\text{observed} - \text{expected}}{\text{standard error}}.$$

The only difference is that the SD of the error box should be estimated by SD^+ of the data instead of the SD. This modified test statistic is usually denoted by "t." Doing the arithmetic, the average of the five measurements is 77.8 ppm., and $SD^+ \approx 8.1$ ppm., so $SE \approx 3.6$ ppm. On the null hypothesis, the expected value is 70 ppm. So

$$t = (77.8 - 70)/3.6 \approx 2.2$$

The average of the five measurements is about 2.2 SEs above the value expected on the null hypothesis.

To compute the observed significance level, Student's curve is used instead of the normal, with

$$\text{degrees of freedom} = \text{number of measurements} - \text{one}$$
$$= 5 - 1 = 4.$$

So the chance of getting a t-value as extreme as (or more extreme than) the observed one is about 5%.

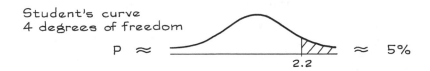

The evidence is running against the null hypothesis, although not very strongly.[5] This completes the example.

Both the z-test and the t-test assume that the data is like the results of drawing at random from a box. The draws are to be made with replacement, or without replacement from a very large box. The t-test involves a further assumption: that the tickets in the box follow the normal curve quite closely. If not, the tail area under Student's curve is only an approximation to the true value of P. Statisticians have shown that this approximation can be very bad when the histogram for the tickets in the box is long-tailed and unsymmetric, especially for small values of P. By comparison, the z-test works, as a good approximation, even when the tickets in the box do not follow the normal curve at all well—provided the number of draws is large enough for the normal approximation to take over.

The t-test is used only when the SD of the box is unknown.

EXERCISE SET C

1. A long series of measurements on a checkweight average out to 253 micrograms above ten grams, with an SD of 7 micrograms. The Gauss model is believed to apply, with negligible bias. At this point, the balance has to be rebuilt; this may introduce bias, as well as changing the SD of the error box. Ten measurements on the checkweight, using the rebuilt scale, show an average of 245 micrograms above ten grams, with an SD of 9 micrograms. Has bias been introduced? Or is this a chance variation? (You may assume that the errors follow the normal curve.)

2. Each (hypothetical) data set below represents some readings on span gas. Assume the Gauss model, with errors following the normal curve: however, bias may be present. In each case, make a t-test to see whether the instrument is properly calibrated or not. In one case, this is impossible. Which one, and why?
   (a) 71, 68, 79
   (b) 71, 68, 79, 84, 78, 85, 69
   (c) 71
   (d) 71, 84

3. A new spectrophotometer is being calibrated. It is not clear whether the errors follow the normal curve, or even whether the Gauss model applies. In two cases, these assumptions should be rejected. Which two, and why? The numbers are replicate measurements on span gas.
    (a) 71, 70, 72, 69, 71, 68, 93, 75, 68, 61, 94, 91
    (b) 71, 73, 69, 74, 65, 67, 71, 69, 70, 75, 71, 68
    (c) 71, 69, 71, 69, 71, 69, 71, 69, 71, 69, 71, 69

4. A company wants to know whether the invoices it gets average less than $100. It takes a random sample of five invoices. The amounts are

$$\$49.76 \qquad \$10.81 \qquad \$207.64 \qquad \$788.95 \qquad \$14.52$$

One person wants to make a t-test. Another objects, on the grounds that the distribution of amounts invoiced has a long right-hand tail. The first person answers that the normal approximation makes it come out right anyway. What does statistical theory say?

The answers to these exercises are on p. A-64.

4. REVIEW EXERCISES

1. A box contains 800 tickets; 400 are drawn at random without replacement, and the SE for their average is computed (using the correction factor) as 5. These tickets are replaced in the box, and another sample of 400 is drawn at random without replacement; the SE for the average of the second group is 5 too. True or false: the SE for the difference between the average of the first 400 draws and the average of the second 400 draws is $\sqrt{5^2 + 5^2}$. Explain your answer.

2. Repeat Exercise 1, if the first sample of 400 is not replaced before drawing the second sample.

3. Box A and box B both contain a large number of red and blue marbles. A hundred marbles are drawn at random from box A, and 54 turn out to be red. Independently, 400 are drawn at random from box B, and 176 turn out to be red. The null hypothesis is that the percentage of red marbles in box A equals the percentage in box B; the alternative is that the percentage in box A is larger than the percentage in box B.
    (a) Estimate the difference between the percentage of red marbles in box A and the percentage of red marbles in box B. Put a give-or-take number on the estimate.
    (b) Compute z and P.
    (c) Which hypothesis do you believe?

4. Is Washington, D.C., different from other cities?[6] A simple random sample of 400 families is taken in Washington; the average of the incomes of the 400 sample families is $21,000 with an SD of $16,000. Another simple random sample of 900 families is taken in Philadelphia; the average of the incomes of these sample families is $17,000 with an SD of $15,000. Are they richer in Washington than in Philadelphia, or is it a chance variation?

(a) Estimate the difference between the average income of all Washington families and the average income of all Philadelphia families. Put a give-or-take number on the estimate.

(b) Formulate the null and alternative hypotheses as statements about a box model.

(c) Compute z and P.

(d) Which hypothesis do you believe?

5. The National Assessment of Educational Progress (NAEP) administers tests in different subjects to a nationwide sample of students each year. General knowledge of science was tested in 1970 and in 1973, and scores were somewhat lower the second time.[7] One question asked of thirteen-year-olds both times was this:

Which of the following diseases is known to be transmitted by an insect:

Cancer   Diabetes   Malaria   Measles   Polio

In 1970, 85% of the thirteen-year-olds correctly identified malaria; however, by 1973 this percentage was down to 77%. Can this drop be explained as a chance variation? You may assume that in each year, NAEP took a nationwide simple random sample of 400 thirteen-year-olds. Formulate the null and alternative hypotheses in terms of a box model before deciding the question.

6. In a 1977 survey, the Gallup poll found that only 10% of Americans believe that women and minorities should get preferential treatment in hiring or college admissions. There was no difference between men and women on this issue, but there was a difference between races: 8% of the whites favored preferential treatment for women and minorities; the corresponding percentage among the blacks was 27%. Is this difference real, or a chance variation? You may assume that Gallup took a simple random sample of 900 whites, and independently a simple random sample of 100 blacks.[8] Formulate the null and alternative hypotheses in terms of a box model before deciding the question.

7. A large university wants to compare the mathematical abilities of its male and female freshmen, using a standardized test.[9] An investigator selects 100 men and 100 women at random from the freshman class. The men average 500 on the test, with an SD of 120. The women average 450, with an SD of 110. The investigator then makes a two-sample z-test:

SE for men's average ≈ 12
SE for women's average ≈ 11
SE for difference $\approx \sqrt{12^2 + 11^2} \approx 16$

$z = (500 - 450)/16 \approx 3,$ P $\approx$ $\approx$ 1 in 1,000.

He concludes that the difference is real, not a chance variation. What does statistical theory say?

8. A large university wants to compare the verbal and mathematical abilities of its male freshmen, on two standardized tests.[10] An investigator selects 100 men at random from the freshman class, and gives each one the two tests. The average score on the verbal test is 430 with an SD of 110. The average score on the mathematics test is 500 with an SD of 120. The investigator then makes a two-sample z-test:

$$\text{SE for verbal average} \approx 11$$
$$\text{SE for mathematics average} \approx 12$$
$$\text{SE for difference} \approx \sqrt{11^2 + 12^2} \approx 16$$

$$z = (430 - 500)/16 \approx -4.4, \quad \text{P} \approx \quad \approx 0$$

He concludes that the difference is real, not a chance variation. What does statistical theory say?

The next two exercises involve the measurement error model.

9. According to the theory of plate tectonics, the California coast (including Los Angeles) is carried on one plate, and the continental United States (including Sacramento) is carried on another plate. These two plates are moving relative to one another in such a way that Los Angeles gets about an inch closer to Sacramento every year. A hypothetical experiment is underway to check this theory, by measuring the distance between Los Angeles and Sacramento once a year for fifty years.[11] The first 25 measurements average out to 31,996,832 inches, with an SD of about 40 inches. The next 25 measurements average out to 31,996,806 inches, with the same SD of 40 inches. The second average is 26 inches less than the first. Is this evidence for plate tectonics, or a chance variation? (You may assume the Gauss model, with no bias.)
   (a) Formulate the null hypothesis as a statement about a box model. (There is no need to formulate the alternative in terms of the box.)
   (b) Compute z and P.
   (c) Would you reject the null hypothesis?

The next exercise also involves material from section 3.

10. A long series of measurements on a checkweight averages out to 508 micrograms above 1 kilogram, with an SD of 47 micrograms. The Gauss model is believed to apply, with negligible bias. The balance is then modified to increase its precision; however, the investigators are afraid that bias may have been introduced. Ten measurements on the checkweight, using the modified balance, show an average of 502 micrograms above 1 kilogram, with an SD of 41 micrograms. Has bias been introduced? Or is this a chance variation? (You may assume the Gauss model, with errors that follow the normal curve.) Be sure to formulate the null and alternative hypotheses in terms of a box model.

5. SUMMARY

1. The standard error for the difference of two independent quantities is $\sqrt{a^2 + b^2}$, where

- a is the SE for the first quantity,
- b is the SE for the second quantity.

This formula does not apply to the difference of two dependent quantities.

2. Many testing problems can be put as follows. Two independent and reasonably large simple random samples are taken from two separate boxes. The null hypothesis is that the two boxes have the same average. The alternative hypothesis is that one box has a bigger average. The appropriate test statistic is

$$z = \frac{\text{average of 1st sample} - \text{average of 2nd sample}}{\text{SE for difference}}$$

Tests based on the statistic are called *two-sample z-tests*.

3. This procedure can handle situations which involve classifying and counting, by putting 0's and 1's in the boxes.

4. Some testing problems can be put as follows. A small number of tickets are drawn at random with replacement from a box whose contents follow the normal curve, with an average of 0 and an unknown SD. Each draw is added to an unknown constant to give a measurement. The null hypothesis is that this unknown constant equals some given value c. An alternative hypothesis is that the unknown constant is bigger than c. The SD of the box can be estimated by the SD^+ of the data. Then the SE for the average of the draws is computed as usual, and the test statistic is

$$t = \frac{\text{average of data} - c}{\text{SE}}$$

The observed significance level is obtained not from the normal curve but from one of Student's curves, with

degrees of freedom = number of measurements − one.

This procedure is a *t-test*.

28

The χ^2-Test

Don't ask what it means, but rather how it is used.

—L. WITTGENSTEIN (1889–1951)

1. INTRODUCTION

How well does it fit the facts? Sooner or later, this question must be asked about any chance model. And in many cases, it can be settled by the χ^2-*test*; invented in 1900 by Karl Pearson.[1] χ is a Greek letter, read like the "ki" in kite; so χ^2 is read as "ki-square." Section 5 of Chapter 26 explained how to test a chance model for a parapsychology experiment. In that experiment, each guess was classified into one of two categories—correct or incorrect. According to the model, each guess had 1 chance in 4 to be correct, so the number of correct guesses was like the sum of draws from the box

$$|\,\boxed{0}\ \boxed{0}\ \boxed{0}\ \boxed{1}\,|$$

In that case, the z-test was appropriate, because only two categories were involved. In many situations, the model will involve more than two categories, and it is then that the χ^2-test should be used.

For instance, suppose a die is being rolled. The model may be that the die is fair—each face has one chance in six to turn up. Now each throw can be classified into one of six categories—either the die landed ⚀, or ⚁, or ⚂, or ⚃, or ⚄, or ⚅. To test the model, you can roll the die, count how many times each face turns up, and make a χ^2-test, as will be indicated by example.

Example 1. A gambler is accused of using a loaded die, but he pleads innocent. A record has been kept of the last sixty throws (Table 1). There is disagreement about how to interpret the data, and a statistician is called in.

Table 1. Sixty rolls of a die, which may be loaded.

4	3	3	1	2	3	4	6	5	6
2	4	1	3	3	5	3	4	3	4
3	3	4	5	4	5	6	4	5	1
6	4	4	2	3	3	2	4	4	5
6	3	6	2	4	6	4	6	3	2
5	4	6	3	3	3	5	3	1	4

The model is that the die is fair. So the numbers in Table 1 are like the results of drawing sixty times (at random, with replacement) from the box

$$\boxed{1}\ \boxed{2}\ \boxed{3}\ \boxed{4}\ \boxed{5}\ \boxed{6}$$

One implication of this model is that each number should turn up about 10 times: its *expected frequency* is 10. To find out how the data compare with these expectations, you have to count and see how many times each number did in fact turn up. These *observed frequencies* are shown in Table 2. As a check on the arithmetic, sum the frequency columns. Each must sum to 60, because that's the total number of entries in Table 1.

Table 2. Observed and expected frequencies for the data in Table 1.

Value	Observed frequency	Expected frequency
1	4	10
2	6	10
3	17	10
4	16	10
5	8	10
6	9	10
sum	60	60

As the table shows, there are too many 3's. The SE for the number of 3's is $\sqrt{60} \times \sqrt{1/6 \times 5/6} \approx 2.9$, so the number of 3's is about 2½ SEs above the expected number. But don't shoot the gambler yet. The statistician won't let you take the table line by line like that, for two reasons.

• Several lines in the table may look suspicious. For example, in Table 2 there are also too many 4's. He wants to take that into account also.

• On the other hand, if the table has many lines, there is a high probability that at least one of them will look suspicious—even if the null hypothesis is true. It's like playing Russian roulette. If you keep on going, sooner or later you lose.

So the problem is to combine the differences from each line of the table into an overall measure of the distance between the observed and expected frequencies. Pearson's solution to this problem was to square each difference, divide by the corresponding expected frequency, and take the sum. As a formula, Pearson's test statistic was

$$\chi^2 = \text{sum of } \frac{(\text{observed frequency} - \text{expected frequency})^2}{\text{expected frequency}}$$

There is one term in the sum for each line in the table.

At first sight, this formula may seem too arbitrary to be worth understanding. And in fact, it is just a definition. However, every statistician uses it because of one very convenient feature, which will be pointed out later. For now, please bear with the formula.

For Table 2, the χ^2-statistic works out as

$$\frac{(4-10)^2}{10} + \frac{(6-10)^2}{10} + \frac{(17-10)^2}{10} + \frac{(16-10)^2}{10} + \frac{(8-10)^2}{10} + \frac{(9-10)^2}{10} = \frac{142}{10} = 14.2$$

When the observed frequency is far from the expected frequency, the corresponding term in the sum is large; when the two are close, this term is small.[2] Large values of χ^2 mean that on the whole, the observed frequencies are far from the expected ones. Small values of χ^2 mean the opposite: on the whole, the observed frequencies are close to the expected ones. So χ^2 does give a measure of the distance between the observed frequencies and the expected frequencies.

Of course, chance variation leads to some range of likely values for χ^2. How does the observed value of 14.2 compare with what could be expected, by chance? To put this another way, even if the data in Table 1 had been generated by rolling a fair die, χ^2 could have come out to 14.2—or more. Consequently, to judge how large 14.2 is, the following probability problem must be solved: find the chance that when a fair die is rolled 60 times and χ^2 is computed from the observed frequencies, it turns out to be 14.2—or more.

Why "or more"? The value 14.2 may be evidence against the model, because it is too big—meaning that the observed frequencies are too far from the expected ones, further than can be explained as a chance variation. If so, values larger than 14.2 would be even stronger evidence against the model. What is the chance that the model will produce such strong evidence against itself? To find out, calculate the chance of getting a χ^2-statistic of 14.2—or more. The chance of getting such a big χ^2-value is called "the observed significance level" and denoted by P, just as before. Computing P looks like a formidable job. In the example, it took an hour on the computer, and the answer was 1.4%.[3]

How is this interpreted? The basic chance experiment consists in rolling a fair die 60 times. Imagine a long series of repetitions of this chance experiment, calculating the χ^2-statistic after each one. About 1.4% of these χ^2-values will be as big as (or bigger than) the value 14.2 computed from Table 1; the other 98.6% will be smaller. At this point, the statistician has finished. Things do not look good for the gambler. If his die is fair, there is only a 1.4% chance for it to produce a χ^2-statistic as big as (or bigger than) the observed one. This completes the discussion of Example 1.

In Pearson's time, there was no computer around to calculate P exactly. So he developed a method for doing it approximately, by hand. This method involves a new curve, called the χ^2-*curve*. More precisely, there is one curve for each number of degrees of freedom.[4] As for Student's t (p. 410),

$$\text{degrees of freedom} = \text{number of terms in sum} - \text{one.}$$

(The differences going into the formula for χ^2 necessarily add up to 0. This constraint eliminates one degree of freedom.) With the die example, there are $6 - 1 = 5$ degrees of freedom. The curves for 5 and 9 degrees of freedom are shown in Figure 1 on the next page.

Pearson's method can be stated as follows:

> For the χ^2-test, P is approximately equal to the area to the right of the observed value for the χ^2-statistic, under the χ^2-curve with the appropriate number of degrees of freedom.

For Example 1,

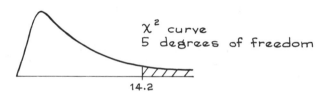

$$P \approx \text{shaded area}$$

This area can be found using tables or a statistical calculator.[5] In principle, there is one table for each curve but this would be so awkward that a different arrangement is used, as shown in Table 3. (This is extracted from a bigger one, on p. A-72.) Areas (in percent) are listed across the top of the table, while degrees of freedom are listed down the left side. For instance, look at the column for 5% and the row for 5 degrees of freedom. In the body of the table there is the entry 11.07, meaning that the area to the right of 11.07 under the curve for 5 degrees of freedom is 5%. The area to the right of 14.2 under the curve for 5 degrees of freedom cannot be read from the table, but it is between

Table 3. A short χ^2 table extracted from the bigger table on p. A-72.

Degrees of freedom	90%	50%	10%	5%	1%
1	0.016	0.46	2.71	3.84	6.64
2	0.21	1.39	4.60	5.99	9.21
3	0.58	2.37	6.25	7.82	11.34
4	1.06	3.36	7.78	9.49	13.28
5	1.61	4.35	9.24	11.07	15.09
6	2.20	5.35	10.65	12.59	16.81
7	2.83	6.35	12.02	14.07	18.48
8	3.49	7.34	13.36	15.51	20.09
9	4.17	8.34	14.68	16.92	21.67
10	4.86	9.34	15.99	18.31	23.21

Figure 1. The χ^2-curves for 5 and 9 degrees of freedom. The curves are not symmetric, having long right-hand tails. The curve for 9 degrees of freedom is flatter, more spread out, and further to the right.

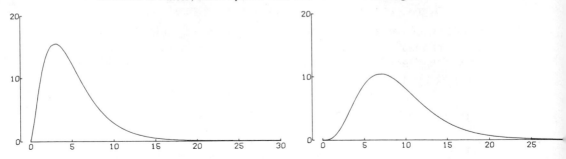

5% (the area to the right of 11.07) and 1% (the area to the right of 15.09). It is reasonable to guess that this area is just a bit more than 1%.

There is one point to watch. The normal table and t-table gave middle areas, but the χ^2-table is different: it gives right-hand tail areas.

How good is Pearson's approximation? Figure 2 shows the probability histogram for the χ^2-statistic with 60 rolls of a fair die.[6] A χ^2-curve with 5 degrees of freedom is superimposed on this histogram, for comparison. The

Figure 2. Pearson's approximation. The top panel shows the probability histogram for the χ^2-statistic with 60 rolls of a fair die, compared with a χ^2-curve (5 degrees of freedom). The bottom panel shows the ratio of the tail areas.

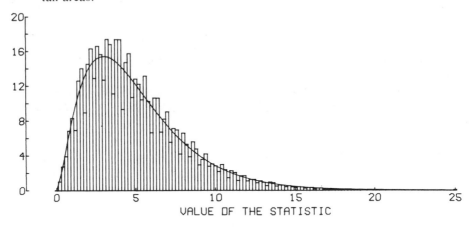

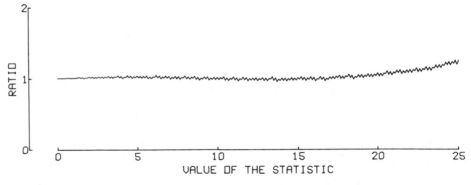

histogram is quite a bit bumpier than the curve, but follows it rather well. In particular, the area to the right of any x under the histogram is going to be very close to the corresponding area under the curve; the ratio of these two tail areas is graphed in the bottom panel, against x. For instance, $x = 14.2$ in Example 1. The area to the right of 14.2 under the histogram (including the box over 14.2) gives the exact value of P. This is 1.4382%. The area to the right of 14.2 under the curve gives Pearson's approximate value for P. This is 1.4388%.

As a rule of thumb, the approximation can be trusted when the expected frequency in each line of the table is ten or more. In Table 2, each expected frequency is 10, and the approximation was excellent. By comparison, the approximation would not be good if the model were drawing 100 times from the box

$$\boxed{1}\ \boxed{2}\ \boxed{3}\quad 96\ \boxed{4}\,\text{'s}$$

How is the χ^2-test related to the z-test for averages? If you are interested in the percentage composition of the box, then use the χ^2-test. If you are only interested in the average of the box, use the z-test. For instance, suppose the model is that of drawing with replacement from a box of tickets numbered 1 through 6; the percentages of the different kinds of tickets are unknown. To test the hypothesis that each value appears on $16\frac{2}{3}\%$ of the tickets, use the χ^2-test. Basically, there is only one box which satisfies this hypothesis:

$$\boxed{1}\ \boxed{2}\ \boxed{3}\ \boxed{4}\ \boxed{5}\ \boxed{6}$$

To test the hypothesis that the average of the box is 3.5, use the z-test. Of course, there are many different boxes which satisfy this hypothesis: for instance,

$$\boxed{1}\ \boxed{2}\ \boxed{3}\ \boxed{4}\ \boxed{5}\ \boxed{6}\quad \text{or}\quad \boxed{1}\ \boxed{1}\ \boxed{2}\ \boxed{3}\ \boxed{4}\ \boxed{5}\ \boxed{6}\ \boxed{6}$$

To sum up:

• The χ^2-test says whether the data are like the result of drawing at random from a box whose contents are specified.

• The z-test says whether the data are like the result of drawing at random from a box whose average is specified.[7]

2. THE STRUCTURE OF THE χ^2-TEST

Section 1 described an application of the χ^2-test. What are the ingredients?

(a) *The basic data.* This consists of some number of observations, usually denoted N. For the die, N was 60 and the basic data were in Table 1 on p. 471.

(b) *The chance model.* So far, only one kind of chance model has been considered in this chapter. There is a box of tickets, whose contents are specified. Then N draws are made at random with replacement from the box: N is the number of observations in the basic data. For the die, N was 60 and the box was

$$\boxed{\,\boxed{1}\ \boxed{2}\ \boxed{3}\ \boxed{4}\ \boxed{5}\ \boxed{6}\,}$$

(c) *The frequency table.*[8] The observed frequency of each value is obtained from the basic data, by counting. The expected frequency is obtained from N and the chance model. For the die, Table 2 on p. 471 is the frequency table.

(d) *The χ^2-statistic.* This is computed from the formula on p. 472, after completing the frequency table.

(e) *The degrees of freedom.* This is one less than the number of terms in the sum for χ^2, when the model is fully specified, as it is here.

(f) *The observed significance level.* This is approximated by the area to the right of the χ^2-statistic, under the χ^2-curve with the right number of degrees of freedom. This can be read from the χ^2-table, or found using a statistical calculator.

This follows the general pattern for making a test, as laid out in section 4 of Chapter 26.

The terminology is complicated. Please keep separate:

- the χ^2-curves, two of which are shown in Figure 1;
- the χ^2-table, which is at the back of the book;
- the χ^2-statistic, which is calculated from the data each time;
- the χ^2-test, which involves going through steps (*a–f*) above.

It is worth noticing that whatever is in the box, the same χ^2-curves and tables can be used to approximate P, provided N is large enough. Pearson's formula was developed to make this so. It is this result which justifies the formula. With almost any other definition, a different curve would be needed in every situation. Notice too that a mathematical theory is needed to connect the value of the χ^2-statistic with the table. The main assumption behind that theory is the box model, that the data are like the results of repeated draws from a box.

EXERCISE SET A

1. Find the area under the χ^2-curve with 5 degrees of freedom to the right of
    (a) 1.00   (b) 1.61   (c) 9.24   (d) 10   (e) 11.78   (f) 13.20.

2. Repeat Exercise 1 with 10 degrees of freedom.

Note: Without a calculator, you will have to guess some of the values in Exercises 1 and 2, using the table on p. A-72.

3. Suppose the observed frequencies in Table 2 had come out as shown in Table 4A, p. 477. Compute the value of χ^2, the number of degrees of freedom, and the observed significance level. What can be inferred? (In later exercises, this will be abbreviated as "Make a χ^2-test.")

4. Suppose the observed frequencies in Table 1 had come out as shown in Table 4B below. Make a χ^2-test.

5. Suppose that Table 1 had 600 entries instead of 60, with observed frequencies as shown in Table 4C. Make a χ^2-test of the model that the observations were generated by rolling a die 600 times.

6. Suppose that Table 1 had 60,000 entries, with the observed frequencies as shown in Table 4D.
   (a) Compute the percentage of times each value showed up.
   (b) Does the die look fair?
   (c) Make a χ^2-test of the model that the observations were generated by rolling a die 60,000 times.

Table 4A.		Table 4B.		Table 4C.		Table 4D.	
Value	Observed frequency	Value	Observed frequency	Value	Observed frequency	Value	Observed frequency
1	5	1	9	1	90	1	10,287
2	7	2	11	2	110	2	10,056
3	17	3	10	3	100	3	9,708
4	16	4	8	4	80	4	10,080
5	8	5	12	5	120	5	9,935
6	7	6	10	6	100	6	9,934

7. Someone wishes to test whether a die is fair. He rolls it 600 times. On each roll, he just records whether the result was even or odd, and large (4, 5, 6) or small (1, 2, 3). His observed frequencies turn out as follows:

	Large	Small
Even	183	113
Odd	88	216

Make a χ^2-test of the fairness of the die.

8. One study of grand juries in Alameda County, California, compared the demographic characteristics of jurors with the general population, to see if the jury panels were representative.[9] Here are the results for age. (Only persons 21 and over are considered; the county age distribution is known from Public Health Department data.)

Age	County-wide percentage	Number of jurors
21 to 40	42	5
41 to 50	23	9
51 to 60	16	19
61 and up	19	33
Total	100	66

One model is that these 66 jurors were selected at random from the population of Alameda County (aged 21 and up). Make a χ^2-test of this model. What can be inferred?

The answers to these exercises are on pp. A-64–A-65.

3. HOW FISHER USED THE χ^2-TEST

Fisher used the χ^2-statistic to show that Mendel's data were fudged (Chapter 25). For each of Mendel's experiments, Fisher computed the χ^2-statistic and its degrees of freedom. These experiments were all independent, for they involved different sets of plants.

> With independent experiments, it is legitimate to pool the results by adding up the values of the χ^2-statistic from each experiment. The degrees of freedom can be found by adding up the numbers of degrees of freedom for all the experiments considered.

For instance, if one experiment gives $\chi^2 = 5.8$ with 5 degrees of freedom, and another independent experiment gives $\chi^2 = 3.1$ with 2 degrees of freedom, the two together have a pooled χ^2 of $5.8 + 3.1 = 8.9$, with $5 + 2 = 7$ degrees of freedom. For Mendel's data, Fisher got a pooled χ^2 value under 42, with 84 degrees of freedom. The area to the left of 42 under the χ^2-curve with 84 degrees of freedom is about 4 in 100,000. This can safely be called decisive proof that the data were massaged.

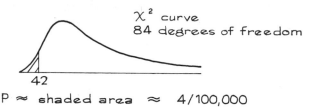

$$\chi^2 \text{ curve}$$
$$84 \text{ degrees of freedom}$$

42

$$P \approx \text{shaded area} \approx 4/100,000$$

At this point, a new principle seems to be involved: P was computed as a left-hand tail area, not a right-hand one. Why?

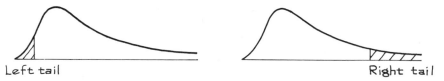

Left tail   Right tail

The basic reason is this. Fisher was not testing Mendel's chance model, he took that for granted. Instead, he was comparing two hypotheses:

- Mendel's were data gathered honestly (the null);
- Mendel's data were fudged to make the "observed" frequencies closer to the expected ones (the alternative).

Small values of χ^2 say the observed frequencies are too close to the expected ones, closer than ordinary chance variations would allow, and argue for the alternative hypothesis. Since it is small values of χ^2 that argue against the null in this problem, P must be computed as a left-hand tail area.

EXERCISE SET B

1. Suppose the same die had been used to generate the data in Exercises 3 and 5 in Set A, rolling it first 60 times for Exercise 3, and then 600 times for Exercise 5. Can you pool the results of the two tests? If so, how?

2. Suppose the same die had been used to generate the data in Exercises 3 and 5 in Set A, rolling it 600 times in all. The first 60 rolls were used for Exercise 3; but Exercise 5 reports the results on all 600 rolls. Can you pool the results of the two tests? If so, how?

3. One of Mendel's breeding trials came out as follows.[10] Make a χ^2-test to see whether these particular data were fudged.

Type of pea	Expected	Observed
Smooth yellow	313	315
Wrinkled yellow	104	101
Smooth green	104	108
Wrinkled green	35	32

The answers to these exercises are on p. A-65.

4. TESTING INDEPENDENCE

The χ^2-statistic is also used to look for dependence, as will be explained in this section. However, the discussion is rather technical, and it can be skipped. The method will be indicated by an example: Are handedness and sex dependent? More precisely, take some population—for instance, all people in the United States in 1962 aged 18 to 79. The question is whether the percentage of right-handers among all the men in this population differs from the percentage of right-handers among the women. If data were available, showing for each man and woman in the population whether they were right-handed or not, it would be possible to settle the issue directly, just by computing the percentages. Such information is not available. However, in 1962 the Health Examination Survey took a probability sample of 6,672 Americans aged 18 to 79. One of the things they found out for each sample person was handedness. The results are shown in Table 5: "other" means left-handed or ambidextrous.

Table 5. Handedness by sex: a 2×2 table from the Health Examination Survey, Cycle I.

	Men	Women
Right-handed	2,780	3,281
Other	311	300

This is called a *2 × 2 table,* because it has 2 rows and 2 columns. A more detailed analysis might require 3 values for the handedness variable:

right-handed, left-handed, ambidextrous. This would lead to a 3 × 2 table, with 3 rows and 2 columns. In general, when studying the relationship between two variables, of which one has *m* values and the other has *n* values, an *m* × *n* table is needed. However, this book will only deal with the simplest case, 2 × 2 tables.[11]

Returning to our example, the numbers in Table 5 are hard to grasp. A first step toward understanding them is to get the row and column totals, as in Table 6.

Table 6. Row and column totals from Table 5.

	Men	Women	Total
Right-handed	2,780	3,281	6,061
Other	311	300	611
Total	3,091	3,581	6,672

Using Table 6, it is possible to compute the percentage of men who were right-handed:

$$\frac{2,780}{3,091} \times 100\% \approx 89.94\%$$

Similarly, the percentage of women who were right-handed is

$$\frac{3,281}{3,581} \times 100\% \approx 91.62\%$$

So women are a bit likelier than men to be right-handed.

According to modern neurophysiology, right-handedness is associated with left-hemisphere dominance in the brain, the rational faculty ruling the emotional.[12] Does the sample show that women are more rational than men? Another interpretation: right-handedness is socially approved, left-handedness is socially deviant. Are women under greater pressure than men to follow the social norm for handedness? Still another interpretation: it's just chance variation. After all, even if the percentage of right-handers among all the men in the population were exactly equal to the corresponding percentage for the women, it could come out differently in the sample. Just by the luck of the draw, there could be too few right-handed men in the sample, or too many right-handed women. The χ^2-test is getting at the question of whether the observed difference in the sample reflects a real difference in the population, or is just the result of chance variation. If you are not worried about chance variation, do not make a χ^2-test.

The HES sampling design is too complicated to analyze by means of the χ^2-test.[13] (This issue came up in sampling, where the formula for the SE was seen to depend on the design, p. 366.) So we are going to pretend that Table 5 reports the results of a simple random sample, choosing 6,672 people at random without replacement from the population. A box model for the data can then be set up as follows. There is one ticket in the box, for each person in the population (Americans aged 18 to 79 in 1962). Each of these millions of tickets is marked in one of the following ways, to describe the corresponding person:

??	right-handed man	??	right-handed woman
??	other-handed man	??	other-handed woman

100 million tickets

The model is that the numbers in Table 5 were generated by drawing 6,672 tickets at random without replacement from this huge box, and counting to see how many tickets there were of each of the four different types. The percentage of each of the four different kinds of tickets in the box is an unknown parameter. So there are four parameters to the model.

Now it is time to formulate some hypotheses about the model. The null hypothesis is that handedness and sex are independent: the percentage of right-handers among all men in the population equals the corresponding percentage among women, and similarly for other-handedness. On this hypothesis, the difference in the sample percentages just reflects chance variation. The alternative hypothesis is dependence: the percentage of right-handers among all men in the population differs from the percentage for women. On this hypothesis, the difference in the sample reflects a difference in the population. To make a χ^2-test of the null hypothesis, we have to compare the observed frequencies (Table 5) with expected frequencies. The technique for computing these expected frequencies may seem a bit complicated and it is explained below. The expected frequencies themselves are shown in Table 7.

Table 7. Comparing observed and expected frequencies from Table 5.

	Observed frequencies		Expected frequencies		Difference	
	Men	Women	Men	Women	Men	Women
Right-handed	2,780	3,281	2,808	3,253	-28	28
Other	311	300	283	328	28	-28

The next step would be to compute the χ^2-statistic using the formula on p. 472:

$$\chi^2 = \frac{(-28)^2}{2,808} + \frac{28^2}{3,253} + \frac{28^2}{283} + \frac{(-28)^2}{328} \approx 5.68$$

How many degrees of freedom are there? Although there are four terms in the sum for χ^2, Table 7 shows there is only one degree of freedom: only one of the differences is free to vary, the others are the same except for sign.

With a 2 × 2 contingency table, there is only one degree of freedom.[14]

Now P can be computed.

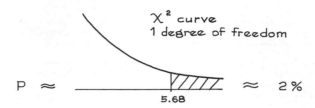

$$P \approx \underline{}\Big|\!\!/\!\!/\!\!/\!\!/\Big| \approx 2\%$$

χ^2 curve
1 degree of freedom

5.68

Thus, for Table 5 the observed significance level P is the area to the right of 5.68 under the χ^2-curve with one degree of freedom, and this is about 2%. There is fairly strong evidence to show that the percentage of right-handers among all the men in the population is different from the corresponding percentage for women. The observed difference in the sample seems to reflect a real difference in the population, rather than a chance variation. That is what the χ^2-test says.

What is left is to compute the expected frequencies in Table 7, and this will take some effort. To get started, suppose the null hypothesis were really specific: not only are handedness and sex independent, but also

- 40% of the population is male, and 60% female;
- 90% of the population is right-handed, and 10% other-handed.

These percentages were chosen to make the arithmetic easy. Using this specific null hypothesis, we can calculate the percentages of each of the four kinds of tickets in the box:

- the percentage of right-handed men is 90% of 40% = 36%
- the percentage of other-handed men is 10% of 40% = 4%
- the percentage of right-handed women is 90% of 60% = 54%
- the percentage of other-handed women is 10% of 60% = 6%

To get the expected frequencies in the sample, on this specific null hypothesis, we would just take the corresponding percentages of the sample size, 6,672. For instance, the expected number of right-handed men would be

$$36\% \text{ of } 6{,}672 \approx 2{,}402$$

The trouble is that the real null hypothesis is much less specific. It only says that handedness and sex are independent. It doesn't specify the percentage of men in the population, or the percentage of right-handers. So, these percentages must be estimated from the sample, before we can compute the expected frequencies. Going back to Table 6,

- the percentage of men in the sample is

$$\frac{3{,}091}{6{,}672} \times 100\% \approx 46.33\%$$

- the percentage of women in the sample is

$$\frac{3{,}581}{6{,}672} \times 100\% \approx 53.67\%$$

• the percentage or right-handed people in the sample is

$$\frac{6{,}061}{6{,}672} \times 100\% \approx 90.84\%$$

• the percentage of other-handed people in the sample is

$$\frac{611}{6{,}672} \times 100\% \approx 9.16\%$$

These four percentages in the sample estimate the corresponding percentages in the population.

Using the assumption of independence, we can now estimate the percentage of each of the four different kinds of tickets in the box:

• the percentage of right-handed men should be around
    90.84% of 46.33% ≈ 42.09%
• the percentage of other-handed men should be around
    9.16% of 46.33% ≈ 4.24%
• the percentage of right-handed women should be around
    90.84% of 53.67% ≈ 48.75%
• the percentage of other-handed women should be around
    9.16% of 53.67% ≈ 4.92%

With 6,672 draws from the box, the expected numbers in the sample can now be worked out, on the basis of independence in the population:

• the expected number of right-handed men in the sample is
    42.09% of 6,672 ≈ 2,808
• the expected number of other-handed men in the sample is
    4.24% of 6,672 ≈ 283
• the expected number of right-handed women in the sample is
    48.75% of 6,672 ≈ 3,253
• the expected number of other-handed women in the sample is
    4.92% of 6,672 ≈ 328

We kept two decimals in the percents, because they were going to be multiplied by the large number of people in the sample. The results were shown in Table 7.

A final, and somewhat technical question must be answered: Why was there only one degree of freedom in Table 7? The basic reason is that the two free parameters of the model, namely the percentage of men and the percentage of right-handers, have been estimated from the data. This makes the model fit the facts better. As can be seen in Table 7, for instance, the expected frequency of men equals the observed frequency, and similarly for right-handers; so the rows and columns of the difference table add up to 0. In consequence, χ^2 is smaller, and to compensate the degrees of freedom must be reduced by the number of parameters which have been estimated.

The other two parameters that were estimated above, namely the percentage of women and the percentage of other-handed people, are redundant:

    percentage of women = 100% − percentage of men
percentage of other-handed = 100% − percentage of right-handed.

EXERCISE SET C

1. (Hypothetical.) In a certain town, there are about one million eligible voters. A simple random sample of size 10,000 was chosen, to study the relationship between sex and participation in the last election. The results:

	Men	Women
Voted	2,792	3,591
Didn't vote	1,486	2,131

Make a χ^2-test of the hypothesis that sex and voting are independent.

2. Subjects in the Health Examination Survey were asked (among other things): "Have you ever felt you were going to have a nervous breakdown?" The results are shown for men and women aged 25 to 34, cross-tabulated by marital status.[15] Suppose these data resulted from a simple random sample.
    (a) Separately for men and for women, make a χ^2-test for independence.
    (b) How do you interpret the results?

	Men aged 25 to 34		Women aged 25 to 34	
	Married	Never married	Married	Never married
Yes	40	9	143	5
No	503	90	431	57

The answers to these exercises are on p. A-65.

5. REVIEW EXERCISES

1. As part of a study on the selection of grand juries in Alameda county, the educational level of grand jurors was compared with the county distribution:[16]

Educational level	County	Number of jurors
Elementary	28.4%	1
Secondary	48.5%	10
Some college	11.9%	16
College degree	11.2%	35
Total	100.0%	62

Could a simple random sample of 62 people from the county show a distribution of educational level that different from the county-wide one? Choose one option and explain why.
    (i)   This is absolutely impossible.
    (ii) This is possible, but fantastically unlikely.
    (iii) This is possible but unlikely—the chance is around 1% or so.
    (iv) This is quite possible—the chance is around 10% or so.
    (v)   This is nearly certain.

2. Each respondent in the Health Examination Survey was classified as "working" or "not at work." The results for men aged 25 to 34 can be cross tabulated by marital status, as follows:[17]

	Married	Never married
Working	506	83
Not at work	11	11

Thus, 98% of the married men were working, compared to 88% for the never-marrieds. Can this difference be explained as the result of chance variation? (Suppose the table results from a simple random sample, and use the method of section 4.)

3. Someone claims to be rolling a pair of fair dice. To test his claim, you make him roll the dice 360 times, and you count up the number of times each sum appears. The results are shown below. (For your convenience, the chance of throwing each sum with a pair of fair dice is shown too.) Should you play craps with this individual?

Sum	Chance	Frequency
2	1/36	11
3	2/36	18
4	3/36	33
5	4/36	41
6	5/36	47
7	6/36	61
8	5/36	52
9	4/36	43
10	3/36	29
11	2/36	17
12	1/36	8

4. The International Rice Research Institute in the Philippines is developing new lines of rice which combine high yields with resistance to disease and insects. The technique involves crossing different lines to get a new line which has the most advantageous combination of genes; detailed genetic modeling is required. One project involved breeding new lines for resistance to an insect called "brown plant hopper"; 374 lines were raised, with the results shown below.[18]

	Number of lines
All plants resistant	97
Mixed: some plants resistant, some susceptible	184
All plants susceptible	93

According to the IRRI model, the lines are independent; each line has a 25% chance to be resistant, a 50% chance to be mixed, and a 25% chance to be susceptible. Are the facts consistent with this model?

5. Two people are trying to decide whether a die is fair. They roll it 100 times, with the results shown below. One person wants to make a z-test, the other wants to make a χ^2-test. What does statistics say?

21 [⚀]'s   15 [⚁]'s   13 [⚂]'s   17 [⚃]'s   19 [⚄]'s   15 [⚅]'s

average of numbers rolled ≈ 3.43,   SD ≈ 1.76
average of list 1, 2, 3, 4, 5, 6 = 3.5,   SD ≈ 1.71

6. SUMMARY

1. The hypothesis that data were generated according to a particular chance model can be tested using the χ^2-statistic.

2. $\chi^2 = $ sum of $\dfrac{(\text{observed frequency} - \text{expected frequency})^2}{\text{expected frequency}}$

3. When the model is fully specified (no parameters to estimate from the data),

degrees of freedom = number of terms − one.

4. The observed significance level P can be approximated as the area under the χ^2-curve to the right of the observed value for χ^2. This gives the chance of the model producing observed frequencies as far from the expected frequencies as those at hand, or even further. Distance is measured by χ^2.

5. Sometimes, the model can be taken as true, and the problem is to decide whether the data have been fudged to make the observed frequencies closer to the expected ones. Then P would be computed as the area to the left of the observed value for χ^2.

6. If experiments are independent, the χ^2-statistics can be pooled by addition. The degrees of freedom are just added too.

7. The χ^2-statistic can also be used to test for independence. This is proper when the data have been obtained from a simple random sample, and it is desired to make an inference about the population. With a 2×2 table, there is only one degree of freedom.

29

A Closer Look at
Tests of Significance

*One of the misfortunes of the law [is that] ideas become encysted in phrases
and thereafter for a long time cease to provoke further analysis.*
—OLIVER WENDELL HOLMES (UNITED STATES, 1809–1894)[1]

1. WAS THE RESULT IMPORTANT?

The previous three chapters explained how tests of significance are
made. This chapter, however, is unrelentingly negative: it explains the limitations of significance tests. The first one is that "significance" is a technical
word. A test can only deal with the question of whether a difference is real, or
just a chance variation. It is not designed to see whether the difference is
important.[2]

This point is easiest to understand in the context of a hypothetical example (based on Exercise 2, p. 462). Suppose that an investigator wants to
compare WISC vocabulary scores for big-city and rural children, aged 6 to
9. He takes a simple random sample of 2,500 big-city children, and an independent simple random sample of 2,500 rural children. The big-city children
average 26 on the test, with an SD of 10 points, while the rural children only
average 25, with the same SD of 10 points. What does this one-point difference mean? To find out, the investigator makes a two-sample z-test. The SE
for the difference can be estimated as 0.3, so $z = (26 - 25)/0.3 \approx 3.3$, and P

is about 5 in 10,000. The difference between big-city children and rural children is highly significant, rural children are lagging behind in the development of language skills, and the investigator launches a crusade to pour money into rural schools.

The common-sense reaction must be: slow down. The z-test is only telling us that the one-point difference between the sample averages is almost impossible to explain as a chance variation. To focus the issue, suppose that the difference observed in the sample applies to the whole population. Let us even suppose that the samples are a perfect image of the population, so that all the big-city children in the United States (not just the ones in the sample) would average 26 points on the WISC vocabulary scale, while the average for all the rural children in the United States would be 25 points. Then what? There is no more chance variation to worry about, so no test of significance can help in judging the difference. All the facts are in, and the problem is to find out what the difference means.

To do that, it is necessary to look at the WISC vocabulary scale itself. There are forty words which the child has to define. Two points are given for a correct definition, and one point for a partially correct definition. So the one-point difference between big-city and rural children only amounts to a partial understanding of one word out of forty. This is not a solid basis for a crusade. Quite the opposite: what this investigator has done is to prove that there is almost no difference between big-city and rural children on the WISC vocabulary scale.

Of course, the sample does not reflect the population perfectly, so a standard error should be attached to his estimate for the difference. Based on the two samples of 2,500 children, the difference in average scores between all the big-city and rural children in the United States would be estimated as one point, give or take 0.3 points or so.

Big samples are good, because they enable investigators to measure small differences quite accurately—with very small SEs. But the z-test compares a difference to its SE. Therefore, with a large sample, even a small difference can lead to an impressive value for z. The z-test can be too sensitive for its own good.

> The P-value of a test depends on the sample size. With a large sample, even a small difference can be "statistically significant," that is, hard to explain by the luck of the draw in sampling. This doesn't necessarily make it important.

The straw-man investigator in the example made a definite blunder. His P-value of 5 in 10,000 looked easier to understand, and much more impressive, than the one-point difference in the sample averages. So he paid attention to the P-value, not the difference itself. But it is the difference which tells you what is going on in the world, not the P-value. The P-value only gets at a statistical issue: whether the difference in the sample averages can be explained as a chance variation. (In the example, that would mean getting too many smart children in the big-city sample, or too few in the rural sample, by the luck of the draw.)

The conclusion can be stated in general. To interpret his findings, an investigator should begin by pretending that his sample is a perfect image of the population, so the percentages, averages, SDs for the population are exactly equal to those observed in the sample. Then he should see what the results mean in practical terms. This is a "test of real significance." The final step is to cope with chance error—the luck of the draw in sampling. One way to do this is to compute P-values. Another is to attach standard errors to the estimates. The standard errors are better, because they are easier to understand.

> In addition to reporting the P-values for their tests, investigators should usually summarize the data. This lets the reader judge the practical importance of a difference.

Example 1. Using data from the Health Examination Survey, a χ^2-test was made of the null hypothesis that handedness and sex are independent (p. 479). The observed significance level turned out to be 2%, so the result was statistically significant. True or false:

(a) Handedness and sex must be very dependent.

(b) This dependence must be very important.

Solution. Both statements are false. The observed significance level doesn't measure the size of the difference or its importance. It only tells you that the difference is hard to explain by the luck of the draw. The HES sample was quite large (6,672 persons), so even small differences can be statistically significant. Going back to the data (Table 5 on p. 479), 91.6% of the women in the sample were right-handed, compared to 89.9% of the men. The difference is 91.6% − 89.9% = 1.7%. It is this difference which you have to look at in order to see whether the dependence is strong or weak, important or unimportant.[3] The χ^2-test has nothing to do with it.

Investigators are sometimes tempted to look at P-values instead of the differences themselves just because the P-values seem easier to understand. This confusion is most visible when they make a test on data for the whole population.

Example 2. In 1960, the population of the United States amounted to 179 million persons, of whom 11.3% were under the age of 5 years. In 1970, the size of the population had increased to 203 million, of whom 8.4% were under the age of 5. These figures are based on a complete count of the population— the Decennial Census.[4] Is the difference between the 1960 and 1970 percentages statistically significant? Or does the question make sense?

Solution. There is nothing to stop you from carrying out the arithmetic for a two-sample z-test, as on p. 459, except that the results would not make any sense. There is no problem about whether, by the luck of the draw, there were too many young children in the sample. There was no sample. The difference may be very important, for instance to school administrators. Or it may not be so important. But the z-test does not help in deciding. In fact, in this example, the z-test is totally irrelevant. A test of significance only makes sense when there is chance variation to worry about.

If an investigator makes a test of significance when he has data for the whole population, watch out.

EXERCISE SET A

1. Say whether each of the following statements are true or false, and explain why.
   (a) A difference which is highly statistically significant must be very important.
   (b) Big samples are bad.
   (c) In the example on the WISC vocabulary scale (p. 487), the z-test automatically checks to see whether the scale is culturally biased or not.

2. A large university wants to compare the performance of male and female undergraduates on a standardized reading test, but can only afford to do this on a sample basis. An investigator chooses 100 male undergraduates at random, and independently 100 females. The men average 49 on the test, with an SD of 10 points. The women average 51 on the test, with the same SD of 10 points. Is the difference real, or a chance variation? Or does the question make sense?

3. Repeat Exercise 2, keeping the averages and SDs the same, but increasing the sample sizes from 100 to 400.

4. In 1960, there were 45 million families, with an average size of 3.7, and an SD of 1.5 persons. In 1970, there were 52 million families; the average size had dropped to 3.6, with the same SD of 1.5 persons. These figures are based on a complete count—the Decennial Census. Is the drop in average family size statistically significant? Or does the question make sense?

5. An investigator makes a study of the selection of freeway routes by the California Division of Highways in San Francisco and Los Angeles from 1958 to 1966.[5] He develops a chance model in order to assess the effects of different variables on the decisions. What he calls "external political and public variables" include the views of other state agencies, school boards, businesses, large property owners, and property owners' associations. To find out how much influence these variables have on the freeway decisions, the investigator makes a test of significance. The null hypothesis is that the external political and public variables do not influence route choice, and the observed significance level is about 3%. Since the result is statistically significant but not highly statistically significant, he concludes that these factors do influence the freeway decisions, but that their impact is relatively weak. Comment briefly.

6. Someone explains the point of a test of significance as follows.[6] "If the null hypothesis is rejected, the difference isn't trivial. It is bigger than what would occur just by chance." Comment briefly.

The answers to these exercises are on pp. A-65–A-66.

2. DOES THE DIFFERENCE PROVE THE POINT?

Usually, the investigator collects data to prove a point. But if the results can be explained by chance—the null hypothesis—they may not prove anything. So the investigator makes a test of significance to show that his difference is real. But that is all the test can do for him. Even if the difference is real, it still may not prove the point that the investigator wants to make. Tests of significance answer the question, "Is the difference real?" They do not answer the question, "What caused the difference?"

For an example, take an ESP experiment in which a die is rolled, and the subject tries to make it land showing six spots.[7] This is repeated 720 times, and the die lands six in 143 of these trials. If the die is fair, and the subject's efforts have no effect, the die has 1 chance in 6 to land six. So in 720 trials, the expected number of sixes is 120. There is a surplus of $143 - 120 = 23$ sixes.

Is this difference real, or a chance variation? This is where a test of significance comes in. The null hypothesis can be formulated in terms of a box model: the number of sixes is like the sum of 720 draws from the box

$$\boxed{0}\ \boxed{0}\ \boxed{0}\ \boxed{0}\ \boxed{0}\ \boxed{1}$$

The SE for the sum of the draws is $\sqrt{720} \times \sqrt{1/6 \times 5/6} = 10$, so $z = (143 - 120)/10 = 2.3$, and P is about 1%. The difference looks real.

But now another question must be asked. Does the difference indicate that ESP exists? After all, that is the point. As it turned out, in another part of the experiment the subject tried to make the die land showing aces—and got too many sixes. In fact, whatever number the subject tried for, he got too many sixes. This doesn't prove that ESP exists. It proves that the die was biased. The cause of the extra sixes was not ESP, but the bias in the die.

A test of significance cannot tell you what the cause of a difference is; it can only tell you that the difference is there. This brings us back to the design issues discussed in Part I. With a well-designed experiment, a real difference proves the investigator's point. With a badly designed study, no difference can prove anything.

> A test of significance does not check the design of a study.

In the ESP experiment, why did the z-test lead us astray? The answer is that it didn't. The test was asked whether there were too many sixes to explain by chance, and it answered, correctly, that there were. But the test was told what "chance" meant: rolling a fair die. The expected value and SE that were substituted into the formula for z were computed on this assumption. All tests of significance have this feature: they have to be told what chances to use. That is what the box model does. If the investigator gets the model wrong, as in the ESP example, it is not fair to blame the test.

In general, many assumptions have to be made about the process generating the data before a statistical test can be done. For example, a box model assumes that there is no dependence in the data, no trends or patterns. Often, the investigator can formulate his null and alternative hypotheses in terms of the box. For instance, the null hypothesis might specify the average of the box. If P turns out to be small, this is usually evidence against the null hypothesis, and for the alternative. However, there is another possible interpretation for a small P-value: the whole box model could be wrong. If so, it would be a serious mistake to think that the results proved an alternative hypothesis about the box.

The procedure for carrying out a test doesn't force the investigator to check the model. This is a real flaw in the method, because usually the investigator sets things up so that what he wants to prove becomes the alternative hypothesis. He leaves its opposite—the null hypothesis—to take the heat and look ridiculous when a difference turns up. This is almost bound to happen if the model is wrong, and the investigator has a large enough sample (section 1). In a way, tests of significance reward investigators for getting the model wrong.

Some investigators adopt a different strategy. They develop a chance model on the basis of a theory that they believe in, and want to test experimentally. This model is taken as the null hypothesis, so that differences between the data and what is expected on the model are explained as chance variations. If the differences are too large to explain this way, the model has failed the test, and must be rejected. If the differences are of the size that would be expected just by chance, the model has passed the test. In this situation, the investigator hopes not to have to reject the null. If the experimental test is harsh, and the model passes, this is very convincing evidence in favor of the theory.[8]

EXERCISE SET B

1. Section 1 of Chapter 26 discussed an experiment in which flex-time was introduced at a plant, and on a sample basis, absenteeism dropped from 6.3 to 5.4 days off work, on the average. A test indicated that this difference was real. Is it fair to conclude that flex-time made the difference? If not, what are some other possible explanations? How could the design of this experiment be improved?

2. Section 5 of Chapter 26 discussed an ESP experiment in which subjects tried to guess which number a machine had picked (from 1 to 4). A test of significance showed that there were many more correct guesses than could be explained by chance. Is it fair to conclude that ESP exists? If not, what are some other possible explanations?

3. Section 5 of Chapter 26 also discussed the Salk vaccine field trial, where there were many fewer polio cases in the vaccine group than in the control group. A test of significance showed that the difference was real. Is it fair to conclude that the vaccine protected the children against polio? If not, what are some other possible explanations?

The answers to these exercises are on p. A-66.

3. FIXED-LEVEL TESTING

How small does P have to get before you reject the null hypothesis? As reported on p. 444, many statisticians draw a line at 5%. If P is less than 5%, the result is "statistically significant," and the "null hypothesis is rejected at the 5% level." However, the question is almost like asking how cold it has to get before you are entitled to say, "It's cold." A temperature of 70°F is balmy, -20°F is cold indeed, and there is no sharp dividing line between the two. Logically, it is the same with testing. There is no sharp dividing line between probable and improbable results. A P-value of 5.1% means just about the same thing as one of 4.9%—especially if both were computed using the normal approximation, which can easily introduce errors bigger than a tenth of a percent. In fact, however, these two P-values would be treated quite differently, because many journals will only publish results which are "statistically significant"—the 5% line. Some of the more prestigious journals will only publish results which are "highly statistically significant"—the 1% line.[9]

These arbitrary lines are taken so seriously that many investigators only report their results as "statistically significant" or "highly statistically significant." They don't even bother telling you what test they used, let alone the value of P. Our recommendation is to summarize the data, say what test was used, and report its P-value, instead of just comparing P to 5% or 1%.

Historical note: Where do the 5% and 1% lines come from? To find out, we have to look at the way statistical tables are laid out. The χ^2-table (p. A-72) is a good example. Part of it is reproduced below as Table 1.

Table 1. A small χ^2-table.

Degrees of freedom	50%	30%	10%	5%	1%
1	0.46	1.07	2.71	3.84	6.64
2	1.39	2.41	4.60	5.99	9.21
3	2.37	3.67	6.25	7.82	11.34
4	3.36	4.88	7.78	9.49	13.28
5	4.35	6.06	9.24	11.07	15.09

How is this table used in testing? Suppose an investigator is making a χ^2-test with 3 degrees of freedom. He is a believer in the 5% line, and therefore wants to know how big his χ^2-statistic has to be in order to achieve "statistical significance"—a P-value below 5%. The table is laid out to make this easy for him. He looks across the row for 3 degrees of freedom and down the column for 5%, finding the entry 7.82 in the body of the table. In other words, the area to the right of 7.82 under the curve for 3 degrees of freedom is 5%. So the result is "statistically significant" as soon as his χ^2 is more than 7.82. In other words, the table gives the cutoff for "statistical significance." Similarly, it gives the cutoff for the 1% line, or for any other significance level listed across the top of the table.

R. A. Fisher was one of the first to publish such tables, and it seems to have been his idea to lay them out this way. There is a limited amount of room across the top of the table. Once the number of levels was limited, 5% and 1% stood out as nice round numbers, and they soon acquired a magical life of their own. With modern calculators, this kind of table is almost obsolete. So are the 5% and 1% levels.[10]

4. TESTING MANY HYPOTHESES

The point of a test of significance is to help distinguish between a real difference and a chance variation. So people sometimes jump to the conclusion that a result which is "statistically significant" cannot possibly be explained as a chance variation. This is false. Even if the null hypothesis is right, there is a 5% chance of getting a difference which the test will call "statistically significant." This 5% chance could have happened to you—an unlikely event, but not impossible. Similarly, on the null hypothesis there is 1% chance to get a difference which is "highly statistically significant," but just a fluke. On the average, an investigator who makes a hundred tests will come up with five results which are "statistically significant," and one which is "highly statistically significant," even if the null hypothesis is right in every case, so that each difference is just a fluke. It is a case of taking the pitcher too often to the well. There is no way to determine, for sure, whether a difference is real or just due to chance.

Investigators often decide which hypotheses to test, and how many tests to run, only after they have seen the data. Statisticians call this *data-snooping*.[11] It is a perfectly reasonable thing to do, but investigators who do it really ought to say how many hypotheses they formulated, and how many tests they ran, before the "statistically significant" difference turned up. And to cut down the chance of being fooled by "statistically significant" flukes, they ought to test their conclusions on an independent batch of data, for instance by replicating the experiment.[12] This good advice is seldom followed.

Much ink has been spilled over one relatively minor aspect of this issue: whether to use *one-tailed* or *two-tailed* tests. The point is easiest to see in a hypothetical example. Suppose it is desired to test whether a coin is fair: does it land heads with probability 50%? The coin is tossed 100 times, and it lands heads on 61 of the tosses. If the coin is fair, the expected number of heads is 50; the null hypothesis is that the difference between 61 and 50 just represents chance variation.

To test this null hypothesis, a box model is needed. The model consists of 100 draws from the box

$$\boxed{?? \boxed{0}\text{'s} \quad ?? \boxed{1}\text{'s}} \quad 0 = \text{tails, } 1 = \text{heads.}$$

The fraction of 1's in this box is an unknown parameter, representing the probability of heads. The null hypothesis is that the fraction of 1's in the box is 1/2—the coin is fair. The test statistic is

$$z = \frac{\text{observed} - \text{expected}}{\text{SE}} = \frac{61 - 50}{5} = 2.20$$

One investigator might formulate the alternative hypothesis that the coin is biased toward heads: in other words, that the fraction of 1's in the box is bigger than 1/2. (This is very reasonable—after you've seen the data.) On this basis, large positive values of z favor the alternative hypothesis, but negative values of z do not. Therefore, the values of z which favor the alternative hypothesis more than the observed one are those which are bigger than 2.20.

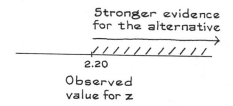

So P is figured as the area to the right of 2.20 under the normal curve:

Before seeing the data, another investigator might formulate a second alternative hypothesis: that the probability of heads differs from 50%, in either direction. In terms of the model, this second alternative hypothesis is that the fraction of 1's in the box differs from 1/2—bigger or smaller. On this basis, large positive values of z favor the alternative, but so do large negative values. If the number of heads is 2.20 SEs above the expected value of 50, that's bad for the null hypothesis. And if the number of heads is 2.20 SEs below the expected value, that's just as bad. The z-values more extreme than the observed 2.20 are:

• 2.20 or more

or

• −2.20 or less.

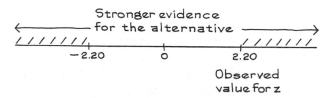

And on this basis, P is figured differently:

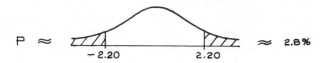

The first way of figuring P is called a *one-tailed z-test,* the second is called a *two-tailed z-test.* The one to use depends on the precise form of the alternative hypothesis. It is just a matter of seeing which z-values argue even more strongly for the alternative hypothesis than the observed one. For example, the two-tailed test is appropriate when the alternative hypothesis is that the average of the box differs from a given value—bigger or smaller. The one-tailed test is appropriate, for instance, when the alternative hypothesis is that the average of the box is bigger than the given value.

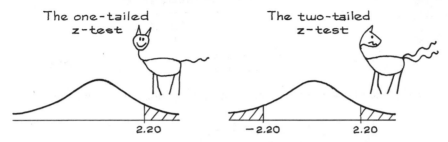

In fact, it doesn't matter very much whether an investigator makes a one-tailed or a two-tailed z-test, as long as he tells you which it was. For instance, if it was one-tailed, and you think it should have been two-tailed, just double the P-value.

To see why such a fuss is made over this issue, suppose an investigator makes a two-tailed z-test, and gets $z = 1.85$, and $P \approx 6\%$.

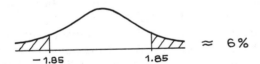

Naturally, he wants to publish. But as it stands, most journals won't touch his report—the result isn't "statistically significant."

What can he do? He could refine his experimental technique, gather more data, use sharper analytical methods. All this is hard. The other possibility is much simpler: he could do a one-tailed test.

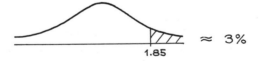

It is the arbitrary lines at 5% and 1% which make the distinction between two-tailed and one-tailed tests loom so large.

With the coin-tossing example, it was easy to double the observed significance level to take account of possible data-snooping—formulating a one-sided alternative hypothesis after seeing that the number of heads was too big. With other kinds of data-snooping, it is much harder to adjust P. One rule of thumb is to multiply P by the number of tests that were run. In general, this gives an overestimate of the adjusted P-value.

EXERCISE SET C

1. An ESP investigator tests a hundred subjects to see if they can guess the results of tossing a coin. Several subjects score significantly above the chance level, and one subject gets a score which is "highly statistically significant." Comment briefly.

2. One hundred draws are made at random from box X. The average of the draws is 51.8, with an SD of 9. The null hypothesis is that the average of the box equals 50, while the alternative hypothesis is that the average of the box differs from 50. Would you make a one-tailed or a two-tailed z-test?

3. Repeat Exercise 2 for the alternative hypothesis that the average of the box is bigger than 50.

4. True or false: "If a result is statistically significant, there are only 5 chances in 100 for it to be due to chance, and 95 chances in 100 for it to be real."

The answers to these exercises are on pp. A-66–A-67.

5. THE ROLE OF THE MODEL

A test of significance answers the question, "Is the difference due to chance?" But the test can't do its job until the word "chance" has been given a precise definition. That is what the box model does.

> Unless there is a clearly defined chance model, a test of significance makes no sense.

This may be hard to understand, because the arithmetic of the test does not seem to use the box model directly. But the fact is that the formulas for the expected value and standard errors, and the statistical tables (normal, t, and χ^2) all make a tacit assumption: that the data are like draws from a box. If this assumption is false, the formulas and the tables do not apply, and may give silly results.

Example 1. An investigator has a list of 100 scores. These average out to 53, with an SD of 11 points. He wants to know whether the 53 is significantly bigger than 50, so he makes a z-test:

$$\text{SE for sum of scores} \approx \sqrt{100} \times 11 = 110$$
$$\text{SE for average of scores} \approx 110/100 = 1.1$$
$$z = (53 - 50)/1.1 \approx 2.7$$

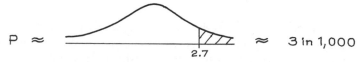

$$P \approx \qquad \approx 3 \text{ in } 1,000$$
$$2.7$$

Is the z-test appropriate here?

Solution. The investigator is confused. If he wants to know whether the difference between 53 and 50 is important, he has to think about what the scores mean; the z-test is irrelevant (section 1). On the other hand, he may want to know whether the difference between 53 and 50 is just due to chance. If so, he implicitly made a whole series of assumptions in calculating the SE and using the normal curve. Basically, he is assuming that his scores are like draws made at random with replacement from a box, and he is asking whether the average of the box equals 50 (the null hypothesis) or is bigger than 50 (the alternative hypothesis). This is the only way to make sense out of this calculation.

The next problem is whether the assumptions, and the question about the box, are reasonable. This depends on where the hundred scores came from, and what box he might be talking about. The investigator would be much better off if he formulated the assumptions explicitly. Then, he could deal with the issues.[13]

Example 2. Graduate division records at the University of California, Berkeley, for fall, 1973, can be used to cross-tabulate admissions by sex, for the largest graduate major on the campus:

	Men	*Women*
Admit	509	89
Deny	316	19

Did this department discriminate against the men? If appropriate, make a χ^2-test for the independence of admissions and sex.

Solution. There is nothing to prevent you from computing χ^2 and P, just as in section 4 of Chapter 28. However, to make sense out of the results, a box model would be needed. And it is hard to invent a plausible one. Chance does seem to enter at two places: first, in recruiting the applicants, and second, in deciding which applicants to admit. However, it is almost impossible to define the pool of potential applicants, and even if you could, the actual applicants were not drawn from this pool by any probability method, so there is no way to calculate chances. Therefore, the first source of variability cannot be dealt with by a box model. Neither can the second source—the selection procedure itself. Departments do not admit candidates by drawing names from a hat, they use their judgment.

There are other ways to set up a chance model, but their relevance to the issue is not clear.[14] The table by itself does not say whether this department discriminates on the basis of sex, and making a χ^2-test is the wrong way to find out.

Statisticians distinguish between samples drawn by probability methods (p. 307) and samples of convenience (p. 366). A sample of convenience consists of whoever is handy—students in freshman psychology class, the first hundred people you bump into, or all the applicants to a given department in a given year. With a sample of convenience, there is no way to calculate chances, no way to interpret the phrase "the difference is due to chance," and no way to make sense out of P-values. The P-value in Example 2 was based on a sample of convenience.

> Be very suspicious of P-values computed from samples of convenience.

Example 3. One month the Current Population Survey (p. 356) sample included 6,000 white men aged 16 to 19 in the civilian labor force, of whom 11% were unemployed. The sample also included 500 black men aged 16 to 19 in the civilian labor force, of whom 23% were unemployed. An investigator wants to know whether this difference in unemployment rates is real or a chance variation.

    (a) Does the question make sense?

    (b) Someone proposes to settle it by making a two-sample z-test, as follows:

SE for number of unemployed whites $\approx \sqrt{6{,}000} \times \sqrt{0.11 \times 0.89} \approx 24$

SE for white unemployment rate $\approx (24/6{,}000) \times 100\% \approx 0.4$ of 1%

SE for number of unemployed blacks $\approx \sqrt{500} \times \sqrt{0.23 \times 0.77} \approx 9$

SE for black unemployment rate $\approx (9/500) \times 100\% \approx 1.8\%$

SE for difference $\approx \sqrt{(0.4)^2 + (1.8)^2} \approx 1.8\%$

$z = (11 - 23)/1.8 \approx -6.7, \quad P \approx \quad \approx 0$

The arithmetic is all in order, but comment briefly on the logic.

Solution. The question makes perfect sense, because the Current Population Survey is based on a probability sample. And the difference is much too large to be a chance variation. However, the calculations of the standard errors are all wrong. They assume two independent simple random samples, whereas in fact a cluster sample is used (p. 364). The SEs would have to be estimated by another method.

It is tempting to think that the layout of the data is what governs the test. With one list of numbers, you make a z-test; with two lists, a two-sample z-test; with a 2×2 table, a χ^2-test, and so on. Unfortunately, it is not that simple. To know what test is appropriate, you have to know how the data were generated. In other words, you need a box model. It is the box model that defines the chances, not the data.

Example 4. Project Follow Through was established by Congress in 1967 to support the academic gains made by minority children in the Headstart pre-school program, when these children enrolled in regular schools. Seven sponsors were given contracts to run project classrooms according to different educational philosophies, and certain other classrooms were used as controls. The Stanford Research Institute (SRI) was hired to evaluate the project, for the Department of Health, Education, and Welfare. The evaluation was reported in *Follow Through Classroom Evaluation 1972–73.*[15] One important question studied in this report was whether the project classrooms really were different from the control classrooms; of course, SRI also studied the effect of the programs on the children.

To see whether or not there were real differences, SRI devised an implementation score to compare project classrooms with control classrooms. This score involved observing the classrooms to determine, for instance, the amount of time children spent playing, working independently, asking questions of the teacher, and so on. (How well this score measures implementation is another problem, which will not be taken up here.) The results for one sponsor, Far West Laboratory, are shown in Table 2.

Table 2. SRI Implementation Scores for twenty Far West Laboratory classrooms. Scores are between 0 and 100.

Site	Classroom scores			
Berkeley	73	79	76	72
Duluth	76	84	81	80
Lebanon	82	76	84	81
Salt Lake City	81	86	76	80
Tacoma	78	72	78	71

The average of these 20 scores is about 78, with an SD of about 4.2. The average score for the control classrooms was about 60, so the difference is 18 points. As far as the SRI Implementation Score is concerned, the Far West classrooms are very different from the control classrooms.

So far, so good. However, SRI was not satisfied. They wished to make a z-test,

to test whether the average implementation score for Follow Through was significantly greater than the average for Non-Follow Through.

The computation is as follows.[16] The SE for the sum of the scores is estimated as $\sqrt{20} \times 4.2 \approx 19$, so the SE for their average is $19/20 \approx 1$, and $z \approx (78 - 60)/1 = 18$. Now

$$P \approx \qquad\qquad\qquad \approx 0$$
$$18$$

The inference is:

the overall Far West classroom average is significantly different from the Non-Follow Through classroom average of 60.

The arithmetic is all in order, and the procedure may seem plausible at first. But there is a real problem, because SRI did not have a chance model for the data. And in fact, it is very hard to invent a plausible model. SRI might be thinking of the twenty treatment classrooms as a sample from the population of all classrooms. But they didn't choose these twenty classrooms by simple random sampling, or even by some more complicated probability method. In fact, no clear procedure for choosing these classrooms was described in the report. It was a sample of convenience, pure and simple.

SRI might be thinking of measurement error. Is there some "exact value" for Far West, which may or may not be different from the one for controls? If so, is this a single number? or does it depend on the site? or on the classroom? the teacher? the students? the year? Or are these part of the error box? If so, isn't the error box different from classroom to classroom, or site to site? Aren't the errors dependent?

The report covers 500 pages, and there isn't one which touches on these problems. It is taken as self-evident that to compare the average of any sample, no matter where it comes from, with an external standard, a test of significance can be used. The whole argument to show that the project classrooms differ from the controls rests on these tests, and the tests rest on nothing. SRI does not have a simple random sample of size 20, or 20 repeated measurements on the same quantity. It has 20 numbers. These numbers have chance components, but almost nothing is understood about the mechanism which generated them. Under these conditions, a test of significance is an act of intellectual desperation.

We went down to SRI to discuss these issues with the investigators. They insisted that they had taken very good statistical advice when designing their study, and were only doing what everybody else did. We pressed our arguments. The discussion went on for several hours. Eventually, the senior investigator said:

Look. When we designed this study, one of our consultants explained that some day, someone would arrive out of the blue and say that none of our statistics made any sense. So you see, everything was very carefully considered.

EXERCISE SET D

1. In fall, 1975, there were 600 students who took the final in Statistics 2 at the University of California, Berkeley. The average score was 65, with an SD of 20 points. At the beginning of the fall quarter in 1976, the 25 teaching assistants assigned to the course that quarter took exactly the same test. The TA's averaged 72, with the same SD of 20 points.[17] Did the TA's do significantly better than the students? If appropriate, make a two-sample z-test. If this isn't appropriate, explain why not.

2. The five planets known to the ancient world may be divided into two groups: the *inner planets* (Mercury and Venus), which are closer to the sun than the Earth is; and the *outer planets* (Mars, Jupiter, and Saturn), which are farther from the sun than the Earth is. The densities of these planets are shown below (the density of the Earth is taken as 1).

Mercury	Venus	Mars	Jupiter	Saturn
0.68	0.94	0.71	0.24	0.12

The two inner planets have an average density of 0.81, while the average density for the three outer planets is 0.36. Is this difference statistically significant?[18] Or does the question make sense?

3. Two researchers studied the relationship between infant mortality and environmental conditions in Dauphin County, Pennsylvania. As a part of the study, the researchers recorded, for each baby born in Dauphin County during the last six months of 1970, in what season the baby was born, and whether or not the baby died before one year of age.[19] If it is appropriate, make a χ^2-test to see whether infant mortality depends on season of birth. If it is not appropriate, explain why not.

	Season of birth	
	July–Aug.–Sept.	*Oct.–Nov.–Dec.*
Died before one year	35	7
Lived one year	958	990

4. In the WISC block design test, subjects are given colored blocks and asked to assemble them to make different patterns shown in pictures. As part of Cycle II of the Health Examination Survey in 1960, this test was given to a nationwide sample of children aged 6 to 9, drawn by probability methods; basically, this was a multistage cluster sample of the kind used by the Current Population Survey (Chapter 22). There were 1,652 children in the sample with family incomes in the range $5,000 to $7,000 a year; these children averaged 14 points on the test, with an SD of 8 points. There were 813 children in the sample with family incomes in the range $10,000 to $15,000 a year; these children averaged 17 points on the test, with an SD of 12 points.[20] Someone asks whether this difference can be explained as a chance variation.

(a) Does this question make sense?

(b) Can it be answered on the basis of the information given? Explain briefly.

The answers to these exercises are on p. A-67.

6. CONCLUSION

When a client is going to be cross-examined, lawyers often give the following advice:

Listen to the question, and answer the question. Don't answer the question they should have asked, or the one you wanted them to ask. Just answer the question they really asked.

Tests of significance follow a completely different strategy. Whatever you ask, they answer one and only one question:

How easy is it to explain the difference between the data and what is expected on the null hypothesis, on the basis of chance variation alone?

Chance variation is defined by a box model. This model is specified (implicitly or explicitly) by the investigator. The test will not check to see whether this model is relevant or plausible. The test will not measure the size of a difference, or its importance. And it will not identify the cause of the difference.

So a test can answer only one very specific question. Often, that's the wrong question to ask. Then, the problem should be resolved not by testing, but by estimation. This involves making a chance model for the data, defining what you want to estimate in terms of the model, estimating it from the data, and attaching a standard error to the estimate.

Today, however, tests of significance are very popular. One reason is that the tests are part of an impressive and well-developed mathematical theory. Another reason is that many investigators just cannot be bothered to set up chance models. The language of testing is such that it is very easy to slide over this point, and talk about "statistically significant" results. This sounds so impressive, and there is so much mathematical machinery clanking around in the background, that tests usually seem truly scientific—even when they are complete nonsense. The Little Prince understood this kind of problem very well:

> When a mystery is too overpowering, one dare not disobey.[21]

7. REVIEW EXERCISES

1. Which of the following questions does a test of significance deal with?
   (i) Is the difference due to chance?
   (ii) Is the difference important?
   (iii) What does the difference prove?
   (iv) Was the experiment properly designed?

2. Can a result which is "highly statistically significant" still be due to chance? Explain briefly.

3. To find out whether the unemployment rate depends on age, an economist looks at white males in the civilian labor force who are in the Current Population Survey one month. He finds 3,000 in the age group 16 to 19, with an unemployment rate of 10%. And he finds 5,000 in the age group 20 to 24, with an unemployment rate of 5%. He wants to know whether this difference in rates can be explained as a chance variation.
   (a) Does the question make sense?
   (b) Can it be answered using the information given?
   Explain your answers briefly.

4. The inner planets (Mercury, Venus) are the ones closer to the sun than the Earth is. The outer planets are farther away. The masses of the planets are shown below, with the mass of the Earth taken as 1.

Mercury	Venus	Mars	Jupiter	Saturn	Uranus	Neptune	Pluto
0.05	0.81	0.11	318	95	15	17	0.8

   The masses of the inner planets average 0.43, while the masses of the outer planets average 74. Is this difference statistically significant?[22] Or does the question make sense?

5. An investigator makes a study of the various factors influencing voting behavior. He develops a chance model, and estimates that in the election of 1976, the issue of inflation contributed about 7 percentage points to the Republican vote. However, the standard error for this estimate is about 5 percentage points. Therefore, the increase is not statistically significant. The investigator concludes that "in fact, and contrary to widely held views, inflation has no impact on voting behavior."[23] Comment briefly.

6. An investigator is trying to decide whether handedness is dependent on sex, or independent. The data are shown below. The sample consists of 215 persons admitted to the department of neurosurgery in a Baltimore hospital during the period 1965–1967. Is it legitimate to make a χ^2-test?[24]

	Male	Female
Right-handed	87	95
Other-handed	19	14

7. The table below shows the population of the twenty largest cities in the United States in 1960 and in 1970. The question is, "Are the cities declining in population?" Two possible procedures are: the sign test (Exercise 9b on p. 452), or the z-test on the differences (Exercise 9c on p. 453).[25] If either test is appropriate, make it: be sure to formulate the null hypothesis in terms of a box model, and to compute the observed significance level. If neither test is appropriate, explain why not.

Population of the twenty largest cities in the United States.

	1970	1960
New York	7,896,000	7,782,000
Chicago	3,369,000	3,550,000
Los Angeles	2,810,000	2,479,000
Philadelphia	1,950,000	2,003,000
Detroit	1,514,000	1,670,000
Houston	1,233,000	938,000
Baltimore	906,000	939,000
Dallas	844,000	680,000
Washington	757,000	764,000
Cleveland	751,000	876,000
Indianapolis	746,000	747,000
Milwaukee	717,000	741,000
San Francisco	716,000	740,000
San Diego	697,000	573,000
San Antonio	654,000	588,000
Boston	641,000	697,000
Memphis	624,000	498,000
St. Louis	622,000	750,000
New Orleans	593,000	628,000
Phoenix	582,000	439,000

8. In a 1977 survey, the Gallup poll found that only 31% of Americans thought that a person should be required to join the union in order to work at a unionized factory or business; this compares to 43% in a 1965 survey on the same topic.[26] Is this difference real, or a chance variation?
   (a) Does the question make sense?
   (b) Can you answer it with the given information if you assume that in each of the two years, Gallup took independent simple random samples of 900 people?
   (c) Can you answer it with the given information if you assume that in each of the two years, Gallup took independent cluster samples of 1,500 people?

9. Two investigators are testing the same null hypothesis about box X:

that its average equals 50. They agree on the alternative hypothesis: that the average differs from 50. They also agree to use a two-tailed z-test. The first one takes 100 tickets at random from the box, with replacement. The second takes 900 tickets at random, also with replacement. Both investigators get the same SD of 10. True or false: The investigator whose average is further from 50 will get the smaller P-value. Explain briefly.[27]

10. Belmont and Marolla conducted a study on the relationship between birth order, family size, and intelligence.[28] The subjects consisted of all Dutch men who reached the age of 19 between 1963 and 1966. These men were required by law to take the Dutch army induction tests, including Raven's intelligence test. The results showed that for each family size, measured intelligence decreased with birth order: first-borns did better than second-borns, second-borns did better than third-borns, and so on. And for any particular birth order, intelligence decreased with family size: for instance, first-borns in two-child families did better than first-borns in three-child families. These results remained true even after controlling for the social class of the parents.

   Taking, for instance, men from two-child families:

   • the first-borns averaged 2.575 on the test;
   • the second-borns averaged 2.678 on the test.

(Raven test scores range from 1 to 6, with 1 being best and 6 worst.) The difference is small, but if it is real, it has interesting implications for genetic theory. To show that the difference was real, Belmont and Marolla made a two-sample z-test. The SD for the test scores was around one point, both for the first-borns and the second-borns, and there were 30,000 of each, so

$$\text{SE for sum} \approx \sqrt{30{,}000} \times 1 \text{ point} \approx 173 \text{ points}$$
$$\text{SE for average} \approx 173/30{,}000 \approx 0.006 \text{ points}$$
$$\text{SE for difference} \approx \sqrt{(0.006)^2 + (0.006)^2} \approx 0.008 \text{ points}.$$

Therefore, $z = (2.575 - 2.678)/0.008 \approx -13$, and P is astonishingly small. Belmont and Marolla concluded:

   Thus the observed difference was highly significant . . . a high level of statistical confidence can be placed in each average because of the large number of cases.

What does statistical theory say?

11. In Title VII litigation, courts have held that there is a *prima facie* case of discrimination against a firm when the percentage of blacks among the employees of the firm is lower than the percentage of blacks in the surrounding geographical region, provided the difference is "statistically significant" by the z-test. Suppose that in one city, 10% of the people are black. Suppose too that every firm in the city hires employees by a process which, as far as race is concerned, is equivalent to simple random sampling. Would any of these firms ever risk being found guilty of discrimination by the z-test? Explain briefly.

8. SUMMARY

1. A test of significance deals with the question of whether a difference is real, or a chance variation. It does not deal with the question of how important the difference is, or what caused the difference. The test does not check the design of the study.

2. The *P*-value of a test depends on the sample size. With a large sample, even a small difference can be "statistically significant," that is, hard to explain as a chance variation. That doesn't necessarily make it important.

3. To decide whether a difference observed in the sample is important, pretend it applies to the whole population, and see what it means in practical terms: this is a "test of real significance." Then, a standard error should be attached to the estimate.

4. Usually, a test of significance does not make sense when data are available for the whole population, because there is no chance variation to screen out.

5. A difference can be "statistically significant," and nevertheless be due to chance. Even if the null hypothesis is right, there is still a 5% chance to get a difference which is "statistically significant," and a 1% chance to get a difference which is "highly statistically significant."

6. To test whether a difference is due to chance, you have to define the chances. That is what the box model does, and a test of significance makes no sense unless there is a clear chance model for the data. In particular, if the test is applied to a sample of convenience, the *P*-value may not mean anything.

A Review of Terms and Equations

Me Tarzan, you Jane.

This is a list of the main terms and equations introduced in the book. If you are unsure of any of them, you may wish to review the corresponding reading.

Part I. Design of Experiments

Terms   Method of comparison: treatment group and control group
Controlled experiment, observational study
Confounding
Randomization, double-blind, placebo

Part II. Descriptive Statistics

Terms   Histogram, density scale
Variable—qualitative, quantitative, discrete, continuous
Average, median
Root-mean-square
Standard deviation, standard units
Normal curve, normal approximation for data
Percentiles
Chance error, bias
Outliers

Equations   SD of list $= \sqrt{\text{average of (deviations from average)}^2}$

individual measurement = exact value + bias + chance error

Part III. Correlation and Regression

Terms   Scatter diagram
Point of averages, SD line
Correlation coefficient
Association and causation

Regression estimates and regression line
Graph of averages
Regression effect and regression fallacy
Residuals: homoscedastic and heteroscedastic
R.m.s. error for regression line
Normal approximation in vertical strips

Equations r = average product of the variables in standard units

If x goes up by one of its SDs, the regression estimate for y only goes up by r times the SD of y.

The r.m.s. error for the regression line estimating y from x is $\sqrt{1 - r^2} \times$ (the SD of y).

Part IV. Probability

Terms   Chance
Compatible and incompatible
Dependent and independent

Equations   Chance of something = 100% − chance of opposite

If two things are incompatible, the chance that at least one will happen is found by adding the two chances. However, it is not legitimate to add the chances of compatible outcomes.

If two things are independent, the chance that both will happen is found by multiplying the two chances. However, it is not legitimate to multiply the chances of dependent outcomes.

The chance of k successes in n independent trials with success probability p at each trial equals

$$\frac{n!}{k!(n-k)!}p^{k}(1-p)^{n-k}$$

Part V. Chance Variability

Terms   The law of averages
Box models
Sum of draws, chance error
Expected value, standard error
Square root law
Classifying and counting zero-one boxes
Probability histograms

Equations   When drawing at random with replacement. from a box, the expected value for the sum of the draws is

(number of draws) × (average of box)

and the SE for the sum of the draws is

$\sqrt{\text{number of draws}}$ × (SD of box)

When there are only 0's and 1's in the box, its SD equals

$\sqrt{(\text{fraction of 1's in box}) \times (\text{fraction of 0's in box})}$

Key principle   Whatever is in the box, with a large enough number of draws, the probability histogram for the sum or average of the draws (when put in standard units) follows the normal curve quite closely.

Part VI. Sampling

Terms   Inference
Sample and population
Parameter and estimate
Bias—selection, nonresponse, response
Probability methods for sampling
Simple random sample
Multistage cluster sample
Quota sampling, sample of convenience
Attaching a standard error to an estimate
Confidence intervals and levels
Ratio estimates, half-sample method

Equations   When drawing at random from a box of 0's and 1's,

percentage of 1's in sample =
percentage of 1's in box + chance error.

To get the SE for the percentage of 1's in the sample, find the SE for the number of 1's, and convert to percent.

When drawing at random from a box of numbered tickets,

$$\text{SE for average of draws} = \frac{\text{SE for sum of draws}}{\text{number of draws}}$$

Part VII. Chance Models

Terms   Gauss model and error box
SD^+, degrees of freedom, Student's curve
Genes: dominant, recessive, additive

Part VIII. Tests of Significance

Terms   Null and alternative hypothesis
Test statistic and observed significance level P
Statistically significant
z-test, two-sample z-test
t-test
χ^2-test,
2×2 table

Equations $z = (\text{observed} - \text{expected})/\text{SE}$

The standard error for the difference between two independent quantities is $\sqrt{a^2 + b^2}$, where a is the standard error for the first quantity and b is the standard error for the second.

$$\chi^2 = \text{sum of } \frac{(\text{observed frequency} - \text{expected frequency})^2}{\text{expected frequency}}$$

Reminder   The simple random sample formulas for standard errors do not apply to other kinds of samples and may give silly results.

Notes

Part I. Design of Experiments

Chapter 1. Controlled Experiments

1. In preparing this chapter, we drew (with permission) on: J. P. Gilbert, R. J. Light and F. Mosteller, *Assessing social innovations: an empirical guide for policy,* Benefit-Cost and Policy Analysis Annual (1974). This is a very useful reference.
2. Thomas Francis, Jr., and others, "An evaluation of the 1954 poliomyelitis vaccine trials—summary report," *Am. J. of Public Health,* Vol. 45 (1955). Also see the article by P. Meier, "The biggest public health experiment ever: the 1954 field trial of the Salk poliomyelitis vaccine," in J. M. Tanur and others, *Statistics: a guide to the unknown* (1972). For more details on the randomization, see note 6 to chapter 26.
3. H. K. Beecher, *Measurement of subjective responses* (New York: Oxford University Press, 1959), pp. 66–67.
4. *J.A.M.A.,* Vol. 180 (1962), 439–444.
5. *New England Journal of Medicine,* Vol. 281 (1969), 16–19. The quote from Ruffin's article is reprinted by permission.
   Also see the article by L. Miao in *Costs, benefits and risks of surgery* (New York: Oxford University Press, 1977).
6. N. D. Grace, H. Muench, and T. C. Chalmers, "The present status of shunts for portal hypertension in cirrhosis," *J. Gastroenterology,* Vol. 50 (1966), 646–691.

Chapter 2. Observational Studies

1. J. Berkson, "Smoking and lung cancer," *J.A.S.A.,* vol. 533 (1958), 28–38. R. A. Fisher, *Smoking: the cancer controversy* (Edinburgh: Oliver and Boyd, 1959).
   In 1977, investigators discovered a substance in cigarette smoke which might be the causal link between smoking and heart attacks. (*New York Times,* September 18.)
2. D. M. Wilner and others, *The housing environment and family life* (Baltimore: Johns Hopkins Press, 1962).
3. The paradox in the Berkeley data was noticed by Eugene Hammel, associate dean of the graduate division. He resolved it with the help of two colleagues. P. Bickel and J. W. O'Connell. We are following their report, "Is there a sex bias in graduate admissions?" *Science* Vol. 187 (1975), 398–404. This article is reprinted in W. B. Fairley, F. Mosteller, *Statistics and public policy* (Reading: Addison-Wesley, 1977).
4. Instituto de Nutrición de Centro America y Panama. The study is supported by the National Institutes of Health, U.S.A. We are grateful to Charles Yarbrough of INCAP for his help.
5. *Atole* and *fresco* are traditional names for different drinks. *Atole* is made from a low-cost diet supplement previously developed by INCAP.
5a. INCAP should not be held responsible for this interpretation, which is based on a regression analysis of the data available through 1975. When more data comes in, or a more sensitive analysis is done, the conclusion may change.
6. See note 2, Chapter 1.
7. Private communication from Peter Renz, formerly of the University of Washington.
8. *San Francisco Chronicle* (Oct. 13–14, 1976).
9. Mathematically, $P(A|B) > P(A)$ implies $P(B|A) > P(B)$.
10. Sheldon and Eleanor Glueck, *Unraveling juvenile delinquency* (Cambridge: 1950), p. 120. This study was called to our attention by Karl Bemesderfer, Montgomery Ward, Chicago. It is discussed in T. Hirschi and H. C. Selvin, *Delinquency research* (New York: 1967), pp. 79 ff.
11. C. E. Ares, A. Rankin, H. Sturz, "The Manhattan Bail Project," *NYU Law Review,* Vol. 38 (1963), 67–95. B. Botein, "The Manhattan Bail Project," *Texas Law Review,* Vol. 43 (1964–65), 319–331.

12. *San Francisco Chronicle* (Dec. 9, 1975). The research report itself is much more sober: D. J. Ullyot, M.D., and others, "Improved survival after coronary artery surgery in patients with extensive coronary artery disease," *Journal of Thoracic and Cardiovascular Surgery,* Vol. 70, No. 3 (1975), 405–413.
13. E. S. Pearson and J. Wishart (eds.), *"Student's" collected papers* (Cambridge University Press, 1942).
14. *Statistical Abstract of the United States* (1971), Table 118.
15. *San Francisco Chronicle* (Oct. 14, 1976).

Part II. Descriptive Statistics

Chapter 3. The Histogram

1. By Antoine de St. Exupéry. Reproduced by permission of the publisher, Harcourt Brace Jovanovich, Inc.
2. *Money income in 1973 of families and persons in the United States,* Current Population Reports, Series P–60, No. 97 (January, 1975). U.S. Department of Commerce.
3. This is exact for class intervals, approximate for other intervals.
4. *Statistical Abstract of the United States* (1971), Table 118.
5. The convention followed in drawing histograms is a bit different for discrete or continuous variables. Many variables can be classified either way, depending on how you view them. Incomes, for instance, can never differ by less than a penny. On a Census data tape, they can never differ by less than $100. Nevertheless, it is convenient to treat income as continuous—because the range is so much larger than the minimum change.
6. With narrow class intervals, the histogram may be so ragged that its shape is impossible to make out. With wider class intervals, the shape of the histogram may be easier to see, even though some information is lost. The two histograms below are for 524 test scores in Statistics 2, fall, 1976, at Berkeley. The top one has class intervals of width 1. The bottom one has class intervals of width 10.

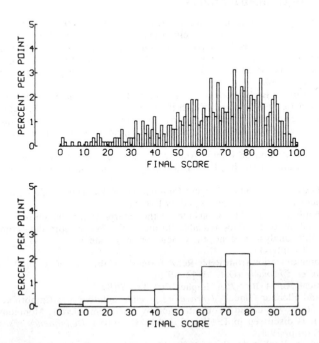

7. Supported by the National Institutes of Health. The director of this study is Dr. Savitri Ramcharan, and we are grateful for her cooperation.
8. I. R. Fisch, S. H. Freedman, A. V. Myatt, "Oral contraceptives, pregnancy, and blood pressure," *Journal of the American Medical Association,* Vol. 222 (1972), 1507–1510.

Our discussion follows this paper. We are grateful to Dr. S. H. Freedman (of the Drug Study) and Dr. Michael Grossman (UCSF Medical Center) for technical advice.

Blood pressure is taken in two phases, called *systolic* and *diastolic*. We are looking at the systolic phase. Results on the diastolic phase are quite similar. In the Contraceptive Drug Study, blood pressures were measured by a machine.

Blood pressure is measured by comparison with the pressure exerted by a column of mercury: it is expressed in terms of the length of the column—hence the units are "mm." or millimeters of mercury.

9. The Drug Study found that four age groups were enough: 17 to 24, 25 to 34, 35 to 44, and 45 to 58. The age distributions of users or nonusers within each of these age groups are quite similar.

10. Excluding women who were pregnant, post-partum, or taking hormonal medication other than the pill.

11. R. C. Tryon, "Genetic differences in maze-learning ability in rats," 39th yearbook, *Nat. Soc. Stud. Educ.,* Part I (1940), 111–119. This article is reprinted in a very nice book of readings: Anne Anastasi, *Individual differences* (New York: Wiley, 1965). This book also includes excerpts by Galton, Pearson, and Spearman. Warning: Tryon uses a non-linear scale for scores in his histograms, so they look quite different from our sketches.

12. *1970 Census of Population,* Vol. 1, Part 1, Sec. 2, Appendix, p. 14. U.S. Department of Commerce. Only persons aged 23 to 99 are counted in the column for 1880, and only persons aged 23 to 82 are counted in the column for 1970.

13. K. Bemesderfer and May, *Social and political inquiry* (Duxbury, 1972), p. 6.

Chapter 4. The Average and the Standard Deviation

1. *Natural inheritance* (London: MacMillan, 1889); reprinted by the American Mathematical Society Press.

2. The highest point of the histogram, called the *mode,* is sometimes used to indicate the center. This is not recommended, as minor changes in the data can cause major shifts in the mode.

3. Tom Alexander, "A revolution called plate tectonics," *Smithsonian Magazine,* Vol. 5, No. 10 (1975). A. Hallam, "Alfred Wegener and the hypothesis of continental drift," *Scientific American,* Vol. 232, No. 2 (1975). Ursula Marvin, *Continental drift* (Smithsonian Press, 1973).

4. In the Department of Health, Education, and Welfare. Data are taken from Series 11 of the *Vital and Health Statistics* publications, and from microdata tapes supplied by the National Center for Health Statistics. We are entirely responsible for all interpretations of the data, right or wrong. We would like to thank Mr. Arthur J. McDowell, Chief of the Division of Health Examination Statistics, for his help.

5. From 18 to 24, from 25 to 34, from 35 to 44, from 45 to 54, from 55 to 64, from 65 to 74, from 75 to 79.

6. This is exact for integer data and class intervals centered at the integers. Otherwise, it is only an approximation.

7. Current Population Reports, Series P–60, No. 97 (January, 1975), Bureau of the Census.

8. The basic reason is called *orthogonality* by statisticians. Briefly, when errors in some situation arise from several independent sources, there is a very simple and exact formula for getting the r.m.s. size of the total error. It turns out that r.m.s. errors combine like the sides of a right-angled triangle. With two independent sources of error

$$c = \sqrt{a^2 + b^2}$$

where a is the r.m.s. size of the errors coming from one source, b is the r.m.s. size of the errors coming from another source, and c is the r.m.s. size of the total error. This fact will be used twice in the book: once in regression (Part III), and once in computing the standard error for a sum (Part V).

Surprisingly, no such formulas are possible for the average absolute value.

9. The figures in the text are rounded off. To two decimals, the average was 63.10 inches, the SD was 2.59 inches, the percent within one SD of the average was 68.00%, the percent within two SDs of the average was 95.95%.

10. This rule works quite well even for many data sets which do not follow the normal curve. Take, for example, the lengths of the reigns of the 61 English monarchs through George VI. These average 18.1 years, with an SD of 15.5 years. Their histogram is shown on the next page, and it is nothing like the normal curve. Still, 42 out of 61, or 69%, were within one SD of average. And 57 out of 61, or 93%, were within two SDs of average. (By definition, the length of a reign is the difference between its first and last

years, as reported on pp. 507–508 of the 1974 *Information please almanac*. This example was contributed by David Lane, Statistics Department, University of Minnesota.)

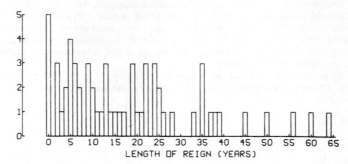

11. The square of the SD is called the *variance*. This is often used as a measure of spread, but we do not recommend it as a descriptive statistic. For instance, the SD of weight for American men is about 30 pounds: individual men are roughly 30 pounds away from average weight. The variance of weight is

$$(30 \text{ pounds})^2 = 900 \text{ square pounds}.$$

12. However, this formula is vulnerable to roundoff error.
13. One good machine is the Monroe 1930.
14. Cycle II of the Health Examination Survey 1963–65; data are from Series 11 of *Vital and Health Statistics*.

Chapter 5. The Normal Approximation for Data

1. Also called standard scores, z-scores, sigma-scores.
2. Sam A. McCandless, "The SAT score decline and its implications for college admissions," 1975 Western Regional Meeting of the College Entrance Examination Board. The numbers have been rounded slightly, the SD being closer to 110. The slide continued into 1977, with the Mathematical SAT dropping to 470 and the Verbal SAT to 429.
3. See note 2.

Chapter 6. Measurement Error

1. We would like to thank Dr. H. H. Ku of the Bureau for his help.
2. *Weight* is used here instead of the more technical word *mass*.
3. Two major sources of chance error in the precision weighing at the Bureau are thought to be:
    • minute amounts of play in the balance mechanism, especially at the knife edge
    • slight variations in the position of the weights on the balance pans.
4. P. E. Pontius, "Measurement philosophy of the pilot program for mass calibration," NBS Technical note #288 (1966). The Bureau rejects outliers only "for cause, such as a door-slam or equipment malfunction."
5. J. N. Morris and J. A. Heady, "Physique of London busmen," *The Lancet* (1956), 569–570. This reference was supplied by Professor Eric Peretz, Statistics Department, Hebrew University, Jerusalem.

Part III. Correlation and Regression

Chapter 8. Correlation

1. There are methods for dealing with more than two variables, but these are quite complicated. Some matrix algebra is needed to follow the discussion. References are: C. Daniel and F. S. Wood, *Fitting equations to data* (New York: Wiley 1971); N. R. Draper and H. Smith, *Applied regression analysis* (New York: Wiley, 1966); H. Scheffé, *The analysis of variance* (New York: Wiley, 1961).
2. *Biometrika* (1903).
3. This term is not standard.

4. These correlations are computed from a data tape supplied by the National Center for Health Statistics (and are rounded off to one decimal). The income variable represents total family income, so only heads of household or men living alone are included: the variable is misleading for others. A few persons with missing data are excluded. This left 529 men aged 25 to 34 and 614 aged 35 to 44. Only income classes are reported on the tape, and were replaced by midpoints: "$15,000 and up" was taken as $18,000. The sample is not exactly self-weighting, but weights were not used. They would increase the correlations, by 10% or so.

5. When the correlation is 0, either slope can be used. "SD line" is not a standard term.

6. However, this formula is quite vulnerable to roundoff error.

7. Dr. Marjorie Honzik was kind enough to supply the data.

Chapter 9. More about Correlation

1. According to Greek legend, Procrustes was a highwayman who used to tie his victims to an iron bed, stretching them or cutting off their legs to make them fit. He was killed by Theseus.

2. T. R. Dawber and others, "Coffee and cardiovascular disease: observations from the Framingham study," *New England Journal of Medicine,* Vol. 291 (1974), 871–874.

3. These statistics are computed from a microdata tape supplied by the National Center for Health Statistics. This correlation was computed for all men aged 25–34, of any marital status, provided the information was complete. There were 621 of them. And $r \approx 0.2214$. An exact test of significance cannot be made, because the sample design is too complicated (Chapter 22). However, a conservative standard error of estimate for this r is 0.05. This correlation is real, not an artifact of sampling.

4. M. Skodak and H. M. Skeels, "A final follow-up study of one hundred adopted children," *Journal of Genetic Psychology,* Vol. 75 (1949), 85–125.

5. Regression to the mean accounts for 7 points, leaving 13 points. There seems to be a difficulty with the environmental explanation, however. At age 2, the average IQ of the adopted children has already gone up to 116, and there is no improvement thereafter. It is very hard to believe that the full effect of the environment is manifested during the first two years of life. There are at least three possible explanations for this anomaly. The likeliest is bad standardization of the tests. The second is bias in testing the mothers (Skeels and Skodak explain that the mothers were under great emotional stress at the time). The third possibility (which Skeels and Skodak deny) is that some aspect of the placement procedures operated to select only the brighter children for adoption.

6. R. Doll, "Etiology of lung cancer," *Advances in Cancer Research,* Vol. 3 (1955), 1–50. Report of the U.S. Surgeon General, *Smoking and health* (1964).

7. Using a 1 in 10,000 sample of records on a data tape supplied by the Census Bureau.

8. W. S. Robinson, *American Sociological Review,* Vol. 15 (1950). Robinson gives the example of literacy and race, based on 1930 Census data. It may be noted that if each cluster is bivariate normal, with a common regression line, then the slope and intercept can be estimated from the averages.

9. E. Durkheim, *Suicide* (New York: Macmillan, 1951), p. 164. We computed the correlation. Durkheim in fact looked at averages of clusters of provinces, for which the correlation was 0.9; his conclusion was "Public Instruction and Suicide are identically distributed."

10. M. P. Rogin and J. L. Shover, *Political change in California* (Greenwood, 1970), p. xvii.

11. This replicates a study by M. and B. Rodin, "Student evaluations of teachers," *Science* (1972), 1164–1166.

Chapter 10. Regression

1. These figures are slightly rounded. The exact figures (unweighted):

   average height = 68.644 inches,   SD = 2.560 inches
   average weight = 158.533 pounds,   SD = 24.893 pounds, $r = 0.36526$

The weight distribution is long-tailed, so many statisticians would recommend taking logs. Next, they would regress log weight on height to get

$$\text{estimated log weight} = \alpha + \beta \text{ height}$$

and then solve, to get

$$\text{estimated weight} = \exp[\alpha + \beta \text{ height}].$$

However, this regression curve coincides almost exactly with the ordinary regression line of weight on height; the average absolute error and r.m.s. error are unchanged.

2. These figures are computed from 1 in 10,000 sample tape provided by the Census Bureau. The exact statistics are:

education: average = 11.15 years,   SD = 3.53 years
income: average = $10,172,   SD = $7,497
$n = 2,155$, $r = 0.430$

The income variable is "personal income," truncated below at $1. About 10% of the data is imputed. Since the spread of incomes increases with education, it is customary to regress log income on education:

estimated log income = $\alpha + \beta$ education
estimated income = $\exp[\alpha + \beta$ education].

However, the correlation between log income and education is only 0.289. Furthermore, the average absolute error from the exponential regression curve is $4,439, compared to $4,338 for the ordinary regression line. The r.m.s. error for the exponential curve is $7,181, compared to $6,768 for the line. The logarithm does not help here.

3. The term "graph of averages" is not standard.
4. The average height of the father was 67.7 inches, with an SD of 2.74 inches; the average height of the sons was 68.7 inches, with an SD of 2.76 inches; r was 0.508.
5. The Institute of Human Development, Berkeley. Dr. Marjorie Honzik was kind enough to supply the data.

Chapter 11. The R.M.S. Error for Regression

1. Edinburgh: Oliver and Boyd, 1958, p. 182.
2. In multiple regression, the residuals can be plotted against: the dependent variable, each independent variable, the fitted values. The last plot is often the most useful.
3. There were 61 families where the father was 64 inches tall (to the nearest inch); the sons averaged 66.7 inches in height, with an SD of 2.0 inches. There were 49 families where the father was 72 inches tall (to the nearest inch); the sons averaged 70.8 inches in height, with an SD of 2.2 inches.
4. The data are taken from the Census tape described in note 2 to Chapter 10. The exact summary statistics are:

average education = 12.3 years,   SD = 2.83 years
average income = $7,420,   SD = $4,210
$r = 0.192$

As explained in note 2 to Chapter 10, it is customary to regress log income on education. But taking logs makes the fit much worse. The correlation drops from 0.19 to 0.15 and the scatter diagram stays heteroscedastic, as shown below: the line of dots at the bottom of the residual plot reflects the people whose nonpositive income was truncated to $1. Regressing log income on education gives

estimated log income = $\alpha + \beta$ education
estimated income = $\exp[\alpha + \beta$ education].

The average absolute error for the exponential regression curve is $3,612, compared to $2,953 for the ordinary regression line. The r.m.s. error for the curve is $4,743, compared to $4,129 for the line.

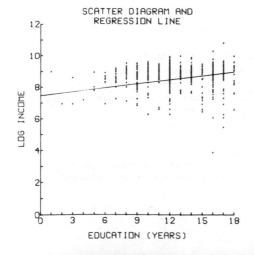

SCATTER DIAGRAM AND
REGRESSION LINE

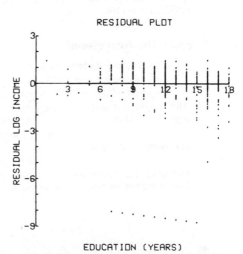

RESIDUAL PLOT

5. The data were supplied by Dr. Marjorie Honzik.

Chapter 12. The Regression Line

1. *Abbandlungen zur Methode der kleinsten Quadrate* (Berlin: 1887), p. 6. We follow the translation by L. Le Cam and J. Neyman, *Bayes-Bernoulli-Laplace Seminar* (New York: Springer, 1965), p. viii.
2. Based on a 1 in 10,000 census sample tape. The exact figures are

$$\text{average education} = 14.04 \text{ years}, \quad \text{SD} = 3.02 \text{ years}$$
$$\text{average income} = \$15,495 \quad \text{SD} = \$9,690$$
$$n = 481, \quad r = 0.2939$$

The income variable is "personal income." "Professional or managerial" is Census recode 1, which does include some lower-status jobs. At work "full time" means the subject worked at least 27 weeks during the previous year, and at least 35 hours during the survey week. The scatter diagram is heteroscedastic, but transforming the variables, for instance by logarithms, worsens the fit. See note 2 to Chapter 10 and note 4 to Chapter 11.
3. See note 2. The exact values are

$$\text{average education} = 13.91 \text{ years}, \quad \text{SD} = 2.87 \text{ years}$$
$$\text{average income} = \$7,425, \quad \text{SD} = \$3,318$$
$$n = 144, \quad r = 0.3799$$

4. J. Mincer, *Schooling, experience and earnings, 1974.* J. Mincer and S. Polachek, "Family Investments in human capital: earnings of women," *Journal of Political Economy,* Vol. 82, No. 2, Part II (March/April 1974), 76–108.
5. This equation is based on rounded values of the summary statistics supplied by IRRI.
6. See note 2. The exact values are

white $\begin{cases} \text{average education} = 11.39 \text{ years}, \quad \text{SD} = 3.38 \text{ years} \\ \text{average income} = \$10,590 \quad \text{SD} = \$7,626 \end{cases}$

$$n = 1955, \quad r = 0.411$$

black $\begin{cases} \text{average education} = 8.55 \text{ years}, \quad \text{SD} = 3.84 \text{ years} \\ \text{average income} = \$5,707 \quad \text{SD} = \$3,636 \end{cases}$

$$n = 180, \quad r = 0.455$$

7. Computed from a microdata tape supplied by the National Center for Health Statistics. The sample is so large that the positive slope is almost impossible to explain as sampling error.
8. Carried out by Professor William Fretter, of the University of California, Physics Department, as a demonstration in his elementary physics course.
9. Regression is appropriate here, in Berkson's case of the *errors-in-variables* model. The nominal values of the weights are fixed by the investigator, the actual value is subject to error; it is the nominal value which goes into the regression. When the value of the weight is measured, subject to measurement error, and the measured value goes into the regression, then the usual regression estimates are biased. A references is G. W. Snedecor and W. G. Cochran, *Statistical methods,* 6th ed. (Iowa: 1973).
10. Data are from *The Economic Report of the President* (1974). As usual, consumption and income are taken per capita and in constant dollars, but this makes little difference to the regression. The possibility of autocorrelation is being ignored. The main catch is that the slope for 1950–59 is appreciably larger than the slope for 1960–69. (Fitting separate regression lines gives slopes of 92% and 85%, respectively.) This could just be chance. Or a shift in the consumption function. Or evidence that the consumption function is not a linear function of income. Or evidence against the existence of the consumption function. It is hard to say from the data. When the mechanism which produced the data is not fully understood (as here) regression can be quite hard to interpret.
11. This study was made by Mr. George Thomas, a student in the University of California, Los Angeles, Statistics Department.
12. See note 7.

Part IV. Probability

Chapter 13. What Are the Chances?

1. For other views of chance, see R. A. Fisher, *Statistical methods and scientific inference,* 2nd ed. (Edinburgh: Oliver and Boyd, 1959). L. J. Savage, *Foundations of statistics.*
2. The third edition was published in 1756, after de Moivre's death. It has been reprinted by Chelsea, New York, 1967.

3. *Statistical Abstract of the United States* (1976), Table 28.
4. This exercise is due to D. Kahnemann and A. Tversky, "Judgment under uncertainty: heuristics and bias," *Science,* Vol. 185 (Sept. 27, 1974), 1124–1131.

Chapter 14. More about Chance

1. From the dedication to the *Doctrine of chances*.
2. For the convenience of the reader, this imaginary dialog has been translated into modern English.
3. See Exercise 7 on p. 257.
4. W. Fairley and F. Mosteller, "A conversation about Collins," *University of Chicago Law Review* (1974).
5. The prosecutor calculated two "chances" for two "events," slipping back and forth between them. The first event was that the accused were guilty. The second event was that no other couple in Los Angeles matched the description. For a frequentist, the concept of chance does not apply to the first event, as argued in the text. Even a Bayesian might find some difficulty here, because there is no reasonable chance model to connect the data with the hypothesis of guilt or innocence. There are similar problems with the second event.

   Were there other couples in Los Angeles matching the description? In principle, this might seem like a statistical issue, which could perhaps be settled by taking a sample. However, a straightforward calculation will show that sampling any fraction of the couples in the city does not settle the issue with any reasonable level of confidence: a complete census is needed. There is no way to estimate the requisite conditional probabilities with satisfactory accuracy from a sample.

Chapter 15. The Binomial Coefficients

1. This exercise was suggested by D. Kahneman and A. Tversky, "Judgment under uncertainty: heuristics and bias," *Science,* Vol. 185 (Sept. 27, 1974), 1124–1131.
2. The model is only approximately correct; there is a slightly better than even chance for a newborn to be male, and successive births in the family are slightly dependent.

Part V. Chance Variability

Chapter 16. The Law of Averages

1. *An experimental introduction to the theory of probability* (Johannesburg, South Africa: University of Witwatersrand Press). Kerrich went to teach there after World War II.
2. We would like to thank Professor Joseph Tussman, Department of Philosophy, University of California, Berkeley, for help with the exposition on this chapter.
3. We are not referring to "absolute value," that is, with sign ignored. Instead, we are trying to distinguish between the difference as a number (in "absolute terms") and the difference as a percent.
4. We are attempting to use "chance process" in a nontechnical sense. A "number generated by a chance process" is the observed value of a random variable.
5. Computer programs are deterministic, and therefore cannot generate numbers in a truly random way. However, a program can generate a sequence of numbers which look as random as can be. One method involves a multiplier M, which is a very big number. A "seed" x is chosen by the programmer; x is between 0 and 1. The computer works out M times x, which has an integer part and a decimal part:

   $$. . . . aaaaaaaaaaaaaaaa \, . \, bbbbbbbbbbb. . . .$$

   The six digits to the left of the decimal point are printed out as the first random number, and the decimal part is used as the seed for the next random number.
6. "Sum of draws from a box" is not standard term but it is lighter on the tongue than "sum of independent, identically distributed, random variables." "Box model" is not standard either.

Chapter 17. The Expected Value and Standard Error

1. Keno is the Las Vegas equivalent of Bingo. There are 80 balls, numbered 1 through 80. On each play, 20 balls are chosen at random without replacement. If you bet on the

single number 17, for example, you are betting that ball number 17 will be among the 20 that are chosen.

2. In this book, we use SD for data and SE for chance quantities (random variables). This distinction in nomenclature is not standard, and the term SD is often used in both situations.

3. E. O. Thorp, *Beat the dealer* (New York: Random House, 1966). Some side bets at baccarat also have positive expected values.

Chapter 18. The Normal Approximation for Probability Histograms

1. Computed on the Monroe 1930.

2. A mathematical discussion can be found in Chapter 7 of Feller, *An introduction to probability theory and its applications,* 3rd ed., I (New York: Wiley, 1968).

3. Computed on the Monroe 1930.

4. A mathematical analysis of the effect of this bias is provided by the Edgeworth expansion; there is a discussion in Chapter 16 of Feller, *An introduction to probability theory and its applications,* 2nd ed., II (New York: Wiley, 1970).

5. The waves can be explained as follows. If the box were $\boxed{1}\ \boxed{1}\ \boxed{9}$, the possible values for the sum would be: 25, 33, 41, . . . , separated by gaps of 8. If the box were $\boxed{2}\ \boxed{2}\ \boxed{9}$, the possible values for the sum would be 50, 57, 64, . . . separated by gaps of 7. The box in Figure 12 is intermediate between these two, and the peak-to-peak distance alternates between 7 and 8.

6. The tacit assumptions: nonzero SD, and a finite number of tickets in the box.

7. Suppose the tickets in two boxes have the same average, and average absolute deviation from average. If they also have the same SD, the asymptotic behavior of the sums will be the same. If not, not. An example would be:

Box A: $\boxed{-1}\ \boxed{1}$ Box B: $\boxed{-2}\ \boxed{0}\ \boxed{0}\ \boxed{+2}$

In both boxes, the tickets average out to 0, and the average absolute deviation from average is +1. But the SD for box A is 1, while the SD for box B is about 1.4: consequently, the sum of 100 draws from box B is about 1.4 times as spread out (by any measure of spread) on the sum of 100 draws from box A. It is the average and SD of the numbers in the box which control the asymptotic distribution of the sum: other measures of location and spread do not.

8. Let *n* denote the number of draws, and *k* the number of repetitions. The implicit condition is that $k/\sqrt{n} \log n \to \infty$. D. Freedman, "A central limit theorem for empirical histograms," *Z. Wahrscheinlichkeitstheorie* (1977).

Part VI. Sampling

Chapter 19. Sample Surveys

1. Some references on sampling:

LESS TECHNICAL

A. Campbell, G. Gurin, and W. Miller, *The voter decides* (Evanston: Row, Peterson, 1954).

George Gallup, *The sophisticated poll-watcher's guide* (1972).

Herbert Hyman and others, *Interviewing in social research* (Chicago: University of Chicago Press, 1954).

Frederick Mosteller and others, *The pre-election polls of 1948* (New York: Social Science Research Council, 1949).

Mildred Parten, *Surveys, polls and samples* (New York: Harper & Row, 1950).

F. F. Stephan and P. J. McCarthy, *Sampling opinions* (New York: Wiley, 1958).

MORE TECHNICAL

M. H. Hansen, W. N. Hurwitz, and W. G. Madow, *Sample survey methods and theory* (New York: Wiley, 1953).

Leslie Kish, *Survey sampling* (New York: Wiley, 1965).

The chapter opening quote is from *The adventure of the copper beeches.* It was found in Don McNeill, *Interactive data analysis* (New York: Wiley, 1977).

2. All quotes are from the *New York Times* (Oct. 1–15, 1936).

3. We would like to thank Paul Perry of the Gallup Poll, who answered all our requests for information.
4. See Parten, p. 393, and Stephan and McCarthy, pp. 241–270 (note 1).
4a. For another discussion, see M. C. Bryson, "The *Literary Digest* poll: making of a statistical myth," *American Statistician* (Nov., 1976).

Bryson agrees that the *Digest* poll was spoiled by nonresponse bias. However, he discounts selection bias as a problem, and questions whether the *Digest* really drew on phone books for its mailing list (that is, the list of people to be polled).

Our primary source of information about the *Digest* poll was George Gallup—a shrewd, interested, and firsthand observer. He maintained that the *Digest* used phone books, lists of automobile owners, and its own subscription lists as the source for the mailing list. This account is confirmed, at least in essentials, by others like Parten, or Stephen and McCarthy (see note 1).

The *Digest* did not publish any very full account of its procedures that we could find. However, something can be learned by reviewing the issues of the *Digest* for the period August 22 through November 14, 1936. For instance, on p. 3 of the issue for August 22, we find:

> The Poll represents thirty years' constant evolution and perfection. Based on "commercial sampling" methods used for more than a century by publishing houses to push book sales, the present mailing list is drawn from *every telephone book in the United States,* from the rosters of clubs and associations, from city directories, lists of registered voters, classified mail-order and occupational data. [Our italics.]

The article goes on to explain that the list was put together for the 1924 election, but was subsequently revised by "trained experts." The bulk of the names on the list were held over from year to year, and the list was used for polls between elections (Aug. 29, p. 6). Drawing on lists of registered voters seems to have been an innovation for 1936, and such lists were used only for certain "big cities" (Oct. 17, p. 7). Which cities, and how many registered voters, the *Digest* does not say.

By modern standards, the *Digest's* mailing list was put together in a somewhat arbitrary way, and it was biased: it excluded substantial, identifiable portions of the community. Bryson suggests that if the *Digest* had somehow managed to get 100-percent response from its list of ten million names, it would have been able to predict the election results. This proposition is remarkably unlikely. Remember, the *Digest* wasn't interested only in calling the winner, it expected to get the percentages right (Aug. 22, p. 3):

> . . . Once again, THE DIGEST was asking more than ten million voters—one out of four, representing every county in the United States—to settle November's election in October.
>
> Next week, the first answers from these ten million will begin the incoming tide of marked ballots, to be *triple-checked,* verified, *five times* cross-classified and totaled. When the last figure has been totted and checked, if past experience is a criterion, the country will know *to within a fraction of 1 per cent* the actual popular vote of forty millions. [Their italics.]

The *Digest* was off by 19 percentage points. As we say in the text, there were two main reasons: selection bias and nonresponse bias.
5. This 75% is typical of four-call probability samples in 1975. The response rate has declined from about 85% in 1960, and this decline is a major worry for polling organizations.
6. This section draws on the book by Mosteller and others (note 1).
7. Stephan and McCarthy, p. 286 (note 1).
8. It is tempting to confuse quota sampling with stratified sampling, but the two are really very different. Suppose, for instance, that it is desired to draw a sample of size 200 from a certain town, controlling for sex, in fact, making the number of men equal to the number of women. A quota sampler could in principle hire two interviewers, one instructed to interview 100 men, the other to interview 100 women. In other respects, the two interviewers would pick whoever they wanted. Clearly, this is a terrible design. By contrast, a stratified sample would be drawn as follows:
   • Take a simple random sample of 100 men.
   • Independently, take a simple random sample of 100 women.
   This is a good design, because human bias is ruled out.
9. The details are discussed in Chapter 22.

We would argue that stratification is needed to draw the sample in a way that keeps the costs reasonable, but in most polls the stratification does very little to reduce the variance. To take a hypothetical example, suppose a country consisted of two regions, East and West. In the East, 60% of the voters are Democrats; in the West, only 40% are. East and West are equal in size, so the overall percentage of Democrats is 50%. Now, two survey organizations take samples to estimate the overall percentage of Democrats. the first one uses a simple random sample of size n. The standard error is

$$50\%/\sqrt{n}.$$

The second one stratifies, taking a simple random sample of size $n/2$ in the East, and an independent simple random sample of size $n/2$ in the West. The standard error is

$$\sqrt{.4 \times .6} \times 100\%/\sqrt{n}.$$

Since $\sqrt{.4 \times .6} \approx 0.49$, the reduction in variance is minimal. Furthermore, in this artificial example, the difference between the regions is much larger than the difference observed in real elections. So the advantage of stratification in predicting real elections is even less.

10. The Gallup organization explains that "This method of selection within the household has been developed empirically to produce an age distribution by men and women separately which compares closely with the age distribution of the population."

11. Paul Perry, "A comparison of the voting preferences of likely voters and likely non-voters," *The Public Opinion Quarterly*, Vol. 37 (1973), 99–109.

12. Kenneth Stampp, Professor of History, University of California, Berkeley. This was a WPA project, and the subjects must have been in their seventies!

13. R. W. Fogel and S. L. Engerman, *Time on the cross* (Boston: Little, Brown, 1974), I, p. 39, and II, p. 37. A careful critique is by Richard Sutch, "The treatment received by American Slaves," *Explorations in Economic History,* Vol. 12 (1975), 335–438.

14. L. Belmont and F. Marolla, "Birth-order, family-size, and intelligence," *Science,* Vol. 182, No. 4117 (Dec., 1973). On the average, intelligence decreases with birth order and family size, even after controlling for family background.

15. L. L. Bairds, *The graduates* (Princeton: ETS, 1973).

16. Based on an example in Parten's book (note 1).

17. This is due to W. E. Deming.

18. See note 13. The phrase "sold down the river" had, as its original meaning, selling a slave into the interstate trade.

19. E. K. Strong, "Japanese in California," *Stanford J. Ed. Psych.,* Vol. I, No. 2 (1933).

20. For a discussion, see Parten's book (note 1).

21. Based on an example in the paper by D. Kahnemen and A. Tversky, "Judgment under uncertainty: heuristics and bias," *Science,* Vol. 185 (Sept. 27, 1974), 1124–1131.

Chapter 20. Chance Errors in Sampling

1. Data for the entire United States are available in the *Statistical Abstract of the United States* (1976), Tables 45 and 65.

2. 380 US 202, 13 L ed 2nd 759, 85 S Ct 824.

Chapter 21. The Accuracy of Percentages

1. Sir Arthur Conan Doyle, *A study in scarlet* (J. B. Lippincott, 1893; rpt., Ballantine Books, 1975), p. 136.

2. See *Statistical Abstract of the United States* (1976), Table 191, for data on school enrollment.

3. See *Statistical Abstract* (1976), Table 1308, for data on sizes of manufacturing establishments.

4. This book takes a strict frequentist view of all things, including confidence intervals. For other views, see R. A. Fisher, *Statistical methods and scientific inference,* 2nd ed. (Edinburgh: Oliver and Boyd, 1959); L. J. Savage, *The foundations of statistics* (New York: Wiley, 1954). Many colleagues will feel that we shut our eyes and walked across an intellectual minefield in this section. Just so. We hope they will be charitable in their judgment.

5. This picture was proposed by Juan Ludlow, CIMASS UNAM, Mexico.

6. For data on smokers, see *Statistical Abstract* (1976), Table 150.

7. *New York Times* (Jan. 2, 1977). National Assessment of Educational Progress, *Education*

for citizenship (Denver: 1976), p. 26. In fact, a cluster sample was used, but the calculations in the exercise give standard errors which are of the right order of magnitude.

8. Sir Arthur Conan Doyle, *The sign of the four* (J. B. Lippincott, 1899; rpt., Ballantine Books, 1974), p. 91.

9. S. W. Polachek, "Occupational segregation: an alternative hypothesis," *J. Contemporary Business* (Winter, 1976).

   For details on the sample design, see "Dual Careers" U.S. Department of Labor, Monograph 21 (1970). The survey ran from 1967 to 1972. The sample size was 5,000, but the sample was so deeply stratified and clustered that the standard erors indicated in the exercise are the right order of magnitude.

10. See note 9.

Chapter 22. Measuring Employment and Unemployment

1. We would like to thank Julius Shiskin of the Bureau of Labor Statistics, and Morton Boisen, Earl Gerson, Charles Jones, Daniel Levine and Margaret Schooley, of the Census Bureau for their help.

   The Bureau of the Census is responsible for the sample design, collection, and production of data, as well as calculation of the estimates and their standard errors. The Bureau of Labor Statistics does the seasonal adjustments, and is responsible for the publication and economic interpretation of the results.

   Some useful references on the Current Population Survey:

   THE BUREAU OF THE CENSUS

   "The Current Population Survey—a report on methodology." Technical Paper No. 7.

   THE BUREAU OF LABOR STATISTICS

   Handbook of methods, Bulletin No. 1711 (1971).
   Employment and earnings, Vol. 23, No. 4 (Oct. 1976).

   THE PRESIDENT'S COMMITTEE TO APPRAISE EMPLOYMENT AND UNEMPLOYMENT STATISTICS

   Measuring employment and unemployment (1962).

   M. THOMPSON AND G. SHAPIRO

   "The current population survey: an overview," *Annals of Economics and Social Measurement,* Vol. 2, No. 2 (1973).

2. There are a few exceptions, in the Northeast.

2a. There are some exceptional, rotating PSUs.

3. There are some exceptional, large USUs which are treated differently.

4. Excluding inmates of penal and mental institutions, and the military.

   The sampling fraction 1 in 1,600 is approximately correct for 1978. However, the sample size is fixed and the population grows, so this fraction changes over time.

5. The total labor force equals the civilian labor force plus the military.

6. The figures used in this section are hypothetical, but close to the real ones.

7. The actual procedure used by the Bureau is a bit more complicated, since they also cross-classify by other demographic variables. Furthermore, they make an adjustment to correct for the known demographic differences between the sample PSUs and the country, using Census data; and they make another adjustment to the current estimates using information from the previous month's sample.

7a. The procedure involves linearizing the estimates first, and computing some of the building-block variances by the half-sample method. It is sketched by R. S. Woodruff, "A simple method for approximating the variance of a complicated estimate," *JASA,* Vol. 66 (June, 1971), 411–414. A complete description will be available in Technical Paper No. 40, to be published by the Bureau of the Census in 1978.

8. The stratification reduces the standard errors, as does the use of ratio estimates. But the clustering has a large effect—it pushes the standard errors up.

9. *Statistical Abstract* (1976), Tables 749–752.

10. For instance, they may be on strike or on holiday.

11. For September, 1976, the estimates were:
    • 16.0 million part-time workers (Table A-8).
    • 4.3 million with a job but not at work (Table A-25).
    • 7.0 million unemployed (Table A-1).

    During the fourth quarter of 1976, for instance, 11,504 people were reinterviewed, and 4,103 were classified—presumably correctly—as full-time workers (in nonagricultural industries) on reinterview. Of these 4,103 people, 4,052 were classified the same way at the original interview. The discrepancy of 51 is under 2% of 4,103. By comparison, 601 people were classified as unemployed on reinterview; of them, only 541 werc classified as unemployed at the original interview; the discrepancy of 60 is almost 10% of 601.

The data used here were supplied by the Basic Surveys Section, Bureau of the Census. The methodology used by them to make such comparisons is outlined in their Technical Paper No. 19.

12. Based on an example in Hyman's book, referenced in note 1 to Chapter 19.

13. A. Jensen, "Environment, heredity, and intelligence," *Harvard Educational Review* (1969), p. 20. The quote was edited slightly.

14. The rules of Keno are explained in note 1 to Chapter 17. The chance for a single number is 20/80, because there are 20 draws. The chance for a double number is
$$(20 \times 19)/ (80 \times 79).$$

Chapter 23. The Accuracy of Averages

1. The draws are to be made with replacement. They can also be made without replacement, provided the number of tickets left in the box is large too.

2. The *Digest of educational statistics* (1975), Table 89, gives figures for the entire United States, based on the Current Population Survey.

3. Census Bureau report, *General housing characteristics*, Part A (1976).

4. According to *Statistical Abstract of the United States* (1976), Table 1285, 99.9% of "wired homes" in the United States have television sets.

5. *Statistical Abstract* (1976), Table 1285.

6. The report cited in note 3 gives figures for the entire United States.

7. See note 3.

8. The figures are reported in the *Digest of educational statistics* (1975). Table 100. A previous phase of the Carnegie survey is discussed in Martin Trow, ed., *Teachers and students* (New York: McGraw-Hill, 1975). A stratified sample was used.

9. See note 8, and pp. 6–7 of Trow's book. Again, the sample was stratified.

Part VII. Chance Models

Chapter 24. A Model for Measurement Error

1. Such equipment is manufactured by Toledo Scale; it uses four load cells, and the cars can move up to 6 mph as they cross the weigh-bridge.

2. W. J. Youden, *Experimentation and measurement* (Washington: 1962).

3. The error box was a bit complicated: 96% of the tickets followed the normal curve, with an average of 0 and an SD of 5 micrograms; the other 4% followed the normal curve with an average of 0 and an SD of 20 micrograms. Two normal curves were needed, one for the middle and one for the outliers.

4. With this many degrees of freedom, there is essentially no difference between the normal curve and Student's *t*, out to the 95% point or so. Confidence levels in this range computed from Student's *t* are quite robust against departures from normality in the error box.

5. The exact weight of NB 10 was taken as 405 micrograms below ten grams, and the error box was described in note 3.

6. The SD of the 100 measurements is actually 6.4 micrograms, so the SD of the group averages should be 3.2 micrograms. In fact, it is 3.4 micrograms.

Dependence between repeated measurements is often caused by observer bias: the person making the measurements subconsciously wants the second measurement to be close to the first one. The Bureau takes elaborate precautions to eliminate this kind of bias. For instance, the value of NB 10 is obtained by comparing the total masses of different sets of weights. These sets are varied according to a design invented by the Bureau. The person who actually makes the measurements does not know how these sets are related to one another, and so cannot form any expectation about what the scales "should" read.

7. With the exception of a passage on the *t*-test (Chapter 27, section 3).

8. Basically, this is just a convention: the factor $\sqrt{n/n - 1}$ could be absorbed into the multiplier derived from Student's curve. Many desk calculators produce SD+ rather than SD: p. 65.

9. The method was invented by W. S. Gossett (England, 1876–1936). He worked as an executive at the Guiness Brewery, where he went after taking his degree at Oxford. He published under the penname "Student" because his employers didn't want the competition to realize how useful the results could be. This anecdote is reported by Youden (note 2).

10. The equation for the curve is

$$y = \text{constant} \left(1 + \frac{t^2}{d}\right)^{-\frac{d+1}{2}}$$

$$\text{constant} = 100\% \frac{\Gamma\left(\frac{d+1}{2}\right)}{\sqrt{\pi d}\ \Gamma\left(\frac{d}{2}\right)}$$

d = degrees of freedom

Γ = Euler's gamma function

11. The exact distribution for the chance error in the average can be found by the methods of Part IV, using what statisticians call the *convolution*.

12. Gossett's procedure was put on a rigorous mathematical footing by Fisher, who also showed that the procedure gave good approximations even when the errors did not follow the normal curve exactly—robustness. Here, the method works quite well despite the outliers in the NB 10 data.

13. By Michelson, Pease, and Pearson at the Irvine Ranch in 1929–33. The results were rounded off a bit in the exercise. Their average value for the speed of light, converted to miles per second, is about 186,270. The measurements were taken in several groups, and there is some evidence to show that the error SD changed from group to group.

14. The quote is from R. D. Tuddenham and M. M. Snyder, *Physical growth of California boys and girls from birth to eighteen years,* (Berkeley: University of California Press, 1954), p. 191. It was edited slightly. As the authors continue,

> With the wisdom of hindsight, we recognized in the later years of the study that a more accurate estimate of the theoretical 'true value' would have been not the first measurement recorded, nor even the 'most representative,' but simply the [average] of the set.

Chapter 25. Chance Models in Genetics

1. We are grateful for expert advice (some of which we took) from Everett Dempster and Michael Freeling of the Genetics Department and Ann Lane of the Botany Department, University of California, Berkeley. G. A. Marx and D. K. Ourecky of the New York State Agricultural Experiment Station were also extremely helpful.

   Two general reference books are:

   I. M. Lerner, *Heredity, evolution, and society* (San Francisco: W. H. Freeman, 1968);

   E. Rosenberg, *Cell and molecular biology* (New York: Holt, Rinehart and Winston, 1971).

   Two intermediate references are:

   W. T. Keeton, *Elements of biological science,* 2nd ed. (New York: Norton, 1973);

   M. W. Strickberger, *Genetics* (New York: Macmillan, 1970).

   Two advanced references are:

   D. S. Falconer, *An introduction to quantitative genetics* (New York: Ronald Press, 1960);

   J. F. Crow and M. Kimura, *An introduction to population genetics theory* (New York: Harper & Row, 1970).

2. Strictly speaking, this refers only to one part of the seeds, the *cotyledons* or first leaves.

3. The term ''gene'' was introduced by the English naturalist Bateson in 1909, and seems to have acquired so many different shades of meaning that many technical writers now scorn it and insist on the more formidable term ''allele.'' We hope our use of ''gene'' will prove inoffensive. In the case of seed color, there are only two alleles or variants of the color gene. In other cases, there may be three or even more alleles.

4. Sperm are carried by the pollen, eggs are in the ovules. Technically these are nuclei not cells.

5. The location of the genes on the chromosomes was worked out by Lamprecht (*Agric. Hortique Genetica,* 1961). There is a more recent discussion in English by Blixt (same journal, 1972).

6. *Experiments in plant hybridisation* (Edinburgh: Oliver and Boyd, 1965), p. 53. This book reprints Mendel's original paper, and some commentaries by Fisher, based on an article in the *Annals of Science,* Vol. 1, No. 2 (1936), 115–137.

7. This experiment used five characteristics, not just the one discussed here. One trial was

repeated, since Mendel thought the fit was poor. He used 100 plants in each trial, making the total of 600 referred to in the text.

8. "On the correlation between relatives on the assumption of Mendelian inheritance." *Trans. Roy. Soc. Edin.*, Vol. 52, 399–433.

9. *Biometrika* (1903). The factor 1.08 more or less adjusts for the sex difference in heights. The equation is rounded off from the one in the paper.

10. Selection can play no role here: the study population consisted of 1,078 established family units.

11. To get equation (5) from equation (4), take conditional expectation given father's height.
    Thus, equation (5) is an empirical fact, as well as a corollary of the additive model—assuming random mating.

12. This discussion ignores complicated phenomena like mutation and what geneticists call *crossing-over*.

13. This exercise is adapted from Strickberger.

14. Rasmusson, *Hereditas,* Vol. 20 (1935). This problem too is from Strickberger.

Part VIII. Tests of Significance

Chapter 26. Tests of Significance

1. From "Of experience," quoted in Jerome Frank, *Courts on trial* (Princeton: Princeton University Press, 1949).

1a. This is only approximate, but the approximation is good for reasonably large samples.

2. Moderate values of P can be taken as evidence for the null. See J. Berkson, "Tests of significance considered as evidence," *J.A.S.A.* (1942), 325–335.

3. We are using ESP loosely to cover PK and clairvoyance as well. The experiment is described in C. Tart, *Learning to use extrasensory perception* (Chicago: University of Chicago Press, 1976).

4. Many statisticians formulate the alternative hypothesis as drawing from a 0–1 box where the fraction of 1's is bigger than $\frac{1}{4}$. For the reasons given in the text, we do not consider this to be a suitable model for ESP. One sensible alternative hypothesis is that the number of correct guesses is stochastically larger than the sum of 7,500 draws.

5. The generator does produce patterns at above the chance level. It is not clear whether this alone explains the results.
    A reference is Martin Gardner, "ESP at random," *New York Review of Books* (July 14, 1977).

6. In fact, the randomization was a bit more complicated. Inoculation required three separate injections over time, and hence the control group was given three injections (of the placebo) too. Vials containing the injection material were packed six to a box: three contained the vaccine, and had a common code number; the other three contained the placebo, with another code number (common to all three vials). Each vial had enough fluid for ten injections. When the time came for the first round of injections, one vial was selected at random from the box, and ten children got their injections from that vial; the investigator recorded its code number against these ten children; these ten children got their second and third injection from the other two vials with the same code number in that box. The next ten children got their first-round injection from one of the three vials of the other group in the box (with a code number different from the first one used); the code number of the vial was recorded against them; and their subsequent injections were from the remaining two vials in the group. In effect, then, the children were blocked into pairs of groups of ten; a coin was tossed for each pair; one whole group went into treatment, and the other group into control, with a 50–50 chance. The calculation in the text is exact, on the plausible assumption that no two polio cases got injections from the same box. Otherwise, the calculation would have to be modified.
    This particular field trial has been analyzed to death (note 2 to Chapter 1) using the two-sample, z-test, even though polio is a contagious disease, so there is dependence in the data.

6a. On the computer, the chance of 57 or fewer heads in 199 tosses of a coin works out to 0.74×10^{-9}. The normal approximation (with continuity correction) to this chance, 1.3×10^{-9}, is almost the right order of magnitude.

7. One such experiment was conducted by Professor W. Meredith, Psychology Department, University of California, Berkeley.

8. See note 8 to Chapter 23.

9. These data originate with the Public Health Department of New York. We got it from Professor Zabell, Department of Statistics, University of Chicago. A reference is A. J. Izenman and S. L. Zabell, "Babies and the blackout: a time series analysis of New York City birth data," Tech. Rep. 38 (Chicago: University of Chicago, Department of Statistics, 1976).

   Apparently, the *New York Times* sent a reporter around to a few hospitals on Monday, August 8, and Tuesday, August 9, nine months after the blackout. The hospitals reported that their obstetrics wards were busier than usual—apparently because of the general pattern that weekends are slow, Mondays and Tuesdays are busy. These "findings" were published in a front-page article on Wednesday, August 10,1966, under the headline "Births Up 9 Months After the Blackout." This seems to be the origin of the baby-boom myth.

10. National Assessment of Educational Progress, "Math fundamentals: selected results from the first national assessment of mathematics" (Washington, D.C.: 1975), p. 14.

    In a more careful treatment, account would be taken of the possibility of error in the NAEP figure (for instance, using the methods of Chapter 27). The California results are hypothetical.

11. A. N. Doob, *et al.*, "Effect of initial selling price on subsequent sales," *J. Pers. and Soc. Psych.*, Vol. 11, No. 4 (1969), pp. 345-350.

    The experiment actually used several different kinds of merchandise, not just cookies, and only about a dozen pairs of stores were involved: they were independently randomized for each different kind of product. However, the results were very similar in each case.

    We would like to thank Professor J. M. Carlsmith, Psychology Department, Stanford University, for his help with this example.

12. M. Rosenzweig, E. L. Bennett, and M. C. Diamond, "Brain changes in response to experience," *Scientific American* (Feb. 1964), pp. 22-29.

    In fact, the experiment used not pairs but triplets, assigned at random to enriched, standard, and deprived environments.

Chapter 27. More Tests for Averages

1. D. N. Kershaw and F. Skidmore, *The New Jersey graduated work incentive experiment* (Princeton: Mathematica, 1974). We would like to thank Mr. Rob Hollister of Mathematica for his help.

2. "Literacy among youths 12–17 years," Vital and Health Statistics, Series 11, no. 131 (Washington, D.C.: 1973).

   The sample design was like that of the Current Population Survey (Chapter 22), and the investigators estimated the standard errors by the half-sample method. Simple random samples of the size indicated in the exercise will have standard errors about equal to the real ones.

3. "Intellectual development of children by demographic and socioeconomic factors," Vital and Health Statistics, Series 11, No. 110 (Washington, D.C.: 1971).

   For a discussion of the standard errors, see note 2. The correlation between the children's test scores and parental education was 0.5, dropping to 0.3 when parental income was held constant. "Big city" means a population of 3 million or more. In fact, children in cities with a population in the range 1 to 3 million did best, averaging around 28 points.

4. The *t*-test is one of the most popular statistical techniques, and we are sorry indeed to have to present it in a context which is both dry and partially hypothetical. (The story in the text is true, up to where they made a *t*-test. In practice, they don't.) But we didn't run across any examples which were simultaneously real, interesting, and plausible.

   With a large sample, only the *z*-test is needed (Chapter 26). In the case of a small sample, our difficulty was the following. The *t*-test is used to compute small tail areas. But departures from normality can throw the computation off by a large factor. To rely on the *t*-test, then, it seems to be necessary to know quite a lot about the distribution of the errors, without having a fair idea about the spread in the errors—in which case the technique would not apply.

   Small departures from independence can throw off both the *z*-test and the *t*-test by quite a lot.

5. A two-tailed test would be appropriate here (pp. 494 ff.).

6. According to the *San Francisco Chronicle* of February 13, 1977, average family income in Washington, D.C., was $23,600, compared to $16,400 in New York and $14,200 in Los Angeles.

7. National Assessment of Educational Progress, "National assessment of science, 1969–1973" (Washington, D.C.: 1975). The average decline in scores between the two tests was about 2 percentage points in all three age groups. This may not seem like much, but it is roughly equivalent to a loss of half a year of schooling.
8. *New York Times* (May 1, 1977).
9. The figures in the exercise are approximately correct for the SAT-Mathematics in 1975–1976. A reference is R. Jackson, "A summary of SAT score statistics for college board candidates" (Princeton: ETS, 1976). We would like to thank Sam McCandless of ETS for his help.
10. The figures in the exercise are approximately correct for the SAT-Mathematics/Verbal in 1975–1976. See note 9.
11. Geophysicists have tried to measure the separation between Great Britain and Europe (which is also supposed to be increasing). However, the measurements were not accurate enough to detect the drift.

Chapter 28. The χ^2-Test

1. "On the criterion that a given system of deviations from the probable in the case of a correlated system of variables in such that it can reasonably be supposed to have arisen from random sampling." *Phil. Mag.,* Series V, Vol. 1, 157–175.
2. If the chance model is right, each term is expected to be a bit less than one; the sum of all the terms is expected to be around one less than the number of terms.
3. The exact distribution was computed on a PDP 11–45, with a BASIC-PLUS program that ran for about 30 minutes in core. The program stepped through all six-tuples of numbers adding up to 60, arranged in decreasing order. It computed the χ^2-statistic for each six-tuple, and the corresponding probability (using the multinomial formula). These probabilities were summed to give the answer. The calculation seemed to be accurate to about 15 decimal places, since the sum of all the probabilities was only 10^{-15} below 1.
4. The equation for the curve is

$$y = \frac{100\%}{\Gamma(d/2)} \left(\tfrac{1}{2}\right)^{\frac{d}{2}} x^{\frac{d}{2}-1} e^{-x/2}$$

   d = degrees of freedom
   Γ = Euler's gamma function

5. With the Monroe 1930, the procedure is to set the II-switch to χ^2. Key in x, press the II-button, key in the number of degrees of freedom, and hit the equal-button. The machine will compute the area to the left of x under the curve.
6. See note 3. Many books recommend the Yates correction (subtracting 0.5 from the absolute difference before squaring, when this difference exceeds 0.5). With one degree of freedom, this is equivalent to the continuity correction (p. 289) and is a good thing to do. With more than one degree of freedom, numerical calculations show that it is a very bad thing to do. The histogram is shifted much too far to the left. Numerical computations also show that with a minimum of five observations expected per cell, and only a few degrees of freedom, the χ^2-curve can be trusted out to the 5% point or so. With a minimum of ten observations expected per cell, it can be trusted well past the 1% point.
7. When there are only two lines in the table (two kinds of tickets in the box), the χ^2-statistic is equal to the square of the z-statistic. Since the square of a normal variable is χ^2-with 1 degree of freedom, the χ^2-test will in this case give exactly the same results as a (two-tailed) z-test.
8. In some cases (for instance, with only a few observations per cell), it is advisable to group the data.
9. *UCLA Law Review,* vol. 20 (1973), 615.
10. Mendel, *Experiments in plant hybridisation,* ed. by R. A. Fisher (London: Oliver and Boyd, 1965).
11. Similar theory is available for $m \times n$ tables; a reference is G. W. Snedecor and W. G. Cochran, *Statistical methods,* 6th ed. (Iowa, 1973), pp. 250–253.
12. A. R. Luria, *The working brain* (New York: Basic Books, 1973).
13. The HES design involved a cluster sample, so there is some dependence in the data, which the χ^2-test would not take into account. For a way out, see note 3 to Chapter 29.
14. The implicit assumption is that the row and column proportions (men, right-handers) are estimated from the data. With an $m \times n$ contingency table, and all row and column proportions estimated from the data, the number of degrees of freedom is

$$(m - 1) \times (n - 1).$$

15. Those subjects who were widowed, divorced or separated, or with missing data were excluded. Also see note 13.
16. *UCLA Law Review,* Vol. 20 (1973), 616.
17. This is based on "usual activity in the last two weeks." Men "with job, not at work" were counted as working. (There were 19 such married men, 7 never-marrieds.) Men who were "widowed, divorced, or separated" were excluded from the table (33 of them), as were men with missing data (26 marrieds, 5 never-marrieds).

    Using another HES variable, it is possible to compare the distribution of occupations for the two groups. The results:

	Married		Never married	
	Number	*Percent*	*Number*	*Percent*
Professional, managerial	137	27	22	27
Other white collar	51	10	16	19
Blue collar	318	63	45	54
Total	506	100	93	100
Unemployed	11		11	
Unknown	26		5	

The percentages look very different—many more of the married men are "blue collar," many fewer "other white collar."
18. This exercise is adapted from data supplied by IRRI.

Chapter 29. A Closer Look at Tests of Significance

1. 225 US 391, quoted from Jerome Frank, *Courts on Trial* (Princeton: Princeton University Press, 1949).
2. The confusion between "statistical significance" and importance gets worse with correlation coefficients. Instead of looking at the value of r, some investigators will test whether $r = 0$, and then use P as the measure of association. Regression coefficients often get the same treatment. However, it is the analysis of variance wich presents the problem in its most acute form: some investigators will report P-values, F-ratios, everything except the magnitude of their effect.
3. With a simple random sample, confidence intervals are easy to compute. For instance, for Table 5 on p. 479, the percentage of men who are right-handed is 89.9%, compared to 91.6% for women, a difference of 1.7%. The standard error for this difference may be estimated as

$$\sqrt{\frac{.899 \times .101}{3091} + \frac{.916 \times .084}{3581}} \times 100\% \approx 0.7 \text{ of } 1\%$$

as may be seen by conditioning on the column totals. Thus, going 2 SE in either direction gives a 95%-confidence interval. With larger contingency tables, the χ^2-statistic can be used to derive a confidence ellipsoid for the population percents, and hence confidence intervals for the contrasts.

    The HES sample was not a simple random sample but a cluster sample, so the standard error would have to be estimated by another method (Chapter 22).
4. In fact, Census data is contaminated by small chance errors. For instance, the Census is estimated to miss between 2% and 5% of the population. However, these errors are too complicated to model by drawing from a box.

    See J. S. Siegel, "Estimates of coverage of the population by sex, race, and age in the 1970 Census," *Demography,* Vol. 11 (1974), 1–24.
5. Daniel McFadden, "The revealed preferences of a government bureaucracy: empirical evidence," *Bell Journal of Economics,* Vol. 7, No. 1 (1976), 66. This reference was supplied by Chris Achen, Professor of Political Science, University of California, Berkeley. Whether the model makes sense is another issue.
6. This is a close paraphrase of a comment (taken out of context) by D. T. Campbell, "Reforms as experiments," *American Psychologist,* Vol. 24 (1969), 409–429. The reference was supplied by J. Merrill Carlsmith, Professor of Psychology, Stanford.
7. This experiment is discussed by C. E. M. Hansel, *ESP: a scientific evaluation* (New York: Scribner, 1966), chap. 11. The numbers have been changed to simplify the arith-

metic. The point of the experiment was to illustrate the fallacy discussed in the text. The reference was supplied by Dr. Charles Yarbrough, INCAP.

8. The null hypothesis is often set up this way in genetic experiments. There is also a famous example in physics—testing the Poisson model for radioactive decay. See Rutherford, Chadwick, and Ellis, *Radiations from radioactive substances* (Cambridge: 1920), p. 172.

9. Arthur Melton, the editor of the *Journal of Experimental Psychology,* defends the practice in these words:

> The next step in the assessment of an article involved a judgment with respect to the confidence to be placed in the findings—confidence that the results of the experiment would be repeatable under the conditions described. In editing the *Journal* there has been a strong reluctance to accept and publish results related to the principal concern of the research when those results were [only] significant at the .05 level, whether by one- or two-tailed test. This has not implied a slavish worship of the .01 levels, as some critics may have implied. Rather, it reflects a belief that it is the responsibility of the investigator in a science to reveal his effect in such a way that no reasonable man would be in a position to discredit the results by saying they were the product of the way the ball bounces.

Many statisticians would advise Melton that there is a better way to make sure results are repeatable: namely, to insist that important experiments be replicated. The quote comes from an editorial in the *Journal of Experimental Psychology,* Vol. 64 (1962), 553–557. We found it in an article by David Bakan, reprinted in Joseph Steger (ed.), *Readings in statistics* (New York: Holt, Rinehart, Winston, 1971).

10. The history is on the authority of G. A. Barnard, Professor of Statistics, Essex. A machine that makes the table obsolete is the Monroe 1930.

11. One example of this strategy is a computer program called PUBPER (= publish or perish), which continues making tests on your data until it finds a significant difference.

12. Another good strategy is to *cross-validate* regressions: fit the equations on half the data, then see how well the fit holds up when the equations are applied to the other half. Replication is a crucial idea, and the text does not do it justice.

13. To paraphrase Keynes, the significance-tester who thinks he doesn't need a box model just has a naive one. J. M. Keynes, *The general theory of employment, interest, and money* (New York: Harcourt, Brace, 1935), pp. 383–384.

> Practical men, who believe themselves to be quite exempt from any intellectual influences, are usually the slaves of some defunct economist.

14. Doing the arithmetic, $\chi^2 = 18$ with 1 degree of freedom, so P is about 2 in 100,000. We can think of only one way to interpret P. Altogether, there were 933 candidates, of whom 825 were men and 108 were women. If you think that sex and admissions were unrelated, comparing admissions rates for men and women is like comparing the admission rate for any random group of 825 people with the admission rate for the complementary group. (After all, there are many other irrelevant splits, based on surnames, fingerprints, whatever.)

So one possible chance model for the table, which captures the idea that sex and admissions are unrelated, might be the following. Put 933 tickets in a box, one for each applicant. Mark 598 of them "admit," and the other 335 "deny." (Because 598 of the applicants were in fact admitted, the rest denied.) Then draw 825 at random to be the "men," and the other 108 are the "women." This model says that the results in the table are like the results that would be obtained by splitting the applicants into two groups at random, where the first group has 825 members (the men), and the second has 108. There are about 7×10^{143} possible splits, all equally likely. What P is saying is that only about 2 in 100,000 of these splits have χ^2-values exceeding 18—the observed value. This interprets P in the framework of a rather unlikely model.

15. Published by SRI at Menlo Park, California. The senior investigator was Jane Stallings. The quotes were edited slightly.

16. This assumes the control average of 60 to be known without error. In fact, SRI made a two-sample t-test. This is peculiarly unwise, because the SRI scoring procedure was bound to introduce dependence between treatment and control scores—it was based on ranks.

17. These are real numbers. About half the TAs had participated in grading the final, and many had graded similar finals in other quarters.

18. F. Mosteller and R. Rourke, *Sturdy statistics* (Reading: Addison-Wesley, 1971), p. 54. They take this question quite seriously, and propose to deal with it by a rank test.

19. T. A. Ryan, B. L. Joiner, and B. F. Ryan, *Minitab student handbook* (Duxbury: 1976), p. 228; they consider the χ^2-test to be appropriate.
20. "Intellectual development of children by demographic and socioeconomic factors," *Vital and Health Statistics,* Series 11, No. 110 (Washington: 1971).
21. Reproduced by permission of the publisher, Harcourt-Brace-Jovanovich, Inc.
22. See p. 68 of the reference in note 18, which suggests a rank test for "the hypothesis of identical distribution" for the inner and outer planets.
23. F. Arcelus and A. H. Meltzer, "The effect of aggregate economic variables on congressional elections," and a reply by S. Goodman and G. H. Kramer, "Comment on Arcelus and Meltzer," *American Political Science Review,* Vol. 69 (1975), 1232–1239 and 1255–1265. This reference too was supplied by Chris Achen. Again, the validity of the model is open to question.
24. Adapted from a paper by Jere Levy.
25. See p. 256 of the reference in note 19. They used the 50 largest cities, and consider both tests appropriate.
26. *San Francisco Chronicle* (May 12, 1977).
27. Based on question used by A. Tversky and D. Kahnemann. Also see p. 298 in the book of readings referenced in note 9.
28. L. Belmont and F. A. Marolla, "Birth order, family size, and intelligence," *Science,* Vol. 182 (1973), 1096–1101.

Answers to Exercises

Part I. Design of Experiments
Chapter 2. Observational Studies
SET A, PAGE 18

1. No. Successful by comparison with what? To judge the effect of the new method, it would be better to test it in competition with the old method, using a control group.

2. The NFIP controls had a whole range of family backgrounds. The controls in the randomized experiment were volunteered by their parents, selecting upper-class (and more vulnerable) children. See p. 4.

3. No, because the experimental areas were selected in those parts of the country most at risk from polio. See p. 4.

4. No. (In fact, there were many new recruits to the program at the beginning of the second year, and these people turned out to have been in worse shape than those who entered at the beginning of the first year.)

5. False. Altogether, 900 out of 2,000 men are admitted—45%; while 360 out of 1,100 women are admitted—33%. This is because the women tend to apply to Department B, which is harder to get into. Please do the arithmetic yourself!

6. To decide whether three out of 24,000 is a lot or a little, you have to compare it to something. For example, the United States national average death rate for persons aged 65 to 74 is about 80 per 100,000 per week (all causes). So three out of 24,000 in a week is low, not high. Suspending the program on the basis of this evidence alone was unwarranted.

7. No. Delinquents tend to come from large families. With large families, most children are middle children anyway. For instance, with six children, four out of six are middle children. The investigator should have controlled for family size.

8. (a) The evidence is not decisive, because this is an observational study. An alternative explanation is that judges parole the innocent and detain the guilty. (It must be said, however, that many criminal lawyers strongly believe that it is to the client's advantage to appear in court as a free man, rather than being led in by warders.)

   (b) The evidence here, based on a randomized controlled experiment, is decisive.

   (c) This is a terrible comparison. The parolees in the treatment group differ from the detainees in the control group in one very important way besides the letter: the first group was released, the second group detained. We know from (a) that detention is associated with conviction or imprisonment.

   Comparing parolees with parolees and detainees with detainees is better, but it's still not good. For example, there were many fewer parolees in the control group, they must have been selected by the judges on some basis or other, and these variables are now confounded with the letter.

(d) We would expect the default rate to go up, and in fact it did, by a factor of three. In general, people are on their best behavior when they are participating in an experiment. (That is why placebos are so important.) Social scientists call this the *Hawthorne effect*.

Part II. Descriptive Statistics

Chapter 3. The Histogram

SET A, PAGE 27

1. (a) 2%   (b) 3%   (c) 5%   (d) 15%   (e) 15%.

2. About the same.

3. There were more earning $10,000 to $11,000.

4. (a) No   (b) B   (c) 20%   (d) 70%.

5. (a) well over 50%   (b) well under 50%   (c) about 50%.

6. Class (b).

7. There were more in the range 90 to 100.

8. (a) (ii). There is one peak for the men, one for the women, and the left-hand part for young children.
   (b) (iii). There are peaks for men and women, but no children.
   (c) (iv).
   (d) (i). Automobiles are around 4 feet tall, with very little spread.
   Histograms (v) and (vi) have the wrong units.

SET B, PAGE 32

1.

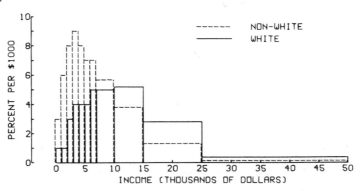

2. The 1970 histogram is shown in Figure 5 on p. 33, and the reason for the spikes is discussed there.

3. Smoothes out the graph between 0 and 8.

4. The educational level went up. For example, more people finished high school and went on to college in 1970 than in 1960. (In this century, there has been a remarkable and steady increase in the educational level of the population. In 1940, only 25% of the population had finished high school. By 1970, this percentage had gone up to 55%, and it is still climbing.)

SET C, PAGE 35

1. 15% per $100.

2. Option (ii) is the answer, because (i) doesn't have units, and (iii) has the wrong units for density.

3. (a) 1.5% per cigarette × 10 cigarettes = 15%.
   (b) 1.5% per cigarette × 20 cigarettes = 30%.
   (c) 30% + 20% = 50%.
   (d) 0.5% per cigarette × 20 cigarettes = 10%.
   (e) 3.5%.

SET D, PAGE 38

1. (a) qualitative.   (b) qualitative.   (c) quantitative, continuous.
   (d) quantitative, continuous.   (e) quantitative, discrete.

2. (a) discrete.   (b)

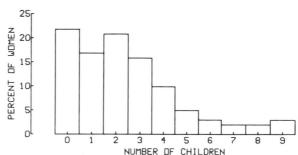

SET E, PAGE 40

1. On the whole, the mothers with four children have higher blood pressures. Causality is not proved, there is the confounding factor of age. The mothers with four children are older. (After controlling for age, the Drug Study found there was no association left between number of children and blood pressure.)

SET F, PAGE 41

1. (a) 7%.   (b) 5%.   (c) The users tend to have higher blood pressures.

2. Use of the pill is associated with an increase in blood pressure of several mm.

3. The younger women have slightly higher blood pressures. (This is a definite anomaly. Most U.S. studies show that systolic blood pressure goes up with age. By comparison with these other studies, the younger women in the Contraceptive Drug Study have blood pressures which are too high, while the older women have blood pressures which are too low. This probably results from a small bias in the procedure used to measure blood pressures at the multiphasic.)

Chapter 4. The Average and the Standard Deviation

SET A, PAGE 51

1. (a)   (b)   (c)

With two numbers, the average is halfway between. If you add bigger numbers to the list, the average moves up. (Smaller numbers move it down.) The average is always somewhere between the smallest and biggest number on the list.

2. If the average is 1, the list consists of ten 1's. If the average is 3, the list consists of ten 3's. The average cannot be 4: it has to be between 1 and 3.

3. The average of (ii) is bigger, for it has the large entry 11.

4. (i)   The average is 2. The histogram is shown in Figure 5 on p. 53.
   (ii)   Average = 3. This list is obtained by adding 1 to each entry in the list (i); this adds 1 to the average, and shifts the histogram over to the right by 1.
   (iii) Average = 4. This list is obtained by doubling each entry in list (i), which doubles the average and spreads the histogram out by a factor of two. The histogram can be drawn in two ways, as follows:

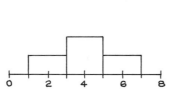 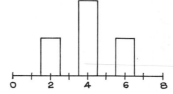

   (iv)   Average = −2. This list is obtained by changing the sign of every entry on the list. This changes the sign of the average, and flips the histogram over.

5. (10 × 66 inches + 77 inches)/11 = 67 inches = 5 feet 7 inches. Or reason this way: the new person is 11 inches taller than the old average. Dividing by 11, he adds 11 inches/11 = 1 inch to the average.

6. 5 feet 6½ inches. As the number of people in the room goes up, each additional person has less of an effect on the average.

7. 5 feet 6 inches + 22 inches = 7 feet 4 inches; it's a giraffe.

NAGY

8. (a) $\dfrac{4 \times 3 + 3 \times 2 + 2 \times 1 + 1 \times 0}{10} = 2$

   (b) $\dfrac{8 \times 3 + 6 \times 2 + 4 \times 1 + 2 \times 0}{20} = 2$

   (c) The average still has to be 2; it only depends on the percentages—the distribution.

9. During recessions, firms tend to lay off the workers with lowest seniority, who are also the lowest paid. This raises the average wage (of those left on the payroll). When the recession ends, these low-paid workers are rehired.

SET B, PAGE 56

1. (a) 50.   (b) 25.   (c) 40.

2. (a) median = average.   (b) median = average.
   (c) average is to the right of the median—long right-hand tail at work.

3. 20.

4. The average has to be bigger than the median, so guess 25. (The exact answer is 27.)

5. The average: long right-hand tail.

6. (a) 1.   (b) 10.   (c) 5.   (d) 5. (''Size'' means: neglecting signs.)

SET C, PAGE 58

1. (a) average = 0,   r.m.s. = 4   (b) average = 0,   r.m.s. = 10.
   On the whole, the numbers in list (b) are bigger in size.

2. (a) 10. (The exact answer is 9.0.)   (b) 20. (The exact answer is 19.8.)
   (c) 1. (The exact answer is 1.3.)

3. smallest, (ii). largest, (i).

4. For both lists, it's 7. All the entries have the same size: 7.

5. The r.m.s. is 3.2.

6. The r.m.s. is 3.1. Note that the r.m.s. is smaller in Exercise 6 than in Exercise 5. There is a reason for this. Suppose we are going to compare each number on a list to some common value. The r.m.s. of the amounts off depends on this common value. For some values it is larger, for some it is smaller. When is it smallest? It can be proved mathematically that the r.m.s. of the amounts off is smallest for the average.

7. Yes. The errors are on the whole much bigger than 3.6. If those ten students are like the rest, there is something wrong with the computer program.

SET D, PAGE 62

1.

   (a) 68% (exact answer from original data—67.1%).
   (b) 95% (exact answer from original data—95.3%).

2. greatest, (iii). least, (ii).
   All three lists have the same range, 0 to 100. But in list (iii), more of the numbers are further away from 50. In list (ii), more of the numbers are closer to 50. There is more to ''spread'' than just the range.

3. greatest, (ii). least, (i).

4. (a) 1, since all deviations from the average of 50 are ± 1.   (b) 2.   (c) 2.
   (d) 2.   (e) 10.

5. 15 years. The average is maybe 45 years, so if 5 years were the answer, many people would be 4 SDs away from the average; with 25 years, most people would be within one SD of the average. The last two options have the wrong units.

6. 4 years. The average is around 10 years, and the reasoning is like Exercise 5. As a rule of thumb, if you take the distance from the highest reasonable value for the variable to the lowest, this range will be from 4 to 6 times the SD, depending on the shape of the histogram.

7. (a) (i).   (b) (ii).   (c) (v).

8. (a) Five 1's, and five 5's.   (b) Five 1's, and five 9's.
   (c) Four 1's, and four 9's, in addition to the two 5's.

9. (a) All 1's. Another possibility: all 5's.
   (b) All 1's. Another possibility: all 9's.   (c) All 5's.

SET E, PAGE 64

1. Guess (ii) is larger. In fact, SD for (i) is 1, SD for (ii) is 2.

2. The SD is different from the average absolute deviation, so the method is wrong.

3. The 0 does count, so the method is wrong.

4. (a) (i) average = 4;   deviations = $-3, -1, 0, 1, 3$;   SD = 2.
      (ii) average = 9;   deviations = $-3, -1, 0, 1, 3$;   SD = 2.
   (b) List (ii) is obtained from list (i) by adding 5 to each entry. This adds 5 to the average, but does not affect the deviations from the average. So, it does not affect the SD.
      Adding the same number to each entry on a list does not affect the SD.

5. (a) (i) average = 4;   deviations = $-3, -1, 0, 1, 3$; SD = 2.
      (ii) average = 12;   deviations = $-9, -3, 0, 3, 9$; SD = 6.
   (b) List (ii) is obtained from list (i) by multiplying each entry by 3. This multiplies the average by 3. It also increases the deviations from the average by a factor of 3, so it multiplies the SD by a factor of 3.
      Multiplying each entry on a list by the same positive number just multiplies the SD by that number.

6. (a) (i) average = 2;   deviations = $3, -6, 1, -3, 5$;   SD = 4.
      (ii) average = -2;   deviations = $-3, 6, -1, 3, -5$; SD = 4.
   (b) List (ii) is obtained from list (i) by changing the sign of each entry. This changes the sign of the average, and all the deviations from the average. This does not affect the SD.

7. (a) This would increase the average by $70, but leave the SD alone.
   (b) This would increase the average and SD by 5%.

8. The r.m.s. size is 17, and the SD is 0.

9. The SD is much smaller than the r.m.s. size. See p. 63 for a discussion of how the SD and the r.m.s. size are related.

10. No.

11. Yes: the list 1, 1, 16 has an average of 6 and an SD of about 7.

12. (a) All 2's.   (b) Five 1's and five 3's.   (c) No.

13. The average is 15, the SD is 5. It only depends on the percent getting each score—the distribution. It does not depend on the number of students in the class.

14. Guess the average, 68 inches; expect to be off by around the SD, 3 inches. Each amount off is a deviation from average. The r.m.s. of the deviations from average is just the SD, so the answer is 3 inches.

15. 3 inches.

Chapter 5. The Normal Approximation for Data

SET A, PAGE 73

1. (a) 60 is 10 above average; that's 1 SD. So 60 is +1 in standard units. Similarly, 45 is −0.5 and 75 is +2.5.

   (b) 0 corresponds to the average, 50. The score which is 1.2 in standard units is 1.2 SDs above average; that's 12 points above average, or 62 points. The score 22 is −2.8 in standard units.

2. The average is 10, with an SD of 2.

   (a) In standard units: 13 is +1.5, and 9 is −0.5, and 11 is +0.5, and 7 is −1.5, and 10 is 0. The converted list is: +1.5, −0.5, +0.5, −1.5, 0.

   (b) The converted list has an average of 0 and an SD of 1. This is always so: when converted to standard units, any list will average out to 0 with an SD of 1.

SET B, PAGE 75

1. (a)

121 mm   147 mm
Ave

$$\frac{147\,mm - 121\,mm}{13\,mm} = 2$$

0   2

   (b)

−2   0

   (c)

−2   0   2

2. (a)

0

   (b)

−1.2   0

   (c)

−1.2   0

3. (a)

−2   0   2

   (b)

−0.5   0   2

   (c)

−0.5   0   0.5

(There is no need to draw this kind of diagram to scale.)

SET C, PAGE 78

1. (a) 11%.   (b) 34%.   (c) 79%.   (d) 80%.   (e) 82%.   (f) 25%.
   (g) 43%.   (h) 23%.   (i) 13%.

2. (a) 1.   (b) 1.15.

3. 1.65.

4. (a) $100\% - 39\% = 61\%$.   (b) impossible without further information.

5. (a) $58\% \div 2 = 29\%$.   (b) $50\% - 29\% = 21\%$.
   (c) impossible without further information.

SET D, PAGE 81

1. (a)

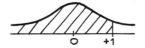

Percent ≈ shaded area
≈ 84%

   (b) 82%.   (c) 98%.

2. (a) 50%.   (b) 42%.   (c) 10%.

3. (a) 89%.   (b) 69%.

SET E, PAGE 82

1. 9%.

2. $4,000.

3. (a) This histogram has fatter tails than the normal curve.   (b) 15.

SET F, PAGE 84

1.

If ⟨curve⟩ = 90% then ⟨curve⟩ = 80%

   Look down the $A(z)$ column till you see percents near 80%:
   • the area between -1.25 and $+1.25$ is 78.87%.
   • the area between -1.30 and $+1.30$ is 80.64%.
   So z is between 1.25 and 1.30; we take $z \approx 1.3$.

2. The score is 1.3 in standard units (from Exercise 1). So it is
   $1.3\,\text{SDs} = 1.3 \times 60 \approx 80$ points above average. That's $650 + 80 = 730$.

3. The first step is to get the score in standard units:

If ⟨curve⟩ = 20% then ⟨curve⟩ = 60%.

   So $z \approx 0.85$. The 20th percentile is 0.85 SDs, or about 50 points, below average;
   so the 20th percentile is 600 points.

4. This GPA is -0.50 in standard units, 2.75 in original units.

Chapter 7. Plotting Points and Lines

SET A, PAGE 100

1. A = (1, 2)   B = (4, 4)   C = (5, 3)   D = (5, 1)   E = (3, 0).

2. *x* up by 3, *y* up by 2.

3. Point D.

SET B, PAGE 100

1. The four points all lie on a line.

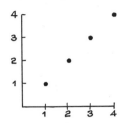

2. The maverick is (1, 2) and it is above the line.

3. The points all lie on a line.

x	*y*
1	3
2	5
3	7
4	9

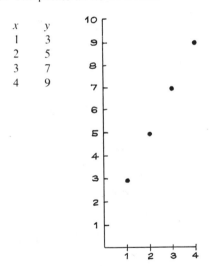

4. (1, 2) is out,   (2, 1) is in.

5. (1, 2) is in,   (2, 1) is out.

6. (1, 2) is in,   (2, 1) is out.

SET C, PAGE 103

1.

	Fig. 16	Fig. 17	Fig. 18
Slope	−1/4 in. per lb.	5	1
Intercept	1 in.	−10	0

Note: In Figure 18, the axes cross at (2, 2).

SET D, PAGE 104

1.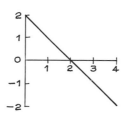

2. on the line.

3. on the line.

4. above the line.

5.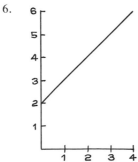

6.

SET E, PAGE 105

1.

	Slope	Intercept	Height at $x = 2$
(a)	2	1	5
(b)	1/2	2	3

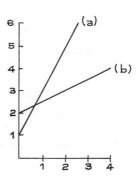

2. (a) $y = \frac{3}{4}x + 1$.   (b) $y = -\frac{1}{4}x + 4$.   (c) $y = -\frac{1}{2}x + 2$.

3. They are all on the line $y = 2x$.

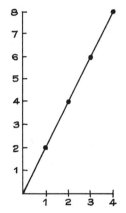

4. They are all on the line $y = x$.

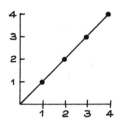

5. (a) on the line.   (b) above the line.   (c) below the line.

6. All three statements are true. If you understand Exercises 4, 5, and 6, you are in good shape for Part III.

Part III. Correlation and Regression

Chapter 8. Correlation

SET A, PAGE 113

1. | x | y |
    |---|---|
    | 1 | 4 |
    | 2 | 3 |
    | 3 | 1 |
    | 4 | 1 |
    | 4 | 2 |

2. (a) ave $x = 3$.   (b) ave $y = 1.5$.   (c) the x-values.

   (d) | x | y |
        |---|---|
        | 0 | 2 |
        | 1 | 1 |
        | 3 | 1 |
        | 4 | 1 |
        | 4 | 2 |
        | 6 | 2 |

   (e) ave $= 3$,   ave $y = 1.5$,   SD $x = 2$,   and SD $y = 0.5$.

3. (a) ave of $x = 1.5$.   (b) SD of $x = 0.5$.   (c) ave of $y = 2$.
   (d) SD of $y \approx 1.5$.

4.

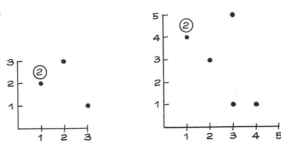

5. (a) shortest father, 59 inches; his son, 65 inches.
   (b) tallest fathers, 75 inches; sons, 69 inches, 70 inches, 71 inches, 73 inches.
   (c) 76 inches, 66 inches. (d) Four: 68 inches, 70 inches, 71 inches, 72 inches.
   (e) ave ≈ 68 inches.   (f) SD ≈ 3 inches.

6. (a) A, B, F.   (b) C, G, H.   (c) ave ≈ 50.   (d) SD ≈ 25.
   (e) ave ≈ 25.   (f) false.   (g) false, the association is negative.

7. (a) 75.   (b) 10.   (c) 20.
   (d) the final: everyone did 50 or better on the midterm.   (e) the final.
   (f) true.

SET B, PAGE 120

1. Negative. The older the car, the lower the price.

2. Left: ave $x = 3.0$,   SD $x = 1.0$,   ave $y = 1.5$,   SD $y = 0.5$,
   positive correlation.
   Right: ave $x = 3.0$,   SD $x = 1.0$,   ave $y = 1.5$,   SD $y = 0.5$,
   negative correlation.

3. The left diagram has correlation closer to 0, it's less like a line.

4. The correlation is about 0.5.

5. The correlation is nearly 0. Psychologists call this "attenuation." If you restrict the range of one variable, that usually cuts the correlation way down.

6. 0.3. Taller men tend to marry taller women, so it's 0.3 or 0.9. But the association is quite loose, ruling out 0.9.

7. All the points on the scatter diagram would lie on a line sloping up, so the correlation is 1.

8. Meaningless.

SET C, PAGE 123

1. He is one SD above average in height so he must weigh
   $$140 + 20 \text{ lbs.} = 160 \text{ lbs.}$$

2. (a) yes.   (b) no.   (c) yes.

3. dashed.

4. (a) dashed.   (b) solid.

SET D, PAGE 126

1. (a) ave of $x = 4$,   SD of $x = 2$
      ave of $y = 4$,   SD of $y = 2$

Standard units		
x	*y*	*Product*
−1.5	1.0	−1.50
−1.0	1.5	−1.50
−0.5	0.5	−0.25
0.0	0.0	0.00
0.5	−0.5	−0.25
1.0	−1.5	−1.50
1.5	−1.0	−1.50

r = average of products $\approx$ −0.93

(b) $r = 0.82$.

(c) $r = -1$. The points all lie on a line sloping down, because in this data set, x and y are related by the equation $y = 8 - x$.

2. No. It doesn't affect the products.

3. No. It washes out in standard units.

4. No. It washes out in standard units.

5. Yes. That will change the last two products in Table 2.

6. Yes. It's just a change of scale.

7. Yes. It's like a diagram in Exercise 6, and $r \approx 0.7$.

8. About 50%.

9. About 25%.

10. About 5%.

SET E, PAGE 129

1. In (a) and (b), $r \approx 0.3$.   In (c), $r \approx 0.7$.

2. Quite a bit higher. It's like Exercise 1—putting the clouds together makes it more linear. (Technically, the distance of points to the SD line does not change when the men and women are put together, but the SD of y goes up. So the points are a smaller fraction of an SD away from the line and r is closer to 1. See pp. 129–130.)

Chapter 9. More about Correlation

SET A, PAGE 138

1. It seems to be due to the association between coffee drinking and cigarette smoking. Coffee drinkers are likelier to smoke, smoking is associated with heart trouble.

2. (a) Positive for both.
   (b) Causality is not suggested. Blood pressure and income are both positively correlated with age. (If you control for age, blood pressure and income are negatively correlated, at least in the HES data.)

3. Going to school has no effect on height. Our interpretation is that height and education are both correlated with a third factor, in family background.

4. This is an observational study, not a controlled experiment, and we would expect chaos if points from the fifties or seventies were plotted on the graph. In this instance, we are right:

The Phillips "Curve" for the period 1949–74.

Source: *Economic Report of the President,* 1975.

5. (a) $r = 1$.   (b) r drops.   (c) r drops.

Measurement error reduces r. Psychologists call this *attenuation* too.

6. No. Religion is a qualitative variable.

SET B, PAGE 142

1. (b) Yes.   (c) No.

2. No. This correlation might well exaggerate the strength of the relationship—it's based on rates.

Chapter 10. Regression

SET A, PAGE 150

1. These men have completed eight years of schooling, which is three years below average. They are $3/4 = 0.75$ SDs below average in schooling. The estimate is that they are below average in income, but not by 0.75 SDs—only by $r \times 0.75 = 0.3$ SDs of income. In dollars, that's $0.3 \times \$8,000 = \$2,400$. Their average income is estimated as
$$\text{overall average} - \$2,400 = \$10,000 - \$2,400 = \$7,600.$$

2. These men are 4 inches above average in height, or 4 inches/2.5 inches = 1.6 SDs. Hence, they are below average in blood pressure (r is negative), by $0.2 \times 1.6 = 0.32$ SDs of blood pressure. In mm., that's 0.32×15 mm. $= 4.8$ mm. Their average blood pressure is estimated as
$$\text{overall average} - 4.8 \text{ mm.} = 120 \text{ mm.} - 4.8 \text{ mm.} = 115.2 \text{ mm.}$$

3. (a) 60.   (b) 67.5.   (c) 45.

4. (a) 70 inches.   (b) 74 inches.   (c) 68.7 inches.
   The regression estimates always lie on a line—the regression line. More about this in Chapter 12.

5. (a) 168 pounds.   (b) 160 pounds.   (c) −8 pounds.   (d) −104 pounds.

This is getting ridiculous, but the Public Health Service didn't run into any little men 2 feet tall, so the regression line doesn't pay much attention to this possibility. The regression line should be trusted less and less the further away it gets from the center of the scatter diagram.

6. No. You would need the averages, SDs, and r for the thirty year olds.

7. False. Think of the scatter diagram for the heights and weights of all the men. Take a vertical strip over 68 inches, representing all the men whose height was just about average. Their average weight should be just about the overall average. But the men aged 45 to 54 are represented by a different collection of points, some of which are in the strip, and many of which aren't. The regression line says how average weight depends on height, not age.

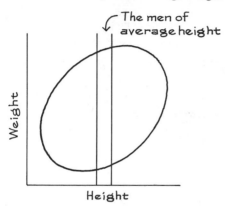
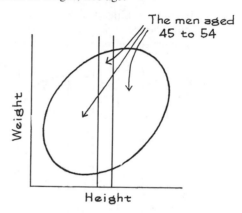

8. False. There are two completely different groups of men here. The ones who are 70.5 inches tall are in the vertical strip, and average 167 pounds in weight, as shown by the cross. The ones who weighed 167 pounds are in the horizontal strip. Their average height is shown by a heavy dot, and it's a lot less than 70.5 inches.

In fact, there are two regression lines—
- one for estimating weight from height,
- one for estimating height from weight.

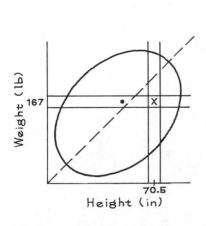
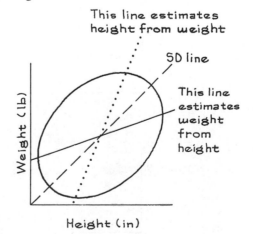

9. The points must all lie on the SD line, which slopes down.

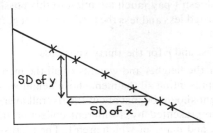

SET B, PAGE 152

1. In each case, the dashed line is the SD line, and the solid line is the regression line. The regression line rises less steeply than the SD line. This is the graphical counterpart of the *regression effect,* to be discussed in the next section.

2.

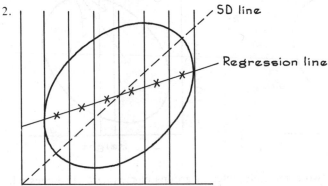

The crosses fall on the solid regression line, the dashed line is the SD line.

3.

	Scatter diagram	Graph of averages and regression line

(a)

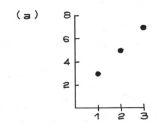

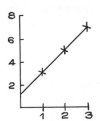

(b)

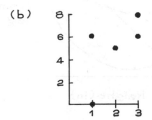

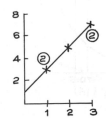

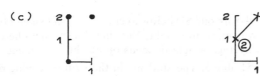

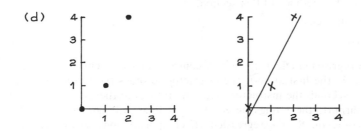

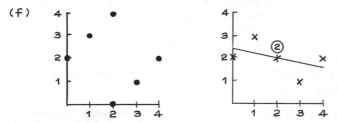

The point of averages isn't shown. (f) shows the regression effect in action. A vicious case. See section 4.

4. (a) yes.   (b) no, nonlinearity.
   (c) no, marital status is a qualitative variable.

SET C, PAGE 157

1. (a) 60.   (b) 60.   (c) 67.5.   (d) 45.
2. (a) 50%.   (b) 50%.   (c) 79%.   (d) 38%.
   Work for (c):

In standard units, his LSAT score was 1.3. The regression estimate for his first-year score is $0.6 \times 1.3 \approx 0.8$ in standard units.

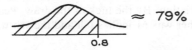

This corresponds to a percentile rank of 79%. In Example 2, the estimated percentile rank was only 69%. Because the correlation was lower in Example 2, the estimated percentile rank was closer to 50%.

3. False. It is true that the men who are one SD below average in height should be on the average 0.4 SDs below average in weight. But that doesn't help here. The regression line says how average weight depends on height, not on age.

4. False. See Exercise 8 on p. 151 above. The students in the 69th percentile of the first-year tests averaged in the 58th percentile on the LSAT, as can be checked using the methods of this section.

5. The SD line—dashed.

SET D, PAGE 162

1. No, it is the regression effect. Imagine a controlled experiment, carried on at two airports. At the first airport, the instructors discuss the ratings with the pilots. At the second, the instructors keep the ratings to themselves. Even at the second airport, the ratings on the two landings will not be identical—differences come in. So the regression effect appears: on the average, the bottom group improves a bit, and the top group falls back. That is all the air force saw at the first airport.

2. No. It looks like the tutoring had an effect—regression would only take them closer to the average, but they got to the other side.

3. No, it would be better to expect the average IQ of their husbands to be around 110. The families where the husband has an IQ of 140 are shown in the vertical strip. The average *y*-coordinate in this strip is 120. The families where the wife has an IQ of 120 are shown in the horizontal strip. This is a completely different set of families. The average *x*-coordinate for points in the horizontal strip is about 110. Basically, there are two regression lines: one for estimating the wife's IQ when husband's IQ is given; the other for estimating the husband's IQ when the wife's is given.

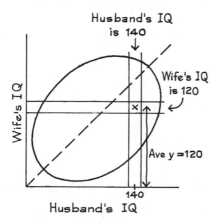

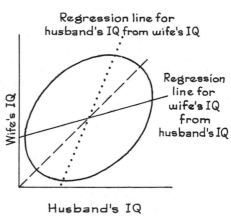

4. Well under 50%—regression effect.

5. The sons of the 61-inch fathers are taller, on the average, than the sons of the 62-inch fathers. This is just a chance variation. By the luck of the draw, Pearson got too many families where the father was 61 inches tall and the son was extra tall.

6. False. The fathers only average 69 inches. See Exercise 3.

7. (a) 68 inches.   (b) 69 inches, 2 inches.
   (c) Individuals do not tend to average. But if you take an extreme group on one test (high or low), their average on the second test will be closer to the overall average.

Chapter 11. The R.M.S. Error for Regression

SET A, PAGE 171

1. (a) 0.2.   (b) 1.   (c) 5.

2. A few thousand dollars.

3. The one with the smaller r.m.s. error should be used, as it will be more accurate overall.

4. (a) one r.m.s. error: 8 points.   (b) two r.m.s. errors: 16 points.

SET B, PAGE 173

1. (a) 65.   (b) r.m.s. = SD of first-year scores = 10 points.   (c) regression.
   (d) $\sqrt{1 - (.6)^2} \times 10 = 8$ points.   (e) The answer to (d) is smaller.
   (f) 8 points—the r.m.s. error. See Figure 3 on p. 170.

2. (a) 68 inches.   (b) r.m.s. = SD = 3 inches.
   (c) regression: if one twin is 6 feet 6 inches tall, the other's height is estimated as 6 feet 5½ inches.
   (d) $\sqrt{1 - (.95)^2} \times 3'' \approx 0.9''$.
   (e) The answer to (d) is quite a bit smaller. When $r = 0.95$, there is quite a large reduction in r.m.s. error.
   (f) 1.8 inches—twice the r.m.s. error. See Figure 3 on p. 170.

SET C, PAGE 176

1. (a) (iii).   (b) (ii).   (c) (i).

2. (a, b) ave of $x \approx 4$,   SD ≈ 1;   ave of $y \approx 4$,   SD ≈ 1.
   (c) $r \approx 0.8$.   (d) 0.   (e) 0.6.   (f) 4.8.   (g) 0.6.

SET D, PAGE 179

1. (a) 2.3 inches—the r.m.s. error.   (b) 68.
   (c) 4.6 inches—two r.m.s. errors.

2. 95%—it's two r.m.s. errors. The r.m.s. error works like the SD.

3. (a) True.   (b) False, because the scatter diagram is heteroscedastic.

SET E, PAGE 183

1. (a)

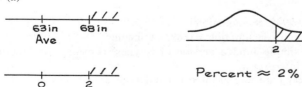

(b) new average ≈ 63.9 inches,   new SD ≈ 2.4 inches

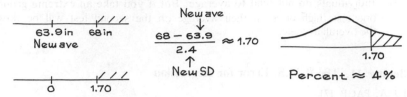

2. (a) 14%.   (b) 33%.
3. (a) 38%.   (b) 60%.

Chapter 12. The Regression Line

SET A, PAGE 191

1. (a) 240 ounces = 15 pounds.   (b) 20 ounces.
   (c) 3 ounces of nitrogen yields $18\frac{3}{4}$ pounds of rice: 4 ounces of nitrogen yields 20 pounds of rice.   (d) controlled.
   (e) Yes. The line fits quite well ($r = 0.95$), and 3 ounces is close to a value that was used.   (f) No. That's too far away from the amounts used.

2. White: $11,960. Black: $7,500. The income distribution for whites and blacks are hard to compare directly, without adjusting for education. The regression does that.

3. With 12 years of education, height is estimated as 68.2 inches; with 16 years, height is estimated as 68.6 inches. Going to college clearly has no effect on height. This observational study picked up a correlation between height and education due to confounding factors.

4. Exercise 2:
   estimated blood pressure = (− 1.2 mm. per in.) × (height) + 201.6 mm.
   Exercise 3: estimated final score = 0.5 × midterm score + 30.
   Exercise 5: estimated weight = (4 lbs. per in.) × height − 104 lbs.

5. Estimated son's height = (0.5) × father's height + 35 inches.

6. Estimated father's height = (0.5) × son's height + 33.5 inches.

SET B, PAGE 196

1. From p. 194, the regression equation is
   estimated length = m × load + b
   = (.05 cm. per kg.) × load + 439.01 cm.
   Substituting 3 kg. and 7 kg. for the load gives estimated lengths of 439.16 cm. and 439.36 cm. The method should not be used for a load of 50 kg. which is much bigger than anything in Table 1. The wire might snap under this load, for example.

2. The correlation is 0.9985, the equation is
   estimated consumption = 0.8937 × income + $42.
   This equation is a simple-minded version of Keynes' consumption function, and we hesitate to interpret it.
   (It would be very hard to plot the original data on any scale which allows you to examine the residuals. The residual plot makes it easy. This regression is not healthy.)

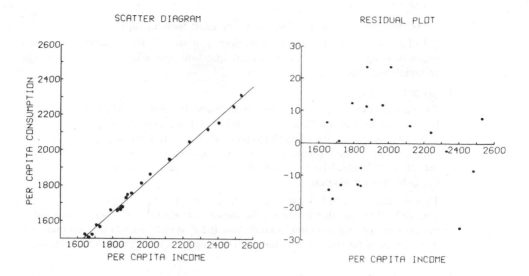

Part IV. Probability

Chapter 13. What Are the Chances?

SET A, PAGE 206

1. (a) (vi).   (b) (iii).   (c) (iv).   (d) (i).   (e) (ii).   (f) (v).
   (g) (vi).

2. about 500.

3. about 1,000.

4. about 14.

SET B, PAGE 208

1. about 300 reds, 200 blues.

2. (a) $308.   (b) $273.   (c) $300.

3. $300 − $200 = $100.

4. 300 × $8 − 200 × $10 = $400.

5. Box (ii), because ⎡3⎤ pays more than ⎡2⎤, and the other ticket is the same.

6. (a) 25%.   (b) 10%.

SET C, PAGE 211

1. 2.1% + 4.6% = 6.7%.

2. True.

3. False. See Example 4 on p. 210.

4. False.

5. False.

SET D, PAGE 216

1. (a) independent.   (b) independent.   (c) dependent.

2. $1/6 \times 1/6 = 1/36$.

3. False, they're dependent. (The cards are dealt without replacement.)

4. "At least one ace" is the better option: you would, for instance, prefer an exam in which you only had to get one question right out of six over an exam in which you had to get all six right.

5. (a) 5%.   (b) 20%.   (c) 17%.

   To figure (a) out, suppose you have 80 men and 20 women in the class. You also have 15 cards marked "freshman" and 85 cards marked "sophomore." You want to give out a card to each student, so that as few women as possible get "sophomore." The strategy is to give a "sophomore" card to each man; you are left with 5, which have to go to five women. The 15 "freshman" cards go to the other fifteen women.

6. This is false. It's like saying someone doesn't have a temperature because you can't find the thermometer. To figure out whether two things are independent or not, you pretend to know how the first one turned out, and then see if the chances for the second change. The emphasis is on the word "pretend."

Chapter 14. More about Chance

SET A, PAGE 222

1. ⚀⚁ ⚁⚅ ⚅⚁ ⚁⚀ . The chance is 4/36.

2. Most often: 7; least often: 2, 12. (Use Figure 1 to get the chance of each total, as in Exercise 1.)

3. There are 25 possible results; the chance is 5/25.

SET B, PAGE 226

1. (a) 3/4.   (b) 3/4.   (c) 9/16.   (d) 9/16.   (e) 7/16.

2. $1 - (5/6)^6 \approx 67\%$. Like de Méré, with 6 rolls instead of 4.

3. $1 - (35/36)^{36} \approx 64\%$.

4. The chance that the point 17 will not come up in 22 throws is
$$(31/32)^{22} \approx 49.7\%$$
The chance that it will come up in 22 throws is therefore
$$100\% - 49.7\% = 50.3\%.$$
So this wager (laid at even money) was also favorable to the Master of the Ball. Poor Adventurers.

5. The chance of surviving 50 missions is $(0.98)^{50} \approx 36\%$.

Chapter 15. The Binomial Coefficients

SET A, PAGE 234

1. The number is 4.

2. The number is 6.

3. (a) $(5/6)^4 = 625/1,296 \approx 48\%$.   (b) $4(1/6)(5/6)^3 = 500/1,296 \approx 39\%$.

   (c) $6(1/6)^2(5/6)^2 = 150/1,296 \approx 12\%$.   (d) $4(1/6)^3(5/6) = 20/1,296 \approx 1.5\%$.

   (e) $(1/6)^4 = 1/1,296 \approx 0.08$ of 1%.   (f) addition rule: $171/1,296 \approx 13\%$.

4. This is the same as Exercise 3a–c: think of 1 as red, and 2 through 6 as green. More explicitly, imagine two people performing different chance experiments:

   (A) rolls a die four times and counts the number of aces.

(B) draws four times at random with replacement from the box R G G G G G
and counts the number of R's.

The equipment is different, but as far as the chance of getting any particular
number of reds is concerned, the two experiments are equivalent.
  • There are four rolls, just as there are four draws.
  • The rolls are independent; so are the draws.
  • Each roll has 1 chance in 6 to contribute one to the count (ace); similarly
    for each draw (red).

5. The chance of getting exactly 4 heads is $\frac{10!}{4!6!}$ $(1/2)^{10} = \frac{210}{1,024} \approx 21\%$. The
chance of getting exactly 6 heads is the same. The chance of getting exactly
5 heads is $\frac{10!}{5!5!}$ $(1/2)^{10} = \frac{252}{1,024} \approx 25\%$. By the addition rule, the chance of getting 4 through 6 heads is $\frac{672}{1,024} \approx 66\%$.

6. $4(1/2)^4 = 25\%$.

7. $\frac{10!}{3!7!}$ $(1/6)^3(5/6)^7 \approx 16\%$.

Part V. Chance Variability

Chapter 16. The Law of Averages

SET A, PAGE 243

1. The error is 50 in absolute terms, 5% in percentage terms.

2. The error is 1,000 in absolute terms, 1/10 of 1% in percentage terms. Comparing this with the previous exercise, the chance error has gone up in absolute
   terms (from 50 to 1,000) but down in percentage terms (5% to 1/10 of 1%).

3. False. The chance stays at 50%. See p. 240.

4. Ten tosses. As the number of tosses goes up, so does the absolute size of the
   chance error. See Figure 1 on p. 241.

5. One hundred tosses: percentagewise, the chance error gets smaller as the
   number of tosses goes up, the percentage of heads is likelier to be closer to
   50%. (With 10 tosses, the chance of getting 4 through 6 heads is about 66%, by
   the method of Chapter 15. With 100 tosses, the chance of getting 40 through 60
   heads is about 95%, as will be seen in Chapter 18.)

6. Ten tosses. It's the opposite of Exercise 5. With more tosses, the percentage
   of heads should be closer to 50%, so there's less of a chance to get over 60%
   heads.

7. Option (ii), the reason is chance error.

8. It's about the same with or without replacement.

9. Same. Both have 50% −1's and 50% +1's.

10. Eventually, the chance error would be large and negative. Then, it would get
    positive again. The swings get wilder and wilder.

SET B, PAGE 246

1. $47 \times 1 + 53 \times 2 = 153$.

2. (a) 100, 200.   (b) 50, 50.   (c) $50 \times 1 + 50 \times 2 = 150$.

3. (a) 100, 900.   (b) 400. Note: 400 isn't halfway between 100 and 900.

4. Guess 500 in all three cases; (iii) is best, (i) worst.

5. Option (ii) cuts down the variability, because the box keeps getting smaller. So (ii) is better. Guess 50 in both cases.

6. Box (i) is better, it has fewer -1's, and the same $+2$.

7. Options (i) and (ii) do it. Your net gain is the sum of your wins and losses, taking signs into account: wins are positive, losses are negative.

SET C, PAGE 251

1. (i) and (ii) are the same. (iii) means that all ten draws are "1," which is worse than (i).

2. The chance for "0" is 1 in 10; the chance for "3 or less" is 4 in 10; the chance for "4 or more" is 6 in 10. If in doubt, read Chapter 13.

3. Option (i) is no good; the sum of the draws is unrelated to the net gain. Option (ii) is no good; it says you may win $17 with 2 chances in 36 on a single play, but your chances are 2 in 38. Option (iii) is right. If in doubt, review Example 1 on p. 250.

4. Your net gain is like the sum of ten draws made at random with replacement from the box

$$\boxed{\ 1 \text{ ticket } \boxed{\$36} \quad 215 \text{ tickets } \boxed{-\$1}\ }\ .$$

This is a terrible game.

5. Ten draws from the box $\boxed{\ \boxed{0}\ \boxed{1}\ }$.

Chapter 17. The Expected Value and Standard Error

SET A, PAGE 257

1. (a) $100 \times 2 = 200$.   (b) -25.   (c) 0.   (d) $66\frac{2}{3}$.

   Note on (d): The sum cannot possibly equal its expected value exactly, but it comes close. The "expected value" need not be one of the possible values. It's like saying that the average family has 2.1 children. This is sensible, even though the "average family" is a statistical fiction.

2. This is the same as the expected value for the sum of two draws from the box $\boxed{\ \boxed{1}\ \boxed{2}\ \boxed{3}\ \boxed{4}\ \boxed{5}\ \boxed{6}\ }$. So the answer is $2 \times 3.5 = 7$ squares.

3. The model is given on pp. 250–251. The average of the box is $(\$35 - \$37)/38 = -\$2/38 \approx -\$0.05$. The expected net gain is around $-\$5$. On the average, you lose around $5.

4. The model is laid out on pp. 248–250. The average of the box is
   $$(\$18 - \$20)/38 = -\$2/38 \approx -\$0.05,$$
   so the expected net gain is $-\$5$. Comparing Exercises 3 and 4 shows that with either bet (number or red-and-black), you can expect to lose 1/19 of your stake on each play. All bets at roulette have this feature.

5. $-\$50$. Moral: the more you play, the more you lose.

6. Average of box $= (18x - \$20)/38$. To be fair, this has to equal 0. The equation is $18x - \$20 = 0$. So $x \approx \$1.11$. They should pay you $1.11.

7. The Master of the Ball should have paid £31, just as the Adventurers thought. Moral: the Adventurers may have the fun, but it is the Master of the Ball who has the profit.

SET B, PAGE 260

1. (a) The average of the numbers in the box is 4, with an SD of 2. So the expected value for the sum is $100 \times 4 = 400$, with an SE of $\sqrt{100} \times 2 = 20$.
   (b) Guess 400, off by 20 or so.

2. The net gain is like the sum of 100 draws from the box  . The average of the box is $0, with an SD of $1.00. The sum of 100 draws has expected value $0, with an SE of $\sqrt{100} \times \$1.00 = \$10$. So your net gain will be around $0, give or take $10 or so.

3. With option (ii), the numbers are too close to 50—no number is more than 5 away. With option (iii), the numbers alternate much too regularly. Option (i) is it.

4. The net gain is like the sum of 100 draws from the box

    .

   The average of this box is $-\$0.25$, with an SD of $1.30. The SE for the net gain is $\sqrt{100} \times \$1.30 = \$13$. In 100 plays, you can expect to lose around $25, give or take $13 or so.

5. Yes. The chance is small, but positive. If you wait long enough, events of small probability do happen.

SET C, PAGE 263

1. The average of the box is 2, with an SD of 1. The sum has an expected value of $100 \times 2 = 200$, with an SE of $\sqrt{100} \times 1 = 10$.
   (a) It will be around 200, give or take 10 or so.
   (b) Smallest, 100; largest, 400.
   (c)

Chance ≈ shaded area
≈ 0%

2. The average of the box is 4, with an SD of 3. The expected value for the sum is $100 \times 4 = 400$, with an SE of $\sqrt{100} \times 3 = 30$.
   (a) largest, 900; smallest, 100.   (b) chance ≈ 68%.

3. (a) The expected value is 0, so the sum is around 0, and your best hope is chance variability. This goes up with the number of draws, choose 100.
   (b) Same as (a).   (c) Now chance variability works against you—choose 10.

4. (i) expected value for sum = 500, SE for sum = 30.
   (ii) expected value for sum = 500, SE for sum ≈ 28.
   Both sums will be around 500, but sum (i) will be a bit further away.
   In (a) and (b), chance variability helps—choose (i).
   In (c), chance variability hurts—choose (ii).

5. 95%.

6. Either they win $25,000 (with chance $20/38 \approx 53\%$) or they lose $25,000 (with chance $18/38 \approx 47\%$). The answer is 50%. Note: The casino is much happier

with a lot of small bets, where the profit is almost guaranteed, than with one big bet, where there is a lot of risk.

7. One number will pay off $35, the other thirty-seven will bomb, so you lose $2.

8. False. The formula only applies when drawing at random with replacement from one box. This person has split the box in two, which will reduce the variability.

SET D, PAGE 269

1. (a) You can't add up words, so box (i) is out. With box (iii), you get 2 chances in 3 to go up each time, and it should only be 1 in 2. Box (ii) is the one.

   (b) Average of box = 0.5 and SD of box = 0.5 too. The sum of 16 draws has an expected value of $16 \times 0.5 = 8$, with an SE of $\sqrt{16} \times 0.5 = 2$. The number of heads will be around 8, give or take 2 or so.

2. (a) False—the formula calls for fractions, not numbers.
   (b) False—it's not a zero-one box.   (c) True.

3. New box: $\boxed{0}\ \boxed{0}\ \boxed{0}\ \boxed{0}\ \boxed{1}$. It's ± 3 SE, chance is about 99.7%.

4. New box: $\boxed{0}\ \boxed{1}$. It's 1 SE or more, chance is about 16%.

5. Expect about 68, see 69.

6. About 99.7%, see Example 2 on p. 267.

7. About 99.7%. Comparing 6 and 7, when the number of tosses goes up from 10,000 to 1,000,000, the percentage of heads gets closer to 50%: the 99.7% interval shrinks from 50% ± 1.5% to 50% ± 0.15%.

8. Put in five 0's and five 1's. Tell it to draw 1,000 times.

9. It's fine. The number of aces isn't supposed to be 16.67 exactly, it's only supposed to be around that.

10. (a) Box (iii) is best (guess 0, have 5 chances in 6 to be right); box (i) is worst.
    (b) Box (iii), least variable; box (i), most variable.
    (c) Box (iii), smallest SD; box (i), largest.
    The SD measures the variability.

Chapter 18. The Normal Approximation for Probability Histograms

SET A, PAGE 283

1. From 70 through 80 inclusive. See p. 278.

2. False. The probability histogram for the sum tells you the chances for the sum. It doesn't tell you how the draws turned out. The shaded area represents the chance that the sum will be in the range from 5 to 10 inclusive.

   (The box had 85 tickets marked "0," 2 tickets marked "1," and 13 tickets marked "2.")

3. The values 70, 76, and 77 came up most often in the first 100 repetitions. The value 72 came up most often in the first 1,000. The value 74 came up most often in the first 10,000 repetitions. The most likely value is 75, the others came out more often just by chance.

4. (a) 75%.   (b) 50%.   (c) 25%.   (d) 1%.

5. (a) 100%.   (b) 1%.

SET B, PAGE 290

1. The histogram gives the exact chance: the normal curve is only an approximation (but a very good one).

2. The expected number of heads is 50, with an SE of 5. You want the area of the rectangle over 60 in the top panel of Figure 8, p. 288.

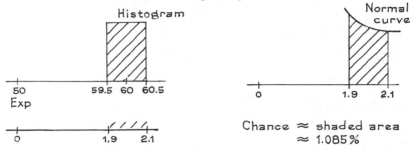

(By computer, the exact chance is 1.084%.)

3. From Exercise 2, about one group in a hundred should have 60 heads. In fact, exactly one group in the hundred does (number 6,901–7,000).

4. The expected number of heads is 5,000, with an SE of 50.
   (a)

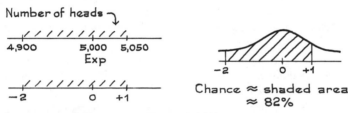

   (b) chance ≈ 2%.   (c) chance ≈ 16%.

5. (a) Yes. The blocks are big. (In fact, keeping track of the edges changes the estimate from 50% to 54%.)
   (b) No. Small blocks.

SET C, PAGE 292

1. You want the area of the rectangle over 40, at the bottom of Figure 10, p. 291. Now the expected number of heads is 40, with an SE of 6.

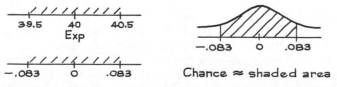

   From the table, this area is between 4% and 8%; actually, it's 6.6%.

2. The normal curve is much lower than the histogram around 1, so the estimate would be quite a bit too low.

3. Yes. Big blocks.

4. (a) Most likely, 105; least likely, 101.   (b) The expected value is 100. There is a trough in this histogram at the expected value. (With 100 draws the trough has disappeared.)

5. No, the histogram will be quite different from the normal curve. If he gets +1 the first time, he is much more likely to get +1 the next time, and so on. So the sum is likely to be +100. If he gets −1 the first time, he is much more likely to get −1 the second time, and so on. So the sum is likely to be − 100. The histogram will look like this:

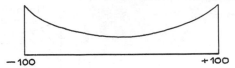

−100   +100

Part VI. Sampling

Chapter 19. Sample Surveys

SET A, PAGE 315

1. Yes. Telephone subscribers are probably different from nonsubscribers, even in the 1970s. (However, the percentage of nonsubscribers is so small that this bias can usually be ignored.)

2. No. You might expect the respondents interviewed by blacks to be much more critical. (And they were.)

3. No, this parish might have been quite different from the rest of the South. (In fact, it was. Plaquemines is sugar country, and sugar requires more highly skilled labor than cotton.)

4. The population and the sample are the same, namely, all men aged 18 in Holland in 1968; so there is no room for chance error. This study is discussed again in Review Exercise 10, Chapter 29.

5. No. First, the ETS judgment about "representative" schools may have been biased. Next, the schools may not have used good methods to draw a sample of their own students. Finally, sampling schools then students is very different from sampling students, and could easily produce too many students in very small schools. (There are about 2,500 institutions of higher learning in the United States, including junior colleges, community colleges, teachers' colleges. About 1,000 of them are very small, altogether enrolling only 10% of the student population. At the other end, there are about 100 schools with enrollments over 20,000—and these account for 30% of the student population.)

6. Choose (ii). See sections 2 and 3.

7. (a) The population consists of all undergraduates registered in the current quarter. The parameter is the percentage of these undergraduates living at home.
   (b) This is a probability method: it is perfectly definite, chance enters in a planned way, and nobody has any discretion as to who gets in the sample. However, it is very different from simple random sampling. For instance, two people whose names are adjacent on the list have no chance to get into the sample together. Simple random samples are defined on p. 308.
   (c) This is not a probability method, and there is no way to calculate the chance that a particular person will get into the sample. For instance, suppose the interviewer goes and stands at a definite place at a definite time. What is the chance that a particular student passes by? It is difficult or impossible to say. (In actual studies of this kind conducted at Berkeley, male interviewers got very high percentages of women in their sample.)

Chapter 20. Chance Errors in Sampling

SET A, PAGE 328

1. The number of red marbles is like the sum of 400 draws from the box $\boxed{0}\,\boxed{1}\,\boxed{1}\,\boxed{1}\,\boxed{1}$. The expected value for the sum is 320, with an SE of 8. So the number of red marbles will be 320 out of 400, give or take 8 or so. Converting to percent, the percentage of red marbles among the draws will be 80%, give or take 2% or so.

2. (a) There should be 100,000 tickets in the box, one for each person in the population, of which 80,000 are marked "1" (married) and 20,000 are marked "0." The SD of the box is $\sqrt{0.8 \times 0.2} = 0.4$. The number of married people in the sample is like the sum of 1,600 draws from the box. The expected value for the sum is $1,600 \times 0.8 = 1,280$, with an SE of $\sqrt{1,600} \times 0.4 = 16$. The number of married people in the sample will be 1,280, give or take 16 or so. Now 1,280 out of 1,600 is 80%, and 16 out of 1,600 is 1%. So 80% of the people in the sample are married, give or take 1% or so. Convert to standard units and use the normal curve:

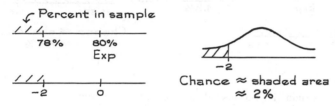

   (b) There should be 100,000 tickets in the box, of which 20,000 are marked "1" (income over $25,000) and the other 80,000 are marked "0." The chance is about 16%.

   (c) The box has 100,000 tickets, of which 10,000 are marked "1" (college degrees) and other 90,000 are marked "0." The chance is about 82%.

3. (a) Around 68%.   (b) 68% of 25 = 17.   (c) Fifteen have 41 to 51 inclusive.

4. The chance is around 95%. As the number of draws goes up, the percentage of men in the sample should get closer to the percentage in the population.

5. (a) Around 8%. See p. 289 for the method.   (b) 8% of 50 = 4.   (c) 2.

6. He is figuring the SE for the number of heads, which is 50. But then, he has to convert to percent: $(50/10,000) \times 100\% = 0.5$ of 1%.

7. False. They forgot to change the box. See pp. 267 and 328. The number of 1's is like the sum of 400 draws from the box $\boxed{0}\,\boxed{0}\,\boxed{0}\,\boxed{1}\,\boxed{0}$. This box has an SD of 0.4, so the SE for the number of 1's is $\sqrt{400} \times 0.4 = 8$, and the SE for the percentage of 1's is $(8/400) \times 100\% = 2\%$.

8. Choose (ii): chance error. Assuming an even split (over 21 or under 21) in the population, which looks about right, the SE for the sample percentage is around 2.5%, so the error in (iii) is way too big.

9. The total distance advanced equals the total number of spots thrown. This is like the sum of 200 draws (at random with replacement) from the box $\boxed{1}\,\boxed{2}\,\boxed{3}\,\boxed{4}\,\boxed{5}\,\boxed{6}$: a hundred throws of a pair of dice is like two hundred throws of a single die. So he can expect to advance around $200 \times 3.5 = 700$ squares, give or take $\sqrt{200} \times 1.7 \approx 25$ squares or so.

10. There should be one ticket in the box for each person in the population, marked "1" if the person is over 6 feet tall, and "0" otherwise. The first step is to compute the percentage of 1's in the box, and this can be done using the normal approximation (Chapter 5):

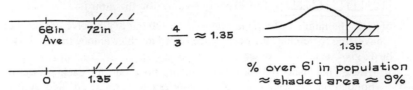

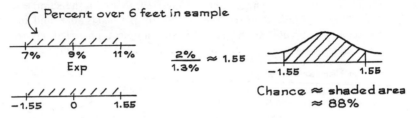

The fraction of 1's in the box is about .09, so the SD of the box is $\sqrt{.09 \times .91} \approx .29$. The SE for the percentage of 1's among 500 draws will be around 1.3%.

SET B, PAGE 333

1. Option (iii) is right. If you missed this, reread the section.

2.

Number of draws	SE for percent of 1's among draws
2,500	1%
25,000	0.27 of 1%
100,000	0

With 2,500 draws, the correction factor is 0.99, and can be ignored. With 25,000 draws, it is 0.87. At 100,000 draws, there is quite a difference between drawing with or without replacement.

3. The sample size should be 2,500.

4. SE with = 20%;   SE without = $20\% \times \sqrt{\dfrac{10 - 4}{10 - 1}} \approx 16\%$.

This is an artificial example where the number of draws is a large fraction of the number of tickets in the box, so the correction factor must be used.

5. (a) The size of the sample is conspicuous by its absence. (Field poll samples run 500 to 1,000 people.)

   (b) It could just be chance error. (As measured by the Field poll, the public's "confidence" in these various institutions went right back up in 1977; *San Francisco Chronicle*, May 12.)

Chapter 21. The Accuracy of Percentages

SET A, PAGE 343

1. The first step is to set up the model. There are 100,000 tickets in the box, some marked "1" (for red) and the others "0" (for blue). Then 1,600 draws are

made from the box to get the sample. The fraction of 1's observed in the sample is $322/1,600 \approx 0.20$, so the SD of the box is estimated as

$$\sqrt{0.20 \times 0.80} = 0.40.$$

The SE for the number of 1's in the sample is then estimated as

$$\sqrt{1,600} \times 0.40 = 16.$$

So the SE for the percentage of 1's is estimated as

$$(16/1,600) \times 100\% = 1\%.$$

This gives the likely size of the difference between the percentage of 1's in the sample and the percentage of 1's in the box. The percentage of red marbles in the box is estimated as 20%, give or take 1% or so.

2. The estimate is: 20% give or take 1% or so. The reasoning is just like Exercise 1.

3. The estimate is 48%, give or take 5% or so.

4. The estimate is 4%, give or take 1% or so.

5. The estimate is 65%, give or take 2.4% or so.

6. No. Most people work for the few large firms.

7. SE = 2%.

8. (a) $18\% \pm 1.9\%$.   (b) $21\% \pm 2.0\%$.   (c) $24.5\% \pm 2.2\%$.
   The third person is off by a couple of SEs in estimating the percentage of 1's in the box; even so, his estimated standard error is only off by 0.2 of 1%. This shows how good the bootstrap method is.

SET B, PAGE 347

1. (a) 18% to 22%.   (b) 17% to 23%.   (c) impossible, see p. 345.

2. 18% to 22%.

3. 18.4% to 21.6%. As the sample size goes up, the confidence interval gets shorter.

4. False. The normal approximation cannot be used here. Basically, this is because the box is so unsymmetric. As best we can estimate from the sample, 1% of the balls in the box are red, and 99% are blue. To illustrate the effect of this unsymmetry, we had the computer work out the probability histogram for the number of red balls in 100 draws made at random from a box with 1% reds and 99% blues. This histogram is shown below, and it is nothing like the normal curve.

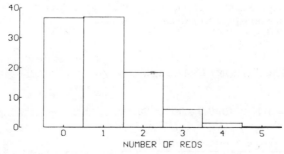

5. The one from the organization making the more modest claim. There are no 100%-confidence intervals.

SET C, PAGE 349

1. (a) 18% ± 3.8%, covers.   (b) 21% ± 4.0%, covers.
   (c) 24.5% ± 4.4%, doesn't cover.

2. False.

3. True.

   It is OK to talk about the chance that the sample percentage will come out one way or the other—it changes from sample to sample. In the frequency theory of probability, it is not OK to talk about the chance that the population percentage is one way or another—are you thinking about one population, or many?

SET D, PAGE 350

1. Theory says, watch out for this man. What population is he talking about? Why are his students like a simple random sample from the population? Until he can answer these questions, don't pay much attention to the calculations. He may be using them just to pull the wool over your eyes.

2. This is not a simple random sample, so the formula can't be used.

   (According to the Gallup poll, as reported in the *San Francisco Chronicle*, May 16, 1977, about 59% of the persons aged 18 to 24 in the United States have smoked marijuana.)

SET E, PAGE 352

1. This isn't a simple random sample, the formulas don't apply.

Chapter 22. Measuring Employment and Unemployment

SET A, PAGE 366

1. This is just fine; it is a simple random sample of households.

2. This is not so good. It is not a simple random sample of people, but a cluster sample. (The household is the cluster.) People living in the same household tend to have similar educational levels, so this sample is not as good as a simple random sample of 144 people. The SE will be more than 2.5%.

3. False. The Bureau would use ratio estimates, and it would come out a bit different. If you missed this, read section 4 again.

4. The one for white males; it is based on a lot more people.

5. 7.0 million ± 0.1 million. See p. 365.

6. The SE for the percentage is only 0.2 of 1%, so the discrepancy 63% − 56% = 7% is almost impossible to explain as a chance error. People like to say they voted, even if they didn't.

Chapter 23. The Accuracy of Averages

SET A, PAGE 374

1. (a) 110/100 = 1.1   (b) 0.9 × 100 = 90.

2.

Number of draws	SE for sum of draws	SE for average of draws
25	10	0.4
100	20	0.2
400	40	0.1

3. The SE for the sum of the draws is $\sqrt{100} \times 20 = 200$. The SE for their average is $200/100 = 2$. The average of the draws will be around 50, give or take 2 or so. This is still true if the draws are made without replacement, because only a small fraction of the tickets in the box are drawn out.

4. The SE for the average is 2, so the answer to (a) is almost 100%. The answer to (b) is 68%. Do not confuse the SE for the average of the draws with the SD of the box.

5. (a) False.   (b) True.
   To repeat, do not mix up the SE for the average of the draw with the SD of the box.

6. The SE for the average of the draws is

   1 from box A   1.4 from box B   2 from box C.

   (a) 203.6 is very unlikely to come from A—it is 3.6 SEs away from the expected value for the average of 100 draws from box A. It is also quite unlikely to come from box B, because $3.6/1.4 \approx 2.6$ is too many SEs. So it comes from box C. Similarly, 198.1 comes from box B, leaving 200.4 for box A by elimination.

   (b) It could be otherwise, but that would be very unlikely.

SET B, PAGE 381

1. The box has 30,000 tickets, one for each registered student, showing his or her age. Then 900 draws are made for the sample. The SD of the box is estimated as 4.5 years, the SE for the sum of the draws is $\sqrt{900} \times 4.5 = 135$ years, the SE for their average is $135/900 \approx 0.15$ years. The interval is 22.3 ± 0.3 years.

2. (a) The SD of the box is estimated at $80, the SE for the sum of the rents is $\sqrt{400} \times \$80 = \$1,600$, the SE for their average is $\$1,600/400 = \$4$. The interval is $\$184 \pm \$4$. Even though the data don't follow the normal curve, the probability histogram for the average of the draws does.

   (b) This is false: $4 is the SE for the average of the draws, not the SD of the box.

3. The shaded area represents the chance that the average of 900 draws from the box will be in the range $14,000 to $14,500.

SET C, PAGE 382

1. The SD of the box is estimated as 0.90; the SE for the total number of TV sets in the sample households is $\sqrt{400} \times 0.90 = 18$; The SE for the average is $18/400 = 0.045$. The 95%-confidence interval is 1.70 ± 0.09.

2. This can't be done with the normal curve. It is like drawing from a box which has 99.75% $\boxed{1}$'s and 0.25 of 1% $\boxed{0}$'s. This box is so skewed that with 400 draws, the probability histogram for the sum still won't be anything like the normal curve. See p. 291, and Exercise 4 on p. 347.

3. This is qualitative data, use the method of Chapter 21. The interval is $37.5\% \pm 2.4\%$.

4. This is a cluster sample—the household is the cluster. So the SE can't be estimated by the method of this chapter. The half-sample method could be used instead (p. 365).

Part VII. Chance Models

Chapter 24. A Model for Measurement Error

SET A, PAGE 392

1. You would have to toss the thumbtack many times, and see whether the percentage of times it landed point down was closer to 50% or to 67%. (This will depend on the surface. In one experiment, the tack landed point down 66% of the time when tossed on linoleum, but only 50% of the time when tossed on a carpet.)

2. No. The rainy days all come close together in the rainy season. If it rains one day, it is more likely to rain the next.

3. Last digits, yes. First digits, no. For instance, in the San Francisco phone book the first digit cannot be 0. Also, many more phone numbers start with 9 than with 2.

4. No, the letters come out in alphabetical order. No box will do that.

5. (a) Tossing a coin is a situation to which the mathematical theory of chance applies directly. Playing a sport isn't. If you tried to develop a chance model for sports, making the successive games independent would be quite silly.
   (b) Like a shot. You have a 50–50 chance to win $5 or lose $4.

SET B, PAGE 394

1. (a) 500 pounds.   (b) 100 pounds. See Example 1 on p. 394.

2. (a) 800 microns.   (b) 80 microns.

SET C, PAGE 398

1. Option (ii) is best. The errors in (i) are too small, those in (iv) are too large, and (v) has the wrong units. In (iii), all the errors are positive, but they have to average out to 0.

2. Option (i) is best. The errors in (ii) are too large, the errors in (iii) don't average out to 0, and the errors in (iv) have the wrong units.

3. The factor is 5.

4. In both cases, the measurement is 504 micrograms above ten grams.

5. No, as the previous exercise shows.

6. The answer is 1.4 micrograms. See Example 1 on p. 397.

7. With $9 on A he gets 3 weighings, SE $\approx$ 0.58 mg.
   With $9 on B he gets 9 weighings, SE $\approx$ 0.67 mg.
   So he's better off with A—smaller chance errors.

8. (a) True: each number on a list is off the average of the list by an SD or so.
   (b) False: the average is 300,007 exactly.
   (c) False: 2 is the SE, not the SD.

9. The SD of the error box is estimated as 50 micrograms.
   (a) 5 micrograms—the SE for the average.
   (b) 50 micrograms—the estimated SD of the error box.
   (c) 95%—two SEs.

10. The answer is 2 inches. Here is the reason. Each measurement equals the exact length, plus a draw from the error box. The measured distance AE is the

sum of the four measurements, and is off the exact length AE by the sum of 4 draws from the box. The average of the box is 0. So the sum of 4 draws will be around 0, give or take an SE or so. It is the SE for the sum which is the right give-or-take number. The SD of the box is 1 inch, so the SE for the sum is $\sqrt{4} \times 1$ inch = 2 inches.

Finding the length AE involved adding the measurements, not averaging them. This problem is about sums, not averages.

SET D, PAGE 403

1. 0.910835 meters ± 18 microns.

2. (a) False. This range is 2 SEs, not 2 SDs, either way from the average.
   (b) False. Same reason as in (a).   (c) True.

SET E, PAGE 408

1. (a) This seems reasonable.
   (b) This is silly. Part of the spread in the 50 measurements represents chance error, but part is monkey-to-monkey variation. These should be kept separate. The Gauss model does not apply, because there are five different values—one for each monkey. The investigator does not have 50 measurements on the same thing.

2. It does not seem at all reasonable that the chance errors for different people behave the same. Also, if the same person takes the test several times, it seems unreasonable to suppose the errors are independent. The Gauss model does not really seem to apply.

3. This is silly. What is this "average response"? If he's talking about the actual data, the average is 4.2; if he's talking about some parameter in a model, let's hear what it is. The Gauss model can't apply, because the successive responses are bound to be dependent.

4. (a) True.   (b) False, the square root law applies to any box.
   (c) False, see p. 93.

5. Can't say: there is no reason to suppose that these 100 scores are like draws from a box.

SET F, PAGE 415

1. (a) 1.25   (b) 1.10   (c) 2.26.
   Comparing (a) and (b): as the number of degrees of freedom goes up, the curve gets more piled up in the middle. Comparing (b) and (c): to get a higher confidence level, a longer interval is needed.

2. (a) Average ± 1.9 micrograms.   (b) Average ± 2.3 micrograms.

3. Student's method gives the longer interval, but is preferred here. The number of measurements is small, and the SD is being estimated from the data, so the normal curve does not apply.

4. (a) 1.0002 ± 0.0013 meters. With so few measurements, a very big multiplier is needed to get 95% confidence.
   (b) 1.0002 ± 0.00008 meters.   (c) 1.0002 ± 0.00002 meters.
   With more measurements, the 95%-confidence interval gets shorter.

5. Area (iii) is largest, (i) is smallest.

6. With only four measurements, the normal curve can't be used because the box doesn't follow the curve. Student's method doesn't apply either, because the SD of the box is given. With 100 measurements, the probability histogram for

the average of the draws will follow the normal curve, and it can be used for confidence levels.

7. Even with four measurements, the normal curve can be used to figure confidence levels, provided the numbers in the box follow the curve (as is given), and their SD is known (as it is here).

Chapter 25. Chance Models in Genetics

SET A, PAGE 425

1. Each seed has a 50% chance to get *y* from the *y/g* parent, and a 50% chance to get *g*. It is bound to get *g* from the *g/g* parent. So the seed has a 50% chance to be *y/g*, and yellow in color; it has a 50% chance to be *g/g*, and green in color. About 50% of the seeds should be yellow.

The number of yellows among 1,600 seeds is like the sum of 1,600 draws from the box $\boxed{0}\boxed{1}$. The expected number of yellows is $1{,}600 \times 1/2 = 800$, with an SE of $\sqrt{1{,}600} \times 1/2 = 20$. Now the normal approximation can be used:

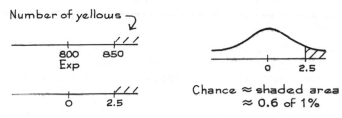

Chance ≈ shaded area ≈ 0.6 of 1%

2. (a) white × red → 100% pink
white × pink → 50% white, 50% pink
pink × pink → 25% red, 50% pink, 25% white.
Work for pink × pink: Each parent is *r/w*, so the offspring's flower color is determined by choosing a row and column at random from the table below.

	r	*w*
r	red	pink
w	pink	white

(b) The expected number of pinks in 400 plants is 200, with an SE of 10. Using the normal approximation

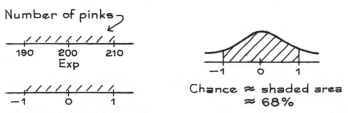

Chance ≈ shaded area ≈ 68%

3. (a) One gene-pair controls leaf width, with variants *w* (wide) and *n* (narrow). The rules are: *w/w* makes wide, *w/n* and *n/w* make medium, and *n/n* makes narrow.
(b) narrow × narrow = *n/n* × *n/n* → 100% *n/n* = narrow.
narrow × medium = *n/n* × *n/w* → 50% *n/n* = narrow, 50% *n/w* = medium.

4. *B* = brown, *b* = blue. Husband is *B/b*, wife is *b/b*. Each child has 1 chance in 2 of having brown eyes. The three children are independent, so the chance that all three will be brown-eyed is $(\frac{1}{2})^3 = \frac{1}{8}$.

Part VIII. Tests of Significance

Chapter 26. Tests of Significance

SET A, PAGE 439

1. Yes, the sample average is only a little over one SE below the old average.

2. If the die is fair, the total number of spots is like the sum of 100 draws from the box | $\boxed{1}$ $\boxed{2}$ $\boxed{3}$ $\boxed{4}$ $\boxed{5}$ $\boxed{6}$ | . The average of the box is 3.5, with an SD of 1.7, so the expected value for the sum is 350, with an SE of 17. The number of spots is a little over one SE above its expected value, which looks like a chance variation.

3. The problem can be set up like Exercise 2, but this time the number of spots is over three SEs above its expected value. This doesn't look like a chance variation. (A more complete test for the fairness of a die will be presented in Chapter 28.)

SET B, PAGE 440

1. False, that's the null.

2. True.

3. The SD of the box can be estimated as 10, so the SE for the average of 100 draws is estimated as 1. If the average of the box is 100, then the average of the draws is 2.7 SEs above its expected value. This isn't plausible.

SET C, PAGE 444

1. SE for average ≈ 0.29, so $z = (5.5 - 6.3)/0.29 \approx -2.8$ and P is approximately the area to the left of -2.8 under the normal curve. From the table, this is 0.255 of 1%, which rounds to 0.3 of 1%. This answer is fine. However, such a small chance may be easier to understand when expressed as a fraction: 0.3 of 1% is

$$0.3 \text{ of } 0.01 = 0.3 \times 0.1 = 0.003 = 3/1,000.$$

2. $z = (5.5 - 6.3)/0.27 \approx -3.0$ and $P \approx 1$ in 1,000.

3. True, see pp. 442–443.

4. False, see p. 442.

5. $z \approx 2.2$ and $P \approx 1\%$. The null hypothesis does not look good.

SET D, PAGE 445

1. True, P is below 5%.

2. False, P has to be below 1%.

3. False. For instance, if P is 3%, and the null hypothesis is true, then 3% of the time the experiment will give an even more extreme value for the test statistic.

4. False, see p. 442.

5. True.

6. True.

SET E, PAGE 448

1. (a) The data consist of a record of the 100 interviews, showing whether the subjects were male (''1'') or female (''0''). On the null hypothesis, this is like 100 draws made at random from a large box, in which 67% of the

tickets are marked ``1`` (for the men), and 33% are ``0`` (for the women). The number of men is like the sum of the draws.

(b) The SD of the box is $\sqrt{0.67 \times 0.33} \approx 0.47$. The expected number of men in the sample is 67, with an SE of 4.7, so $z = (53 - 67)/4.7 \approx -3$, and $P = 1$ in 1,000.

(c) He has too few men in the sample to explain as a chance variation. Choosing people haphazardly, without a definite plan, isn't at all like choosing a simple random sample (Chapter 19).

2. (a) This is like drawing at random with replacement 10,000 times from a 0–1 box, with 0 = tails and 1 = heads. The fraction of 1's in the box is an unknown parameter. The null hypothesis is that this fraction equals ½, the alternative is that it's bigger than ½. The number of heads is like the sum of the draws. (This problem involves a zero-one box with an alternative hypothesis.)

(b) $z \approx 3.3$, $P \approx 5$ in 10,000.

(c) There are too many heads to explain as chance variation.

3. (a) Same as 2(a).   (b) $z \approx 1.35$, $P \approx 9\%$.

(c) The evidence against the null doesn't look strong. The coin really could be fair.

4. (a) The data are like 400 draws from a box, with one ticket in the box for each six year old in Maine, showing its score. The null hypothesis is that the average of the box equals 50, the alternative is that the average is bigger than 50.

(b) The SD of the box is estimated as 12 (no reason it should be the same as for the nation), the SE for the average of 400 draws is 0.6, so
$$z = (51.3 - 50)/0.6 \approx 2.2 \text{ and } P \approx 1\%.$$

(c) There is good evidence that the average for Maine is bigger than 50.

5. The data consist of the 25 weights. On the null hypothesis, this is like 25 draws made at random from a box, with one ticket in the box for each animal in the colony, showing its weight. The average of the box is 30 grams with an SD of 5 grams, so the SE for the average of 25 draws is 1 gram. Now
$$z = (33 - 30)/1 = 3, \text{ and } P \approx 1 \text{ in } 1,000.$$
Choosing haphazardly is not like taking a simple random sample! See Chapter 19 for a discussion.

   One theory is that when you reach into the cage to pick an animal, it is the tamer ones who come to your hand, and they are a bit heavier than the others.

6. (a) and (b) According to the Gauss model (p. 396), each of the 100 measurements equals the exact weight plus a draw from the error box. The tickets in the error box average out to 0, and their SD can be estimated from the past data as 50 micrograms—the error box belongs to the equipment. The unknown parameter is the exact weight. The null hypothesis is that this is still 512 micrograms above a kilogram. The alternative is that the exact weight is less.

(c) The SE for the average of 100 measurements is 5 micrograms, so
$z = (508 - 512)/5 = -0.8$ and $P \approx 21\%$.

(d) The drop in weight looks like a chance variation.

Chapter 27. More Tests for Averages

SET A, PAGE 456

1. The expected value is $100 - 50 = 50$, with an SE of $\sqrt{2^2 + 3^2} \approx 3.6$.

2. The expected value for each percent is 50%; the SEs are 2.5 and 5 percentage points. The expected value for the difference is 0, with an SE of

$\sqrt{2.5^2 + 5^2} \approx 5.6$ percentage points. The square root law applies because the two percents are independent.

3. True. If the draws were made with replacement, the two averages would be independent, and the SE for the difference would equal $\sqrt{3^2 + 3^2}$ exactly. The box is so large that there is no practical difference between drawing with or without replacement.

4. False. With only 200 tickets in the box, the two averages are completely dependent—when you know one average, you automatically know the other. The square root law would not apply.

5. The SD of box X can be estimated as 3, so the SE for the average of 100 draws from box X is 0.3; similarly, the SE for the average of 400 draws from box Y is estimated as 0.4; the averages are independent so the SE for the difference is $\sqrt{(0.3)^2 + (0.4)^2} = 0.5$. If the two boxes have the same average, the observed difference $51 - 48 = 3$ is 6 SEs away from the expected value of 0. Not a likely story.

SET B, PAGE 461

1. The data consist of 1,600 0's and 1's for the men (1 = illiterate), and another 1,600 0's and 1's for the women. The model has two boxes, M and F. Box M has a ticket for every male youth in the country, marked "1" for the illiterates and "0" for the literates. Box F is similar, for the females. The data are like the result of drawing 1,600 tickets at random from each box. The null hypothesis is that the percentage of 1's is the same in the two boxes, the alternative being that the percentage of 1's is bigger in box M. The SE for the percentage of 1's in the male sample can be estimated as 0.6 of 1%; for the female sample, the SE is 0.4 of 1%. So the SE for the difference is $\sqrt{(0.6)^2 + (0.4)^2} \approx 0.7$ of 1%. Then $z = (7 - 3)/0.7 = 5.7$ and P is almost 0. This difference is almost impossible to explain as a chance variation.

   In this exercise, a two-sample z-test is called for. The reason is that two percentages—of 1's in box M and 1's in box F—are unknown and are estimated from the two samples. For comparison, look at Review Exercise 6 on p. 451. There, a one-sample z-test was appropriate, because only one percentage—of 1's in the California box—was estimated from a sample. The other percentage—the national standard of 66%—was given.

   When comparing two percentages estimated from two simple random samples, use the two-sample z-test. When comparing a percentage estimated from a simple random sample to a given standard, use a one-sample z-test.

2. The data consist of 400 scores for the big-city children, and 400 scores for the rural children. There are two boxes. One has a ticket for each big-city child, showing its score; the other has a ticket for every rural child, showing its score. The data are like 400 draws from each of the two boxes. The null hypothesis is that average of the two boxes are equal, the alternative that the average of the big-city box is bigger. The SE for the difference of the two averages can be estimated as $\sqrt{(0.5)^2 + (0.5)^2} \approx 0.7$, because the two averages are independent. So $z = (26 - 25)/0.7 \approx 1.4$, and $P \approx 8\%$. The difference could well be due to chance.

3. $z = 1/0.45 \approx 2.25$ and $P \approx 1\%$. The observed significance level depends on the sample size. With large samples, even small differences will be highly statistically significant. More about this in Chapter 29.

4. The treatment and control averages are dependent, because the rats came in pairs from the same litter, so if one rat has a heavy cortex, the other one in the pair is likely to also. This analysis is not legitimate.

5. Statistical theory says, watch out. First, he doesn't have a random sample. Second, and worse, children in the same family are apt to have similar test scores, so the two averages are dependent, and the square root law cannot be used to figure the SE for the difference. It would be better to analyze the data as in Exercise 9 on pp. 451–453.

SET C, PAGE 465

1. Use the Gauss model as in the text. The bias is an unknown parameter.
     Null: bias is 0.
     Alt.:   bias is negative.
   Now, degrees of freedom $= 10 - 1 = 9$

   $$SD^+ = \sqrt{\frac{10}{9}} \times SD \approx 9.5 \text{ micrograms}, \quad SE \approx 3 \text{ micrograms}$$

   $$t \approx \frac{245 - 253}{3} \approx -2.7, \quad P \approx 1\%$$

   Inference: bias was introduced.

2. (a) degrees of freedom $= 2$,   ave ≈ 72.7, $SD^+ \approx 5.7$, $SE \approx 3.3$

   $$t \approx \frac{72.7 - 70}{3.3} \approx 0.8, \quad P \approx 25\%$$

   Inference: the calibration is fine.
   (b) $P \approx 3\%$, recalibrate.   (c) One measurement is never enough.
   (d) $P \approx 23\%$. (Two measurements are better than one, but more would be even better.)

3. In (a), the numbers 93, 94, and 91 are quite large, suggesting an error box with a long right-hand tail. The approximation (using Student's curve) might not work very well. In (c), the numbers are just switching back and forth between 69 and 71. This is not good for the Gauss model.

4. The first person is all wet. The normal approximation has nothing to do with it. See p. 465.

Chapter 28. The χ^2-Test

SET A, PAGE 476

1. (a) 96%.   (b) 90%.   (c) 10%.   (d) 7.5%.   (e) 3.8%.   (f) 2.2%.

2. (a) 99.98%.   (b) 99.85%.   (c) 51%.   (d) 44%.   (e) 30%.
   (f) 21%.
   As the degrees of freedom go up, the curve shifts to the right and spreads out.

3. $\chi^2 = 13.2$, $d = 5$, $P \approx 2.2\%$.   Note: d = degrees of freedom.
   The data do not fit the model at all well.

4. $\chi^2 = 1$, $d = 5$, $P \approx 96\%$.

5. $\chi^2 = 10$, $d = 5$, $P \approx 7.5\%$.
   The data do not fit the model especially well. Comparing Exercises 4 and 5, the observed frequencies just got multiplied by 10; this doesn't change the per-cents. So the result of the χ^2-test depends on the sample size. With very large samples, the χ^2-test will reject very reasonable models: more about this in Exercise 6 and in section 1 of Chapter 29.

6. $\chi^2 \approx 18.6$, $d = 5$, $P \approx 0.2$ of 1%—although for most purposes, the die is as fair as could be wanted. More about this in section 1 of Chapter 29.

7.

	Ways	Chance	Expected	Observed	
Even, large	4, 6	2/6	200	183	$\chi^2 \approx 5.9$
Even, small	2	1/6	100	113	$d = 3$
Odd, large	5	1/6	100	88	$P \approx 12\%$
Odd, small	1, 3	2/6	200	216	inconclusive

8. $\chi^2 \approx 61$, $d = 3$, $P \approx 0$.

With simple random sampling, it is almost impossible for a jury to differ this much from the county age distribution. The inference is that grand juries are not selected at random. (No surprise, grand jurors are nominated by judges.)

SET B, PAGE 479

1. Pooled $\chi^2 \approx 13.2 + 10 = 23.2$, $d = 5 + 5 = 10$, $P \approx 1\%$.

2. No, dependent experiments.

3. $\chi^2 \approx 0.5$, $d = 3$, $P \approx 8\%$. Inconclusive, but points to fudging.

SET C, PAGE 484

1. $\chi^2 \approx 6.6$, $d = 1$, $P \approx 1\%$.

Notice that 65% of the men voted, compared to 63% of the women. This is a small difference, but with a large sample it is accurately estimated. All P tells you is whether the difference can be explained by chance. More about this in section 1 of Chapter 29.

2. For the men

Observed			Expected		Difference	
40	9	49	41	8	−1	+1
503	90	593	502	91	+1	−1
543	99	642				

$\chi^2 \approx 0.2$, $d = 1$, $P \approx 65\%$.

(The approximation may be somewhat off, because one of the expected frequencies is below 10.)

For the men, marital status does not seem to be related to mental health.

For the women,

Observed			Expected		Difference	
143	5	148	134	14	+9	−9
431	57	488	440	48	−9	+9
574	62	636				

$\chi^2 = {}^{81}/_{134} + {}^{81}/_{440} + {}^{81}/_{14} + {}^{81}/_{48} \approx 8.3$, $d = 1$, $P \approx 0.4$ of 1%.

For the women, marital status does seem to be related to mental health, and the married ones seem to be under much more stress.

Discussion: The HES had 12 measures of psychological distress, only one is considered here. However, the others turn out very much the same way. Married women (controlling for age and even education) show much more stress than unmarried women. Interestingly, among the married women, there is little difference between those with or without children.

Chapter 29. A Closer Look at Tests of Significance

SET A, PAGE 490

1. (a) False. With a large sample, even a trivial difference will be "highly statistically significant."

(b) False. Big samples are good, because they let you estimate things very accurately.

(c) False. You can do the arithmetic without knowing anything about the WISC.

2. The question makes perfect sense, because we are dealing with simple random samples, and it can be answered by a two-sample z-test:

$$\text{SE for men's average} \approx 1, \quad \text{SE for women's average} \approx 1$$

$$\text{SE for difference} \approx \sqrt{1^2 + 1^2} \approx 1.4, \quad z \approx 2/1.4 \approx 1.4, \quad P \approx 8\%.$$

This could easily be a chance variation.

3. The SE for each average is 0.5, so the SE for the difference is $\sqrt{(0.5)^2 + (0.5)^2} \approx 0.7$, and $z \approx 2/0.7 \approx 2.85$. Now P drops to 2 in 1,000.

The observed significance level depends on the sample size. With the smaller sample, the difference was estimated as 2 ± 1.4 points; with the larger, 2 ± 0.7 points.

4. This question does not make sense, because the data are for the whole population.

5. The P-value does not measure the size of the difference, so there is no way of telling just from P whether the impact is weak or strong.

6. The null hypothesis can be rejected on the basis of a trivial difference—if the sample is large. With a small sample, however, quite a large difference may fail to be statistically significant. Remember, the z-test compares a difference to its SE, and the SE for a percent or average goes down as the sample size goes up.

SET B, PAGE 492

1. There is no real proof that flex-time made the difference. It could be mild weather, or a general improvement in health, or more interesting work, to name just a few possibilities. A better design, if employees work independently, would be to choose 100 people at random for treatment (flex-time) and 100 at random for control (fixed hours). Then compare the two groups at the end of the year.

2. If the "random number generator" actually generates numbers which more or less follow a pattern, a subject might spot that and do very well at guessing. Other possibilities are mentioned in the text.

3. This experiment was very well designed. It is fair to conclude that the vaccine protected the children against polio.

SET C, PAGE 497

1. Even if nothing is going on except chance variation, one subject in a hundred should get a score which is "highly statistically significant," and several scores should be "statistically significant."

2. Two-tailed.

3. One-tailed.

4. This is false. The explanation is a bit complicated. Suppose, for instance, that you are making a two-sample z-test. To say that the difference between the sample averages is real means that the averages of the two boxes are different. And in the frequency theory, it does not make sense to talk about the probability that the boxes are one way or another. The two boxes either have

the same average or they don't, so the "probability" of the difference being real is 0% or 100%.

One correct assertion would be this: If the null hypothesis is right, there are only 5 chances in 100 to get a difference between the sample averages which is large enough to be "statistically significant." This assigns chances to a statement about how the data will come out. By contrast, the quote in the exercise assigns chances to a statement about the boxes.

SET D, PAGE 501

1. A two-sample z-test is not appropriate because there are no simple random samples here. Either you think of the 600 students and 25 TAs as the total population, in which case there is no "luck of the draw" to worry about, or you think of them as a sample of convenience. In neither case is a test relevant. Actually, the students seem to have done very well by comparison with the TAs.

2. The question is nonsense. The two inner planets do not constitute a random sample of size 2 from the population of inner planets. They are the inner planets. Similarly for the outer ones.

   (The problem, as stated, misses the interesting feature of the data: Mercury, Venus, the Earth, and Mars are all relatively dense; Jupiter and Saturn have much lower density—they are largely made up of gas.)

3. A χ^2-test is inappropriate. This is not a simple random sample.

4. The question makes perfect sense, because we are dealing with a probability sample. However, it cannot be answered on the basis of the information given. This is a cluster sample, so the simple random sample formulas do not apply. Due to the clustering, the SEs for this sample are about the same as for a simple random sample of half the size. Of course, this kind of sample is much cheaper to draw and interview than a simple random sample! (The SEs can be estimated by the half-sample method discussed on p. 365.)

Tables

A Square Root Table

N	$\sqrt{N}$	N	$\sqrt{N}$	N	$\sqrt{N}$	N	$\sqrt{N}$	N	$\sqrt{N}$
0.1	0.3162	3.1	1.7607	6.1	2.4698	9.1	3.0166	31	5.5678
0.2	0.4472	3.2	1.7889	6.2	2.4900	9.2	3.0332	32	5.6569
0.3	0.5477	3.3	1.8166	6.3	2.5100	9.3	3.0496	33	5.7446
0.4	0.6324	3.4	1.8439	6.4	2.5298	9.4	3.0659	34	5.8310
0.5	0.7071	3.5	1.8708	6.5	2.5495	9.5	3.0822	35	5.9161
0.6	0.7746	3.6	1.8974	6.6	2.5690	9.6	3.0984	36	6.0000
0.7	0.8367	3.7	1.9235	6.7	2.5884	9.7	3.1145	37	6.0828
0.8	0.8944	3.8	1.9494	6.8	2.6077	9.8	3.1305	38	6.1644
0.9	0.9487	3.9	1.9748	6.9	2.6269	9.9	3.1464	39	6.2450
1.0	1.0000	4.0	2.0000	7.0	2.6458	10.0	3.1623	40	6.3246
1.1	1.0488	4.1	2.0248	7.1	2.6646	11	3.3166	41	6.4031
1.2	1.0954	4.2	2.0494	7.2	2.6833	12	3.4641	42	6.4807
1.3	1.1402	4.3	2.0736	7.3	2.7019	13	3.6056	43	6.5574
1.4	1.1832	4.4	2.0976	7.4	2.7203	14	3.7417	44	6.6332
1.5	1.2247	4.5	2.1213	7.5	2.7386	15	3.8730	45	6.7082
1.6	1.2649	4.6	2.1448	7.6	2.7568	16	4.0000	46	6.7823
1.7	1.3038	4.7	2.1679	7.7	2.7749	17	4.1231	47	6.8557
1.8	1.3416	4.8	2.1909	7.8	2.7928	18	4.2426	48	6.9282
1.9	1.3784	4.9	2.2136	7.9	2.8107	19	4.3589	49	7.0000
2.0	1.4142	5.0	2.2361	8.0	2.8284	20	4.4721	50	7.0711
2.1	1.4491	5.1	2.2583	8.1	2.8460	21	4.5826	51	7.1414
2.2	1.4832	5.2	2.2804	8.2	2.8636	22	4.6904	52	7.2111
2.3	1.5166	5.3	2.3022	8.3	2.8810	23	4.7958	53	7.2801
2.4	1.5492	5.4	2.3238	8.4	2.8983	24	4.8990	54	7.3485
2.5	1.5811	5.5	2.3452	8.5	2.9155	25	5.0000	55	7.4162
2.6	1.6124	5.6	2.3664	8.6	2.9326	26	5.0990	56	7.4833
2.7	1.6432	5.7	2.3875	8.7	2.9496	27	5.1962	57	7.5498
2.8	1.6733	5.8	2.4083	8.8	2.9665	28	5.2915	58	7.6158
2.9	1.7029	5.9	2.4290	8.9	2.9833	29	5.3852	59	7.6811
3.0	1.7320	6.0	2.4495	9.0	3.0000	30	5.4772	60	7.7460

N	$\sqrt{N}$	N	$\sqrt{N}$	N	$\sqrt{N}$	N	$\sqrt{N}$	N	$\sqrt{N}$
61	7.8102	71	8.4261	81	9.0000	91	9.5394	101	10.0499
62	7.8740	72	8.4853	82	9.0554	92	9.5917	102	10.0995
63	7.9373	73	8.5440	83	9.1104	93	9.6436	103	10.1489
64	8.0000	74	8.6023	84	9.1652	94	9.6954	104	10.1980
65	8.0623	75	8.6603	85	9.2195	95	9.7468	105	10.2470
66	8.1240	76	8.7178	86	9.2736	96	9.7980	106	10.2956
67	8.1854	77	8.7750	87	9.3274	97	9.8490	107	10.3440
68	8.2462	78	8.8318	88	9.3808	98	9.8995	108	10.3923
69	8.3066	79	8.8882	89	9.4340	99	9.9499	109	10.4403
70	8.3666	80	8.9443	90	9.4868	100	10.0000	110	10.4881

A Normal Table

A(z) is this area (percent)   h(z) is this height (percent)

z	h(z)	A(z)	z	h(z)	A(z)	z	h(z)	A(z)
0.00	39.89	0	1.50	12.95	86.64	3.00	0.443	99.730
0.05	39.84	3.99	1.55	12.00	87.89	3.05	0.381	99.771
0.10	39.69	7.97	1.60	11.09	89.04	3.10	0.327	99.806
0.15	39.45	11.92	1.65	10.23	90.11	3.15	0.279	99.837
0.20	39.10	15.85	1.70	9.40	91.09	3.20	0.238	99.863
0.25	38.67	19.74	1.75	8.63	91.99	3.25	0.203	99.885
0.30	38.14	23.58	1.80	7.90	92.81	3.30	0.172	99.903
0.35	37.52	27.37	1.85	7.21	93.57	3.35	0.146	99.919
0.40	36.83	31.08	1.90	6.56	94.26	3.40	0.123	99.933
0.45	36.05	34.73	1.95	5.96	94.88	3.45	0.104	99.944
0.50	35.21	38.29	2.00	5.40	95.45	3.50	0.087	99.953
0.55	34.29	41.77	2.05	4.88	95.96	3.55	0.073	99.961
0.60	33.32	45.15	2.10	4.40	96.43	3.60	0.061	99.968
0.65	32.30	48.43	2.15	3.96	96.84	3.65	0.051	99.974
0.70	31.23	51.61	2.20	3.55	97.22	3.70	0.042	99.978
0.75	30.11	54.67	2.25	3.17	97.56	3.75	0.035	99.982
0.80	28.97	57.63	2.30	2.83	97.86	3.80	0.029	99.986
0.85	27.80	60.47	2.35	2.52	98.12	3.85	0.024	99.988
0.90	26.61	63.19	2.40	2.24	98.36	3.90	0.020	99.990
0.95	25.41	65.79	2.45	1.98	98.57	3.95	0.016	99.992
1.00	24.20	68.27	2.50	1.75	98.76	4.00	0.013	99.9937
1.05	22.99	70.63	2.55	1.54	98.92	4.05	0.011	99.9949
1.10	21.79	72.87	2.60	1.36	99.07	4.10	0.009	99.9959
1.15	20.59	74.99	2.65	1.19	99.20	4.15	0.007	99.9967
1.20	19.42	76.99	2.70	1.04	99.31	4.20	0.006	99.9973
1.25	18.26	78.87	2.75	0.91	99.40	4.25	0.005	99.9979
1.30	17.14	80.64	2.80	0.79	99.49	4.30	0.004	99.9983
1.35	16.04	82.30	2.85	0.69	99.56	4.35	0.003	99.9986
1.40	14.97	83.85	2.90	0.60	99.63	4.40	0.002	99.9989
1.45	13.94	85.29	2.95	0.51	99.68	4.45	0.002	99.9991

A *t*-Table

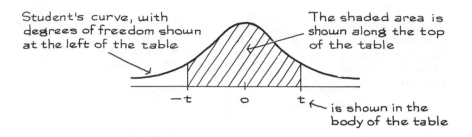

Student's curve, with degrees of freedom shown at the left of the table

The shaded area is shown along the top of the table

−t   o   t ← is shown in the body of the table

Degrees of freedom	50%	70%	90%	95%	99%
1	1.00	1.96	6.31	12.71	63.66
2	0.82	1.39	2.92	4.30	9.92
3	0.76	1.25	2.35	3.18	5.84
4	0.74	1.19	2.13	2.78	4.60
5	0.73	1.16	2.02	2.57	4.02
6	0.72	1.13	1.94	2.45	3.71
7	0.71	1.12	1.90	2.37	3.50
8	0.71	1.11	1.86	2.31	3.36
9	0.70	1.10	1.83	2.26	3.25
10	0.70	1.09	1.81	2.23	3.17
11	0.70	1.09	1.80	2.20	3.11
12	0.70	1.08	1.78	2.18	3.06
13	0.69	1.08	1.77	2.16	3.01
14	0.69	1.08	1.76	2.14	2.98
15	0.69	1.07	1.75	2.13	2.95
16	0.69	1.07	1.75	2.12	2.92
17	0.69	1.07	1.74	2.11	2.90
18	0.69	1.07	1.73	2.10	2.88
19	0.69	1.07	1.73	2.09	2.86
20	0.69	1.06	1.72	2.09	2.84

Source: Adapted from p. 174 of R. A. Fisher, *Statistical Methods for Research Workers* (Edinburgh, Oliver and Boyd, 1958).

A Chi-Square Table

The chi-square curve, with degrees of freedom shown along the left of the table ⟶

The shaded area is shown along the top of the table

is shown in the body of the table

Degrees of freedom	99%	95%	90%	70%	50%	30%	10%	5%	1%
1	0.00016	0.0039	0.016	0.15	0.46	1.07	2.71	3.84	6.64
2	0.020	0.10	0.21	0.71	1.39	2.41	4.60	5.99	9.21
3	0.12	0.35	0.58	1.42	2.37	3.67	6.25	7.82	11.34
4	0.30	0.71	1.06	2.20	3.36	4.88	7.78	9.49	13.28
5	0.55	1.14	1.61	3.00	4.35	6.06	9.24	11.07	15.09
6	0.87	1.64	2.20	3.83	5.35	7.23	10.65	12.59	16.81
7	1.24	2.17	2.83	4.67	6.35	8.38	12.02	14.07	18.48
8	1.65	2.73	3.49	5.53	7.34	9.52	13.36	15.51	20.09
9	2.09	3.33	4.17	6.39	8.34	10.66	14.68	16.92	21.67
10	2.56	3.94	4.86	7.27	9.34	11.78	15.99	18.31	23.21
11	3.05	4.58	5.58	8.15	10.34	12.90	17.28	19.68	24.73
12	3.57	5.23	6.30	9.03	11.34	14.01	18.55	21.03	26.22
13	4.11	5.89	7.04	9.93	12.34	15.12	19.81	22.36	27.69
14	4.66	6.57	7.79	10.82	13.34	16.22	21.06	23.69	29.14
15	5.23	7.26	8.55	11.72	14.34	17.32	22.31	25.00	30.58
16	5.81	7.96	9.31	12.62	15.34	18.42	23.54	26.30	32.00
17	6.41	8.67	10.09	13.53	16.34	19.51	24.77	27.59	33.41
18	7.00	9.39	10.87	14.44	17.34	20.60	25.99	28.87	34.81
19	7.63	10.12	11.65	15.35	18.34	21.69	27.20	30.14	36.19
20	8.26	10.85	12.44	16.27	19.34	22.78	28.41	31.41	37.57

Source: Adapted from p. 112 of Sir R. A. Fisher, *Statistical Methods for Research Workers* (Edinburgh: Oliver and Boyd, 1958).

Index